DISCRETE MATHEMATICS WITH APPLICATIONS

Instructor's Manual

2nd Edition

Susanna S. Epp

DePaul University

PWS Publishing Company

I(T)P **An International Thomson Publishing Company**

Boston • Albany • Bonn • Cincinnati • Detroit • London • Madrid • Melbourne • Mexico City
New York • Paris • San Francisco • Singapore • Tokyo • Toronto • Washington

PWS PUBLISHING COMPANY
20 Park Plaza, Boston, MA 02116-4324

I(T)P™
International Thomson Publishing
The trademark ITP is used under license.

Copyright © 1995 by PWS Publishing Company.

All rights reserved. Instructors of classes using DISCRETE MATHEMATICS, SECOND EDITION by Susanna S. Epp may reproduce materials for classroom use. Otherwise, no part of this book may be reproduced, stored in a retrieval system, or transcribed, in any form or by any means, electronic, mechanical, photocopying, recording, or otherwise--without the prior written permission of PWS Publishing Company, 20 Park Plaza, Boston, Massachusetts 02116-4324.

PWS Publishing Company is a division of International Thomson Publishing Inc.

Printed and bound in the United States of America by Malloy Lithographing.
02 03 – 9 8 7 6

Contents

	Introduction	iv
Chapter 1	The Logic of Compound Statements	1
Chapter 2	The Logic of Quantified Statements	23
Chapter 3	Elementary Number Theory and Methods of Proof	30
Chapter 4	Sequences and Mathematics Induction	62
Chapter 5	Set Theory	88
Chapter 6	Counting	109
Chapter 7	Functions	135
Chapter 8	Recursion	167
Chapter 9	Functions and the Efficiency of Algorithms	198
Chapter 10	Relations	230
Chapter 11	Graphs and Trees	257

Introduction

This manual contains complete solutions to all exercises not fully solved in Appendix B of the second edition of *Discrete Mathematics with Applications*. It also contains some suggestions for how to teach the material in the book. These are directed primarily toward those who have not previously taught discrete mathematics at the freshman-sophomore level. Most of the suggestions are based on my own experience and are offered modestly and with apologies to those whose pedagogical insights are deeper than my own. Comments and suggestions from users of this manual are always welcome.

The primary aim of *Discrete Mathematics with Applications* is to help students attain the knowledge and reasoning skills they need to be successful in upper-level computer science and mathematics courses. Three parallel threads run through the book: exposition of standard facts of discrete mathematics, incremental development of mathematical reasoning skills, and discussion of applications. Almost every section contains exercises designed to help students explore facts, theory, and applications. You can assign whatever mix you wish of exercises of the three types.

The book contains unusually complete explanations and a very large number of illustrative examples on the theory that it is easier for a student who has caught on to an idea to skim over a passage than for a student who is still mystified to fill in the reason for something that is not understood. The extensiveness of the exposition should also make it possible to use the book in courses that use class time primarily for collaborative learning and problem solving rather than lecturing. Each section contains exercises of several different levels of difficulty so that you can choose those most appropriate for your students.

Like many other skills, the skill of mathematical reasoning, once acquired, becomes such a fundamental part of a person's being that it passes from consciousness to instinct. It is natural for mathematicians to think that things that seem totally obvious and trivial to them are equally obvious and trivial to their students. However, the habit of reasoning according to standard logical principles is thought to be innate in only about 4% of the population. Large numbers of students confuse a statement with its converse, do not intuitively understand the equivalence between a statement and its contrapositive (which means that the idea of proof by contradiction is foreign to them), think that the negation of "All mathematicians wear glasses" is "No mathematicians wear glasses," and have the dim impression that an irrational number is any decimal that goes on forever (such as 0.333333. . .). Of course, the extent to which students have such misconceptions varies from school to school, but no institution is immune and at many the problem is epidemic.

Because of the wide variety of backgrounds and abilities of students typically enrolled in a discrete mathematics course, it is important to stay in close touch with how your students are doing. The following are a number of techniques that have been found helpful:

- Increasing the discussion part of a lecture-discussion class by encouraging students to interrupt lectures with questions and by asking frequent, non-rhetorical questions of students. To counteract the problem of having only a small group of students respond to questions, you might call on students by name, using a class list if necessary. If you find that students are unable to answer simple questions on the material being presented, you can backtrack immediately to find common ground. A side benefit of increased class participation is that it improves students' ability to use mathematical language correctly.

- Having students present their solutions to selected homework exercises at the board, following each presentation with class discussion and (tactful) critique. This technique is especially useful when covering topics that involve proof and disproof and is an effective means both of giving feedback on how students are doing and of conveying to students the standards of exposition required in the course.

- Giving frequent quizzes and grading them promptly. This technique gives both you and your students feedback on the kinds of problems students are having at a point in the course when both you and they can take measures to correct them.

- Assigning a few problems to be done on an "until correct" basis, or allowing students to hand in "draft solutions" that are read by you and are handed back for redoing if they are not entirely correct.

It is likely that some, perhaps even many, of your students will do excellent work. But it is also likely that some will have considerable difficulty, especially if asked to write proofs or justify answers to questions. If you are teaching discrete mathematics for the first time, you may be appalled by the awkwardness and illogic of some of your students' efforts. But the very fact that students have difficulty expressing mathematics logically and coherently attests to the value and importance of trying to teach them how to do it. It takes months for a child to learn to walk, days or weeks to learn to ride a bicycle, swim, or play tennis, and years to learn to read at an advanced level. We should not be discouraged if students catch on slowly to new modes of thinking. Prospective employers of students of mathematics and computer science are looking for brain power in those whom they hire, and students enjoy the feeling that they are increasing their mental capacity. It is in everyone's interest to help students develop as much reasoning capability as possible.

Students often come to a discrete mathematics skeptical of its practical usefulness, especially of those parts that seem abstract. One reason for including the large number of applications in the book is to overcome such skepticism. You can also encourage students to broaden their perspective by referring, whenever possible, to relationships between the topics and modes of thought under discussion and actual uses in computer science. You do not need to be an expert to convey these ideas effectively. Much can be learned by spending some time looking through computer science texts in such areas as data structures, design and analysis of algorithms, relational database theory, theory of computation, and artificial intelligence. It is also desirable, if possible, to coordinate the topics of the course with those of a computer science course that students take concurrently. Hardly anything more effectively convinces students of the utility of discrete mathematics than to hear references to it in their computer science courses.

Acknowledgement

I would like to thank my husband, Helmut Epp, for converting the text of the first edition of this manual from T^3 to TeX for assisting me in dealing with the intricacies of TeX for the second edition, and for transforming my rough sketches into computer graphics.

Chapter 1: The Logic of Compound Statements

Skill in handling mathematical logic correctly is essential for students to be able to solve problems in abstract mathematics and computer science. Because a large number of students come to college very shaky in such skill, a primary aim of both Chapters 1 and 2 is to help students develop an inner voice that speaks with logical precision. Consequently, the various rules used in logical reasoning are developed both symbolically and in the context of their somewhat limited but very important use in everyday language. Exercise sets for Sections 1.1–1.3 and 2.1–2.3 all contain sentences for students to negate, write the contrapositive for, and so forth. Virtually all students benefit from doing these exercises. Another aim of Chapters 1 and 2 is to teach students the rudiments of symbolic logic as a foundation for a variety of upper-level courses. Symbolic logic is used in, among others, the study of digital logic circuits, relational databases, artificial intelligence, and program verification.

Suggestions

1. In Sections 1.1 and 1.4, students have more difficulty than you might expect simplifying statement forms and circuits. Only through trial and error can you learn the extent to which this is the case at your institution. If it is, you could either assign only the easier exercises or build in extra time to teach students how to do the more complicated ones. Discussion of simplification techniques occurs again in Chapter 5 in the context of set theory. At this later point in the course most students are able to deal with it successfully.

2. When covering the translations of "only if" and "necessary and sufficient conditions" to if-then form in Section 1.2, you might try to get the class to agree to the translations of certain concrete statements and suggest that students recall the general translation rules by recalling these particular translations. For instance, when discussing "only if" you could use a statement such as "The Cubs will win the pennant only if they win tomorrow's game," first getting students to agree and then suggesting that they remember that this statement translates to "If the Cubs do not win tomorrow's game then they will not win the pennant."

With students of lower ability, you may find yourself tying up excessive amounts of class time discussing "only if" and "necessary and sufficient conditions" with little success. You might just assign the easier exercises or you might assign exercises on these topics to be done for extra credit (putting corresponding extra credit problems on exams) and use the results to help distinguish A from B students. It is probably best not to omit these topics altogether because the language of "only if" and "necessary and sufficient conditions" is a standard part of the technical vocabulary of textbooks used in upper-level courses.

3. In Section 1.3, many students mistakenly conclude that an argument is valid if when they compute the truth table, they find a single row in which both the premises and the conclusion are true. To help counteract this misconception, you could give examples of both valid and invalid arguments whose truth tables have eight rows, several of which are critical. The source of students' difficulty understanding the concept of validity appears to be their tendency to ignore quantification and to misinterpret if-then statements as "and" statements. Since the definition of validity includes both a universal quantifier and if-then, it is helpful to go back over the definition and the procedures for testing for validity and invalidity after discussing the general topic of universal conditional statements in Section 2.1.

4. Also in Section 1.3, you might suggest that students just familiarize themselves with but not memorize all the forms of valid arguments covered in Section 1.3. It is wise, however, to ask students to learn modus ponens and modus tollens and converse and inverse errors.

Section 1.1

2. common form: If p then q.
   $\sim q$
   Therefore, $\sim p$.

   b. all prime numbers are odd; 2 is odd

4. common form: If p then q.
   If q then r.
   Therefore, if p then r.

   b. x equals 0; the guard condition for the **while** loop is false; program execution moves to the next instruction following the loop

5. b. The truth or falsity of this sentence depends on the reference for the pronoun "she." Considered on its own, the sentence cannot be said to be either true or false, and so it is not a statement.

   c. This sentence is true; hence it is a statement.

   d. This is not a statement because its truth or falsity depends on the value of x.

7. $m \wedge \sim c$

8. b. $\sim w \wedge (h \wedge s)$

   c. $\sim w \wedge \sim h \wedge \sim s$

9. $(n \vee k) \wedge \sim (n \wedge k)$

10. b. $p \wedge \sim q$ d. $(\sim p \wedge q) \wedge \sim r$ e. $\sim p \vee (q \wedge r)$

11. It is used in an inclusive sense.

13.

p	q	$p \wedge q$	$p \vee q$	$\sim (p \vee q)$	$(p \wedge q) \vee \sim (p \vee q)$
T	T	T	T	F	T
T	F	F	T	F	F
F	T	F	T	F	F
F	F	F	F	T	T

15.

p	q	r	$\sim p$	$\sim r$	$q \vee \sim r$	$\sim p \wedge (q \vee \sim r)$
T	T	T	F	F	T	F
T	T	F	F	T	T	F
T	F	T	F	F	F	F
T	F	F	F	T	T	F
F	T	T	T	F	T	T
F	T	F	T	T	T	T
F	F	T	T	F	F	F
F	F	F	T	T	T	T

16.

p	q	r	$\sim p$	$\sim r$	$\sim p \vee q$	$q \wedge \sim r$	$\sim (q \wedge \sim r)$	$p \vee (\sim p \vee q)$	$(p \vee (\sim p \vee q)) \wedge \sim (q \wedge \sim r)$
T	T	T	F	F	T	F	T	T	T
T	T	F	F	T	T	T	F	T	F
T	F	T	F	F	F	F	T	T	T
T	F	F	F	T	F	F	T	T	T
F	T	T	T	F	T	F	T	T	T
F	T	F	T	T	T	T	F	T	F
F	F	T	T	F	T	F	T	T	T
F	F	F	T	T	T	F	T	T	T

18.

p	q	$p \vee q$	$\sim p$	$\sim q$	$\sim (p \vee q)$	$\sim p \wedge \sim q$
T	T	T	F	F	F	F
T	F	T	F	T	F	F
F	T	T	T	F	F	F
F	F	F	T	T	T	T

$\underbrace{\qquad\qquad\qquad\qquad}_{\text{same truth values}}$

Since $\sim (p \vee q)$ and $\sim p \wedge \sim q$ have the same truth values, they are logically equivalent.

20.

p	c	$p \vee c$	p
T	F	T	T
F	F	F	F

$\underbrace{\qquad\qquad}_{\text{same truth values}}$

Since $p \vee c$ and p have the same truth values, they are logically equivalent.

22.

p	q	r	$q \vee r$	$p \wedge q$	$p \wedge r$	$p \wedge (q \vee r)$	$(p \wedge q) \vee (p \wedge r)$
T	T	T	T	T	T	T	T
T	T	F	T	T	F	T	T
T	F	T	T	F	T	T	T
T	F	F	F	F	F	F	F
F	T	T	T	F	F	F	F
F	T	F	T	F	F	F	F
F	F	T	T	F	F	F	F
F	F	F	F	F	F	F	T

$\underbrace{\qquad\qquad\qquad}_{\text{same truth values}}$

Since $p \wedge (q \vee r)$ and $(p \wedge q) \vee (p \wedge r)$ have the same truth values, they are logically equivalent.

24.

p	q	r	$p \vee q$	$p \wedge r$	$(p \vee q) \vee (p \wedge r)$	$(p \vee q) \wedge r$
T	T	T	T	T	T	T
T	T	F	T	F	T	F
T	F	T	T	T	T	T
T	F	F	T	F	T	F
F	T	T	T	F	T	T
F	T	F	T	F	T	F
F	F	T	F	F	F	F
F	F	F	F	F	F	F

$$\underbrace{\qquad\qquad\qquad\qquad\qquad\qquad}_{\text{different truth values}}$$

Since $p \vee q$ $p \wedge r$ $(p \vee q) \vee (p \wedge r)$ and $(p \vee q) \wedge r$ have different truth values in rows 2, 3, and 6, they are not logically equivalent.

26.

p	q	r	$r \vee p$	$\sim r$	$p \wedge q$	$\sim r \vee (p \wedge q)$	$r \vee q$	$(\sim r \vee (p \wedge q)) \wedge (r \vee q)$	$(r \vee p) \wedge ((\sim r \vee (p \wedge q)) \wedge (r \vee q))$	$p \wedge q$
T	T	T	T	F	T	T	T	T	T	T
T	T	F	T	T	T	T	T	T	T	T
T	F	T	T	F	F	F	T	F	F	F
T	F	F	T	T	F	T	F	F	F	F
F	T	T	T	F	F	F	T	F	F	F
F	T	F	F	T	F	T	T	T	F	F
F	F	T	T	F	F	F	T	F	F	F
F	F	F	F	T	F	T	F	F	F	F

$$\underbrace{\qquad\qquad\qquad\qquad}_{\text{same truth values}}$$

Since $(r \vee p) \wedge ((\sim r \vee (p \wedge q)) \wedge (r \vee q))$ and $p \wedge q$ have the same truth values, they are logically equivalent.

28. Sam does not swim on Thursdays or Kate does not play tennis on Saturdays.

30. This computer program has no logical errors in the first ten lines and it is not being run with an incomplete data set.

31. The dollar is not at an all-time high or the stock market is not at a record low.

32. The train is not late and my watch is not fast.

34. $x \leq -4$ or $x \geq -1$

36. $x > 0$ or $x \leq -5$

39.

p	q	r	$\sim p$	$\sim q$	$\sim p \wedge q$	$q \wedge r$	$((\sim p \wedge q) \wedge (q \wedge r))$	$((\sim p \wedge q) \wedge (q \wedge r))$ $\wedge$ $\sim q$
T	T	T	F	F	F	T	F	F
T	T	F	F	F	F	F	F	F
T	F	T	F	T	F	F	F	F
T	F	F	F	T	F	F	F	F
F	T	T	T	F	T	T	T	F
F	T	F	T	F	T	F	F	F
F	F	T	T	T	F	F	F	F
F	F	F	T	T	F	F	F	F

all $F's$

Since all the truth values of $((\sim p \wedge q) \wedge (q \wedge r)) \wedge \sim q$ are F, $((\sim p \wedge q) \wedge (q \wedge r)) \wedge \sim q$ is a contradiction.

40.

p	q	$\sim p$	$\sim q$	$\sim p \vee q$	$p \wedge \sim q$	$(\sim p \vee q) \vee (p \wedge \sim q)$
T	T	F	F	T	F	T
T	F	F	T	F	T	T
F	T	T	F	T	F	T
F	F	T	T	T	F	T

all $T's$

Since all the truth values of $(\sim p \vee q) \vee (p \wedge \sim q)$ are T, $(\sim p \vee q) \vee (p \wedge \sim q)$ is a tautology.

42. *a.* the commutative law for $\vee$ *b.* the distributive law *c.* the negation law for $\wedge$ *d.* the identity law for $\vee$

44. $\quad p \wedge (\sim q \vee p) \equiv p \wedge (p \vee \sim q) \quad$ by the commutative law for $\vee$
$\qquad\qquad\qquad\quad\;\; \equiv p \qquad\qquad\quad\;$ by the absorption law

45. $\sim (p \vee \sim q) \vee (\sim p \wedge \sim q) \equiv (\sim p \wedge q) \vee (\sim p \wedge \sim q) \quad$ by De Morgan's law and
$\qquad\qquad\qquad\qquad\qquad\qquad\qquad\qquad\qquad\qquad\qquad\qquad$ the double negative law
$\qquad\qquad\qquad\qquad\qquad\qquad \equiv \sim p \wedge (q \vee \sim q) \quad$ by the distributive law
$\qquad\qquad\qquad\qquad\qquad\qquad \equiv \sim p \wedge t \qquad\qquad$ by the negation law for $\vee$
$\qquad\qquad\qquad\qquad\qquad\qquad \equiv \sim p \qquad\qquad\quad$ by the identity law for $\wedge$

47. $(p \wedge (\sim (\sim p \vee q))) \vee (p \wedge q) \equiv (p \wedge (p \wedge \sim q)) \vee (p \wedge q) \quad$ by De Morgan's law and
$\qquad\qquad\qquad\qquad\qquad\qquad\qquad\qquad\qquad\qquad\qquad\qquad\qquad$ the double negative law
$\qquad\qquad\qquad\qquad\qquad\qquad\qquad \equiv ((p \wedge p) \wedge \sim q) \vee (p \wedge q) \quad$ by the associative law for $\wedge$
$\qquad\qquad\qquad\qquad\qquad\qquad\qquad \equiv (p \wedge \sim q) \vee (p \wedge q) \qquad\;\;$ by the idempotent law $\wedge$
$\qquad\qquad\qquad\qquad\qquad\qquad\qquad \equiv p \wedge (\sim q \vee q) \qquad\qquad\quad\;$ by the distributive law
$\qquad\qquad\qquad\qquad\qquad\qquad\qquad \equiv p \wedge (q \vee \sim q) \qquad\qquad\quad\;$ by the commutative law for $\vee$
$\qquad\qquad\qquad\qquad\qquad\qquad\qquad \equiv p \wedge t \qquad\qquad\qquad\qquad\;\;\;$ by the negation law for $\vee$
$\qquad\qquad\qquad\qquad\qquad\qquad\qquad \equiv p \qquad\qquad\qquad\qquad\qquad$ by the identity law $\wedge$

48. b. Yes.

p	q	r	$p \oplus q$	$q \oplus r$	$(p \oplus q) \oplus r$	$p \oplus (q \oplus r)$
T	T	T	F	F	T	T
T	T	F	F	T	F	F
T	F	T	T	T	F	F
T	F	F	T	F	T	T
F	T	T	T	F	F	F
F	T	F	T	T	T	T
F	F	T	F	T	T	T
F	F	F	F	F	F	F

$\underbrace{\qquad\qquad\qquad\qquad}_{\text{same truth values}}$

Since $(p \oplus q) \oplus r$ and $p \oplus (q \oplus r)$ have the same truth values, they are logically equivalent.

c. Yes.

p	q	r	$p \oplus q$	$p \wedge r$	$q \wedge r$	$(p \oplus q) \wedge r$	$(p \wedge r) \oplus (q \wedge r)$
T	T	T	F	T	T	F	F
T	T	F	F	F	F	F	F
T	F	T	T	T	F	T	T
T	F	F	T	F	F	F	F
F	T	T	T	F	T	T	T
F	T	F	T	F	F	F	F
F	F	T	F	F	F	F	F
F	F	F	F	F	F	F	F

$\underbrace{\qquad\qquad\qquad\qquad}_{\text{same truth values}}$

Since $(p \oplus q) \wedge r$ and $(p \wedge r) \oplus (q \wedge r)$ have the same truth values, they are logically equivalent.

50. Because of slight ambiguity in the rules, there are two possible answers: $(p \vee q) \wedge \sim (r \wedge (s \wedge t))$ and $p \vee (q \wedge \sim (r \wedge (s \wedge t)))$, either of which may be expressed in a logically equivalent form.

Section 1.2

2. If I catch the 8:05 bus, then I am on time for work.

4. If you do not fix my ceiling, then I won't pay my rent.

6.

p	q	$\sim p$	$\sim p \wedge q$	$p \vee (\sim p \wedge q)$	$p \vee (\sim p \wedge q) \to q$
T	T	F	F	T	T
T	F	F	F	T	F
F	T	T	T	T	T
F	F	T	F	F	T

8.

p	q	r	$\sim p$	$\sim p \vee q$	$\sim p \vee q \to r$
T	T	T	F	T	T
T	T	F	F	T	F
T	F	T	F	F	T
T	F	F	F	F	T
F	T	T	T	T	T
F	T	F	T	T	F
F	F	T	T	T	T
F	F	F	T	T	F

6

10.

p	q	r	$p \to r$	$q \to r$	$(p \to r) \leftrightarrow (q \to r)$
T	T	T	T	T	T
T	T	F	F	F	T
T	F	T	T	T	T
T	F	F	F	T	F
F	T	T	T	T	T
F	T	F	T	F	F
F	F	T	T	T	T
F	F	F	T	T	T

11.

p	q	r	$q \to r$	$p \to (q \to r)$	$p \wedge q$	$p \wedge q \to r$	$(p \to (q \to r)) \leftrightarrow (p \wedge q \to r)$
T	T	T	T	T	T	T	T
T	T	F	F	F	T	F	T
T	F	T	T	T	F	T	T
T	F	F	T	T	F	T	T
F	T	T	T	T	F	T	T
F	T	F	F	T	F	T	T
F	F	T	T	T	F	T	T
F	F	F	T	T	F	T	T

13. a.

p	q	$\sim p$	$p \to q$	$\sim p \vee q$
T	T	F	T	T
T	F	F	F	F
F	T	T	T	T
F	F	T	T	T

$\underbrace{\qquad\qquad\qquad}_{\text{same truth values}}$

Since $p \to q$ and $\sim p \vee q$ have the same truth values, they are logically equivalent.

b.

p	q	$\sim q$	$p \to q$	$\sim (p \to q)$	$p \wedge \sim q$
T	T	F	T	F	F
T	F	T	F	T	T
F	T	F	T	F	F
F	F	T	T	F	F

$\underbrace{\qquad\qquad\qquad}_{\text{same truth values}}$

Since $\sim (p \to q)$ and $p \wedge \sim q$ have the same truth values, they are logically equivalent.

14. a.

p	q	r	$\sim q$	$\sim r$	$q \vee r$	$p \wedge \sim q$	$p \wedge \sim r$	$p \to q \vee r$	$p \wedge \sim q \to r$	$p \wedge \sim r \to q$
T	T	T	F	F	T	F	F	T	T	T
T	T	F	F	T	T	F	T	T	T	T
T	F	T	T	F	T	T	F	T	T	T
T	F	F	T	T	F	T	T	F	F	F
F	T	T	F	F	T	F	F	T	T	T
F	T	F	F	T	T	F	F	T	T	T
F	F	T	T	F	T	F	F	T	T	T
F	F	F	T	T	F	F	F	T	T	T

$\underbrace{\qquad\qquad\qquad\qquad\qquad}_{\text{same truth values}}$

7

Since $p \to q \vee r$, $p \wedge \sim q \to r$, and $p \wedge \sim r \to q$ all have the same truth values, they are all logically equivalent.

   b. If n is prime and n is not odd, then n is 2.

   And: If n is prime and n is not 2, then n is odd.

16. b. Today is Thanksgiving and tomorrow is not Friday.

    c. r is rational and the decimal expansion of r is not repeating.

    e. x is nonnegative and x is not positive and x is not 0.

    Or: x is nonnegative but x is not positive and x is not 0.

    Or: x is nonnegative and x is neither positive nor 0.

    g. n is divisible by 6 and either n is not divisible by 2 or n is not divisible by 3.

17. Note that $p \to q$ is false if, and only if, p is true and q is false. Under these circumstances, (b) $p \vee q$ is true and (c) $q \to p$ is true.

18. b. If tomorrow is not Friday, then today is not Thanksgiving.

    c. If the decimal expansion of r is not repeating, then r is not rational.

    e. If x is not positive and x is not 0, then x is not nonnegative.

    Or: If x is neither positive nor 0, then x is negative.

    g. If n is not divisible by 2 or n is not divisible by 3, then n is not divisible by 6.

19. b. *converse*: If tomorrow is Friday, then today is Thanskgiving.

    inverse: If today is not Thanksgiving, then tomorrow is not Friday.

    c. *converse*: If the decimal expansion of r is repeating, then r is rational.

    inverse: If r is not rational, then the decimal expansion of r is not repeating.

    e. *converse*: If x is positive or x is 0, then x is nonnegative.

    inverse: If x is not nonnegative, then both x is not positive and x is not 0.

      Or: If x is negative, then x is neither positive nor 0.

21.

p	q	$\sim p$	$\sim q$	$p \to q$	$\sim p \to \sim q$
T	T	F	F	T	T
T	F	F	T	F	T
F	T	T	F	T	F
F	F	T	T	T	T

different truth values

Since $p \to q$ and $\sim p \to \sim q$ have different truth values in rows 2 and 3, they are not logically equivalent.

23.

p	q	$\sim p$	$\sim q$	$q \to p$	$\sim p \to \sim q$
T	T	F	F	T	T
T	F	F	T	T	T
F	T	T	F	F	F
F	F	T	T	T	T

same truth values

Since $q \to p$ and $\sim p \to \sim q$ have the same truth values, they are logically equivalent.

24. The if-then form of "I say what I mean" is "If I mean something, then I say it." The if-then form of "I mean what I say" is "If I say something, then I mean it." Thus "I mean what I say" is the converse of "I say what I mean." The two statements are not logically equivalent.

26. If Sam is not an expert sailor, then he will not be allowed on Signe's racing boat.

   If Sam is allowed on Signe's racing boat, then he is an expert sailor.

27. The Personnel Director did not lie. By using the phrase "only if," the Personnel Director set forth conditions that were necessary but not sufficient for being hired: if you did not satisfy those conditions then you would not be hired. The Personnel Director's statement said nothing about what would happen if you did satisfy those conditions.

28. b. If a security code is not entered, then the door will not open.

30. a. $\sim p \vee q \to r \vee \sim q \equiv \sim (\sim p \vee q) \vee (r \vee \sim q)$ [an acceptable answer]
    $\equiv (p \wedge \sim q) \vee (r \vee \sim q)$ by De Morgan's law
    [another acceptable answer]

   b. $\sim p \vee q \to r \vee \sim q \equiv (p \wedge \sim q) \vee (r \vee \sim q)$ by part (a)
    $\equiv \sim [\sim (p \wedge \sim q) \wedge \sim (r \vee \sim q)]$ by De Morgan's law
    $\equiv \sim [\sim (p \wedge \sim q) \wedge (\sim r \wedge \sim (\sim q))]$ by De Morgan's law
    $\equiv \sim [\sim (p \wedge \sim q) \wedge (\sim r \wedge q)]$ by the double negative law

32. a. $(p \to (q \to r)) \leftrightarrow ((p \wedge q) \to r) \equiv [\sim p \vee (q \to r)] \leftrightarrow [\sim (p \wedge q) \vee r]$
    $\equiv [\sim p \vee (\sim q \vee r)] \leftrightarrow [\sim (p \wedge q) \vee r]$
    $\equiv \sim [\sim p \vee (\sim q \vee r)] \vee [\sim (p \wedge q) \vee r]$
    $\wedge \sim [\sim (p \wedge q) \vee r] \vee [\sim p \vee (\sim q \vee r)]$

   b. By part (a), De Morgan's law, and the double negative law,
   $(p \to (q \to r)) \leftrightarrow ((p \wedge q) \to r) \equiv \sim [\sim p \vee (\sim q \vee r)] \vee [\sim (p \wedge q) \vee r]$
   $\wedge \sim [\sim (p \wedge q) \vee r] \vee [\sim p \vee (\sim q \vee r)]$
   $\equiv \sim ([\sim p \vee (\sim q \vee r)] \wedge \sim [\sim (p \wedge q) \vee r])$
   $\wedge \sim (\sim [(p \wedge q) \wedge \sim r] \wedge \sim [\sim p \vee (\sim q \vee r)])$
   $\equiv \sim (\sim [p \wedge \sim (\sim q \vee r)] \wedge [(p \wedge q) \wedge \sim r])$
   $\wedge \sim (\sim [(p \wedge q) \wedge \sim r] \wedge [p \wedge \sim (\sim q \vee r)])$
   $\equiv \sim (\sim [p \wedge (q \wedge \sim r)] \wedge [(p \wedge q) \wedge \sim r])$
   $\wedge \sim (\sim [(p \wedge q) \wedge \sim r] \wedge [p \wedge (q \wedge \sim r)]).$

33. Yes. As in exercises 29–32, the logical equivalences $p \to q \equiv \sim p \vee q$, $p \leftrightarrow q \equiv (\sim p \vee q) \wedge (\sim q \vee p)$, $p \vee q \equiv \sim (\sim p \wedge \sim q)$, and $\sim (\sim p) \equiv p$ can be used to rewrite any statement form in a logically equivalent way using only $\sim$ and $\wedge$. Similarly, the logical equivalence $p \wedge q \equiv \sim (\sim p \vee \sim q)$ can be used to rewrite any statement form in a logically equivalent way using only $\sim$ and $\vee$.

35. If this triangle has two 45° angles, then it is a right triangle.

37. If Jim does not do his homework regularly, then Jim will not pass the course.

   If Jim passes the course, then he will have done his homework regularly.

39. If this computer program produces error messages during translation, then it is not correct.

   If this computer program is correct, then it does not produce error messages during translation.

40. c. must be true   d. not necessarily true   e. must be true   f. not necessarily true

   Note: To solve this problem, it may be helpful to imagine a compound whose boiling point is greater than 250°. For concreteness, suppose it is 300°. Then the given statement would be true for this compound, but statements a, d, and f would be false.

Section 1.3

2. This is a **while** loop.

4. This polygon is not a triangle.
5. They did not telephone.

9.

				premises			conclusion
p	q	r	$q \wedge r$	$p \to q$	$p \to r$	$p \to q \wedge r$	
T	T	T	T	T	T	T ←	critical row
T	T	F	F	T	F	F	
T	F	T	F	F	T	F	
T	F	F	F	F	F	F	
F	T	T	T	T	T	T ←	critical row
F	T	F	F	T	T	T ←	critical row
F	F	T	F	T	T	T ←	critical row
F	F	F	F	T	T	T ←	critical row

The first, fifth, sixth, seventh, and eighth rows are the critical rows (all the premises are true in these rows). In each critical row the conclusion is also true. Hence the argument form is valid.

10.

					premises			conclusion	
p	q	r	$\sim q$	$p \wedge \sim q$	$p \wedge \sim q \to r$	$p \vee q$	$q \to p$	r	
T	T	T	F	F	T	T	T	T ←	critical row
T	T	F	F	F	T	T	T	F ←	critical row
T	F	T	T	T	T	T	T	T ←	critical row
T	F	F	T	T	F	T	T	F	
F	T	T	F	F	T	T	F	T	
F	T	F	F	F	T	T	F	F	
F	F	T	T	F	T	F	T	T	
F	F	F	T	F	T	F	T	F	

In the second row all the premises are true but the conclusion is false. Hence the argument is invalid.

11.

		premises		conclusion	
p	q	$p \to q$	$\sim q$	$\sim p$	
T	T	T	F	F	
T	F	F	T	F	
F	T	T	F	T	
F	F	T	T	T ←	critical row

The only critical row is the fourth, and in this row the conclusion is true. Therefore, the argument is valid.

12. b.

		premises		conclusion	
p	q	$p \to q$	$\sim p$	$\sim q$	
T	T	T	F	F	
T	F	F	F	T	
F	T	T	T	F ←	critical row
F	F	T	T	T ←	critical row

In the third row the premises are both true but the conclusion is false. Therefore, this argument form is invalid.

14.

		premise	conclusion	
p	q	q	$p \vee q$	
T	T	T	T ←	critical row
T	F	F	T	
F	T	T	T ←	critical row
F	F	F	F	

10

There are two critical rows (in which the premise is true), and in these rows the conclusion is also true. Therefore, the argument is valid.

15.

		premise	conclusion
p	q	p ∧ q	p
T	T	T	T ← critical row
T	F	F	T
F	T	F	F
F	F	F	F

The only critical row is the first, and in this row the conclusion is true. Therefore, the argument is valid.

16.

		premise	conclusion
p	q	p ∧ q	q
T	T	T	T ← critical row
T	F	F	F
F	T	F	T
F	F	F	F

The only critical row is the first, and in this row the conclusion is true. Therefore, the argument is valid.

18.

		premises		conclusion
p	q	p ∨ q	∼p	q
T	T	T	F	T
T	F	T	F	F
F	T	T	T	T ← critical row
F	F	F	T	F

The only critical row is the third, and in this row the conclusion is true. Therefore, the argument is valid.

19.

			premises		conclusion
p	q	r	p → q	q → r	p → r
T	T	T	T	T	T ← critical row
T	T	F	T	F	F
T	F	T	F	T	T
T	F	F	F	T	F
F	T	T	T	T	T ← critical row
F	T	F	T	F	T
F	F	T	T	T	T ← critical row
F	F	F	T	T	T ← critical row

The first, fifth, seventh, and eighth rows are all the critical rows, and in each of these rows the conclusion is true. Therefore, the argument form is valid.

20.

				premises		conclusion
p	q	r	p ∨ q	p → r	q → r	r
T	T	T	T	T	T	T ← critical row
T	T	F	T	F	F	F
T	F	T	T	T	T	T ← critical row
T	F	F	T	F	T	F
F	T	T	T	T	T	T ← critical row
F	T	F	T	T	F	F
F	F	T	F	T	T	T
F	F	F	F	T	T	F

The first, third, and fifth rows are all the critical rows, and in each of these rows the conclusion is true. Therefore, the argument form is valid.

22. form: $p \vee q$
    $p \rightarrow q$
    $\therefore q \vee \sim r$

				premises		conclusion	
p	q	r	$\sim r$	$p \vee q$	$p \rightarrow r$	$q \vee \sim r$	
T	T	T	F	T	T	T	← critical row
T	T	F	T	T	F	T	
T	F	T	F	T	T	F	← critical row
T	F	F	T	T	F	T	
F	T	T	F	T	T	T	← critical row
F	T	F	T	T	T	T	← critical row
F	F	T	F	F	T	F	
F	F	F	T	F	T	T	

In the third row the premises are both true but the conclusion is false, and so the argument is invalid.

25. form: $p \rightarrow q$ valid, hypothetical syllogism
    $q \rightarrow r$
    $\therefore p \rightarrow r$

27. form: $p \rightarrow q$ invalid, converse error
    q
    $\therefore p$

28. form: $p \rightarrow q$ invalid, inverse error
    $\sim p$
    $\therefore \sim q$

29. form: $p \rightarrow q$ invalid, converse error
    q
    $\therefore p$

30. form: $p \wedge q$ valid, conjunctive simplification
    $\therefore q$

31. form: $p \rightarrow r$ valid, dilemma (proof by division into cases)
    $q \rightarrow r$
    $\therefore p \vee q \rightarrow r$

34. A correct answer should indicate that for a valid argument, any argument of the same form that has true premises has a true conclusion, whereas for an invalid argument, it is possible to find an argument of the same form that has true premises and a false conclusion.

37. b. 1. Suppose C is a knight.
    2. $\therefore C$ is a knave (because what C said was true).
    3. $\therefore C$ is both a knight and a knave (by (1) and (2)), which is a contradiction.
    4. $\therefore C$ is not a knight (because by the contradiction rule the supposition is false).
    5. $\therefore$ What C says is false (because since C is not a knight he is a knave and knaves always speak falsely).
    6. $\therefore$ At least one of C or D is a knight (by De Morgan's law).
    7. $\therefore D$ is a knight (by (4) and (6) and disjunctive syllogism).
    8. $\therefore C$ is a knave and D is a knight (by (4) and (7)).
    To check that the problem situation is not inherently contradictory, note that if C is a knave and D is a knight, then each could have spoken as reported.

c. There is one knave. E and F cannot both be knights because then both would also be knaves (since each would have spoken the truth), which is a contradiction. Nor can E and F both be knaves because then both would be telling the truth which is impossible for knaves. Hence, the only possible answer is that one is a knight and the other is a knave. But in this case both E and F could have spoken as reported, without contradiction.

d. The following is one of many solutions.

1. The statement made by U must be false because if it were true then U would not be a knight (since none would be a knight), but since he spoke the truth he would be a knight and this would be a contradiction.

2. ∴ there is at least one knight, and U is a knave (since his statement that there are no knights is false).

3. Suppose Z spoke the truth. Then so did W (since if there is exactly one knight then it is also true that there are at most three knights). But this implies that there are at least two knights, which contradicts $Z's$ statement. Hence Z cannot have spoken the truth.

4. ∴ there are at least two knights, and Z is a knave (since his statement that there is exactly one knight is false). Also $X's$ statement is false because since both U and Z are knaves it is impossible for there to be exactly five knights. Hence X also is a knave.

5. ∴ there are at least three knaves (U, Z, and X), and so there are at most three knights.

6. ∴ $W's$ statement is true, and so W is a knight.

7. Suppose V spoke the truth. Then V, W, and Y are all knights (otherwise there would not be at least three knights because U, Z, and X are known to be knaves). It follows that Y spoke the truth. But Y said that exactly two were knights. This contradicts the result that V, W, and Y are all knights.

8. ∴ V cannot have spoken the truth, and so V is a knave.

9. ∴ U, Z, X, and V are all knaves, and so there are at most two knights.

10. Suppose that Y is a knave. Then the only knight is W, which means that Z spoke the truth. But we have already seen that this is impossible. Hence Y is a knight.

11. By 6, 9, and 10, the only possible solution is that U, Z, X, and V are knaves and W and Y are knights. Examination of the statements shows that this solution is consistent: in this case, the statements of U, Z, X, and V are false and those of W and Y are true.

39. Suppose Socko is telling the truth. Then Fats is also telling the truth because if Lefty killed Sharky then Muscles didn't kill Sharky. Consequently, two of the men were telling the truth, which contradicts the fact that all were lying except one. Therefore, Socko is not telling the truth: Lefty did not kill Sharky. Hence Muscles is telling the truth and all the others are lying. It follows that Fats is lying, and so Muscles killed Sharky.

41. (1) $q \to r$ premise
         $\sim r$ premise
    ∴ $\sim q$ by modus tollens

(2) $p \lor q$ premise
        $\sim q$ by (1)
    ∴ p by disjunctive syllogism

(3) $\sim q \to u \land s$ premise
        $\sim q$ by (1)
    ∴ $u \land s$ by modus ponens

(4) $u \land s$ by (3)
    ∴ s by conjunctive simplification

(5) p by (2)
        s by (4)
    ∴ $p \land s$ by conjunctive addition

(6) $p \wedge s \to t$ premise
     $p \wedge s$ by (5)
     $\therefore t$ by modus ponens

43. (1) $\sim q \vee s$ premise
     $\sim s$ premise
     $\therefore \sim q$ by disjunctive syllogism

(2) $p \to q$ premise
     $\sim q$ by (1)
     $\therefore \sim p$ by modus tollens

(3) $r \vee s$ premise
     $\sim s$ premise
     $\therefore r$ by disjunctive syllogism

(4) $\sim p$ by (2)
     r by (3)
     $\therefore \sim p \wedge r$ by conjunctive addition

(5) $\sim p \wedge r \to u$ premise
     $\sim p \wedge r$ by (4)
     $\therefore u$ by modus ponens

(6) $\sim s \to \sim t$ premise
     $\sim s$ premise
     $\therefore \sim t$ by modus ponens

(7) $w \vee t$ premise
     $\sim t$ by (6)
     $\therefore w$ by disjunctive syllogism

(8) u by (5)
     w by (7)
     $\therefore u \wedge w$ by conjunctive addition

Section 1.4

2. $R = 1$

4. $S = 0$

6. The input/output table is as follows.

Input		Output
P	Q	R
1	1	1
1	0	1
0	1	0
0	0	1

8. The input/output table is as follows.

Input			Output
P	Q	R	S
1	1	1	1
1	1	0	1
1	0	1	1
1	0	0	1
0	1	1	1
0	1	0	0
0	0	1	0
0	0	0	0

10. $(P \wedge Q) \vee \sim Q$

12. $(P \vee Q) \wedge (P \vee R)$

14.

15.

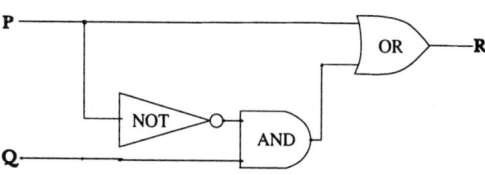

17.

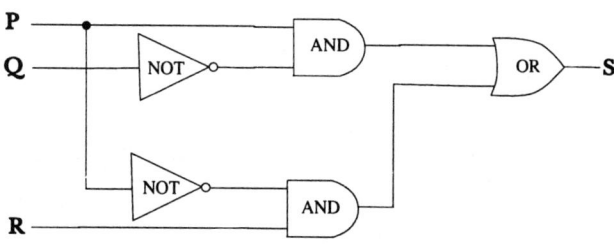

19. a. $(P \wedge \sim Q \wedge R) \vee (P \wedge \sim Q \wedge \sim R) \vee (\sim P \wedge Q \wedge \sim R)$

b. One circuit having the given input/output table is the following.

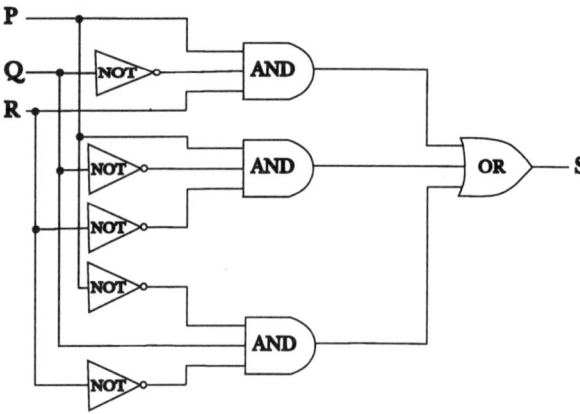

21. a. $(P \wedge Q \wedge \sim R) \vee (\sim P \wedge Q \wedge R) \vee (\sim P \wedge \sim Q \wedge R)$

b. One circuit having the given input/output table is the following.

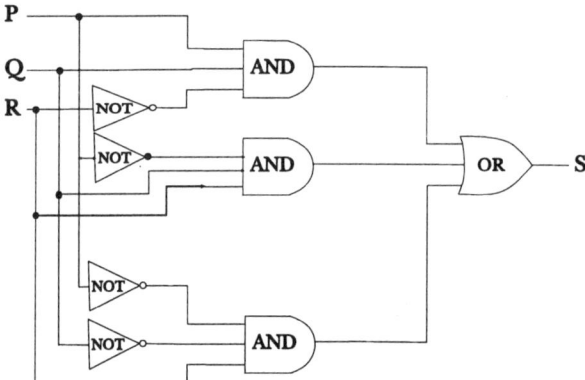

23. (The following is the answer for the second and subsequent printings of the text.)

The input/output table is as follows.

Input			Output
P	Q	R	S
1	1	1	1
1	1	0	0
1	0	1	0
1	0	0	0
0	1	1	0
0	1	0	0
0	0	1	0
0	0	0	1

One circuit (among many) having this input/output table is the following.

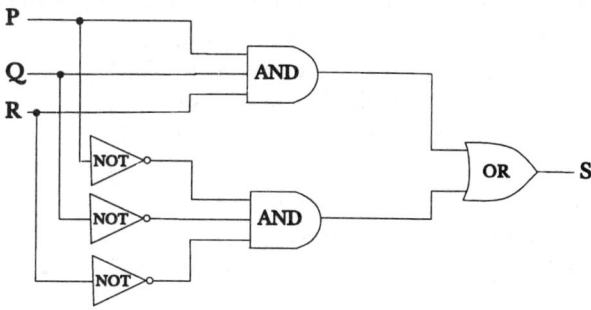

25. Let P, Q, and R indicate the positions of the driver's side door, the passenger's side door, and the ceiling switch, respectively. For P and Q, let 0 indicate a closed door and 1 and open door. For R, let 0 indicate that the switch is in the back position and 1 indicate that it is in the front position. Let S indicate whether the light is off (0) or on (1). Initially, $P = Q = R = 0$ and $S = 0$. Because each movement of any door or switch turns the light on if it is off and off it it is on, the complete input/output table is as follows.

Input			Output
P	Q	R	S
1	1	1	1
1	1	0	0
1	0	1	0
1	0	0	1
0	1	1	0
0	1	0	1
0	0	1	1
0	0	0	0

One circuit (among many) having this input/output table is the following.

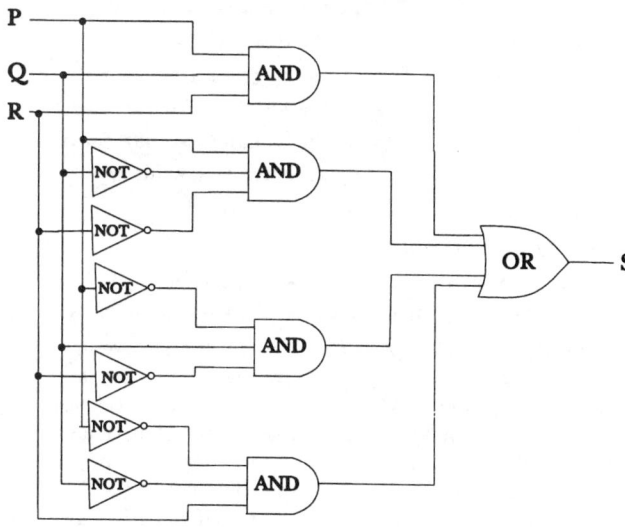

Note: An alternate answer would interchange the 1's and 0's.

27. The Boolean expression for (a) is $\sim(P \wedge Q) \wedge P$ and for (b) it is $P \wedge \sim Q$. We must show that if these expressions are regarded as statement forms, then they are logically equivalent. But

$$\begin{aligned}
\sim(P \wedge Q) \wedge P &\equiv (\sim P \vee \sim Q) \wedge P && \text{by De Morgan's law} \\
&\equiv P \wedge (\sim P \vee \sim Q) && \text{by the commutative law for } \wedge \\
&\equiv (P \wedge \sim P) \vee (P \wedge \sim Q) && \text{by the distributive law} \\
&\equiv c \vee (P \wedge \sim Q) && \text{by the negation law for } \wedge \\
&\equiv (P \wedge \sim Q) \vee c && \text{by the commutative law for } \vee \\
&\equiv P \wedge \sim Q && \text{by the identity law for } \vee.
\end{aligned}$$

29. The Boolean expression for (a) is $(P \wedge Q) \vee (\sim P \wedge Q) \vee (P \wedge \sim Q)$ and for (b) it is $P \vee Q$. We must show that if these expressions are regarded as statement forms, then they are logically equivalent. But

$$\begin{aligned}
(P \wedge Q) &\vee (\sim P \wedge Q) \vee (P \wedge \sim Q) \\
&\equiv ((P \wedge Q) \vee (\sim P \wedge Q)) \vee (P \wedge \sim Q) && \text{by inserting parentheses (which is legal by the} \\
& && \text{associative law for } \vee) \\
&\equiv ((Q \wedge P) \vee (Q \wedge \sim P)) \vee (P \wedge \sim Q) && \text{by the commutative law for } \wedge \\
&\equiv (Q \wedge (P \vee \sim P)) \vee (P \wedge \sim Q) && \text{by the distributive law} \\
&\equiv (Q \wedge t) \vee (P \wedge \sim Q) && \text{by the negation law for } \vee \\
&\equiv Q \vee (P \wedge \sim Q) && \text{by the identity law for } \wedge \\
&\equiv (Q \vee P) \wedge (Q \vee \sim Q) && \text{by the distributive law} \\
&\equiv (Q \vee P) \wedge t && \text{by the negation law for } \vee \\
&\equiv Q \vee P && \text{by the identity law for } \wedge \\
&\equiv P \vee Q && \text{by the commutative law for } \vee.
\end{aligned}$$

31. $(\sim P \wedge \sim Q) \vee (\sim P \wedge Q) \vee (P \wedge \sim Q)$

$$\begin{aligned}
&\equiv ((\sim P \wedge \sim Q) \vee (\sim P \wedge Q)) \vee (P \wedge \sim Q) && \text{by inserting parentheses (which is legal by the} \\
& && \text{associative law for } \vee) \\
&\equiv (\sim P \wedge (\sim Q \vee Q)) \vee (P \wedge \sim Q) && \text{by the distributive law} \\
&\equiv (\sim P \wedge (Q \vee \sim Q)) \vee (P \wedge \sim Q) && \text{by the commutative law for } \vee \\
&\equiv (\sim P \wedge t) \vee (P \wedge \sim Q) && \text{by the negation law for } \vee \\
&\equiv \sim P \vee (P \wedge \sim Q) && \text{by the identity law for } \wedge \\
&\equiv (\sim P \vee P) \wedge (\sim P \vee \sim Q) && \text{by the distributive law} \\
&\equiv (P \vee \sim P) \wedge (\sim P \vee \sim Q) && \text{by the commutative law for } \vee \\
&\equiv t \wedge (\sim P \vee \sim Q) && \text{by the negation law for } \vee \\
&\equiv (\sim P \vee \sim Q) \wedge t && \text{by the commutative law for } \wedge \\
&\equiv \sim P \vee \sim Q && \text{by the identity law for } \wedge \\
&\equiv \sim(P \vee Q) && \text{by De Morgan's law.}
\end{aligned}$$

32. $(P \wedge Q \wedge R) \vee ((P \wedge \sim Q \wedge R) \vee (P \wedge \sim Q \wedge \sim R)$

$$\begin{aligned}
&\equiv ((P \wedge (Q \wedge R)) \vee (P \wedge (\sim Q \wedge R))) \vee (P \wedge (\sim Q \wedge \sim R)) \vee (P \wedge \sim Q) \\
& \quad \text{by inserting parentheses (which is legal by the associative law for } \vee) \\
&\equiv (P \wedge [(Q \wedge R) \vee (\sim Q \wedge R)]) \vee (P \wedge [\sim Q \wedge \sim R]) && \text{by the distributive law} \\
&\equiv P \wedge ([(Q \wedge R) \vee (\sim Q \wedge R)] \vee [\sim Q \wedge \sim R]) && \text{by the distributive law} \\
&\equiv P \wedge ([(R \wedge Q) \vee (R \wedge \sim Q)] \vee [\sim Q \wedge \sim R]) && \text{by the commutative law for } \wedge \\
&\equiv P \wedge ([(R \wedge (Q \vee \sim Q)] \vee [\sim Q \wedge \sim R]) && \text{by the distributive law} \\
&\equiv P \wedge ([(R \wedge t] \vee [\sim Q \wedge \sim R]) && \text{by the negation law for } \vee \\
&\equiv P \wedge (R \vee [\sim Q \wedge \sim R]) && \text{by the identity law for } \wedge \\
&\equiv P \wedge ((R \vee \sim Q) \wedge (R \vee \sim R)) && \text{by the distributive law} \\
&\equiv P \wedge ((R \vee \sim Q) \wedge t) && \text{by the negation law for } \vee \\
&\equiv P \wedge (R \vee \sim Q) && \text{by the identity law for } \wedge.
\end{aligned}$$

33. *a.*

$$\begin{aligned}(p \mid q) \mid (p \mid q) &\equiv \sim[(p \mid q) \wedge (p \mid q)] &&\text{by definition of } \mid\\ &\equiv \sim(p \mid q) &&\text{by the idempotent law for } \wedge\\ &\equiv \sim[\sim(p \wedge q)] &&\text{by definition of } \mid\\ &\equiv p \wedge q &&\text{by the double negative law.}\end{aligned}$$

b.

$$\begin{aligned}p \wedge (\sim q \vee r) &\equiv (p \mid (\sim q \vee r)) \mid (p \mid (\sim q \vee r)) &&\text{by part (a)}\\ &\equiv (p \mid [(\sim q \mid \sim q) \mid (r \mid r)]) \mid (p \mid [(\sim q \mid \sim q) \mid (r \mid r)]) &&\text{by Example 1.4.7(b)}\\ &\equiv (p \mid [((q \mid q) \mid (q \mid q)) \mid (r \mid r)]) \mid (p \mid [((q \mid q) \mid (q \mid q)) \mid (r \mid r)]) &&\text{by Example 1.4.7(a)}.\end{aligned}$$

34. *b.*

$$\begin{aligned}p \vee q &\equiv \sim(\sim(p \vee q)) &&\text{by the double negative law}\\ &\equiv \sim(p \downarrow q) &&\text{by definition of } \downarrow\\ &\equiv (p \downarrow q) \downarrow (p \downarrow q) &&\text{by part (a).}\end{aligned}$$

c.

$$\begin{aligned}p \wedge q &\equiv \sim(\sim p \vee \sim q) &&\text{by De Morgan's law and the double negative law}\\ &\equiv \sim p \downarrow \sim q &&\text{by definition of } \downarrow\\ &\equiv (p \downarrow p) \downarrow (q \downarrow q) &&\text{by part (a).}\end{aligned}$$

d.

$$\begin{aligned}p \rightarrow q &\equiv \sim p \vee q &&\text{by Exercise 13(a) of Section 1.2}\\ &\equiv (\sim p \downarrow q) \downarrow (\sim p \downarrow q) &&\text{by part (b)}\\ &\equiv ((p \downarrow p) \downarrow q) \downarrow ((p \downarrow p) \downarrow q) &&\text{by part (a).}\end{aligned}$$

e.

$$\begin{aligned}p \leftrightarrow q &\equiv (p \rightarrow q) \wedge (q \rightarrow p) &&\text{by the truth table on page 24}\\ &\equiv ([[(p \downarrow p) \downarrow q] \downarrow [(p \downarrow p) \downarrow q]]) \wedge ([[(q \downarrow q) \downarrow p] \downarrow [(q \downarrow q) \downarrow p]]) &&\text{by part (d)}\\ &\equiv ((([(p \downarrow p) \downarrow q] \downarrow [(p \downarrow p) \downarrow q])) \downarrow ([(p \downarrow p) \downarrow q] \downarrow [(p \downarrow p) \downarrow q])))\\ &\quad \downarrow ((([(q \downarrow q) \downarrow p] \downarrow [(q \downarrow q) \downarrow p]) \downarrow ([(q \downarrow q) \downarrow p] \downarrow [(q \downarrow q) \downarrow p]))) &&\text{by part (c).}\end{aligned}$$

Section 1.5

2. $43 = 32 + 8 + 2 + 1 = 101011_2$
3. $287 = 256 + 16 + 8 + 4 + 2 + 1 = 100011111_2$
5. $1297 = 1024 + 256 + 16 + 1 = 10100010001_2$
7. $1 + 4 + 16 = 21$
8. $2 + 4 + 16 + 32 = 54$
10. $1 + 2 + 4 + 64 = 71$
12. 11000_2
14. 10000010101_2
16. 1101_2
18. 1011101_2
19. *b.* $S = 0, T = 1$ *c.* $S = 0, T = 0$
21. $67_{10} = (64 + 2 + 1)_{10} = 01000011_2 \longrightarrow 10111100 \longrightarrow 10111101$. So the answer is 10111101.
23. $108_{10} = (64 + 32 + 8 + 4)_{10} = 01101100_2 \longrightarrow 10010011 \longrightarrow 10010100$. So the answer is 10010100.

25. $10001001 \longrightarrow 01110110 \longrightarrow 01110111_2 = (64+32+16+4+2+1)_{10} = 119_{10}$. So the answer is 119.

27. $10111010 \longrightarrow 01000101 \longrightarrow 01000110_2 = (64+4+2)_{10} = 70_{10}$. So the answer is 70.

31. $79_{10} = (64+8+4+2+1)_{10} = 1001111_2 \longrightarrow 01001111$

    $-43_{10} = -(32+8+2+1)_{10} = -101011_2 \longrightarrow 00101011 \longrightarrow 11010100 \longrightarrow 11010101$

    So the 8-bit representations of 79 and -43 are 01001111 and 11010101. Adding the 8-bit representations gives

    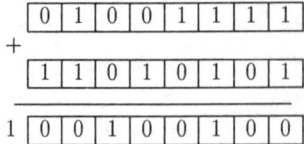

    Truncating the 1 in the 2^8th position gives 00100100. Since the leading bit of this number is a 0, the answer is positive. Converting back to decimal form gives

    $$00100100 \longrightarrow 100100_2 = (32+4)_{10} = 36_{10}.$$

    So the answer is 36.

32. $123_{10} = (64+32+16+8+2+1)_{10} = 1111011_2 \longrightarrow 01111011$

    $-94_{10} = -(64+16+8+4+2)_{10} = -1011110_2 \longrightarrow 01011110 \longrightarrow 10100001 \longrightarrow 10100010$

    So the 8-bit representations of 123 and -94 are 01111011 and 10100010. Adding the 8-bit representations gives

    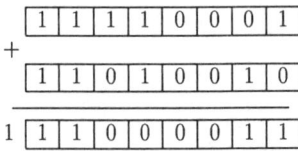

    Truncating the 1 in the 2^8th position gives 00011101. Since the leading bit of this number is a 0, the answer is positive. Converting back to decimal form gives

    $$00011101 \longrightarrow 11101_2 = (16+8+4+1)_{10} = 29_{10}.$$

    So the answer is 36.

33. $-15_{10} = -(8+4+2+1)_{10} = -1111_2 \longrightarrow 00001111 \longrightarrow 11110000 \longrightarrow 11110001$

    $-46_{10} = -(32+8+4+2)_{10} = -101110_2 \longrightarrow 00101110 \longrightarrow 11010001 \longrightarrow 11010010$

    So the 8-bit representations of -15 and -46 are 11110001 and 10100010. Adding the 8-bit representations gives

    Truncating the 1 in the 2^8th position gives 11000011. Since the leading bit of this number is a 1, the answer is negative. Converting back to decimal form gives

    $$11000011 \longrightarrow 00111100 \longrightarrow 00111101 \longrightarrow -111101_2 = -(32+16+8+4+1)_{10} = -61_{10}.$$

    So the answer is -61.

34. Suppose a and b are two integers in the range from 1 through 128 whose sum $a+b$ is also in this range. Since
$$1 \le a+b \le 128$$
then
$$-1 \ge -(a+b) \ge -128 \quad \text{by multiplying through by } -1.$$
Adding 2^9 to all parts of the inequality gives
$$2^9 - 1 \ge 2^9 - (a+b) \ge 2^9 - 128 > 2^8,$$
and so
$$2^8 < 2^9 - (a+b) < 2^9. \quad (*)$$
Now observe that
$$(2^8 - a) + (2^8 - b) = 2 \cdot 2^8 - (a+b) = 2^9 - (a+b).$$
Hence by substitution into $(*)$,
$$2^8 < (2^8 - a) + (2^8 - b) < 2^9.$$
Consequently,
$$(2^8 - a) + (2^8 - b) = 2^8 + \text{ smaller powers of } 2,$$
and so the binary representation of $(2^8 - a) + (2^8 - b)$ has a leading 1 in the 2^8th position.

36. $E0D_{16} = 14 \cdot 16^2 + 0 + 13 = 3597$
37. $29FB_{16} = 2 \cdot 16^3 + 9 \cdot 16^2 + 15 \cdot 16 + 11 = 10{,}747$
39. $B53DF8_{16} = 1011\ 0101\ 0011\ 1101\ 1111\ 1000_2$
40. $2ACF93_{16} = 0010\ 1010\ 1100\ 1111\ 1001\ 0011_2$
42. $1011011111000101_2 = B7C5_{16}$
43. $0011101000111100_2 = 3A3C_{16}$
44. a. $61502_8 = (6 \cdot 8^4 + 1 \cdot 8^3 + 5 \cdot 8^2 + 0 \cdot 8 + 2)_{10} = 25410_{10}$

To convert an integer from octal to binary notation:

1. Write each octal digit of the integer in fixed 3-bit binary notation (including leading zeros as needed);
2. Juxtapose the results.

As an example, consider converting 61502_8 to binary notation:
$$6_8 = 110_2 \quad 1_8 = 001_2 \quad 5_8 = 101_2 \quad 0_8 = 000_2 \quad 2_8 = 010_2.$$
So in binary notation the integer should be 110001101000010_2. To check this result, write this integer in decimal notation and compare it to the result of part (a):
$$110001101000010_2 = (1 \cdot 2^{14} + 1 \cdot 2^{13} + 1 \cdot 2^9 + 1 \cdot 2^8 + 1 \cdot 2^6 + 1 \cdot 2)_{10} = 25410_{10}.$$
It agrees.

b. To convert an integer from binary to octal notation:

1. Group the digits of the binary number into sets of three, starting from the right and adding leading zeros as needed;
2. Convert the binary numbers in each set of three into octal digits;
3. Juxtapose those octal digits.

As an example consider converting 1101011101_2 to octal notation. Grouping the binary digits in sets of three and adding two leading zeros gives

$$001\ 101\ 011\ 101.$$

  c. To convert each group into an octal digit, note that

$$001_2 = 1_8 \quad 101_2 = 5_8 \quad 011_2 = 3_8 \quad 101_2 = 5_8.$$

So the octal version of the integer should be 1535_8. To check this result, observe that

$$1101011101_2 = (1 \cdot 2^9 + 1 \cdot 2^8 + 1 \cdot 2^6 + 1 \cdot 2^4 + 1 \cdot 2^3 + 1 \cdot 2^2 + 1)_{10} = 861_{10}$$

and

$$1535_8 = (1 \cdot 8^3 + 5 \cdot 8^2 + 3 \cdot 8 + 5)_{10} = 861_{10}$$

also.

Chapter 2: The Logic of Quantified Statements

Ability to use the logic of quantified statements correctly is necessary for doing mathematics because mathematics is, in a very broad sense, about quantity. The main purpose of this chapter is to familiarize students with the language of universal and existential statments. The various facts about quantified statements developed in this chapter are used extensively in Chapter 3 and are referred to throughout the rest of the book. Experience with the formalism of quantification is especially useful to students planning to study LISP or Prolog, program verification, or relational databases.

Many students come to college with inconsistent interpretations of quantified statements. In tests made at DePaul University, over 60% of students chose the statement "No firetrucks are red" as the negation of "All firetrucks are red." Yet, through guided discussion, these same students came fairly quickly to accept that "Some firetrucks are not red" conveys the negation more accurately, and most learned to take negations of general statements of the form "$\forall x$, if $P(x)$ then $Q(x)$," "$\forall x$, $\exists y$ such that $P(x,y)$," and so forth with reliable accuracy.

Suggestions

1. The exercises in Sections 2.1 and 2.2 are designed to try to imprint new language patterns on students' minds. Because it takes time to develop new habits, it is helpful to continue assigning exercises from these sections for several days after covering them in class. To prepare for Chapter 3, universal conditional statements should especially be emphasized. As in Section 1.2, care may need to be taken not to spend excessive class time going over the more difficult exercises on necessary and sufficient conditions.

2. The most important idea of Section 2.3 is also the simplest: the rule of universal instantiation. Yet to a very great extent this inference rule drives mathematical reasoning. If you wish to move rapidly through Chapter 2, you could focus on this rule and its immediate consequences in Section 2.3 and omit the discussion of how to use diagrams to check validity of arguments.

Section 2.1

1. *c.* False   *d.* True   *e.* False   *f.* True

2. *b.* $\{n \in \mathbf{Z} \mid n^2 \leq 30\} = \{n \in \mathbf{Z} \mid -\sqrt{30} \leq n \leq \sqrt{30}\} = \{-5, -4, -3, -2, -1, 0, 1, 2, 3, 4, 5\}$

3. *b.* True   *d.* True: $x^2 > 4 \Leftrightarrow x > 2$ or $x < -2 \Leftrightarrow |x| > 2$

5. *Counterexample*: Let $a = 1$ or $a = -1$, and note that both $(1-1)/1 = 0$ and $(-1-1)/(-1) = 2$ are integers.

7. *Counterexample*: Let $x = 1$ and $y = 1$, and note that $\sqrt{1+1} = \sqrt{2}$, $\sqrt{1} + \sqrt{1} = 1 + 1 = 2$, and $2 \neq \sqrt{2}$. (This is one counterexample among many.)

10. *a. Some acceptable answers:* All squares are rectangles. If a figure is a square then that figure is a rectangle. Every square is a rectangle. All figures that are squares are rectangles. Any figure that is a square is a rectangle.

    b. Some acceptable answers: There is a set with sixteen subsets. Some set has sixteen subsets. Some sets have sixteen subsets. There is at least one set that has sixteen subsets.

11. *b.* ∀ real numbers x, x is positive, negative, or zero.

    d. ∀ logicians x, x is not lazy.

12. *b.* ∃ a real number x such that x is rational.

14. *Some acceptable answers*: If a student is in CSC 310, then that student has taken MAT 140. All students in CSC 310 have taken MAT 140. Every student in CSC 310 has taken MAT 140. Each student who is in CSC 310 has taken MAT 140. Given any student in CSC 310, that student has taken MAT 140.

15. *b.* ∀x, if x is a valid argument with true premises, then x has a true conclusion.

    Or: ∀ valid arguments x, if x has true premises then x has a true conclusion.

    d. ∀ integers m and n, if m and n are odd then mn is odd.

16. *b.* ∀x, if x is a computer science student then x needs to take assembly language programming.

17. *b.* ∃ a question x such that x is easy. ∃x such that x is a question and x is easy.

20. *b. Some acceptable answers*: There is a real number that is not positive, negative, or zero. There is a real number that is neither positive, negative, nor zero. ∃ a real number x such that x is not positive and x is not negative and x is not zero.

    d. Some acceptable answers: Some logicians are lazy. ∃ a logician x such that x is lazy.

21. *b. Some acceptable answers*: ∀ real numbers x, x is not rational. No real numbers are rational. All real numbers are irrational.

22. *b. Some acceptable answers*: There is a valid argument with true premises that has a false conclusion. ∃ a valid argument x such that x has true premises and x does not have a true conclusion. Some valid arguments with true premises do not have true conclusions.

    d. Some acceptable answers: There are two odd integers whose product is even. ∃ integers m and n such that m and n are odd and mn is even. ∃ odd integers m and n such that mn is even. For some odd integers m and n, mn is even.

23. *Some acceptable answers:* Some computer science student does not need to take assembly language programming. ∃ a computer science student x such that x does not need to take assembly language programming. Some computer science students do not need to take assembly language programming. There is at least one computer science student who does not need to take assembly language programming.

25. The proposed negation is not correct. *Correct negation:* There are an irrational number x and a rational number y such that $x \cdot y$ is rational. Or: There are an irrational number and a rational number whose product is rational.

27. The proposed negation is not correct. *Correct negation:* ∃ real numbers x_1 and x_2 such that $x_1^2 = x_2^2$ and $x_1 \neq x_2$.

28. *b.* True   *d.* True   *e.* False: $x = 36$ is a counterexample.

30. ∃ a computer program P such that P is correct and P does not compile without error messages.

32. ∃ an integer n such that n is prime and both n is not odd and $n \neq 2$.

    Or: ∃ an integer n such that n is prime and neither is n odd nor does n equal 2.

34. ∃ an animal x such that x is a cat and either x does not have whiskers or x does not have claws.

36. There is an integer n such that n^2 is even but n is not even.

37. The statement is not existential. An informal negation of the statement is "There are some easy questions on the exam" or "Some of the questions on the exam are easy." A formal version of the statement is "∀ questions on the exam x, x is not easy" or "∀x, if x is a question on the exam then x is not easy." In informal language, these can be written as "None of the questions on the exam are easy."

38. *b. One possible answer:* Let $P(x)$ be "$x^2 \neq 2$." The statements "$\forall x \in \mathbf{Z}, x^2 \neq 2$" and "$\forall x \in \mathbf{Q}, x^2 \neq 2$" are true, but the statement "$\forall x \in \mathbf{R}, x^2 \neq 2$" is false.

Section 2.2

In 2-17 there are other correct answers besides those shown.

2. *a.* Let $n = 16$. *b.* Let $n = 10^8 + 1$. *c.* Let $n = 10^{10^{10}} + 1$.

4. *a.* There is a book that everyone has read.

   b. Given any book, there is a person who has not read that book.

6. *a.* Every rational number is equal to a ratio of some two integers.

   b. There is a rational number that is not equal to a ratio of any two integers.

8. *a.* There is a real number whose sum with any real number equals zero.

   b. Given any real number x, there is a real number y such that $x + y \neq 0$.

11. *a.* ∀ people x, ∃ a person y such that x trusts y.

    b. ∃ a person x such that ∀ people y, x does not trust y.

12. *a.* ∃ a person x such that ∀ people y, x trusts y.

    b. ∀ people x, ∃ a person y such that x does not trust y.

14. *a.* ∀ truth tables T, ∃ an integer n such that the number of rows in T equals 2^n.

    b. ∃ a truth table T such that ∀ integers n, the number of rows in T is not equal to 2^n.

15. *a.* ∀ actions A, ∃ a reaction R such that R is equal and opposite to A.

    b. ∃ an action A such that ∀ reactions R, R is not equal and opposite to A.

17. *a.* ∀ integers n, ∃ a prime number p such that $n < p < 2n$.

    b. ∃ an integer n such that ∀ prime numbers p, $p \leq n$ or $p \geq 2n$.

19. *a.* ∀ real numbers x, ∃ a negative real number y such that $x > y$.

    b. The given statement is true and the version with interchanged quantifiers is also true.

20. *b.* This statement says that every student chose a salad. This is false: Yuen did not choose a salad.

    c. This statement says that some particular dessert was chosen by every student. This is true: every student chose pie.

    d. This statement says that some particular beverage was chosen by every student. This is false: There is no beverage that was chosen by every student.

    e. This statement says that some partucular item was not chosen by any student. This is false: every item was chosen by at least one student.

    f. This statement says that there is a station from which every student made a selection. This is true: every student chose a main course, every student chose a dessert, and every student chose a beverage.

21. *b. One solution:* Present all the students in the class with a list of residence halls and ask them to check off all residence halls containing a person they have dated. If some residence hall is checked off by every student in the class, then (assuming the students are all truthful) the statement is true. Otherwise, the statement is false.

   c. One solution: Present all the students in the class with a list of residence halls and ask them to write the number of people they have dated from each hall next to the name of that hall. If no number written down is a 1, then (assuming the students are all truthful) the statement is true. Otherwise, the statement is false.

23. *contrapositive:* ∀ computer programs P, if P does not compile without error messages, then P is not correct.

   converse: ∀ computer programs P, if P compiles without error messages then P is correct.

   inverse: ∀ computer programs P, if P is not correct, then P does not compile without error messages.

25. *contrapositive:* If an integer is not even, then its square is not even.

   converse: If an integer is even, then its square is even.

   inverse: If the square of an integer is not even, then the integer is not even.

27. *contrapositive:* ∀ integers n, if n is not odd and $n \neq 2$ then n is not prime.

   converse: ∀ integers n, if n is odd or $n = 2$, then n is prime.

   inverse: ∀ integers n, if n is not prime, then both n is not odd and $n \neq 2$.

   Or: ∀ integers n, if n is not prime, then neither is n odd nor is n equal to 2.

29. *contrapositive:* ∀ animals A, if A does not have whiskers or A does not have claws then A is not a cat.

   converse: ∀ animals A, if A has whiskers and A has claws then A is a cat.

   inverse: ∀ animals A, if A is not a cat, then either A does not have whiskers or A does not have claws.

30. Consider the statement: ∀ real numbers x, if $x > 0$ then $x^2 > 0$. This statement is true. But its inverse is "∀ real numbers x, if $x \not> 0$ then $x^2 \not> 0$," which is false. (One counterexample is $x = -1$ because $-1 \not> 0$ but $(-1)^2 > 0$.)

32. If an integer is divisible by 6, then it is divisible by 3.

34. If a person does not have a grade-point average of at least 3.7, then that person cannot graduate with honors. Or: If a person graduates with honors then that person earned a grade-point average of at least 3.7.

36. There is a person who does not have a large income and is happy.

38. There is a function that is continuous but not differentiable.

40. *b.* No   *c.* No   *e.* $X = g$ *g.* $X = b_1$, $X = w_1$

41. ∃ a real number $\epsilon > 0$ such that ∀ integers N, ∃ an integer n with $n > N$ and either $L - \epsilon \geq a_n$ or $a_n \geq L + \epsilon$.

42. *b.* This statement is false: there are two distinct integers x such that $1/x$ is an integer, namely $x = 1$ and $x = -1$.

   c. This statement is true: given any real number x, if $x + y = 0$ then $y = -x$ and so y is unique.

43. $\exists! x \in D$ such that $P(x) \equiv \exists x \in D$ such that $(P(x) \land (\forall y \in D, \text{if } P(y) \text{ then } y = x))$

44. *b.* These statements do not necessarily have the same truth values. For instance, let $D = \mathbf{R}$, the set of all real numbers, let $P(x)$ be "x is positive," and let $Q(x)$ be "x is negative." Then "$\exists x \in D, (P(x) \wedge Q(x))$" can be written "$\exists$ a real number x such that x is both positive and negative," which is false. On the other hand, "$(\exists x \in D, \ P(x)) \wedge (\exists x \in D, Q(x))$" can be written "$\exists$ a real number that is positive and $\exists$ a real number that is negative," which is true.

c. These statements do not necessarily have the same truth values. For example, let $D = \mathbf{Z}$, the set of all integers, let $P(x)$ be "x is even," and let $Q(x)$ be "x is odd." Then the statement "$\forall x \in D, (P(x) \vee Q(x))$" can be written "$\forall$ integers x, x is even or x is odd," which is true. On the other hand, "$(\forall x \in D, P(x)) \vee (\forall x \in D, \ Q(x))$" can be written "All integers are even or all integers are odd," which is false.

d. These statements have the same truth values for all domains D and predicates $P(x)$ and $Q(x)$.

If the statement "$\exists x \in D, (P(x) \vee Q(x))$" is true, then by definition of the truth values for $\exists$, the predicate $P(x) \vee Q(x)$ is true for at least one element x in D. Let's call such an element x_0. Then $P(x_0) \vee Q(x_0)$ is true, and so by definition of the truth values for $\vee$, at least one of $P(x_0)$ or $Q(x_0)$ is true. In case $P(x_0)$ is true, then the statement "$\exists x \in D, \ P(x)$" is true. In case $Q(x_0)$ is true, then the statement "$\exists x \in D, \ Q(x)$" is true. Since at least one of these cases must occur, the statement "$\exists x \in D, (P(x) \vee Q(x))$" is true by definition of truth values for $\vee$.

If the statement "$(\exists x \in D, \ P(x)) \vee (\exists x \in D, \ Q(x))$" is true, then by definition of truth values for $\vee$, at least one of the statements "$\exists x \in D, \ P(x)$" or "$\exists x \in D, \ Q(x)$" is true. In case "$\exists x \in D, \ P(x)$" is true, then by definition of truth values for $\exists$, there exists an element, say x_1, in D such that $P(x_1)$ is true. Then by definition of the truth values for $\vee$, $P(x_1) \vee Q(x_1)$ is true, and so by definition of the truth values for $\exists$, "$\exists x, (P(x) \vee Q(x))$" is true. Similarly, in case "$\exists x \in D, \ Q(x)$" is true, then by definition of the truth values for $\exists$, there exists an element, say x_2, in D such that $Q(x_2)$ is true. It follows by definition of the truth values for $\vee$ that $P(x_2) \vee Q(x_2)$ is true, and so by definition of the truth values for $\exists$, "$\exists x, (P(x) \vee Q(x))$" is true. Since one of the two cases must occur, we can conclude that the statement "$\exists x \in D, (P(x) \vee Q(x))$" is true.

Section 2.3

1. *a.* $(x+y)^2 = x^2 + 2xy + y^2$
   e. $(\log(t_1) + \log(t_2))^2 = (\log(t_1))^2 + 2(\log(t_1))(\log(t_2)) + (\log(t_2))^2$

4. $(5^{\frac{1}{2}})^4 = 5^{\frac{1}{2} \cdot 4}$

6. This computer program is not correct.

11. invalid, converse error

12. invalid, inverse error

13. valid, universal modus ponens

14. invalid, inverse error

15. invalid, converse error

16. invalid, converse error

17. invalid, converse error

18. valid, universal modus tollens

19. *c.* valid, universal modus tollens
    d. invalid, inverse error

20. *a.* Either of the following diagrams could represent the given premises.

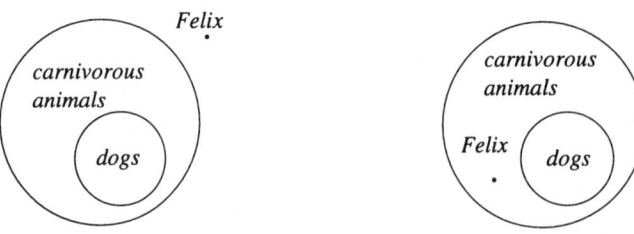

In both, the premises are true, but in (1) the conclusion is true whereas in (2) the conclusion is false.

b. The answer to (a) shows that there is an argument of the given form with true premises and a false conclusion. Hence the argument form is invalid.

22. Invalid. Let D be the set of all discrete mathematics students, T the set of all thoughtful people, and V the set of all people who can tell a valid from an invalid argument. Any one of the following diagrams could represent the given premises.

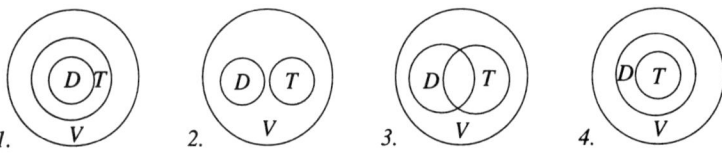

Only in drawing (1) is the conclusion true. Hence it is possible for the premises to be true while the conclusion is false, and so the argument is invalid.

24. Invalid. Let D represent the set of all college dormitory food, G the set of all good food, and W the set of all wasted food. Then any one of the following diagrams could represent the given premises.

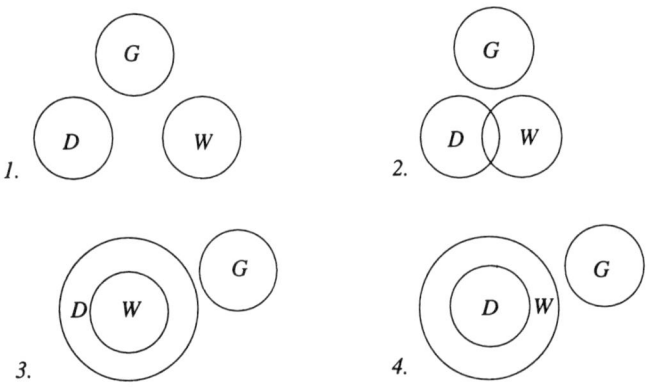

Only in drawing (1) is the conclusion true. Hence it is possible for the premises to be true while the conclusion is false, and so the argument is invalid.

25. Valid. The only drawing representing the truth of the premises also represents the truth of the conclusion.

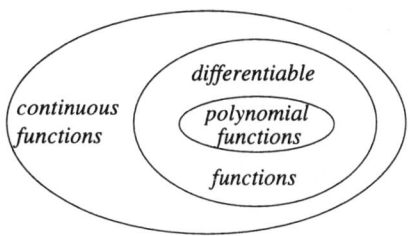

26. Valid. The only drawing representing the truth of the premises also represents the truth of the conclusion.

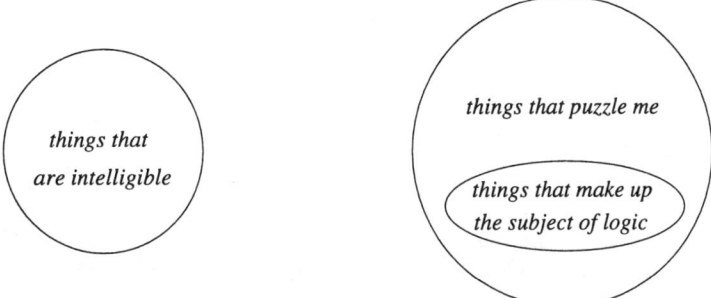

28. 2. These arguments are not arranged in regular order like the ones I am used to.

    4. If examples are not arranged in regular order like the ones I am used to, then I cannot understand them.

    1. (*contrapositive form*) If I can't understand a logic example, then I grumble when I work it.

    5. If I grumble at an example, then it gives me a headache.

    3. (*contrapositive form*) If an example makes my head ache, then it is not easy.

∴ These arguments are not easy.

30. 3. Shakespeare wrote *Hamlet*.

    5. If a person wrote *Hamlet*, then that person was a true poet.

    2. If a person is a true poet, then he can stir the hearts of men.

    4. If a writer can stir the hearts of men, then that writer understands human nature.

    1. If a writer understands human nature, then that writer is clever.

∴ Shakespeare was clever.

31. The law of universal modus tollens says that the following form of argument is valid:

$\forall x$ in D, if $P(x)$ then $Q(x)$.   ← *major premise*
$\sim Q(c)$ for a particular c in D.   ← *minor premise*
∴ $\sim P(c)$.

Proof of Validity: Suppose the major and minor premises of the above argument form are both true. *[We must show that the conclusion $\sim P(c)$ is true.]* By the minor premise, $\sim Q(c)$ is true for a particular value of c in D. By the major premise and the rule of universal instantiation, the statement "If $P(c)$ then $Q(c)$" is true for that particular c. But by modus tollens, since the statements "If $P(c)$ then $Q(c)$" and "$\sim Q(c)$" are both true, it follows that $\sim P(c)$ is also true. *[This is what was to be shown.]*

Chapter 3:
Elementary Number Theory and Methods of Proof

One aim of this chapter is to introduce students to methods for evaluating whether a given mathematical statement is true or false. Throughout the chapter the emphasis is on learning to prove and disprove statements of the form "$\forall x$ in D, if $P(x)$ then $Q(x)$." To prove such a statement directly, one supposes one has a particular but arbitrarily chosen element x in D for which $P(x)$ is true and one shows that $Q(x)$ must also be true. To disprove such a statement, one shows that there is an element x in D (a counterexample) for which $P(x)$ is true and $Q(x)$ is false. To prove such a statement by contradiction, one shows that no counterexample exists, that is, one supposes that there is an x in D for which $P(x)$ is true and $Q(x)$ is false and one shows that this supposition leads to a contradiction. Direct proof, disproof by counterexample, and proof by contradiction can, therefore, all be viewed as three aspects of one whole. One arrives at one or the other by a thoughtful examination of the given statement, knowing what it means for a statement of that form to be true or false.

Another aim of this chapter is to provide students with fundamental knowledge about numbers that is needed in mathematics and computer science. Surprisingly many college students have little intuition for numbers, even integers. Many claim not to be familiar with how to write down a prime factorization, and even very good students often do not know that a rational number is a ratio of integers (having been taught to think of rational numbers as certain kinds of decimals). To accomodate a wide range of student backgrounds and abilities, the exercise sets contain problems of varying difficulty. It is especially important in this chapter to keep in close touch with how students are doing so as to assign problems that are appropriate.

Suggestions

1. The careful use of definitions is stressed throughout the chapter. To bring the idea alive in class, you might try the following technique. Each time you write the definition for a new term, go through a few examples, phrasing each as a question. For instance, immediately after defining rational, write "Is 0.873 rational?" and simultaneously ask the question out loud. To a student's answer of "yes," write "Yes, because" and look expectantly at the student. The student may be surprised that you seem to expect additional words but is generally able to supply the reason without difficulty (or other students may help out). You move on to slightly more complicated examples (Is $-(5/3)$ rational? Is 0 rational? Is $0.252525\ldots$ rational?), each time acting as if you take it for granted that the student answering the question will give a reason. Soon students learn to give the reference to the definition without prompting and gradually they come to understand the value of using the definition to answer such questions. By the way, if your students dispute that 0 is divisible by, say, 2 (a common occurrence), you can use the occasion to emphasize that it is the definition and only the definition that determines the answer.

2. On the due date for the first assignment asking for proofs, it is extremely helpful to recruit a member of the class to present his or her proof to the class as a whole. If the student's proof is perfect, it serves as a model for the rest of the class. If it is less than perfect, the class benefits from analyzing it together. It is especially helpful at this early stage for students to see what kinds of things the instructor finds to both correct and to praise in the proof. Corrections must, of course, be made with regard for the feelings of the student making the presentation. But normally you can balance criticism with praise in a way that students find encouraging.

3. A number of exercises in the book are phrased as open-ended questions, requiring students to find an answer and justify it with either a proof or a counterexample. After you have assigned one of these exercises, students will often ask that the answer be shown in class. When this happens,

you might leave the question open for a while for the class to discuss as a group, perhaps suggesting that students imagine they are the mathematical problem-solving group of a large company and that the answer to the question carries consequences of considerable importance to the company. This works best if you act as a leader but stay somewhat in the background, identifying students who hold opposing points of view and inviting them to come to the board to present their answers for the rest of the class to critique, but feigning ignorance as to the answer and indicating that it is the responsibility of the group as a whole to come to a consensus. (The reason you need to lead the discussion is that if, for instance, the statement is false, it is important to arrange for a false "proof" to be shown before a counterexample.) Determining collectively which proofs are valid and which are not and which counterexamples work and which do not can be an informative demonstration of the nature of mathematical truth and a good advertisement for the usefulness of some of the logic studied in the course.

In cases when the given statement is false, you might ask some of the students who found a counterexample to try to explain to the other students what reasoning they used to discover it. The resulting discussion can be quite worthwhile.

4. At some point during this chapter, you might mention proof by handwaving, proof by intimidation, and so forth, contrasting the feeling of unease produced in the hearer by these methods with the feeling of "mathematical certainty" produced by careful use of the methods discussed in the chapter. The question of what is mathematical certainty has been debated recently in connection with mathematical results such as the proof of the four-color theorem. At the frontiers of mathematical research this question does not have a simple answer. But to understand the issues in the current debate, students need a background of experience in understanding and appreciating simpler proofs, such as those discussed in this chapter.

5. One of the trickiest issues you will face in teaching proof is what style to use for the proofs you write in class. Because students catch onto the idea of the proof at dramatically different rates, it seems best to be careful and complete in class. For instance, you might always start with the words "let" or "suppose," state the full supposition, and even include a bracketed [We must show that ...], pointing out at appropriate times how being aware of what is to be shown helps guide the steps of the argument. Part of the reason for taking care in this chapter is to encourage students to develop habits that will serve them well when they encounter mathematics that is more abstract in later chapters and subsequent courses. For instance, the habit of identifying the supposition and the conclusion to be shown in a proof of a statement is very helpful to students when they confront problems in set theory, functions, and relations (*e.g.*, the proof that a composition of one-to-one functions is one-to-one or that a given binary relation is transitive). Exercises 17-20 of Section 3.1 are designed to lay a groundwork for student appreciation of the power and generality of the method of generalizing from the generic particular. Although many students come to a full appreciation only late in the course, pointing out at an early stage that the method depends only on the form of the statement to be proved creates a point of reference for later discussion.

6. A related and equally tricky issue in teaching proof is how to specify an acceptable range of proof styles for students' work. Students need reassurance that acceptable proofs may be written in many different styles, but they also need encouragement to write coherently. To motivate students to write in sentences and in adequate detail, you might suggest that they imagine writing their proofs for an intelligent classmate who has missed the last few days of the course or in a style that they themselves would have been able to understand when they were first learning the subject matter. To emphasize the importance of precision, you might compare writing a mathematical proof with writing a computer program. You might also suggest that a proof is fundamentally a work of communication and point out that employers in business and industry are demanding ever better communication skills from those whom they hire.

Particular requirements, such as that all variables used in a proof be introduced or that all proofs be written in complete sentences, will undoubtedly vary from one instructor to another. In the answers given in Appendix *B*, two different versions of the first proof of Section 3.1 are written out to illustrate some of the variety that is possible. The text itself contains a note to the effect that students should expect to find out from their instructor what the requirements are to be in their class. In one of my classes I happened to write "p.b.a.c." for "particular but arbitrarily chosen" and found that a number of students enthusiastically incorporated this abbreviation in their written work. Although I tell my students that writing these words is optional, they seem to like to use the abbreviation because it reminds them of the idea of the generic paricular with very little writing.

Comments on Exercises

Exercise Set 3.1: #43 and #47: The solutions use indirect reasoning which is not introduced formally until Section 3.6. #46(b): The solution uses the fact (proved formally in Section 3.4) that any integer is either even or odd.
Exercise Set 3.3: #23: To show that "$a \mid b$ and $a \mid c$" is false, it suffices to show that either $a \nmid b$ or $a \nmid c$. #24: This exercise is especially effective at stimulating lively class discussion. Occasionally, a very good student will come up with an ingenious "proof," and often students will propose counterexamples that are not really counterexamples.
Exercise Set 3.4: #14: Students doing this exercise will discover that January 1, 2000 is a Saturday, which means that New Year's Eve is a Friday and the whole week-end can be devoted to celebration. #29: An indirect argument is needed for part of the proof.
Exercise Set 3.7: #5 and #6 are designed to counter student misgeneralization that any ratio of two numbers is rational and that any square root is irrational.

Section 3.1

2. *a.* Yes: $4rs = 2 \cdot (2rs)$ and $2rs$ is an integer because r and s are integers and products of integers are integers.

   b. Yes: $6r + 4s^2 + 3 = 2(3r + 2s^2 + 1) + 1$ and $3r + 2s^2 + 1$ is an integer because r and s are integers and products and sums of integers are integers.

   c. Yes: $r^2 + 2rs + s^2 = (r+s)^2$ and $r+s$ is an integer that is greater than or equal to 2 because both r and s are positive integers, and so each is greater than or equal to 1.

4. For example, let $a = 1$ and $b = 0$. Then $\sqrt{a+b} = \sqrt{1} = 1$ and $\sqrt{a} + \sqrt{b} = \sqrt{1} + \sqrt{0} = 1$ also. Hence $\sqrt{a+b} = \sqrt{a} + \sqrt{b}$ for these values of a and b.

5. For example, let $x = 60$. Note that to four significant digits $2^{60} \cong 1.153 \times 10^{18}$ and $60^{10} \cong 6.047 \times 10^{17}$, and so $2^x \geq x^{10}$.

7. For example, let $n = 25$. Then n is a perfect square because $25 = 5^2$ and since $25 = 9 + 16 = 3^2 + 4^2$, n can be written as a sum of two other perfect squares.

9. $1^2 - 1 + 11 = 11$, which is prime. $2^2 - 2 + 11 = 13$, which is prime.
   $3^2 - 3 + 11 = 17$, which is prime. $4^2 - 4 + 11 = 23$, which is prime.
   $5^2 - 5 + 11 = 31$, which is prime. $6^2 - 6 + 11 = 41$, which is prime.
   $7^2 - 7 + 11 = 53$, which is prime. $8^2 - 8 + 11 = 67$, which is prime.
   $9^2 - 9 + 11 = 83$, which is prime. $10^2 - 10 + 11 = 101$, which is prime.

12. *Proof:* Suppose m and n are any *[particular but arbitrarily chosen]* odd integers. *[We must show that $m+n$ is even.]* By definition of odd, there exist integers r and s such that $m = 2r+1$ and $n = 2s+1$. Then $m + n = (2r+1) + (2s+1) = 2r + 2s + 2 = 2(r + s + 1)$. But $r + s + 1$ is an integer because r, s, and 1 are integers and a sum of integers is an integer. Hence $m + n$ equals twice an integer, and so by definition of even, $m+n$ is even *[as was to be shown]*.

14. *Proof:* Suppose n is any odd integer. *[We must show that $(-1)^n = -1$.]* By definition of odd, $n = 2k+1$ for some integer k. By substitution and the laws of exponents, $(-1)^n = (-1)^{2k+1} = (-1)^{2k} \cdot (-1) = ((-1)^2)^k \cdot (-1)$. But $(-1)^2 = 1$, and since 1 raised to any power equals 1, $((-1)^2)^k = 1^k = 1$. Hence, by substitution, $(-1)^n = ((-1)^2)^k \cdot (-1) = 1 \cdot (-1) = -1$ *[as was to be shown]*.

16. *Counterexample*: Let $a = -4$ and $b = -1$. Then $a^2 = 16$ and $b^2 = 1$, and so $a < b$ but $a^2 \not< b^2$.

18. *Start of Proof*: Suppose x is any *[particular but arbitrarily chosen]* real number such that $x > 1$. *[We must show that $x^2 > x$.]*

20. *Start of Proof*: Suppose x is any *[particular but arbitrarily chosen]* real number such that $0 < x < 1$. *[We must show that $x^2 < x$.]*

24. This incorrect proof begs the question. The second sentence of the proof states a conclusion that follows from the assumption that $m \cdot n$ is even. The next-to-last sentence of the proof states this conclusion as if it were known to be true. But it is not known to be true. In fact, it is the main task of the proof to derive this conclusion, not from the assumption that it is true but from the hypothesis of the theorem.

26. *Proof*: Suppose m is any even integer and n is any odd integer. *[We must show that $m + n$ is odd.]* By definition of even and odd, there exist integers r and s such that $m = 2r$ and $n = 2s+1$. By substitution and algebra, $m + n = 2r + (2s + 1) = 2(r + s) + 1$. Let $k = r + s$. Then k is an integer because r and s are integers and sums of integers are integers. By substitution, $m + n = 2k + 1$ where k is an integer, and so by definition of odd, $m + n$ is odd *[as was to be shown]*.

27. *Counterexample*: Let $m = 5$ and $n = 3$. Then m and n are odd but $m - n = 2$, which is even.

28. *Proof*: Suppose m is any even integer and n is any integer. *[We must show that mn is even.]* By definition of even, there exists an integer k such that $m = 2k$. By substitution and algebra, $mn = (2k)n = 2(kn)$. But $2(kn)$ is even because kn is an integer (being a product of integers). Hence mn is even *[as was to be shown]*.

30. *Proof*: Let m and n be any even integers. By definition of even, $m = 2r$ and $n = 2s$ for some integers r and s. By substitution, $m - n = 2r - 2s = 2(r - s)$. Since $r - s$ is an integer (being a difference of integers), then $m - n$ equals twice some integer, and so $m - n$ is even by definition of even.

31. *Proof*: Let m and n be any odd integers. By definition of odd, $m = 2r + 1$ and $n = 2s + 1$ for some integers r and s. By substitution, $m - n = (2r + 1) - (2s + 1) = 2(r - s)$. Since $r - s$ is an integer (being a difference of integers), then $m - n$ equals twice some integer, and so $m - n$ is even by definition of even.

32. *Proof*: Suppose m and n are any *[particular but arbitrarily chosen]* integers such that $n - m$ is even. *[We must show that $n^3 - m^3$ is even.]* Note that $n^3 - m^3 = (n - m)(n^2 + nm + m^2)$. Now $n - m$ is even by supposition and $n^2 + nm + m^2$ is an integer (being a sum of products of integers). Thus $(n - m)(n^2 + nm + m^2)$ is the product of an even integer and an integer, and so, by exercise 28, it is even. Hence, by substitution, $n^3 - m^3$ is even *[as was to be shown]*.

(Alternatively, the given statement can be proved by direct use of the definition of even, without reference to exercise 28.)

33. *Counterexample*: Let $n = 2$. Then n is prime but $(-1)^n = (-1)^2 = 1 \neq -1$.

34. *Counterexample*: Let $m = 3$. Then $m^2 - 4 = 9 - 4 = 5$, which is not composite.

35. *Counterexample*: Let $n = 11$. Then $n^2 - n + 11 = 11^2 - 11 + 11 = 11^2$, which is not a prime number.

37. *Counterexample*: The number 28 cannot be expressed as a sum of three or fewer perfect squares. The only perfect squares that could be used are those that are smaller than 28: 1, 4, 9, 16, and 25. The method of exhaustion can be used to show that there are just three ways to express 28 as a sum of four or fewer of these numbers: $28 = 25 + 1 + 1 + 1 = 16 + 4 + 4 + 4 = 9 + 9 + 9 + 1$. In none of these ways are only three perfect squares used.

38. *Proof*: Consider any product of four consecutive integers. Call the second smallest of the four n. Then the product is $(n-1)n(n+1)(n+2)$. Let $m = n^2 - n + 1$. Note that m is an intger because sums, products, and differences of integers are integers. Also

$$\begin{aligned} m^2 - 1 &= (n^2 - n + 1)^2 - 1 = (n^4 + 2n^3 - n^2 - 2n + 1) - 1 \\ &= n^4 + 2n^3 - n^2 - 2n = (n-1)n(n+1)(n+2). \end{aligned}$$

Hence the given product of four consecutive integers is one less than a perfect square.

39. *Proof*: Suppose a and b are any nonnegative real numbers. Then

$$\sqrt{a} = \text{that unique nonnegative real number u such that } u^2 \text{ equals } a$$

and

$$\sqrt{b} = \text{that unique nonnegative real number v such that } v^2 \text{ equals } b$$

By substitution and the laws of exponents, $ab = u^2v^2 = (uv)^2$. So uv is that unique nonnegative real number such that $(uv)^2 = ab$. Hence $\sqrt{ab} = uv = \sqrt{a}\sqrt{b}$.

40. *Counterexample*: Let $a = 1$ and $b = 1$. Then $\sqrt{a+b} = \sqrt{1+1} = \sqrt{2}$ and $\sqrt{a}+\sqrt{b} = \sqrt{1}+\sqrt{1} = 2$, and $\sqrt{2} \neq 2$.

41. *Counterexample*: Let $m = n = 3$. Then $mn = 3 \cdot 3 = 9$, which is a perfect square, but neither m nor n is a perfect square.

42. If m and n are perfect squares, then $m = a^2$ and $n = b^2$ for some integers a and b. We may take a and b to be nonnegative because for any real number x, $x^2 = (-x)^2$ and if x is negative then $-x$ is nonnegative. By substitution,

$$\begin{aligned} m + n + 2\sqrt{mn} &= a^2 + b^2 + 2\sqrt{a^2b^2} \\ &= a^2 + b^2 + 2ab \quad \text{since } a \text{ and } b \text{ are nonnegative} \\ &= (a+b)^2. \end{aligned}$$

But $a + b$ is an integer (since a and b are), and so $m + n + 2\sqrt{mn}$ is a perfect square.

43. If an integer n greater than 1 is not prime, then $n = rs$ where r and s are integers and $1 < r < n$ and $1 < s < n$. Note that one of r or s must be less than or equal to $\sqrt{n}$ because, by the order properties of the real numbers, if both were greater than $\sqrt{n}$ then $rs > \sqrt{n}\sqrt{n} = n$, which is not the case. So n is divisible by some integer that is greater than 1 and less than or equal to $\sqrt{n}$. Let p be any prime factor of this number. [A proof that any integer greater than 1 has a prime factor is given in Section 3.3.] Then n is divisible by p and $p \leq \sqrt{n}$. Hence for any integer n, if n is not prime then n is divisible by a prime number p with $p \leq \sqrt{n}$. By taking the contrapositive of this statement, we conclude that if n is not divisible by any prime number p with $p \leq \sqrt{n}$, then n is prime. [Various of the methods of argument used informally in this justification are introduced formally in Section 3.6.]

44. *Counterexample*: Let $p = 11$. Then $2^p - 1 = 2^{11} - 1 = 2047 = 89 \cdot 23$, and so $2^p - 1$ is not prime.

45. *Counterexample*: Let $n = 5$. Then $2^{2^n} + 1 = 2^{32} + 1 = 4,294,967,297 = (641) \cdot (6,700,417)$, and so $2^{2^n} + 1$ is not prime.

46. a. Note that $(x-r)(x-s) = x^2 - (r+s)x + rs$. If both r and s are odd integers, then $r+s$ is even and rs is odd (by exercises 12 and 25). If both r and s are even integers, then both $r+s$ and rs are even (by exercises 10 and 28).

b. It follows from part(a) that $x^2 - 1253x + 255$ cannot be factored over the integers as $(x-r)(x-s)$ because for none of the possible cases (both r and s odd, both r and s even, and one of r and

s odd and the other even) are both $r+s$ and rs odd integers. *[Strictly speaking, we have not established in Section 3.1 that any integer is either even or odd. We do this formally in Section 3.4.]*

47. The polynomial $15x^3 + 7x^2 - 8x - 27$ cannot be factored over the integers. The reason is that if it could be so factored, then there would exist integers a, b, c, d, and e so that

$$\begin{aligned}15x^3 + 7x^2 - 8x - 27 &= (ax^2 + bx + c)(dx + e)\ . \\ &= adx^3 + (ae + bc)x^2 + (be + cd)x + ce.\end{aligned}$$

Equating coefficients gives

$$ad = 15, \quad ae + bc = 7, \quad be + cd = -8, \quad \text{and} \quad ce = -27.$$

Now since $ad = 15$ and $ce = -27$ and 15 and -27 are both odd integers, then a, d, c, and e are all odd *[because by exercise 28 if one of these integers were even, its product with any other integer would also be even]*. If b were also odd then because a, e, and c are odd, bc and ae would also be odd (by exercise 25) and so $ae + bc$ would be even (by exercise 12). But this is impossible because $ae + bc = 7$ which is odd. Hence b must be even. It follows that be must also be even and cd must be odd, and so $be + cd$ must be odd (by exercise 26). But this is impossible because $be + cd = -8$, which is even. Hence no such integers a, b, c, d, and e can be found, or, in other words, the polynomial $15x^3 + 7x^2 - 8x - 27$ cannot be factored over the integers.

Section 3.2

2. $3.9602 = 39602/10000$

5. Let $x = 0.30303030\ldots$. Then $100x = 30.30303030\ldots$, and so $100x - x = 99x = 30$. Hence $x = 30/99$.

7. Let $x = 6.3215215215\ldots$. Then $10000x = 63215.215215\ldots$ and $10x = 63.215215215\ldots$, and so $10000x - 10x = 9990x = 63215 - 63 = 63152$. Hence $x = 63215/9990$.

8. *a.* $\forall$ real numbers x and y, if $xy = 0$ then $x = 0$ or $y = 0$.

   c. If neither of two real numbers is zero, then their product is nonzero.

10. Because p and q are integers, $2p + 3q$ and $5q$ are both integers (since products and sums of integers are integers). Also by the zero product property, $5q \neq 0$ because $5 \neq 0$ and $q \neq 0$. Hence $(2p + 3q)/5q$ is a quotient of integers with a nonzero denominator, and so it is rational.

11. *Proof:* Suppose n is a *[particular but arbitrarily chosen]* integer. Then $n = n \cdot 1$, and so $n = n/1$ by the laws of algebra. Now n and 1 are both integers and $1 \neq 0$. Hence n can be written as a quotient of integers with a nonzero denominator, and so n is rational.

14. This statement is false.

    Counterexample: Both 1 and 0 are rational numbers (by exercise 11) but $1/0$ is not a rational number. (It is not even a number.)

    Modified Statement: The quotient of any rational number and any nonzero rational number is rational.

    Proof of Modified Statement: Suppose r and s are *[particular but arbitrarily chosen]* rational numbers with $s \neq 0$. By definition of rational, $r = a/b$ and $s = c/d$ for some integers a, b, c, and d with $b \neq 0$ and $d \neq 0$. Furthermore, $c \neq 0$ because $s \neq 0$. By substitution and the laws of algebra,

$$\frac{r}{s} = \frac{\frac{a}{b}}{\frac{c}{d}} = \frac{a}{b} \cdot \frac{d}{c} = \frac{ad}{bc}.$$

Now ad and bc are integers because a, b, c, and d are integers and products of integers are integers. Also $bc \neq 0$ by the zero product property because $b \neq 0$ and $c \neq 0$. Thus r/s can be written as a quotient of integers with a nonzero denominator, and so r/s is rational.

15. *Proof*: Suppose r and s are *[particular but arbitrarily chosen]* rational numbers. *[We must show that $r - s$ is rational.]* By definition of rational, $r = a/b$ and $s = c/d$ for some integers a, b, c, and d with $b \neq 0$ and $d \neq 0$. Then by substitution and the laws of algebra, $r - s = a/b - c/d = (ad - bc)/bd$. But $ad - bc$ and bd are both integers because a, b, c, and d are integers and products and sums of integers are integers and $bd \neq 0$ by the zero product property. Hence $r - s$ is a quotient of integers with a nonzero denominator, and so by definition of rational number $r - s$ is rational *[as was to be shown]*.

16. Suppose r is a *[particular but arbitrarily chosen]* rational number. By definition of rational, $r = a/b$ for some integers a and b with $b \neq 0$. Then $-r = -(a/b) = (-a)/b$ by substitution and the laws of algebra. But since a is an integer, so is -a (being the product of -1 and a). Hence $-r$ is a quotient of integers with a nonzero denominator, and so $-r$ is rational.

17. *Proof*: Suppose r and s are any two distinct rational numbers. By definition of rational, $r = a/b$ and $s = c/d$ for some integers a, b, c, and d with $b \neq 0$ and $d \neq 0$ and $r \neq s$. Without loss of generality assume that $r < s$. (For if $s < r$, we can just interchange the names of r and s.) Let $x = (r + s)/2$. By substitution

$$x = \frac{\frac{a}{b} + \frac{c}{d}}{2} = \frac{\frac{ad+bc}{bd}}{2} = \frac{ad+bc}{2bd}.$$

Now $ad + bc$ and $2bd$ are integers because a, b, c, and d are integers and products and sums of integers are integers. And $2bd \neq 0$ by the zero product property. Hence x is a quotient of integers with a nonzero denominator, and so x is rational. Furthermore, since $r < s$, adding s to both sides gives $(r + s) < 2s$, and so $(r + s)/2 < s$. Also since $r < s$, adding r to both sides gives $2r < (r + s)$, and so $r < (r + s)/2$. By combining the inequalities, $r < (r + s)/2 < s$, and so by substitution $r < x < s$.

19. *Proof*: Suppose n is any odd integer. Then $n^2 = n \cdot n$ is a product of odd integers and hence is odd by exercise 25 of Section 3.1.

20. *Proof*: Suppose n is any even integer. Then $n^2 = n \cdot n$ is a product of an even integer and an integer, and hence is even by exercise 28 of Section 3.1.

22. *Proof*: Suppose r is any rational number. Then $r^2 = r \cdot r$ is a product of rational numbers and hence is rational by exercise 12. Also 2, which is an integer, is rational by exercise 11. Thus $2r^2$ is rational by exercise 13. By exercise 15, $2r^2 - r$ is rational (because it is a difference of two rational numbers). Finally, 1, which is an integer, is rational by exercise 11. So by Theorem 3.2.2, $2r^2 - r + 1 = (2r^2 - r) + 1$ is rational.

24. Yes. Since $\dfrac{ax+b}{cx+d} = 1$, then $ax + b = cx + d$, and so $ax - cx = d - b$, or, equivalently, $(a - c)x = d - b$. Thus $x = (d - b)/(a - c)$. Now $d - b$ and $a - c$ are integers because a, b, c, and d are integers and differences of integers are integers. Also $a - c \neq 0$ because it is given that $a \neq c$. Thus x can be written as a quotient of integers with a nonzero denominator, and so x is rational.

25. Yes.

   a. Solve $yz/(y+z) = c$ for z: Given that $yz/(y+z) = c$, then $yz = c(y+z) = cy + cz$, and so $yz - cz = cy$, or equivalently, $(y-c)z = cy$. Consequently, if $y \neq c$, then $z = cy/(y-c)$. But $y \neq c$ (because if $y = c$, then $(y-c)z = 0$, and so $cy = 0$. This would imply that $y = c = 0$, which was excluded by the problem statement). Hence $z = cy/(y-c)$.

   b. Substitute the result of (a) into the equation $xz/(x+z) = b$ to obtain an equation in x and y:
   $$\frac{x \cdot \frac{cy}{y-c}}{x + \frac{cy}{y-c}} = \frac{cxy}{xy - xc + cy} = b.$$

   c. Solve $xy/(x+y) = a$ for y: Given that $xy/(x+y) = a$, then $xy = a(x+y) = ax + ay$, and so $xy - ay = ax$, or equivalently, $(x-a)y = ax$. Consequently, if $x \neq a$, then $y = ax/(x-a)$. But $x \neq a$ (because if $x = a$, then $(x-a)y = 0$, and so $ax = 0$. This would imply that $x = a = 0$, which was excluded by the problem statement). Hence $y = ax/(x-a)$.

   d. Substitute the result of (c) into the result of (b):
   $$b = \frac{cx \cdot \frac{ax}{x-a}}{x \cdot \frac{ax}{x-a} - xc + c \cdot \frac{ax}{x-a}}$$
   $$= \frac{acx^2}{ax^2 - cx^2 + acx + acx}$$
   $$= \frac{acx}{ax - cx + 2ac}.$$

   Multiply both sides by the denominator to obtain $abx - bcx + 2abc = acx$, or equivalently, $x(ac + bc - ab) = 2abc$. Thus $x = 2abc/(ac + bc - ab)$.

   e. Now $2abc$ and $ac + bc - ab$ are integers because a, b, and c are integers and products and sums of integers are integers. Thus x has been expressed as a ratio of integers.

   Note: An alternate and elegant way to solve this exercise is to start with the observation that $\frac{1}{a} = \frac{x+y}{xy} = \frac{1}{x} + \frac{1}{y}$, $\frac{1}{b} = \frac{1}{x} + \frac{1}{z}$, and $\frac{1}{c} = \frac{1}{y} + \frac{1}{z}$, and then compute $\frac{1}{x} - \frac{1}{z}$ and $\frac{1}{x} + \frac{1}{z}$ and solve for x.

26. Let the quadratic equation be $x^2 + bx + c = 0$ where b and c are rational numbers. Suppose one solution, r, is rational. Call the other solution s. Then $x^2 + bx + c = (x-r)(x-s) = x^2 - (r+s)x + rs$. By equating the coefficients of x, $b = -(r+s)$. Solving for s yields $s = -r - b = -(r+b)$. Since s is the negative of a sum of two rational numbers, s also is rational (by Theorem 3.2.2 and exercise 16).

28. *Proof:* Suppose c is a real number that is a root of a polynomial $p(x) = r_n x^n + r_{n-1} x^{n-1} + \ldots + r_1 x + r_0$ where n is a nonnegative real number, $r_n \neq 0$, and $r_0, r_1, \ldots, r_n$ are all rational numbers. By definition of rational, there exist integers $a_0, a_1, \ldots, a_n$ and $b_0, b_1, \ldots, b_n$ such that $r_i = a_i/b_i$ and $b_i \neq 0$ for all integers i with $0 \leq i \leq n$. Since c is a root of $p(x)$, $r_n c^n + r_{n-1} c^{n-1} + \ldots + r_1 c + r_0 = 0$. By substitution,

$$(*) \qquad \frac{a_n}{b_n} \cdot c^n + \frac{a_{n-1}}{b_{n-1}} \cdot c^{n-1} + \ldots + \frac{a_1}{b_1} \cdot c + \frac{a_0}{b_0} = 0.$$

For each $i = 0, 1, 2, \ldots, n$, let m_i be the product of a_i and all the b_j except b_i. Then each m_i is an integer (being a product of integers). Multiplying both sides of (*) by $b_0 \cdot b_1 \cdot \ldots \cdot b_n$ gives $m_n c^n + m_{n-1} c^{n-1} + \ldots + m_1 c + m_0 = 0$. Hence c satisfies the equation $m_n x^n + m_{n-1} x^{n-1} + \ldots + m_1 x + m_0 = 0$ where $m_0, m_1, \ldots, m_n$ are all integers. Thus c satisfies a polynomial with integer coefficients.

29. This incorrect proof just shows the theorem to be true in the one case where one of the rational numbers is 1/4 and the other is 1/2. A correct proof must show the theorem is true for *any* two rational numbers.

32. This incorrect proof begs the question. The second sentence asserts that a certain conclusion follows if $r + s$ is rational, and the rest of the proof uses that conclusion to deduce that $r + s$ is rational. Thus this incorrect proof assumes what is to be proved.

Section 3.3

2. Yes, $51 = 17 \cdot 3$.

4. Yes, $2m(2m + 2) = 4[m(m + 1)]$ and $m(m + 1)$ is an integer because m is an integer and sums and products of integers are integers.

8. Yes, $6a \cdot 10b = 4(15ab)$ and $15ab$ is an integer because a and b are integers and sums and products of integers are integers.

10. No, $61/11 = 5.45454545\ldots$, which is not an integer.

12. Yes, $n^2 - 1 = (4k + 3)^2 - 1 = (16k^2 + 24k + 9) - 1 = 16k^2 + 24k + 8 = 8(2k^2 + 3k + 1)$, and $2k^2 + 3k + 1$ is an integer because k is an integer and sums and products of integers are integers.

15. *Proof*: Suppose a, b, and c are integers and $a \mid b$ and $a \mid c$. *[We must show that $a \mid (b - c)$.]* By definition of divisibility, there exist integers r and s such that $b = ar$ and $c = as$. Then $b - c = ar - as = a(r - s)$ by substitution and the distributive law. But $r - s$ is an integer since it is a difference of two integers. Hence $a \mid (b - c)$ *[as was to be shown]*.

17. *Proof*: Let m and n be any two even integers. By definition of even, $m = 2r$ and $n = 2s$ for some integers r and s. Then $mn = (2r)(2s) = 4(rs)$. Since rs is an integer (being a product of integers), mn is a multiple of 4 (by definition of divisibility).

19. *Proof*: Suppose n is any integer that is divisible by 16. By definition of divisibility, $n = 16k$ for some integer k. Then $n = 16k = 8 \cdot (2k)$. But $2k$ is an integer (being a product of integers), and so n is divisible by 8 (by definition of divisibility).

21. *Proof*: Suppose a, b, and c are any integers such that $a \mid b$. By definition of divisibility, $b = ak$ for some integer k. Multiplying both sides of this equation by c gives $bc = (ak)c = a(kc)$. But kc is an integer because it is a product of integers, and so $a \mid (bc)$ by definition of divisibility.

20. *Proof*: Suppose n is any integer. By basic algebra, $n(6n + 3) = 3[n(2n + 1)]$. But $n(2n + 1)$ is an integer because n is an integer and sums and products of integers are integers. Hence by definition of divisibility, $n(6n + 3)$ is divisible by 3.

22. *Counterexample*: Let $a = c = 3$ and $b = 2$. Then $a \mid c$ (because $3 \mid 3$) but $ab \nmid c$ (because $6 \nmid 3$).

23. *Counterexample*: Let $a = 2$, $b = 3$, and $c = 1$. Then $a \mid (b + c)$ because $2 \mid 4$ but $a \nmid b$ because $2 \nmid 3$.

24. *Counterexample*: Let $a = 6$, $b = 2$, and $c = 3$. Then $a \mid bc$ because $6 \mid 6$ but $a \nmid b$ and $a \nmid c$ because $6 \nmid 2$ and $6 \nmid 3$.

25. *Proof*: Let a and b be integers such that $a \mid b$. By definition of divisibility, $b = a \cdot k$ for some integer k. Squaring both sides of this equation gives $b^2 = (a \cdot k)^2 = a^2 \cdot k^2$. But k^2 is an integer (being a product of the integer k times itself). Hence by definition of divisibility, $a^2 \mid b^2$.

26. *Counterexample*: Let $a = 25$ and $b = 5$. Then $a \mid 10b$ because $25 \mid 50$ but $a \nmid b$ because $25 \nmid 5$.

28. Let n be the number of minutes past 4 p.m. when the athletes first return to the start together. Then n is the smallest multiple of 8 that is also a multiple of 10. This number is 40. Hence the first time the athlethes will return to the start together will be 4:40 p.m.

29. After crossing out all multiples of 2, 3, 5, and 7 (the prime numbers less than $\sqrt{100}$), the remaining numbers are prime. They are circled below.

> ② ③ 4̸ ⑤ 6̸ ⑦ 8̸ 9̸ 1̸0̸ ⑪ 1̸2̸ ⑬ 1̸4̸ 1̸5̸
> 1̸6̸ ⑰ 1̸8̸ ⑲ 2̸0̸ 2̸1̸ 2̸2̸ ㉓ 2̸4̸ 2̸5̸ 2̸6̸ 2̸7̸ 2̸8̸ 2̸9̸
> 3̸0̸ ㉛ 3̸2̸ 3̸3̸ 3̸4̸ 3̸5̸ 3̸6̸ ㊲ 3̸8̸ 3̸9̸ 4̸0̸ ㊶ 4̸2̸ ㊸
> 4̸4̸ 4̸5̸ 4̸6̸ ㊼ 4̸8̸ 4̸9̸ 5̸0̸ 5̸1̸ 5̸2̸ ㊽ 5̸4̸ 5̸5̸ 5̸6̸ 5̸7̸
> 5̸8̸ ㊾ 6̸0̸ ㉼ 6̸2̸ 6̸3̸ 6̸4̸ 6̸5̸ 6̸6̸ ㊼ 6̸8̸ 6̸9̸ 7̸0̸ ㊼
> 7̸2̸ ㊼ 7̸4̸ 7̸5̸ 7̸6̸ 7̸7̸ 7̸8̸ ㊼ 8̸0̸ 8̸1̸ 8̸2̸ ㊼ 8̸4̸ 8̸5̸
> 8̸6̸ 8̸7̸ 8̸8̸ ㊼ 9̸0̸ 9̸1̸ 9̸2̸ 9̸3̸ 9̸4̸ 9̸5̸ 9̸6̸ ㊼ 9̸8̸ 9̸9̸

30. b. Let $N = 12,858,306,120,312$. The sum of the digits of N is 42, which is divisible by 3 but not by 9. Therefore, N is divisible by 3 but not by 9. The right-most digit of N is neither 5 nor 0, and so N is not divisible by 5. The two right-most digits of N are 12, which is divisible by 4. Therefore, N is divisible by 4.

c. Let $N = 517,924,440,926,512$. The sum of the digits of N is 61, which is not divisible by 3 (and hence not by 9 either). Therefore, N is not divisible either by 3 or by 9. The right-most digit of N is neither 5 nor 0, and so N is not divisible by 5. The two right-most digits of N are 28, which is divisible by 4. Therefore, N is divisible by 4.

d. Let $N = 14,328,083,360,232$. The sum of the digits of N is 45, which is divisible by 9 and hence also by 3. Therefore, N is divisible by 9 and by 3. The right-most digit of N is neither 5 nor 0, and so N is not divisible by 5. The two right-most digits of N are 32, which is divisible by 4. Therefore, N is divisible by 4.

31. b. $4851 = 3^2 \cdot 7^2 \cdot 11$ c. $8925 = 3 \cdot 5^2 \cdot 7 \cdot 17$

32. Yes. The number 151 is prime and is a factor of the right-hand side of the given equation. By the unique factorization theorem, it must be a factor of the left-hand side also. But 151 does not equal any of the prime factors of 2, 3, 4, 5, 6, 7, 8, or 9. Therefore, 151 must be a factor of m.

33. Yes. The number 29 is prime and is a factor of the right-hand side of the given equation. By the unique factorization theorem, it must be a factor of the left-hand side also. But 29 does not equal any of the prime factors of 2, 3, 4, or 5. Therefore, 29 must be a factor of n.

35. $p_1^{3e_1} \cdot p_2^{3e_2} \cdot \ldots \cdot p_k^{3e_k}$

36. a.

$$\begin{aligned}(20!)^2 &= (20 \cdot 19 \cdot 18 \cdot 17 \cdot 16 \cdot 15 \cdot 14 \cdot 13 \cdot 12 \cdot 11 \cdot 10 \cdot 9 \cdot 8 \cdot 7 \cdot 6 \cdot 5 \cdot 4 \cdot 3 \cdot 2 \cdot 1)^2 \\ &= (2^2 \cdot 5 \cdot 19 \cdot 2 \cdot 3^2 \cdot 17 \cdot 2^4 \cdot 3 \cdot 5 \cdot 2 \cdot 7 \cdot 2^2 \cdot 3 \cdot 11 \cdot 2 \cdot 5 \cdot 3^2 \cdot 2^3 \cdot 7 \cdot 2 \cdot 3 \cdot 5 \cdot 2^2 \cdot 3 \cdot 2)^2 \\ &= (2^{18} \cdot 3^8 \cdot 5^4 \cdot 7^2 \cdot 11 \cdot 13 \cdot 17 \cdot 19)^2 \\ &= 2^{36} \cdot 3^{16} \cdot 5^8 \cdot 7^4 \cdot 11^2 \cdot 13^2 \cdot 17^2 \cdot 19^2\end{aligned}$$

b. When $(20!)^2$ is written in ordinary decimal form, there are as many zeros at the end of it as there are factors of the form $2 \cdot 5$ ($= 10$) in its prime factorization. Since the prime factorization of $(20!)^2$ contains eight 5's and more than eight 2's, $(20!)^2$ contains eight factors of 10 and hence eight zeros.

38. *Proof:* Suppose n is a nonnegative integer whose decimal representation ends in 5. By exercise 32, $n = 10m + 5$ for some integer m. By factoring out a 5, $n = 10m + 5 = 5(2m + 1)$, and $2m + 1$ is an integer since m is an integer. Hence $10m$ is divisible by 5.

39. *Proof:* Suppose the decimal representation of a nonnegative integer n ends in $d_1 d_0$. Then $n = d_k 10^k + d_{k-1} 10^{k-1} + \ldots + d_2 10^2 + d_1 10 + d_0$ where k is a nonnegative integer and all the d_i are integers from 0 to 9 inclusive.

    Case 1 ($0 \le k < 2$): In this case, $n = d_1 10 + d_0$. Let $s = 0$. Then $n = 100s + 10d_1 + d_0$.

    Case 2 ($k \ge 2$): In this case, $k - 2 \ge 0$. Factoring out a 100 from all but the two right-most terms of n and using the laws of exponents gives $n = 100(d_k 10^{k-2} + d_{k-1} 10^{k-1} + \ldots + d_2) + 10d_1 + d_0$. Let $s = d_k 10^{k-2} + d_{k-1} 10^{k-1} + \ldots + d_2$. Then s is an integer (being a sum of products of integers) and $n = 100s + 10d_1 + d_0$.

    The above argument shows that regardless of whether case 1 or case 2 holds for n, $n = 100s + 10d_1 + d_0$ for some integer s.

    Suppose in addition that $4 \mid (10d_1 + d_0)$. By definition of divisibility, $10d_1 + d_0 = 4r$ for some integer r. Then by substitution, $n = 100s + 4r = 4(25s + r)$. But $25s + r$ is an integer (being a sum of products of integers). Therefore by definition of divisibility, $4 \mid n$.

40. *Proof:* Suppose n is any nonnegative integer for which the sum of the digits of n is divisible by 9. By definition of decimal representation, n can be written in the form
    $n = d_k 10^k + d_{k-1} 10^{k-1} + \ldots + d_2 10^2 + d_1 10 + d_0$ where k is a nonnegative integer and all the d_i are integers from 0 to 9 inclusive. Then

    $$\begin{aligned}
    n &= d_k(\underbrace{99\ldots9}_{k\text{ 9's}} + 1) + d_{k-1}(\underbrace{99\ldots9}_{(k-1)\text{ 9's}} + 1) + \ldots + d_2(99 + 1) + d_1(9 + 1) + d_0 \\
    &= d_k \cdot \underbrace{99\ldots9}_{k\text{ 9's}} + d_{k-1} \cdot \underbrace{99\ldots9}_{(k-1)\text{ 9's}} + \ldots + d_2 \cdot 99 + d_1 \cdot 9 + (d_k + d_{k-1} + \ldots + d_2 + d_1 + d_0) \\
    &= d_k \cdot \underbrace{11\ldots1}_{k\text{ 1's}} \cdot 9 + d_{k-1} \cdot \underbrace{11\ldots1}_{(k-1)\text{ 1's}} \cdot 9 + \ldots + d_2 \cdot 11 \cdot 9 + d_1 \cdot 9 + (d_k + d_{k-1} + \ldots + d_2 + d_1 + d_0) \\
    &= 9(d_k \cdot \underbrace{11\ldots1}_{k\text{ 1's}} + d_{k-1} \cdot \underbrace{11\ldots1}_{(k-1)\text{ 1's}} + \ldots + d_2 \cdot 11 + d_1) + (d_k + d_{k-1} + \ldots + d_2 + d_1 + d_0) \\
    &= (\text{an integer divisible by 9}) + (\text{the sum of the digits of } n).
    \end{aligned}$$

    Since the sum of the digits of n is divisible by 9, n can be written as a sum of two integers each of which is divisible by 9. It follows from exercise 14 that n is divisible by 9.

41. *Proof:* Suppose n is any nonnegative integer for which the sum of the digits of n is divisible by 3. By the same reasoning as in the answer to exercise 40,

    $$\begin{aligned}
    n &= 9(d_k \cdot \underbrace{11\ldots1}_{k\text{ 1's}} + d_{k-1} \cdot \underbrace{11\ldots1}_{(k-1)\text{ 1's}} + \ldots + d_2 \cdot 11 + d_1) + (d_k + d_{k-1} + \ldots + d_2 + d_1 + d_0) \\
    &= 3[3(d_k \cdot \underbrace{11\ldots1}_{k\text{ 1's}} + d_{k-1} \cdot \underbrace{11\ldots1}_{(k-1)\text{ 1's}} + \ldots + d_2 \cdot 11 + d_1)] + (d_k + d_{k-1} + \ldots + d_2 + d_1 + d_0) \\
    &= (\text{an integer divisible by 3}) + (\text{the sum of the digits of } n).
    \end{aligned}$$

    Since the sum of the digits of n is divisible by 3, n can be written as a sum of two integers each of which is divisible by 3. It follows from exercise 14 that n is divisible by 3.

42. What follows is a rather formal justification for the given statement that uses only the mathematics developed in this chapter. A student might appropriately give a much less formal partial justification.

Lemma: For all integers r, if r is even then $10^r - 1$ is divisible by 11, and if r is odd then $10^r + 1$ is divisible by 11.

Proof: Suppose r is an even integer. Then

$11 \cdot [9(10^{r-2} + 10^{r-4} + 10^{r-6} + \cdots + 10^2 + 1)]$

$\begin{aligned}
&= (10 + 1)[(10 - 1)(10^{r-2} + 10^{r-4} + 10^{r-6} + \cdots + 10^2 + 1)] \\
&= (10^2 - 1)(10^{r-2} + 10^{r-4} + 10^{r-6} + \cdots + 10^2 + 1) \\
&= (10^r - 10^{r-2}) + (10^{r-2} - 10^{r-4}) + (10^{r-4} - 10^{r-6}) + \cdots + ((10^4 - 10^2) + (10^2 - 1) \\
&= 10^r - 1.
\end{aligned}$

Hence if r is even, then $10^r - 1$ is divisible by 11.

Suppose r is an odd integer. Then

$11 \cdot [9 \cdot (10^{r-2} + 10^{r-4} + 10^{r-6} + \cdots + 10^3 + 10) + 1]$

$\begin{aligned}
&= (10 + 1)[(10 - 1)(10^{r-2} + 10^{r-4} + 10^{r-6} + \cdots + 10^3 + 10) + 1] \\
&= (10^2 - 1)(10^{r-2} + 10^{r-4} + 10^{r-6} + \cdots + 10^3 + 10) + (10 + 1) \\
&= (10^r - 10^{r-2}) + (10^{r-2} - 10^{r-4}) + (10^{r-4} - 10^{r-6}) + \cdots + ((10^5 - 10^3) + (10^3 - 10) + (10 + 1) \\
&= 10^r + 1.
\end{aligned}$

Hence if r is odd, then $10^r + 1$ is divisible by 11.

Proof of Exercise Statement: Suppose n is any integer for which the alternating sum of the digits is divisible by 11. By definition of decimal representation, n can be written in the form

$$n = d_k 10^k + d_{k-1} 10^{k-1} + \cdots + d_2 10^2 + d_1 10 + d_0$$

where k is a nonnegative integer and all the d_i are integers from 0 to 9 inclusive.

Case 1 (k is even): In this case,

$n = [d_k(10^k - 1) + d_k] + [d_{k-1}(10^{k-1} - 1) - d_{k-1}] + \ldots + [d_2(10^2 - 1) + d_2] + [d_1(10 - 1) - d_1] + d_0$

$= [d_k(10^k - 1) + d_{k-1}(10^{k-1} + 1) + \ldots + d_2(10^2 - 1)d_1(10 + 1)] + [d_0 - d_1 + d_2 - \cdots - d_{k-1} + d_k]$

Case 2 (k is odd): In this case,

$n = [d_k(10^k + 1) - d_k] + [d_{k-1}(10^{k-1} - 1) + d_{k-1}] + \ldots + [d_2(10^2 - 1) + d_2] + [d_1(10 - 1) - d_1] + d_0$

$= [d_k(10^k + 1) + d_{k-1}(10^{k-1} - 1) + \ldots + d_2(10^2 - 1)d_1(10 + 1)] + [d_0 - d_1 + d_2 - \cdots + d_{k-1} - d_k]$

Observe that in each case each term of the first bracketed sum is divisible by 11 (by the lemma), and the second bracketed sum is divisible by 11 (by hypothesis). Thus in each case n is divisible by 11.

Section 3.4

2. $q = 11, r = 1$

4. $q = 0, d = 5$

6. $q = -5, r = 8$

8. *a.* 3 *b.* 7

9. *a.* 30 *b.* 1

10. *a.* 51 *b.* 3

11. *b.* When today is Sunday, 7 days from today is Sunday also. Hence $DayN$ should be 0. Substituting $DayT = 0$ (Sunday) and $n = 7$ into the formula gives $DayN = (DayT + N) \bmod 7 = (0 + 7) \bmod 7 = 0$, which agrees.

    c. When today is Thursday, ten days from today is one week (which is Thursday) plus three days (which is Sunday). Hence $DayN$ should be 0. Substituting $DayT = 4$ (Thursday) and $N = 10$ into the formula gives $DayN = (DayT + N) \bmod 7 = (4 + 10) \bmod 7 = 14 \bmod 7 = 0$, which agrees.

12. Let the days of the week be numbered from 0 (Sunday) through 6 (Saturday) and let $DayT$ and $DayN$ be variables representing the day of the week today and the day of the week N days from today. By the quotient-remainder theorem, there exist unique integers q and r such that $DayT + N = 7q + r$ and $0 \le r < 7$. Now $DayT + N$ counts the number of days to the day N days from today starting last Sunday (where "last Sunday" is interpreted to mean today if today is a Sunday). Thus $DayN$ is the day of the week that is $DayT + N$ days from last Sunday. Because each week has seven days, $DayN$ is the same as the day of the week $DayT + N - 7q$ days from last Sunday. But $DayT + N - 7q = r$ and $0 \le r < 7$. Therefore, $DayN = r = (DayT + N) \bmod 7$.

14. The number of days from January 1, 1990 to January 1, 2000 is 10 years times 365 days per year plus 2 leap year days (in 1992 and 1996), which gives a total of 3652 days. Substituting into the formula $DayN = (DayT + N) \bmod 7$ with $DayT = 1$ (Monday) and $N = 3652$ gives $DayN = 3653 \bmod 7 = 6$, which is a Saturday.

15. *b.*

A	83	8	8
q	3		
d		0	
n			1
p			3

16. *b.* a_{ij} is stored in location $8304 + 4(i - 1) + (j - 1)$. Thus $n = 4(i - 1) + (j - 1)$.

    c. $r = n \text{ div } 4 + 1$, $s = n \bmod 4 + 1$

    Note that this answer can be checked for consistency with the result of part (b) by using the formulas $a \bmod b = a - \lfloor a/b \rfloor b$ and $a \text{ div } b = \lfloor a/b \rfloor$:

    $$\begin{aligned} 4(r-1) + (s-1) &= 4[(n \text{ div } 4 + 1) - 1] + [(n \bmod 4 + 1) - 1] \\ &= 4 \cdot \left\lfloor \frac{n}{4} \right\rfloor + \left(n - \left\lfloor \frac{n}{4} \right\rfloor \cdot 4\right) \\ &= n \end{aligned}$$

17. $r = k \text{ div } n + 1$, $s = k \bmod m + 1$

18. *Proof:* Consider any two consecutive integers. Call the smaller one n. By the parity theorem, either n is even or n is odd.

    Case 1 (n is even): In this case $n = 2k$ for some integer k. Then $n(n + 1) = 2k(2k + 1) = 2[k(2k + 1)]$. But $k(2k + 1)$ is an integer (because it is a product of sums of integers), and so $n(n + 1)$ is even.

    Case 2 (n is odd): In this case $n = 2k + 1$ for some integer k. Then

    $$n(n+1) = (2k+1)[(2k+1)+1] = (2k+1)(2k+2) = 2[(2k+1)(k+1)].$$

    But $(2k + 1)(k + 1)$ is an integer (because it is a product of sums of integers), and so $n(n + 1)$ is even.

    Hence in either case the product $n(n + 1)$ is even *[as was to be shown]*.

19. *Proof:* Suppose n is any odd integer. By definition of odd, $n = 2k+1$ for some integer k. Then $n^2 - 1 = (n-1)(n+1) = [(2k+1)-1][(2k+1)+1] = 2k(2k+1) = 4k(k+1)$. But $k(k+1)$ is even by exercise 18. Hence $k(k+1) = 2m$ for some integer m. By substitution, $n^2 - 1 = 4 \cdot 2m = 8m$, or, equivalently, $n^2 = 8m+1$ where m is an integer.

20. *Proof:* Suppose n is any integer. By the quotient-remainder theorem with $d = 3$, there exist integers q and r such that $n = 3q + r$ and $0 \leq r < 3$. But the only nonnegative integers r that are less than 3 are $0, 1$, and 2. Therefore $n = 3q + 0 = 3q$, or $n = 3q + 1$, or $n = 3q + 2$ for some integer q.

21. *b.* For any integer n, $n(n+1)(n+2) \bmod 3 = 0$.

22. *Proof:* Suppose n is any integer. By exercise 20, $n = 3q$, or $n = 3q + 1$, or $n = 3q + 2$ for some integer q.

    Case 1 ($n = 3q$ for some integer q): In this case, $n^2 = (3q)^2 = 3(3q^2)$. Let $k = 3q^2$. Then k is an integer because it is a product of integers. Hence $n^2 = 3k$ for some integer k.

    Case 2 ($n = 3q+1$ for some integer q): In this case, $n^2 = (3q+1)^2 = 9q^2 + 6q + 1 = 3(3q^2 + 2q) + 1$. Let $k = 3q^2 + 2q$. Then k is an integer because it is a sum of products of integers. Hence $n^2 = 3k + 1$ for some integer k.

    Case 3 ($n = 3q + 2$ for some integer q): In this case, $n^2 = (3q+2)^2 = 9q^2 + 12q + 4 = 9q^2 + 12q + 3 + 1 = 3(3q^2 + 4q + 1) + 1$. Let $k = 3q^2 + 4q + 1$. Then k is an integer because it is a sum of products of integers. Hence $n^2 = 3k + 1$ for some integer k.

    In all three cases, either $n^2 = 3k$ or $n^2 = 3k + 1$ for some integer k. This is what was to be shown.

23. *a.*

    Proof 1: Suppose m and n are integers.

    Case 1 (both m and n are even): In this case both $m+n$ and $m-n$ are even (by exercises 10 and 30 of Section 3.1).

    Case 2 (one of m and n is even and the other is odd): In this case both $m+n$ and $m-n$ are odd (by exercises 11 and 26 of Section 3.1).

    Case 3 (both m and n are odd): In this case both $m+n$ and $m-n$ are even (by exercises 12 and 31 of Section 3.1).

    Thus in all three possible cases, either both $m+n$ and $m-n$ are even or both $m+n$ and $m-n$ are odd *[as was to be shown]*.

    Proof 2: Suppose m and n are integers.

    Case 1 ($m-n$ is even): In this case, $m+n = (m-n) + 2n$, and so $m+n$ is a sum of two even integers and is therefore even.

    Case 2 ($m-n$ is odd): In this case also, $m+n = (m-n) + 2n$. Hence $m+n$ is the sum of an odd integer and an even integer, and is therefore odd.

    Now either $m-n$ is even or $m-n$ is odd *[by the quotient-remainder theorem]*, and thus either case 1 or case 2 must apply. In each case, $m-n$ and $m+n$ are either both even or both odd *[as was to be shown]*.

    c. If $m^2 - n^2 = 40$, then $40 = (m+n)(m-n)$. Now $40 = 2^3 \cdot 5$, and by the unique factorization theorem this factorization is unique up to the order in which the factors are written down. It follows that the only representations of 40 as a product of two positive integers are $40 = 40 \cdot 1 = 8 \cdot 5 = 4 \cdot 10 = 2 \cdot 20$. By part (a), m and n must both be odd or both be even. Thus the only solutions are either $m+n = 10$ and $m-n = 4$ or $m+n = 20$ and $m-n = 2$. This gives either $m = 7$ and $n = 3$ or $m = 11$ and $n = 9$.

24. *Proof:* Suppose n is any integer. *[We must show that $8 \mid n(n+1)(n+2)(n+3)$.]* By the quotient-remainder theorem, $n = 4k$ or $n = 4k+1$ or $n = 4k+2$ or $n = 4k+3$ for some integer k.

Case 1 ($n = 4k$ for some integer)k: In this case,
$$n(n+1)(n+2)(n+3) = 4k(4k+1)(4k+2)(4k+3) = 8[k(4k+1)(2k+1)(4k+3)],$$
which is divisible by 8 (because k is an integer and sums and products of integers are integers).

Case 2 ($n = 4k+1$ for some integer k): In this case,
$$n(n+1)(n+2)(n+3) = (4k+1)(4k+2)(4k+3)(4k+4) = 8[(4k+1)(2k+1)(4k+3)(k+1)],$$
which is divisible by 8 (because k is an integer and sums and products of integers are integers).

Case 3 ($n = 4k+2$ for some integer)k: In this case,
$$n(n+1)(n+2)(n+3) = (4k+2)(4k+3)(4k+3)(4k+5) = 8[(2k+1)(4k+3)(k+1)(4k+5)],$$
which is divisible by 8 (because k is an integer and sums and products of integers are integers).

Case 4 ($n = 4k+3$ for some integer k): In this case,
$$n(n+1)(n+2)(n+3) = (4k+3)(4k+4)(4k+5)(4k+6) = 8[(4k+3)(k+1)(4k+5)(2k+3)],$$
which is divisible by 8 (because k is an integer and sums and products of integers are integers).

Hence in all four possible cases, $8 \mid n(n+1)(n+2)(n+3)$ *[as was to be shown]*.

25. *Proof:* Let n be any integer. *[We must show that $n^2 = 4k$ or $n^2 = 4k+1$ for some integer k.]* By the quotient-remainder theorem, $n = 4q$ or $n = 4q+1$ or $n = 4q+2$ or $n = 4q+3$ for some integer q.

Case 1 ($n = 4q$ for some integer q): In this case, $n^2 = (4q)^2 = 4(4q^2)$. Let $k = 4q^2$. Then k is an integer because it is a product of integers. Hence $n^2 = 4k$ for some integer k.

Case 2 ($n = 4q+1$ for some integer q): In this case, $n^2 = (4q+1)^2 = 16q^2 + 8q + 1 = 4(4q^2 + 2q) + 1$. Let $k = 4q^2 + 2q$. Then k is an integer because it is a sum of products of integers. Hence $n^2 = 4k+1$ for some integer k.

Case 3 ($n = 4q+2$ for some integer q): In this case, $n^2 = (4q+2)^2 = 16q^2 + 16q + 4 = 4(4q^2 + 4q + 1)$. Let $k = 4q^2 + 2q + 1$. Then k is an integer because it is a sum of products of integers. Hence $n^2 = 4k$ for some integer k.

Case 4 ($n = 4q+3$ for some integer q): In this case, $n^2 = (4q+3)^2 = 16q^2 + 24q + 9 = 16q^2 + 24q + 8 + 1 = 4(4q^2 + 6q + 2) + 1$. Let $k = 4q^2 + 6q + 2$. Then k is an integer because it is a sum of products of integers. Hence $n^2 = 4k+1$ for some integer k.

It follows that in all four possible cases, $n^2 = 4k$ or $n^2 = 4k+1$ for some integer k *[as was to be shown]*.

26. Note that this result can be proved directly by dividing into four cases as is done in the proof for exercise 25. But it can also be deduced as follows from the result of exercise 25.

Proof: Let n be any integer with $n \geq 2$. According to the quotient-remainder theorem, when $n^2 - 3$ is divided by 4, the quotient and remainder are unique (provided the remainder r satisfies the condition $0 \leq r < 4$). If $n^2 - 3$ were divisible by 4, then the remainder would be 0. Now by exercise 25, $n^2 = 4k$ or $n^2 = 4k+1$ for some integer k. In the first case, $n^2 - 3 = 4k - 3 = 4(k-1) + 1$, and in the second case $n^2 - 3 = (4k+1) - 3 = 4k - 2 = 4(k-1) + 2$. Thus in the first case the remainder obtained when $n^2 - 3$ is divided by 4 is 1, and in the second case it is 2. In neither case is the remainder 0, as it would have to be if $n^2 - 3$ were divisible by 4. Therefore, $n^2 - 3$ is not divisible by 4.

27. *Proof:* Consider any four consecutive integers. Call the smallest n. Then the sum of the four integers is $n + (n+1) + (n+2) + (n+3) = 4n + 6 = 4(n+1) + 2$. Consequently,

$$\frac{n + (n+1) + (n+2) + (n+3)}{4} = \frac{4(n+1) + 2}{4} = (n+1) + \frac{1}{2}$$

which is not an integer. So the sum of the four integers is not divisible by 4.

28. *Proof:* Let n be any integer and observe that $n(n^2 - 1)(n+2) = (n-1)n(n+1)(n+2)$, which is a product of four consecutive integers. By exercise 24, this product is divisible by 8, and hence by transitivity of divisibility (Theorem 3.3.1) the product is divisible by 4 *[as was to be shown]*. *Note:* The statement can also be proved directly without using exercise 24 by dividing into four cases as is done in the proof for exercise 24.

29. *Proof:* Let p be any prime number except 2 or 3. By the quotient-remainder theorem, p can be written as $6k$ or $6k + 1$ or $6k + 2$ or $6k + 3$ or $6k + 4$ or $6k + 5$ for some integer k. Since p is prime and $p \neq 2$, p is not divisible by 2. Consequently, $p \neq 6k$, $p \neq 6k + 2$, and $p \neq 6k + 4$ for any integer k *[because all of these numbers are divisible by 2]*. Furthermore, since p is prime and $p \neq 3$, p is not divisible by 3. Thus $p \neq 6k + 3$ *[because this number is divisible by 3]*. Therefore, $p = 6k + 1$ or $p = 6k + 5$ for some integer k.

30. *Proof:* Let n be any odd integer. By Theorem 3.4.4, $n^2 = 8m + 1$ for some integer m. Then $n^4 = (8m+1)^2 = 64m^2 + 16m + 1 = 16(4m^2 + m) + 1$. But $4m^2 + m$ is an integer (because it is a sum of products of integers), and so by the quotient-remainder theorem, the remainder obtained when n^4 is divided by 16 is 1. Hence by definition of mod, $n^4 \ mod \ 16 = 1$.

31. *Proof:* Let n be any integer that is not divisible by 2 or 3. By the quotient-remainder theorem n can be written in one of the forms $n = 6q$ or $6q + 1$ or $6q + 2$ or $6q + 3$ or $6q + 4$ or $6q + 5$ for some integer q. Since n is not divisible by 2, $n \neq 6q$, $n \neq 6q + 2$, and $n \neq 6q + 4$ for any integer q *[because all of these numbers would be divisible by 2]*. Furthermore, since n is not divisible by 3, $n \neq 6q + 3$ for any integer q *[because such a number would be divisible by 3]*. Therefore, $p = 6q + 1$ or $p = 6q + 5$ for some integer q.

Case 1 ($n = 6q + 1$ for some integer q): In this case, $n^2 = (6q + 1)^2 = 36q^2 + 12q + 1 = 12(3q^2 + q) + 1$. Let $k = 3q^2 + q$. Then k is an integer because it is a sum of products of integers, and so $n^2 = 12k + 1$ for some integer k.

Case 2 ($n = 6q + 5$ for some integer q): In this case, $n^2 = (6q + 5)^2 = 36q^2 + 60q + 25 = 12(3q^2 + 5q + 2) + 1$. Let $k = 3q^2 + 5q + 2$. Then k is an integer because it is a sum of products of integers, and so $n^2 = 12k + 1$ for some integer k.

Thus in both possible cases, $n^2 = 12k + 1$ for some integer k, or, equivalently, $n^2 \ mod \ 12 = 1$ *[as was to be shown]*.

33. They are equal.

Proof: Suppose m, n, and d are integers and $d \mid (m-n)$. By definition of divisibility, $m - n = dk$ for some integer k. Therefore, $m = n + dk$. Let $r = n \ mod \ d$. Then by definition of mod, $n = qd + r$ where q and r are integers and $0 \leq r < d$. By substitution, $m = n + dk = (qd + r) + dk = d(q + k) + r$. Since $q + k$ is an integer and $0 \leq r < d$, the integral quotient of the division of n by d is $q + k$ and the remainder is r. Hence $m \ mod \ d = r$ also, and so $n \ mod \ d = m \ mod \ d$.

34. Answer to the first question: not necessarily

Counterexample: Let $m = n = 3$, $d = 2$, $a = 1$, and $b = 1$. Then $m \ mod \ d = n \ mod \ d = 3 \ mod \ 2 = 1 = a = b$. But $a + b = 1 + 1 = 2$, whereas $(m + n) \ mod \ d = 6 \ mod \ 2 = 0$.

Answer to the second question: yes.

Proof: Suppose m, n, a, b, and d are integers and $m \bmod d = a$ and $n \bmod d = b$. By definition of mod, $m = dq_1 + a$ and $n = dq_2 + b$ for some integers q_1 and q_2. By substitution, $m + n = (dq_1 + a) + (dq_2 + b) = d(q_1 + q_2) + (a + b)$. Apply the quotient-remainder theorem to $a+b$ to obtain unique integers q_3 and r such that $a+b = dq_3 + r$ and $0 \leq r < d$. By definition of mod, $r = (a+b) \bmod d$. By substitution, $m+n = d(q_1+q_2)+(a+b) = d(q_1+q_2)+(dq_3+r) = d(q_1+q_2+q_3)+r$ where $q_1+q_2+q_3$ and r are integers and $0 \leq r < d$. Hence by definition of mod, $r = (m+n) \bmod d$, and so $(m+n) \bmod d = (a+b) \bmod d$.

35. Answer to the first question: not necessarily

    Counterexample: Let $m = n = 2$, $d = 3$, $a = 2$, and $b = 2$. Then $m \bmod d = n \bmod d = 2 \bmod 3 = 2 = a = b$. But $a \cdot b = 2 \cdot 2 = 4$, whereas $(m \cdot n) \bmod d = 4 \bmod 3 = 1$.

    Answer to the second question: yes.

    Proof: Suppose m, n, a, b, and d are integers and $m \bmod d = a$ and $n \bmod d = b$. By definition of mod, $m = dq_1 + a$ and $n = dq_2 + b$ for some integers q_1 and q_2. By substitution, $m \cdot n = (dq_1 + a) \cdot (dq_2 + b) = d^2 q_1 q_2 + d(aq_1 + bq_2) + ab$. Apply the quotient-remainder theorem to $a \cdot b$ to obtain unique integers q_3 and r such that $a \cdot b = dq_3 + r$ and $0 \leq r < d$. By definition of mod, $r = (a \cdot b) \bmod d$. By substitution, $m \cdot n = d^2 q_1 q_2 + d(aq_1 + bq_2) + ab = d^2 q_1 q_2 + d(aq_1 + bq_2) + dq_3 + r = d(dq_1 q_2 + aq_1 + bq_2 + q_3) + r$ where $dq_1 q_2 + aq_1 + bq_2 + q_3$ and r are integers and $0 \leq r < d$. Hence by definition of mod, $r = (m \cdot n) \bmod d$, and so $(m \cdot n) \bmod d = (a \cdot b) \bmod d$.

36. a. *Proof:* Let x and y be any real numbers.

    Case 1 (x and y are both nonnegative): In this case $|x| = x$, $|y| = y$, and xy is also nonnegative. So $|xy| = xy = |x| \cdot |y|$.

    Case 2 (x is nonnegative and y is negative): In this case $|x| = x$, $|y| = -y$ and $xy \leq 0$. So $|xy| = -(xy) = x(-y) = |x| \cdot |y|$.

    Case 3 (x is negative and y is nonnegative): In this case $|x| = -x$, $|y| = y$ and $xy \leq 0$. So $|xy| = -(xy) = (-x)y = |x| \cdot |y|$.

    Case 4 (x and y are both negative): In this case $|x| = -x$, $|y| = -y$, and $xy > 0$. So $|xy| = (-x)(-y) = |x| \cdot |y|$.

    Therefore in all four possible cases, $|xy| = |x| \cdot |y|$ *[as was to be shown]*.

    c. *Proof:* Let c be any positive real number and let x be any real number.

    Part 1 (Proof that if $-c \leq x \leq c$ then $|x| \leq c$): Suppose that $-c \leq x \leq c$.(*) By the trichotomy law (see Appendix A, T16), either $x \geq 0$ or $x < 0$.

    Case 1 ($x \geq 0$): In this case $|x| = x$, and so by substitution into (*), $-c \leq |x| \leq c$. In particular, $|x| \leq c$.

    Case 2 ($x < 0$): In this case $|x| = -x$, and so $x = -|x|$. Hence by substitution into (*), $-c \leq -|x| \leq c$. In particular, $-c \leq -|x|$. Multiplying both sides by -1 gives $c \geq |x|$, or, equivalently, $|x| \leq c$.

    Therefore, regardless of whether $x \geq 0$ or $x < 0$, $|x| \leq c$ *[as was to be shown]*.

    Part 2 (Proof that if $|x| \leq c$ then $-c \leq x \leq c$): Suppose that $|x| \leq c$.(**) By the trichotomy law, either $x \geq 0$ or $x < 0$.

    Case 1($x \geq 0$): In this case $|x| = x$, and so by substitution into (**), $x \leq c$. Since $x \geq 0$ and $0 \geq -c$, then by transitivity of order (Appendix A, T17), $x \geq -c$. Hence $-c \leq x \leq c$.

    Case 2 ($x < 0$): In this case $|x| = -x$, and so by substitution into (**) $-x \leq c$. Multiplying both sides of this inequality by -1 gives $x \geq -c$. Also since $x < 0$ and $0 < c$, then $x \leq c$. Thus $-c \leq x \leq c$.

    Therefore, regardless of whether $x \geq 0$ or $x < 0$, we conclude that $-c \leq x \leq c$ *[as was to be shown]*.

d. *Proof:* Let any real numbers x and y be given. By part (b)
$$-|x| \le x \le |x|$$
and
$$-|y| \le y \le |y|.$$
Hence by the order properties of the real numbers (Appendix A),
$$(-|x|) + (-|y|) \le x + y \le (|x| + |y|),$$
or, equivalently,
$$-(|x| + |y|) \le x + y \le (|x| + |y|)$$
It follows immediately from part (c) that $|x + y| \le |x| + |y|$.

Section 3.5

2. $\lfloor 10/3 \rfloor = \lfloor 3\ 1/3 \rfloor = 3$, $\lceil 10/3 \rceil = \lceil 3\ 1/3 \rceil = 4$

4. $\lfloor -57/2 \rfloor = \lfloor -28.5 \rfloor = -29$, $\lceil -57/2 \rceil = \lceil -28.5 \rceil = -28$

5. $589\ div\ 12 = \lfloor 589/12 \rfloor = 49$, $589\ mod\ 12 = 589 - 12\lfloor 589/12 \rfloor = 589 - 12 \cdot 49 = 589 - 588 = 1$

6. If k is an integer, then $\lceil k \rceil = k$ because $k - 1 < k \le k$ and $k - 1$ and k are integers.

7. If k is an integer, then $\lceil k + 1/2 \rceil = k + 1$ because $k < k + 1/2 \le k + 1$ and k and $k + 1$ are both integers.

8. When the ceiling notation is used, the answer is either $\lceil n/7 \rceil - 1$ if $n/7$ is not an integer or $\lceil n/7 \rceil$ if $n/7$ is an integer.

9. The number of boxes that is needed is $\lceil n/24 \rceil$. The ceiling notation is more appropriate for this problem. If the floor notation is used, the answer is more complicated: it is either $\lfloor n/24 \rfloor + 1$ if $n/24$ is not an integer or $\lfloor n/24 \rfloor$ if n is an integer.

10. *b.* When the year $n - 1$ is a leap year, then, because leap years contain an extra day, January 1 of the year n is one day of the week later than it would otherwise be. If leap years occurred exactly every four years, then there would be $\lfloor (n-1)/4 \rfloor$ extra leap year days from year 1 to year n. So the day of the week of January 1 of year n would be pushed forward $\lfloor (n-1)/4 \rfloor\ mod\ 7$ days from its value in year 1. But leap years do not occur exactly every four years. Every century year (year that is a multiple of 100), except those that are multiples of 400, the leap year day is not added. So instead of $\lfloor (n-1)/4 \rfloor$ leap year days from year 1 to year n, there are $\lfloor (n-1)/4 \rfloor - \lfloor (n-1)/100 \rfloor + \lfloor (n-1)/400 \rfloor$ leap year days, where $\lfloor (n-1)/100 \rfloor$ is the number of century years from year 1 to year n and $\lfloor (n-1)/400 \rfloor$ is the number of those that are multiples of 400. Hence the day of the week of January 1 of year n is actually pushed forward $(\lfloor (n-1)/4 \rfloor - \lfloor (n-1)/100 \rfloor + \lfloor (n-1)/400 \rfloor)\ mod\ 7$ days from its value in year 1.

11. A necessary and sufficient condition for the floor of a real number to equal the number is that the number be an integer.

13. *a. Proof:* Suppose n and d are integers with $d \ne 0$ and $d \mid n$. Then $n = d \cdot k$ for some integer k. By substitution and algebra, $\lfloor n/d \rfloor = \lfloor d \cdot k/d \rfloor = \lfloor k \rfloor$, and $\lfloor k \rfloor = k$ because $k \le k < k + 1$ and both k and $k + 1$ are integers. But since $n = d \cdot k$, then $k = n/d$. Hence $\lfloor n/d \rfloor = k = n/d$. Therefore, $n = \lfloor n/d \rfloor \cdot d$ *[as was to be shown]*.

b. Proof: Suppose n and d are integers with $d \ne 0$ and $n = \lfloor n/d \rfloor \cdot d$. By definition of floor, $\lfloor n/d \rfloor$ is an integer. Hence, $n = d \cdot (some\ integer)$, and so by definition of divisibility, $d \mid n$.

c. A necessary and sufficient condition for an integer n to be divisible by an integer d is that $n = \lfloor n/d \rfloor \cdot d$.

14. *Counterexample*: Let $x = 2.1$ and $y = 1.9$. Then $\lfloor x - y \rfloor = \lfloor 2.1 - 1.9 \rfloor = \lfloor 0.2 \rfloor = 0$. On the other hand, $\lfloor x \rfloor - \lfloor y \rfloor = \lfloor 2.1 \rfloor - \lfloor 1.9 \rfloor = 2 - 1 = 1$.

15. *Proof 1*: Suppose x is any *[particular but arbitrarily chosen]* real number. Then $\lfloor x - 1 \rfloor$ is some integer: say $\lfloor x - 1 \rfloor = n$. By definition of floor, $n \leq x - 1 < n + 1$. Adding 1 to all parts of this inequality gives $n + 1 \leq x < n + 2$, and thus by definition of floor $\lfloor x \rfloor = n + 1$. Solving this equation for n gives $n = \lfloor x \rfloor - 1$. But $n = \lfloor x - 1 \rfloor$ also. Hence $\lfloor x - 1 \rfloor = \lfloor x \rfloor - 1$ *[as was to be shown]*.

    Proof 2: Suppose x is any *[particular but arbitrarily chosen]* real number. Apply Theorem 3.5.1 with $m = -1$. Then $\lfloor x + m \rfloor = \lfloor x + (-1) \rfloor = \lfloor x - 1 \rfloor$, $\lfloor x \rfloor + m = \lfloor x \rfloor + (-1) = \lfloor x \rfloor + 1$, and so $\lfloor x - 1 \rfloor = \lfloor x \rfloor - 1$ *[as was to be shown]*.

16. *Counterexample*: Let $x = 3/2$. Then $\lfloor x^2 \rfloor = \lfloor (3/2)^2 \rfloor = \lfloor 9/4 \rfloor = 2$, whereas $\lfloor x \rfloor^2 = \lfloor 3/2 \rfloor^2 = 1^2 = 1$.

17. *Proof*: Let n be any integer. By the quotient-remainder theorem and the definition of *mod*, either $n \bmod 3 = 0$ or $n \bmod 3 = 1$ or $n \bmod 3 = 2$.

    Case 1 ($n \bmod 3 = 0$): In this case, $n = 3q$ for some integer q by definition of *mod*. By substitution and algebra, $\lfloor n/3 \rfloor = \lfloor 3q/3 \rfloor = \lfloor q \rfloor$ and $\lfloor q \rfloor = q$ because q is an integer and $q \leq q < q + 1$. But solving $n = 3q$ for q gives $q = n/3$. Thus $\lfloor n/3 \rfloor = q = n/3$ *[as was to be shown]*.

    Case 2 ($n \bmod 3 = 1$): In this case, $n = 3q + 1$ for some integer q by definition of *mod*. By substitution and algebra, $\lfloor n/3 \rfloor = \lfloor (3q + 1)/3 \rfloor = \lfloor q + 1/3 \rfloor$ and $\lfloor q + 1/3 \rfloor = q$ because q is an integer and $q \leq q + 1/3 < q + 1$. But solving $n = 3q + 1$ for q gives $q = (n - 1)/3$. Thus $\lfloor n/3 \rfloor = q = (n - 1)/3$ *[as was to be shown]*.

    Case 3 ($n \bmod 3 = 2$): The proof for this case is included in the answers in Appendix B.

18. *Counterexample*: Let $x = y = 1.5$. Then $\lceil x + y \rceil = \lceil 1.5 + 1.5 \rceil = \lceil 3 \rceil = 3$, whereas $\lceil x \rceil + \lceil y \rceil = \lceil 1.5 \rceil + \lceil 1.5 \rceil = 2 + 2 = 4$.

19. *Proof*: Let x be any *[particular but arbitrarily chosen]* real number. Then $\lceil x + 1 \rceil$ is some integer: say $\lceil x + 1 \rceil = n$. By definition of ceiling, $n - 1 < x + 1 \leq n$. Subtracting 1 from all parts of this inequality gives $n - 2 < x \leq n - 1$, and thus by definition of ceiling $\lceil x \rceil = n - 1$. Solving this equation for n gives $n = \lceil x \rceil + 1$. But $n = \lceil x + 1 \rceil$ also. Hence $\lceil x + 1 \rceil = \lceil x \rceil + 1$ *[as was to be shown]*.

20. *Counterexample*: Let $x = y = 1.1$. Then $\lceil x \cdot y \rceil = \lceil (1.1) \cdot (1.1) \rceil = \lceil 1.21 \rceil = 2$. On the other hand, $\lceil x \rceil \cdot \lceil y \rceil = \lceil 1.1 \rceil \cdot \lceil 1.1 \rceil = 2 \cdot 2 = 4$.

21. *Proof*: Let n be any odd integer. *[We must show that $\lceil n/2 \rceil = (n + 1)/2$.]* By definition of odd, $n = 2k + 1$ for some integer k. Substituting into the left-hand side of the equation to be proved gives

$$\left\lceil \frac{n}{2} \right\rceil = \left\lceil \frac{2k + 1}{2} \right\rceil = \left\lceil k + \frac{1}{2} \right\rceil = k + 1,$$

where $\left\lceil k + \dfrac{1}{2} \right\rceil = k + 1$ by definition of ceiling because $k < k + 1/2 < k + 1$ and k is an integer. On the other hand, substituting into the right-hand side of the equation to be shown gives

$$\frac{n + 1}{2} = \frac{(2k + 1) + 1}{2} = \frac{2k + 2}{2} = \frac{2(k + 1)}{2} = k + 1$$

also. Thus both the left- and right-hand sides of the equation to be proved equal $k + 1$, and so both are equal to each other. In other words, $\lceil n/2 \rceil = (n + 1)/2$ *[as was to be shown]*.

22. *Counterexample*: Let $x = y = 1.9$. Then $\lceil xy \rceil = \lceil (1.9)(1.9) \rceil = \lceil 3.61 \rceil = 4$, whereas $\lceil 1.9 \rceil \cdot \lfloor 1.9 \rfloor = 2 \cdot 1 = 2$.

24. *Proof:* Suppose m is any integer and x is any real number that is not an integer. By definition of floor, $\lfloor x \rfloor = n$ where n is an integer and $n \leq x < n+1$. Since x is not an integer, $x \neq n$, and so $n < x < n+1$. Multiply all parts of this inequality by -1 to obtain $-n > -x > -n-1$. Then add m to all parts to obtain $m-n > m-x > m-n-1$, or, equivalently, $m-n-1 < m-x < m-n$. But $m-n-1$ and $m-n$ are both integers, and so by definition of floor, $\lfloor m-x \rfloor = m-n-1$. By substitution, $\lfloor x \rfloor + \lfloor m-x \rfloor = n + (m-n-1) = m-1$ *[as was to be shown]*.

25. *Proof:* Let x be any *[particular but arbitrarily chosen]* real number and let $n = \lfloor x/2 \rfloor$. Then by definition of floor, $n \leq x/2 < n+1$.

    Case 1 (n is even): In this case, $n/2$ is an integer, and we divide all parts of the inequality $n \leq x/2 < n+1$ by 2 to obtain $n/2 \leq x/4 < (n+1)/2$. But $(n+1)/2 = n/2 + 1/2 < n/2 + 1$. Hence $n/2 \leq x/4 < n/2 + 1$. Since n is even, $n/2$ is an integer, and so by definition of floor, $\lfloor x/4 \rfloor = n/2$. Since $n = \lfloor x/2 \rfloor$, then $\lfloor \lfloor x/2 \rfloor / 2 \rfloor = \lfloor x/4 \rfloor$.

    Case 2 (n is odd): In this case $(n-1)/2$ is an integer, and by Theorem 3.5.2 $\lfloor n/2 \rfloor = (n-1)/2$. We divide all parts of the inequality $n \leq x/2 < n+1$ by 2 to obtain $n/2 \leq x/4 < (n+1)/2$. But $n/2 > (n-1)/2$. Thus $(n-1)/2 \leq x/4 < (n+1)/2$. Now $(n-1)/2$ is an integer and $(n+1)/2 = (n-1)/2 + 1$. Hence by definition of floor, $\lfloor x/4 \rfloor = (n-1)/2$. Since $(n-1)/2 = \lfloor n/2 \rfloor$ and $n = \lfloor x/2 \rfloor$, then $\lfloor \lfloor x/2 \rfloor / 2 \rfloor = \lfloor n/2 \rfloor = \lfloor x/4 \rfloor$.

27. *Proof:* Suppose x is any real number such that $x - \lfloor x \rfloor \geq 1/2$. Multiply both sides by 2 to obtain $2x - 2\lfloor x \rfloor \geq 1$, or equivalently, $2x \geq 2\lfloor x \rfloor + 1$. Now by definition of floor, $x < \lfloor x \rfloor + 1$. Hence $2x < 2\lfloor x \rfloor + 2$. Put the two inequalties involving x together to obtain $2\lfloor x \rfloor + 1 \leq 2x < 2\lfloor x \rfloor + 2$. By definition of floor, then, $\lfloor 2x \rfloor = 2\lfloor x \rfloor + 1$.

28. *Proof:* Let n be any odd integer. *[We must show that $\left\lfloor \dfrac{n^2}{4} \right\rfloor = \left(\dfrac{n-1}{2}\right)\left(\dfrac{n+1}{2}\right)$.]* By definition of odd, $n = 2k+1$ for some integer k. Substituting into the left-hand side of the equation to be proved gives

$$\left\lfloor \frac{n^2}{4} \right\rfloor = \left\lfloor \frac{(2k+1)^2}{4} \right\rfloor = \left\lfloor \frac{4k^2 + 4k + 1}{4} \right\rfloor = \left\lfloor k^2 + k + \frac{1}{4} \right\rfloor = k^2 + k.$$

where $\left\lfloor k^2 + k + \dfrac{1}{4} \right\rfloor = k^2 + k$ by definition of floor because $k^2 + k$ is an integer and $k^2 + k < k^2 + k + \dfrac{1}{4} < k^2 + k + 1$. On the other hand, substituting into the right-hand side of the equation to be proved gives

$$\left(\frac{n-1}{2}\right)\left(\frac{n+1}{2}\right) = \left(\frac{(2k+1)-1}{2}\right)\left(\frac{(2k+1)+1}{2}\right) = \left(\frac{2k}{2}\right)\left(\frac{2k+2}{2}\right) = k(k+1) = k^2 + k$$

also. Thus the left- and right-hand sides of the equation to be proved both equal $k^2 + k$, and so the two sides are equal to each other. In other words, $\left\lfloor \dfrac{n^2}{4} \right\rfloor = \left(\dfrac{n-1}{2}\right)\left(\dfrac{n+1}{2}\right)$ *[as was to be shown]*.

29. *Proof:* Let n be any odd integer. *[We must show that $\left\lceil \dfrac{n^2}{4} \right\rceil = \dfrac{n^2+3}{4}$.]* By definition of odd, $n = 2k+1$ for some integer k. Substituting into the left-hand side of the equation to be proved gives

$$\left\lceil \frac{n^2}{4} \right\rceil = \left\lceil \frac{(2k+1)^2}{4} \right\rceil = \left\lceil \frac{4k^2 + 4k + 1}{4} \right\rceil = \left\lceil k^2 + k + \frac{1}{4} \right\rceil = k^2 + k + 1$$

where $\left\lceil k^2 + k + \dfrac{1}{4} \right\rceil = k^2 + k + 1$ by definition of floor because $k^2 + k + 1$ is an integer and $k^2 + k < k^2 + k + \dfrac{1}{4} < k^2 + k + 1$. On the other hand, substituting into the right-hand side of the equation to be proved gives

$$\frac{n^2+3}{4} = \frac{(2k+1)^2+3}{4} = \frac{4k^2+4k+1+3}{4} = \frac{4k^2+4k+4}{4} = k^2+k+1$$

also. Thus the left- and right-hand sides of the equation to be proved both equal k^2+k+1, and so the two sides are equal to each other. In other words, $\left\lceil\frac{n^2}{4}\right\rceil = \frac{n^2+3}{4}$ [as was to be shown].

Section 3.6

2. *Negation of statement*: There is a greatest negative real number.

    Proof of statement: Suppose not. Suppose there is a greatest negative real number a. [*We must deduce a contradiction.*] Then $a < 0$ and $a \geq x$ for every negative real number x. Let $b = a/2$. Then b is a real number because b is a quotient of two real numbers (with a nonzero denominator). Also $a < a/2 < 0$. [*The reason is that adding a to both sides of $a < 0$ gives $2a < a$, and so $2a < a < 0$. Dividing all parts of the inequality by 2 gives $a < a/2 < 0$.*] By substitution, then, $a < b < 0$. Thus b is a negative real number that is greater than a. This contradicts the supposition that a is the greatest negative real number. [*Hence the supposition is false and the statement is true.*]

4. *Proof*: Suppose not. That is, suppose there is a least positive rational number. Call it r. Then r is a real number such that $r > 0$, r is rational, and for all positive rational numbers x, $r \leq x$. Let $s = r/2$. Note that if we divide both sides of the inequality $0 < r$ by 2, we obtain $0 < r/2 = s$, and if we add r to the inequality $0 < r$ and then divide by 2, we obtain $\frac{0+r}{2} < \frac{r+r}{2}$, or, equivalently, $s = \frac{r}{2} < r$. Hence $0 < s < r$. Note also that since r is rational, $r = a/b$ for some integers a and b with $b \neq 0$, and so $s = \frac{r}{2} = \frac{a/b}{2} = \frac{a}{2b}$. Since a and $2b$ are integers and $2b \neq 0$, s is rational. Thus we have found a positive rational number s such that $s < r$. This contradicts the supposition that r is the *least* positive rational number. Therefore, there is no least positive rational number.

8. *Proof (by contraposition)*: Suppose a and b are [*particular but arbitrarily chosen*] real numbers such that $a \geq 25$ and $b \geq 25$. Then $a + b \geq 25 + 25 = 50$. Hence if $a + b < 50$, then $a < 25$ or $b < 25$.

10. a. *Proof by contraposition*: Suppose n is a [*particular but arbitrarily chosen*] integer that is not even. [*We must show that n^2 is not even.*] By the parity theorem, since n is not even, then n is odd. So $n^2 = n \cdot n$ is also odd (by exercise 17 of Section 3.1). Hence by the parity theorem, n^2 is not even [*as was to be shown.*]

    b. *Proof by contradiction*: Suppose not. Suppose $\exists$ an integer n such that n^2 is odd and n is even. [*We must derive a contradiction.*] By definition of even, $n = 2k$ for some integer k. Then $n^2 = (2k)^2 = 2(2k^2)$ by the laws of algebra. Let $m = 2k^2$. Then m is an integer because it is a product of integers. Thus $n^2 = 2m$ for some integer m, and so by definition of even, n^2 is even. This contradicts the supposition that n^2 is odd. [*Hence the supposition is false and the statement is true.*]

11. a. *Proof by contraposition*: Suppose n is an integer and p is a prime number such that n is not divisible by p. [*We must show that n^2 is not divisible by p.*] By the unique factorization theorem, n has a unique representation as $p_1^{e_1} \cdot p_2^{e_2} \cdot \ldots \cdot p_k^{e_k}$ where $p_1, p_2 \ldots p_k$ are prime numbers, $e_1, e_2, \ldots, e_k$ are positive integers, and $p_1 < p_2 < \ldots < p_k$. Since $p \nmid n$, $p \neq p_i$ for any $i = 1, 2, \ldots, k$. By the laws of exponents, $n^2 = p_1^{2e_1} \cdot p_2^{2e_2} \cdot \ldots \cdot p_k^{2e_k}$, and so by the uniqueness part of

the unique factorization theorem no prime number can divide n^2 except for $p_1, p_2, \ldots, p_k$. But $p \neq p_i$ for any $i = 1, 2, \ldots, k$. Thus n^2 is not divisible by p *[as was to be shown]*.

b. Proof by contradiction: Suppose not. Suppose $\exists$ an integer n and a prime number p such that n^2 is divisible by p and n is not divisible by p. *[We must derive a contradiction.]* Since n is not divisible by p, p does not appear in the prime factorization of n. But the prime factorization of n^2 consists of the product of all the prime factors of n each written twice. By the uniqueness part of the unique factorization theorem, no other prime factors than these can occur in the prime factorization of n^2. Hence p cannot occur in the prime factorization of n^2, and so n^2 is not divisible by p. This contradicts the supposition that n^2 is divisible by p. *[Hence the supposition is false and the statement is true.]*

13. *a. Proof (by contraposition)*: Suppose m and n are integers such that one of m and n is even and the other is odd. By exercise 18 of Section 3.1, the sum of any even integer and any odd integer is odd. Hence $m + n$ is odd. *[This is what was to be shown]*.

    b. Proof by contradiction: Suppose not. Suppose $\exists$ integers m and n such that $m + n$ is even and either m is even and n is odd or m is odd and n is even. By exercise 18 in Section 3.1, the sum of any even integer and any odd integer is odd. Thus both when m is even and n is odd and when m is odd and n is even, the sum $m + n$ is odd. This contradicts the supposition that $m + n$ is even. *[Hence the supposition is false and the statement is true.]*

14. *a. Proof (by contraposition)*: By De Morgan's law, we must show that for all integers a, b, and c, if $a \mid (b + c)$ then $a \nmid b$ or $a \mid c$. But by the logical equivalence of $p \rightarrow q \vee r$ and $p \wedge \sim q \rightarrow r$, it suffices to show that for all integers a, b, and c, if $a \mid (b + c)$ and $a \mid b$, then $a \mid c$. So suppose a, b, and c are any integers with $a \mid (b + c)$ and $a \mid b$. *[We must show that $a \mid c$.]* By definition of divisibility, $b + c = as$ and $b = ar$ for some integers s and r. By substitution, $c = (b+c) - b = as - ar = a(s-r)$. But $s - r$ is an integer (because it is a difference of integers). Hence by definition of divisibility, $a \mid c$ *[as was to be shown]*.

    b. Proof by contradiction: Suppose not. Suppose $\exists$ integers a, b, and c such that $a \mid b$ and $a \nmid c$ and $a \mid (b + c)$. *[We must derive a contradiction.]* By definition of divisibility, $\exists$ integers r and s such that $b = ar$ and $b + c = as$. By substitution, $ar + c = as$. Subtracting ar from both sides gives $c = as - ar = a(s - r)$. But $s - r$ is an integer because r and s are integers. Hence by definition of divisibility, $a \mid c$. This contradicts the supposition that $a \nmid c$. *[Hence the supposition is false and the statement is true.]*

15. *Counterexample*: Let $a = 4$ and $n = 2$. Then $a \mid n^2$ because $4 \mid 4$ (since $4/4$ is an integer), but $a \nmid n$ because $4 \nmid 2$ (since $2/4$ is not an integer).

18. *Proof*: Suppose not. Suppose $\exists$ a nonzero rational number x and an irrational number y such that xy is rational. *[We must derive a contradiction.]* By definition of rational, $x = a/b$ and $xy = c/d$ for some integers a, b, c and d with $b \neq 0$ and $d \neq 0$. Also $a \neq 0$ because x is nonzero. By substitution, $xy = (a/b)y = c/d$. Solving for y gives $y = bc/ad$. Now bc and ad are integers (being products of integers) and $ad \neq 0$ (by the zero product property). Thus by definition of rational, y is rational, which contradicts the supposition that y is irrational. *[Hence the supposition is false and the statement is true.]*

19. *Counterexample*: Both $\sqrt{2}$ and $2 - \sqrt{2}$ are positive (because $2 > \sqrt{2}$) and both are irrational (by exercises 13 and 4). Furthermore, $\sqrt{2} + (2 - \sqrt{2}) = 2$, which is rational (by Theorem 3.2.1). Thus $\exists$ positive irrational numbers whose sum is rational.

20. *Counterexample*: $\sqrt{2}$ is irrational. Also $\sqrt{2} - \sqrt{2} = 0$ and 0 is rational (by Theorem 3.2.1). Thus $\exists$ irrational numbers whose difference is rational.

21. *Counterexample*: $\sqrt{2}$ is irrational. Also $\sqrt{2} \cdot \sqrt{2} = 2$ and 2 is rational (by Theorem 3.2.1). Thus $\exists$ irrational numbers whose product is rational.

22. *Proof:* Suppose not. Suppose a and b are rational numbers, $b \neq 0$, r is an irrational number, and $a + br$ is rational. *[We must derive a contradiction.]* By definition of rational, $a = \dfrac{i}{j}, b = \dfrac{k}{l}$, and $a + br = \dfrac{m}{n}$ where i, j, k, l, m, and n are integers and $j \neq 0, l \neq 0$, and $n \neq 0$. Since $b \neq 0$, we also have that $k \neq 0$. By substitution $a + br = \dfrac{i}{j} + \dfrac{k}{l} \cdot r = \dfrac{m}{n}$, or, equivalently, $\dfrac{k}{l} \cdot r = \dfrac{m}{n} - \dfrac{i}{j}$. Solving for r gives $r = \dfrac{mj - in}{nj} \cdot \dfrac{l}{k} = \dfrac{mjl - inl}{njk}$. Now $mjl - inl$ and njk are both integers *[because products and differences of integers are integers]* and $njk \neq 0$ because $n \neq 0, j \neq 0$, and $k \neq 0$. Hence by definition of rational, r is a rational number. But this contradicts the supposition that r is irrational. *[Hence the supposition is false and the statement is true.]*

23. *Counterexample:* Let $r = 0$ and $s = \sqrt{2}$. Then r is rational and s is irrational and $r/s = 0/\sqrt{2} = 0$, which is rational (by Theorem 3.2.1).

24. Yes. *Proof:* Suppose not. That is, suppose $\exists$ integers a, b, and c such that a and b are odd and $a^2 + b^2 = c^2$. Since any product of two odd integers is odd (exercise 25, Section 3.1) both a^2 and b^2 are odd, and since any sum of two odd integers is even (exercise 12, Section 3.1) $a^2 + b^2$ is even. Thus c^2 is even (because $c^2 = a^2 + b^2$), and so (by Proposition 3.6.3) c is even. Hence $c = 2r$ for some integer r, and consequently $c^2 = 4r$, or, equivalently, $c^2 \bmod r = 0$. On the other hand, since a^2 and b^2 are both odd, by exercise 10 both a and b are odd. Hence $a = 2s + 1$ and $b = 2t + 1$ for some integers s and t. It follows that

$$c^2 = a^2 + b^2 = (2s+1)^2 + (2t+1)^2 = (4s^2 + 4s + 1) + (4t^2 + 4t + 1) = 4(s^2 + s + t^2 + t) + 2.$$

Since $s^2 + s + t^2 + t$ is an integer (because sums and products of integers are integers), then $c^2 \bmod 4 = 2$. We have shown that both $c^2 \bmod 4 = 0$ and $c^2 \bmod 4 = 2$ which is impossible by the uniqueness condition of the quotient-remainder theorem. Thus the supposition is false, and we conclude that if a, b, and c are integers and $a^2 + b^2 = c^2$, then at least one of a and b must be even.

Section 3.7

1. We can deduce that $p = 3$.
   Proof: Let a be any integer, and let p be any prime number such that $p \mid a$ and $p \mid (a + 3)$. By definition of divisibility, $a = pr$ and $a + 3 = ps$ for some integers r and s. Then $3 = (a+3) - a = ps - pr = p(s - r)$. But $s - r$ is an integer (because it is a difference of integers), and so by definition of divisibility $p \mid 3$. But since 3 is a prime number, its only positive divisors are 1 and itself. So $p = 1$ or $p = 3$. However, since p is a prime number, $p \neq 1$. Hence $p = 3$.

3. *Proof:* Suppose not. Suppose $4\sqrt{2} - 9$ is rational. *[We must derive a contradiction.]* By definition of rational, $\exists$ integers a and b with $4\sqrt{2} - 9 = a/b$ and $b \neq 0$. Solving for $\sqrt{2}$ gives $\sqrt{2} = (a/b + 9)/4 = (a + 9b)/4b$. But $a + 9b$ and $4b$ are integers (because products and sums of integers are integers) and $4b \neq 0$ (by the zero product property). Therefore by definition of rational, $\sqrt{2}$ is rational. This contradicts Theorem 3.7.1 which states that $\sqrt{2}$ is irrational. *[Hence the supposition is false and the statement is true.]*

4. *Proof:* Suppose not. Suppose $\sqrt{3}$ is rational. *[We must derive a contradiction.]* By definition of rational, $\exists$ integers m and n with $\sqrt{3} = m/n$ and $n \neq 0$. Without loss of generality, we may assume that m and n have no common divisors. *[by dividing m and n by any common divisors if necessary.]* Squaring both sides of the equation above gives $3 = m^2/n^2$, or, equivalently, $m^2 = 3n^2$. Since n^2 is an integer, $3 \mid m^2$ by definition of divisibility. It follows that $3 \mid m$ (by exercise 6 with $p = 3$), and so by definition of divisibility there exists an integer k such that $m = 3k$. Substituting into $m^2 = 3n^2$ gives $(3k)^2 = 3n^2$, and so $9k^2 = 3n^2$, which implies that $3k^2 = n^2$. Since k^2 is an integer, $3 \mid n^2$ by definition of divisibility. It follows that $3 \mid n$

(by exercise 6 with $p = 3$). Thus 3 is a common divisor of m and n. But this contradicts the assumption that m and n have no common divisors. *[Hence the supposition is false and the statement is true.]*

6. False. $\sqrt{2}/4 = (1/4) \cdot \sqrt{2}$, which is a product of a nonzero rational number and an irrational number. By exercise 18 of Section 3.6, such a product is irrational.

7. *Proof:* Suppose not. Suppose $\exists$ an integer a such that $4 \mid (a^2 - 2)$. *[We must derive a contradiction.]* By definition of divisibility, $a^2 - 2 = 4b$ for some integer b. Then $a^2 = 4b + 2 = 2(2b+1)$, and so a^2 is even. Hence a is even (by Proposition 3.6.3). By definition of even, $a = 2c$ for some integer c. Substituting into the equation $a^2 - 2 = 4b$ gives $(2c)^2 - 2 = 4c^2 - 2 = 4b$, and so $4c^2 = 4b + 2$. Dividing by 2 gives $2c^2 = 2b + 1$. Since c^2 is an integer, this implies that $2b + 1$ is even, which it is not. *[Hence the supposition is false and the statement is true.]*

8. *Proof:* Suppose not. Suppose $\exists$ an integer n such that n is not a perfect square and $\sqrt{n}$ is rational. *[We must derive a contradiction.]* By definition of rational, there exist integers a and b such that $\sqrt{n} = a/b$ and $b \neq 0$. Without loss of generality, we may assume that a and b have no common divisors *[by dividing a and b by any common divisors if necessary]*. Squaring both sides of the equation above gives $n = a^2/b^2$ and multiplying by b^2 gives $b^2 n = a^2$. By the unique factorization theorem, a, b, and n have representations as products of primes that are unique except for the order in which the prime factors are written down. By the laws of exponents, a^2 and b^2 are products of the same prime numbers, each written twice. Consequently, each prime factor in a^2 and in b^2 occurs an even number of times. Since n is not a perfect square, some prime factor in n occurs an odd number of times (again by the laws of exponents). It follows that this same prime factor occurs an odd number of times in the product $n \cdot b^2$ (because all prime factors in b^2 occur an even number of times). Since $n \cdot b^2 = a^2$, a^2 contains a prime factor that occurs an odd number of times. This contradicts the fact that every prime factor of a^2 occurs an even number of times. *[Hence the supposition is false and the statement is true.]*

9. *Proof:* Suppose not. Suppose $\sqrt{2} + \sqrt{3}$ is rational. *[We must derive a contradiction.]* By definition of rational, $\sqrt{2} + \sqrt{3} = a/b$ for some integers a and b with $b \neq 0$. Squaring both sides gives $2 + 2 \cdot \sqrt{2} \cdot \sqrt{3} + 3 = a^2/b^2$. Then $2\sqrt{6} = a^2/b^2 - 5$, and so $\sqrt{6} = (a^2 - 5b^2)/2b^2$. Now $a^2 - 5b^2$ and $2b^2$ are both integers (because products and differences of integers are integers) and $2b^2 \neq 0$ (by the zero product property). Therefore, $\sqrt{6}$ is rational by definition of rational. But $\sqrt{6}$ is irrational by exercise 8 (since 6 is not a perfect square), which is a contradiction. *[Hence the supposition is false and the statement is true.]*

10. *Lemma:* For all integers n, if n^3 is even then n is even.

    Proof of lemma (by contraposition): Let n be any integer such that n is not even. *[We must show that n^3 is not even.]* By the parity theorem, n is odd, and so n^3 is also odd (by exercise 25 of Section 3.1 applied twice). Thus (again by the parity theorem), n^3 is even *[as was to be shown]*.

    Proof of statement: Suppose the statement is false. That is, suppose $\sqrt[3]{2}$ is rational. By definition of rational, $\sqrt[3]{2} = a/b$ for some integers a and b with $b \neq 0$. By canceling any common factors if necessary, we may assume that a and b have no common factors. Cubing both sides of the equation $\sqrt[3]{2} = a/b$ gives $2b^3 = a^3$, and so a^3 is even. By the lemma, a is even, and thus $a = 2k$ for some integer k. By substitution $a^3 = (2k)^3 = 8k^3 = 2b^3$, and so $b^3 = 4k^3 = 2(2k^3)$. It follows that b^3 is even, and hence (by the lemma) b is even. Thus both a and b are even which contradicts the assumption that a and b have no common factor. Therefore, the supposition is false, and $\sqrt[3]{2}$ is irrational.

11. *Counterexample:* Let $n = 64$. Then n is a perfect square because $64 = 8^2$, but $\sqrt[3]{64} = 4$ which is not irrational.

12. *Proof:* Suppose not. That is, suppose ∃ an irrational number x such that $\sqrt{x}$ is rational. By definition of rational, $\sqrt{x} = a/b$ for some integers a and b with $b \neq 0$. Then $x = (\sqrt{x})^2 = (a/b)^2 = a^2/b^2$. But a^2 and b^2 ae both integers (being products of integers), and $b^2 \neq 0$ by the zero product property. Hence, x is rational (by definition of rational). This contradicts the supposition that x is irrational, and so the supposition is false. Therefore, the square root of an irrational number is irrational.

13. *Proof:* Suppose not. That is, suppose $\sqrt{2}$ is rational. By definition of rational, we may write $\sqrt{2} = a/b$ for some integers a and b with $b \neq 0$. Then $2 = a^2/b^2$, and so $a^2 = 2b^2$. Consider the prime factorizations for a^2 and for $2b^2$. By the unique factorization theorem, these factorizations are unique except for the order in which the factors are written down. Now because every prime factor of a occurs twice in the prime factorization of a^2, the prime factorization of a^2 contains an even number of 2's. If 2 is a factor of a, then this even number is positive, and if 2 is not a factor of a, then this even number is 0. On the other hand, because every prime factor of b occurs twice in the prime factorization of b^2, the prime factorization of $2b^2$ contains an odd number of 2's. Therefore, the equation $a^2 = 2b^2$ cannot be true. So the supposition is false, and hence $\sqrt{2}$ is irrational.

14. The answer is no.
    Proof: Let a, b, and c be any odd integers, and suppose $ax^2 + bx + c = 0$ has a rational solution. Then ∃ integers r and s with $s \neq 0$ such that
    $$a\left(\frac{r}{s}\right)^2 + b\left(\frac{r}{s}\right) + c = 0, \quad \text{or, equivalently,} \quad ar^2 + brs + cs^2 = 0. (*)$$
    By the parity theorem, either r is even or r is odd.

    Case 1 (r is even): In this case, we solve equation (*) for cs^2 to obtain $cs^2 = -ar^2 - brs = r(-ar - bs)$. Since r is even, by exercise 28 of Section 3.1, $r(-ar - bs)$ is even, and so cs^2 is even. Now because c is odd we must have that s^2 is even (since otherwise cs^2 would be odd by exercise 25 of Section 3.1). Hence by Proposition 3.6.3, s is even. Thus both r and s have a common factor of 2, which contradicts the assumption that they have no common factor.

    Case 2 (r is odd): In this case, we solve equation (*) for ar^2 to obtain $ar^2 = -brs - cs^2$. Also we note that since a and r are both odd, ar^2 is also odd (by exercise 25 of Section 3.1 applied twice). By the parity theorem, either s is even or s is odd.

    Subcase 1 (s is even): In this case $ar^2 = -brs - cs^2 = s(-br - cs)$, which is a product of an even integer times an integer and so is even (by exercise 12 of Section 3.1). Hence ar^2 is even.

    Subcase 2 (s is odd): In this case because b, c, and r are also odd, $-brs$ and $-cs^2$ are both odd (by exercise 25 of Section 3.1 applied several times). It follows by exercise 12 of Section 3.1 that $(-brs) + (-cs^2)$ is even. Hence $ar^2 = -brs - cs^2$ is even.

    Thus regardless of whether s is even or odd, when r is odd, ar^2 is even. But we noted above that when r is odd, ar^2 is odd. So we have a contradiction.

    The above arguments show that both in case r is even and in case r is odd, a contradiction can be deduced. Therefore, the supposition is false. We conclude that if a, b, and c are any odd integers, then all solutions to $ax^2 + bx + c = 0$ are irrational.

15. *Proof:* Suppose that $\log_2(3)$ is rational. Then $\log_2(3) = a/b$ for some intgers a and b with $b \neq 0$. Since logarithms are always positive, we may assume that a and b are both positive. By definition of logarithm, $2^{a/b} = 3$. Raising both sides to the bth power gives $2^a = 3^b$. Let $N = 2^a = 3^b$, and consider the prime factorization of N. Since $N = 2^a$, the prime factors of N are all 2. On the other hand, since $N = 3^b$, the prime factors of N are all 3. This contradicts the unique factorization theorem which states that the prime factors of any integer greater than 1 are unique except for the order in which they are written. Hence the supposition is false, and so $\log_2(3)$ is irrational.

16. *Proof:* Let n be any integer greater than 11. Then each of the numbers $n-4$, $n-6$, and $n-8$ is greater than 3. Observe that $n = (n-4)+4 = (n-6)+6 = (n-8)+8$ and that all three of 4, 6, and 8 are composite. Thus, if at least one of $n-4$, $n-6$, and $n-8$ is composite, then n can be written as a sum of two composite numbers. In particular, in case either $n-6$ or $n-8$ is composite, we are done. Suppose, therefore, that neither $n-6$ nor $n-8$ is composite. *[We will show that $n-4$ is composite.]* Observe that $n-8$, $n-7$, and $n-6$ constitute a sequence of three consecutive integers. A proof almost identical to the proof for exercise 21 of Section 3.4 shows that in any such sequence one of the integers is divisible by 3. Now neither of $n-6$ nor $n-8$ is divisible by 3 (because since $n > 11$, $n-6$ and $n-8$ are both greater than 3, and each of $n-6$ and $n-8$ is prime). Consequently, $n-7$ is divisible by 3, and so $n-7 = 3k$ for some integer k. Thus $n-4 = (n-7)+3 = 3k+3 = 3(k+1)$. Therefore $n-4$ is composite because it is a product of 3 and $k+1$. *[We know that $k+1 > 1$ because since $n > 11$, $3k = n-7 > 4$, and so $k > 4/3$.]* Hence in case neither $n-6$ nor $n-8$ is composite, then $n-4$ is composite, and so in this case also n is a sum of two composite numbers, namely $n-4$ and 4.

17. *Proof (by contraposition):* Suppose that for some integer $n > 2$ that is not a power of 2, $x^n + y^n = z^n$ has a positive integer solution, say it is $x = x_0$, $y = y_0$, and $z = z_0$. If n is prime, then for some prime number p (namely $p = n$), $x^p + y^p = z^p$ has a positive integer solution and we are done. If n is not prime, then n is divisible by a prime (by Theorem 3.3.2), and so, since n is not a power of 2, there exist a prime number $p > 2$ and an integer k such that $n = kp$. Also since $x = x_0$, $y = y_0$, and $z = z_0$ is an integer solution to $x^n + y^n = z^n$, then $x_0^n + y_0^n = z_0^n$. But $n = kp$, and so $x_0^{kp} + y_0^{kp} = z_0^{kp}$, or, equivalently (by the laws of exponents), $(x_0^k)^p + (y_0^k)^p = (z_0^k)^p$. Now x_0^k, y_0^k, and z_0^k are all integers (because they are integer powers of integers). Consequently, the equation $x^p + y^p = z^p$ has an integer solution (namely, x_0^k, y_0^k, and z_0^k).

18. The following are all prime numbers: $2 \cdot 3 + 1 = 7$, $2 \cdot 3 \cdot 5 + 1 = 31$, $2 \cdot 3 \cdot 5 \cdot 7 + 1 = 211$, $2 \cdot 3 \cdot 5 \cdot 7 \cdot 11 + 1 = 2311$. However, $2 \cdot 3 \cdot 5 \cdot 7 \cdot 11 \cdot 13 + 1 = 30031 = 59 \cdot 509$. Thus the smallest non-prime integer of the given form is 30,031.

19. *Proof:* Suppose $p_1, p_2, \ldots, p_n$ are distinct prime numbers with $p_1 = 2$ and $n > 1$. *[We must show that $p_1 \cdot p_2 \cdot \ldots \cdot p_n + 1 = 4k+3$ for some integer k.]* Let $N = p_1 \cdot p_2 \cdot \ldots \cdot p_n + 1$. By the quotient-remainder theorem, N can be written in one of the forms $4k$, $4k+1$, $4k+2$, or $4k+3$ for some integer k. Now N is odd (because $p_1 = 2$); hence N equals either $4k+1$ or $4k+3$ for some integer k. Suppose $N = 4k+1$ for some integer k. *[We will show that this supposition leads to a contradiction.]* By substitution, $4k+1 = p_1 \cdot p_2 \cdot \ldots \cdot p_n + 1$, and so $4k = p_1 \cdot p_2 \cdot \ldots \cdot p_n$. Hence $4 \mid p_1 \cdot p_2 \cdot \ldots \cdot p_n$. But $p_1 = 2$ and all of $p_2, p_3, \ldots, p_n$ are odd (being prime numbers that are greater than 2). Consequently, there is only one factor of 2 in the prime factorization of $p_1 \cdot p_2 \cdot \ldots \cdot p_n$, and so $4 \nmid p_1 \cdot p_2 \cdot \ldots \cdot p_n$. This is a contradiction. Hence the supposition that $N = 4k+1$ for some integer k is false, and so *[by disjunctive syllogism!]* $N = 4k+3$ for some integer k *[as was to be shown]*.

20. *Proof:* Suppose n is any integer that is greater than 2. Then $n! - 1$ is an integer that is greater than 2, and so by Theorem 3.3.2 there is a prime number p that divides $n! - 1$. Thus $p \le n! - 1 < (n!)$. Now either $p > n$ or $p \le n$. Suppose $p \le n$. *[We will show that this supposition leads to a contradiction.]* Since $p \le n$, then $p \mid (n!)$. So $p \mid (n!)$ and also $p \mid (n!+1)$, and thus (by exercise 14 of Section 3.3) p divides $((n!+1) - n!)$, which equals 1. But the only divisors of 1 are 1 and -1, and so $p = 1$ or $p = -1$. However since p is prime, $p > 1$. This is a contradiction. Hence the supposition that $p \le n$ is false, and so $p > n$. Therefore, p is a prime number such that $n < p < (n!)$.

22. *Existence Proof:* When $n = 2$, then $n^2 + 2n - 3 = 2^2 + 2 \cdot 2 - 3 = 5$, which is prime. Thus there is a prime number of the form $n^2 + 2n - 3$, where n is a positive integer.

Uniqueness Proof (by contradiction): Suppose not. By the existence proof above, we know that when $n = 2$, then $n^2 + 2n - 3$ is prime. Suppose there is another positive integer m, not equal to 2, such that $m^2 + 2m - 3$ is prime. *[We must derive a contradiction.]* By factoring, we see that $m^2 + 2m - 3 = (m+3)(m-1)$. Now $m \neq 1$ because otherwise $m^2 + 2m - 3 = 0$, which is not prime. Also $m \neq 2$ by supposition. Thus $m > 2$. Consequently, $m + 3 > 5$ and $m - 1 > 1$, and so $m^2 + 2m - 3$ can be written as a product of two positive integers neither of which is 1 (namely $m+3$ and $m-1$). This contradicts the supposition that $m^2 + 2m - 3$ is prime. Hence the supposition is false: there is no integer m other than 2 such that $m^2 + 2m - 3$ is prime.

Uniqueness Proof (direct): Suppose m is any positive integer such that $m^2 + 2m - 3$ is prime. *[We must show that $m = 2$.]* By factoring, $m^2 + 2m - 3 = (m+3)(m-1)$. Since $m^2 + 2m - 3$ is prime, either $m + 3 = 1$ or $m - 1 = 1$. Now $m + 3 \neq 1$ because m is positive (and if $m + 3 = 1$ then $m = -2$). Thus $m - 1 = 1$, which implies that $m = 2$ *[as was to be shown]*.

24. *Proof (by contradiction)*: Suppose not. Suppose there exist two distinct real number b_1 and b_2 such that for all real numbers r, (1) $b_1 r = r$ and (2) $b_2 r = r$. Then $b_1 b_2 = b_2$ (by (1) with $r = b_1$) and $b_2 b_1 = b_1$ (by (2) with $r = b_2$). Consequently, $b_2 = b_1 b_2 = b_2 b_1 = b_1$ by substitution and the commutative law of multiplication. But this implies that $b_1 = b_2$, which contradicts the supposition that b_1 and b_2 are distinct. *[Thus the supposition is false and there exists at most one real number b such that $br = r$ for all real numbers r.]*

Proof (direct): Suppose b_1 and b_2 are real numbers such that (1) $b_1 r = r$ and (2) $b_2 r = r$ for all real numbers r. By (1) $b_1 b_2 = b_2$, and by the commutative law for multiplication and (2), $b_1 b_2 = b_2 b_1 = b_1$. Since both b_1 and b_2 are equal to $b_1 b_2$, we conclude that $b_1 = b_2$.

Section 3.8

2. $z = 1$

3. b. $z = 4$

5. $e = 41/24$

7.
a	54				
d	11				
q	0	1	2	3	4
r	54	43	32	21	10

10. $\gcd(5, 10) = 5$

11. $\gcd(42, 63) = \gcd(2 \cdot 3 \cdot 7, 3 \cdot 3 \cdot 7) = 21$

13.
$$544 \overline{)1001}^{1}$$
$$\underline{544}$$
$$457$$

So $1001 = 544 \cdot 1 + 457$, and hence $\gcd(1001, 544) = \gcd(544, 457)$

$$457 \overline{)544}^{1}$$
$$\underline{457}$$
$$87$$

So $1 = 457 \cdot 544 + 87$, and hence $\gcd(544, 457) = \gcd(457, 87)$

$$87 \overline{)457}^{5}$$
$$\underline{435}$$
$$22$$

So $457 = 87 \cdot 5 + 22$, and hence $\gcd(457, 87) = \gcd(87, 22)$

$$22 \overline{)87}^{3}$$
$$\underline{66}$$
$$21$$

So $87 = 22 \cdot 3 + 21$, and hence $\gcd(87, 22) = \gcd(22, 21)$

$$21 \overline{)22}^{1}$$
$$\underline{21}$$
$$1$$

So $22 = 21 \cdot 1 + 1$, and hence $\gcd(22, 21) = \gcd(21, 1)$

$$1 \overline{)21}^{1}$$
$$\underline{21}$$
$$0$$

So $21 = 1 \cdot 21 + 0$, and hence $\gcd(21, 1) = \gcd(1, 0)$

But $\gcd(1, 0) = 1$. So $\gcd(1001, 544) = 1$.

14.
$$672 \overline{)3510}^{5}$$
$$\underline{3360}$$
$$150$$

So $3510 = 672 \cdot 5 + 150$, and hence $\gcd(3510, 672) = \gcd(672, 150)$

$$150 \overline{)672}^{4}$$
$$\underline{600}$$
$$72$$

So $672 = 150 \cdot 4 + 72$, and hence $\gcd(672, 150) = \gcd(150, 72)$

$$72 \overline{)150}^{2}$$
$$\underline{144}$$
$$6$$

So $150 = 72 \cdot 2 + 6$, and hence $\gcd(150, 72) = \gcd(72, 6)$

$$6 \overline{)72}^{12}$$
$$\underline{72}$$
$$0$$

So $72 = 6 \cdot 12 + 0$, and hence $\gcd(72, 6) = \gcd(6, 0)$

But $\gcd(6, 0) = 6$. So $\gcd(3510, 672) = 6$.

16.

A	2628							
B	738							
r	738	414	324	90	54	36	18	0
a	2628	738	414	324	90	54	36	18
b	738	414	324	90	54	36	18	0
gcd								18

17. *Proof:* Let a and b be any positive integers.

   Part 1 (proof that if $\gcd(a,b) = a$ then $a \mid b$): Suppose that $\gcd(a,b) = a$. By definition of greatest common divisor, $\gcd(a,b) \mid b$, and so by substitution, $a \mid b$.

   Part 2 (proof that if $a \mid b$ then $\gcd(a,b) = a$): Suppose that $a \mid b$. Then since it is also the case that $a \mid a$, a is a common divisor of a and b. Thus by definition of greatest common divisor, $a \leq \gcd(a,b)$. On the other hand, since no integer greater than a divides a, the greatest common divisor of a and b is less than or equal to a. In symbols, $\gcd(a,b) \leq a$. Therefore, since $a \leq \gcd(a,b)$ and $\gcd(a,b) \leq a$, then $\gcd(a,b) = a$.

18. *Lemma:* If a and b are integers, not both zero, and $d = \gcd(a,b)$, then a/d and b/d are integers with no common divisor that is greater than 1.

   Proof: Let a and b be integers, not both zero, and let $d = \gcd(a,b)$. By definition of gcd, $d \mid a$ and $d \mid b$. Hence a/d and b/d are integers. Suppose a/d and b/d have a common divisor c that is greater than 1. *[We will derive a contradiction.]* Then $a/d = cr$ and $b/d = cs$ for some integers r and s. It follows that $a = (cd)r$ and $b = (cd)s$, and so $cd \mid a$ and $cd \mid b$. But $c > 1$, and so $cd > d$. Thus cd is a common divisor of a and b that is greater than the greatest common divisor of a and b. This is a contradiction. Hence the supposition that a/d and b/d have a common divisor that is greater than 1 is false, and so a/d and b/d have no common divisor that is greater than 1.

   Algorithm: Reducing a Fraction

   [Given two integers A and B with $B \neq 0$, this algorithm finds integers C and D so that $A/B = C/D$ and C and D have no common divisor that is greater than 1. The algorithm first adjusts for the fact that A/B may be negative by setting variables $a = |A|$ and $b = |B|$. Then it calls the Euclidean algorithm to compute $\gcd(a,b)$ and sets $c = a$ div $\gcd(a,b)$ and $d = b$ div $\gcd(a,b)$. By the lemma above, c and d are integers that are divisible by $\gcd(a,b)$. Consequently,

   $$\frac{c}{d} = \frac{a \ \text{div} \ \gcd(a,b)}{b \ \text{div} \ \gcd(a,b)} = \frac{\frac{a}{\gcd(a,b)}}{\frac{b}{\gcd(a,b)}} = \frac{a}{b}.$$

   It also follows from the lemma that c and d have no common factor that is greater than 1. Thus c/d is the reduced form of a/b. Finally the sign of the reduced fraction is adjusted: C is set equal to $-c$ if A/B is less than 0 and to c if A/B is greater than or equal to 0, and D is set equal to d.]

   Input: A, B *[integers with $B \neq 0$]*

   Algorithm Body:

   if $A/B < 0$ then *sign* $:= -1$ else *sign* $:= 1$

   if $A < 0$ then a $:= -A$ else a $:= A$

   if $B < 0$ then b $:= -B$ else b $:= B$

   $gcd := \gcd(a,b)$

   [The value of gcd can be computed by calling the Euclidean algorithm.]

$c := a \text{ div } \gcd, \quad d := b \text{ div } \gcd$

[The values of c and d can be computed by calling the Division algorithm.]

[When execution reaches this point, $a/b = c/d$ and c and d have no common divisors that are greater than 1.]

$C := \text{sign} \cdot c, D := d$

[When execution reaches this point, $A/B = C/D$ and C and D have no common divisors that are greater than 1.]

Output: C, D *[integers with $D \neq 0$]*

end Algorithm

19. *Proof*: Suppose a and b are any positive integers and q and r are any integers such that $a = bq+r$ and $0 \leq r < b$. *[We must show that $\gcd(b, r) \leq \gcd(a, b)$.]*

    a. We will first show that any common divisor of b and r is also a common divisor of a and b.

    Let c be a common divisor of b and r. Then $c \mid b$ and $c \mid r$, and so by definition of divisibility, $b = n \cdot c$ and $r = m \cdot c$ for some integers n and m. Now substitute into the equation $a = bq + r$ to obtain $a = (n \cdot c) \cdot q + m \cdot c = c \cdot (n \cdot q + m)$. But $n \cdot q + m$ is an integer, and so by definition of divisibility $c \mid a$. Now we already know that $c \mid b$; hence c is a common divisor of a and b *[as was to be shown]*.

    b. Next we show that $\gcd(b, r) \leq \gcd(a, b)$.

    By part (a), every common divisor of b and r is a common divisor of a and b. It follows that the greatest common divisor of b and r is a common divisor of a and b. But then $\gcd(b, r)$ (being one of the common divisors of a and b) is less than or equal to the greatest common divisor of a and b: $\gcd(b, r) \leq \gcd(a, b)$ *[as was to be shown]*.

21. a. Suppose a and d are positive integers and q and r are integers such that $a = dq + r$ and $0 < r < d$. Then $-a = -dq - r = -dq - d + d - r = d(-q-1) + (d-r) = d(-(q+1)) + (d-r)$. Also since $0 < r < d$, then $0 > -r > -d$ (by multiplying all parts of the inequality by -1). Adding d to all parts of the inequality gives $d + 0 > d + (-r) > d + (-d)$, and hence $d > d - r > 0$.

    b. If the input a is negative, then use the current algorithm with input $-a$ to obtain a quotient q and a remainder r so that $-a = dq + r$ and $0 \leq r < d$. If $r = 0$, then $a = -(-a) = -dq = d(-q)$, and thus the quotient of the division of a by d is $-q$ and the remainder is 0. If $r > 0$, then by part (a), $a = -(-a) = d(-(q+1)) + (d-r)$ and $0 < d - r < d$, and thus the quotient of the division of a by d is $-(q+1)$ and the remainder is $d-r$. Hence Algorithm 3.8.1 can be modified as follows.

    Algorithm Body:

        if $a < 0$ **then** $sign := -1, a := -a$ **else** $sign := 1$

        [Same steps as the body of Algorithm 3.8.1.]

        if $sign = -1$ **then if** $r = 0$

                         then $q := -q$

                         else $q := -(q+1), r := d - r$

22. a. *Proof*: Suppose a, d, q, and r are integers such that $a = d \cdot q + r$ and $0 \leq r < d$. *[We must show that $q = \lfloor a/d \rfloor$ and $r = a - \lfloor a/d \rfloor \cdot d$.]* Solving $a = d \cdot q + r$ for r gives $r = a - d \cdot q$, and substituting into $0 \leq r < d$ gives $0 \leq a - d \cdot q < d$. Then $d \cdot q \leq a < d + d \cdot q = d \cdot (q+1)$, and so $q \leq a/d < q + 1$. Thus by definition of floor, $q = \lfloor a/d \rfloor$, and by substitution into $r = a - d \cdot q$, we have $r = a - \lfloor a/d \rfloor \cdot d$ *[as was to be shown]*.

    b. $r := B, a := A, b := B$

        while $(b \neq 0)$

$$r := a - \lfloor a/b \rfloor \cdot b$$
$$a := b$$
$$b := r$$
end while
$$gcd := a$$

23. *a. Proof:* Suppose a and b are integers and $a \geq b > 0$. *[We will show that every common divisor of a and b is a common divisor of b and $a - b$, and conversely.]*

    Part 1 (proof that every common divisor of a and b is a common divisor of b and $a - b$):

    Suppose $d \mid a$ and $d \mid b$. Then $d \mid (a - b)$ by exercise 14 of Section 3.3. Hence d is a common divisor of a and $a - b$.

    Part 2 (proof that every common divisor of b and $a - b$ is a common divisor of a and b):

    Suppose $d \mid b$ and $d \mid (a - b)$. Then by exercise 13 of Section 3.3, $a \mid [b + (a - b)]$. But $b + (a - b) = a$, and so $d \mid a$. Hence d is a common divisor of a and b.

    Because every common divisor of a and b is a common divisor of b and $a - b$, the greatest common divisor of a and b is a common divisor of b and $a - b$ and so is less than or equal to the greatest common divisor of a and $a - b$. Thus $\gcd(a,b) \leq \gcd(b, a-b)$. By similar reasoning, $\gcd(b, a-b) \leq \gcd(a,b)$. Therefore, $\gcd(a,b) = \gcd(b, a-b)$.

    b.

A	630								
B	336								
a	630	294		252	210	168	126	84	42	0
b	336		42						42
gcd									42

24. b. $\text{lcm}(2 \cdot 3^2 \cdot 5, 2^3 \cdot 3) = 2^3 \cdot 3^2 \cdot 5 = 360$

    c. $\text{lcm}(3500, 1960) = \text{lcm}(2^2 \cdot 5^3 \cdot 7, 2^3 \cdot 5 \cdot 7^2) = 2^3 \cdot 5^3 \cdot 7^2 = 49,000$

26. *Proof:* Let a and b be any positive integers.

    Part 1 (proof that if $\text{lcm}(a,b) = b$ then $a \mid b$): Suppose that $\text{lcm}(a,b) = b$. By definition of least common multiple, $a \mid \text{lcm}(a,b)$, and so by substitution, $a \mid b$.

    Part 2 (proof that if $a \mid b$ then $\text{lcm}(a,b) = b$): Suppose that $a \mid b$. Then since it is also the case that $b \mid b$, b is a common multiple of a and b. Thus by definition of least common multiple, $\text{lcm}(a,b) \leq b$. On the other hand, since no integer less than b divides b, the least common multiple of a and b is greater than or equal to b. In symbols, $\text{lcm}(a,b) \geq b$. Therefore, since $\text{lcm}(a,b) \leq b$ and $\text{lcm}(a,b) \geq b$, then $\text{lcm}(a,b) = b$.

27. *Proof:* Let a and b be any integers. By definition of greatest common divisor, $\gcd(a,b) \mid a$, and by definition of least common multiple, $a \mid \text{lcm}(a,b)$. Hence by transitivity of divisibility, $\gcd(a,b) \mid \text{lcm}(a,b)$.

28. *Proof:* Let a and b be any positive integers.

    Part 1 (proof that $\gcd(a,b) \cdot \text{lcm}(a,b) \leq ab$): By definition of greatest common divisor, $\gcd(a,b) \mid a$. Hence by definition of divisibility, $a = \gcd(a,b) \cdot k$ for some integer k. Multiplying both sides by b gives $ab = \gcd(a,b) \cdot k \cdot b$, and so $\dfrac{ab}{\gcd(a,b)} = bk$. It follows by definition of divisibility that $b \mid \left[\dfrac{ab}{\gcd(a,b)}\right]$. An almost exactly identical sequence of steps shows that $a \mid \left[\dfrac{ab}{\gcd(a,b)}\right]$. Thus by definition of least common multiple, $\text{lcm}(a,b) \leq \dfrac{ab}{\gcd(a,b)}$, or, equivalently, $\gcd(a,b) \cdot \text{lcm}(a,b) \leq ab$.

Part 2(proof that $ab \leq \gcd(a,b) \cdot \text{lcm}(a,b)$): By definition of least common multiple, $a \mid \text{lcm}$. Hence by definition of divisibility, $\text{lcm}(a,b) = ak$ for some integer k. Multiplying both sides by b gives $b \cdot \text{lcm}(a,b) = ak \cdot b$, and so $b = \left[\dfrac{ab}{\text{lcm}(a,b)}\right] \cdot k$. It follows by definition of divisibility that $\left[\dfrac{ab}{\text{lcm}(a,b)}\right] \mid b$. An almost exactly identical sequence of steps shows that $\left[\dfrac{ab}{\text{lcm}(a,b)}\right] \mid a$. Thus by definition of greatest common divisor, $\dfrac{ab}{\text{lcm}(a,b)} \leq \gcd(a,b) \cdot \text{lcm}(a,b) \leq ab$, or, equivalently, $ab \leq \gcd(a,b) \cdot \text{lcm}(a,b)$.

By part 1, $\gcd(a,b) \cdot \text{lcm}(a,b) \leq ab$, and by part 2, $ab \leq \gcd(a,b) \cdot \text{lcm}(a,b)$. Therefore, $\gcd(a,b) \cdot \text{lcm}(a,b) = ab$

Chapter 4: Sequences and Mathematical Induction

The first section of this chapter introduces the notation for sequences, summations, products, and factorial. Students have two main difficulties with this material. One is learning to recognize patterns so as to be able, for instance, to write summations in closed form, and the other is learning how to handle subscripts, particularly to change variables for summations and to distinguish index variables from variables that are constant with respect to a summation.

The second, third, and fourth sections of this chapter treat mathematical induction. The ordinary form is discussed in Sections 4.2 and 4.3 and the strong form in Section 4.4. Because of the importance of mathematical induction in discrete mathematics, a wide variety of examples is given so that students will become comfortable with using the technique in many different situations.

Students may find it helpful if you relate the logic of ordinary mathematical induction to the logic discussed in Chapters 1 and 2. The main point is that the inductive step establishes the truth of a sequence of if-then statements. Together with the basis step, this sequence gives rise to a chain of inferences that lead to the desired conclusion. More formally:

Suppose
1. $P(1)$ is true; and
2. for all integers $k \geq 1$, if $P(k)$ is true then $P(k+1)$ is true.

The truth of statement (2) implies, according to the law of universal instantiation, that no matter what particular integer $k \geq 1$ is substituted in place of k, the statement "If $P(k)$ then $P(k+1)$" is true. The following argument, therefore, has true premises, and so by modus ponens it has a true conclusion:

If $P(1)$ then $P(2)$. by 2 and universal instantiation
$P(1)$ by 1
∴ $P(2)$ by modus ponens

Similar reasoning gives the following chain of arguments, each of which has a true conclusion by modus ponens:

If $P(2)$ then $P(3)$.
$P(2)$
∴ $P(3)$
If $P(3)$ then $P(4)$.
$P(3)$
∴ $P(4)$
If $P(4)$ then $P(5)$.
$P(4)$
∴ $P(5)$
And so forth.

Thus no matter how large a positive integer n is specified, the truth of $P(n)$ can be deduced as the final conclusion of a (possibly very long) chain of arguments continuing those shown above.

Note that in Section 4.2 the formula for the sum of the geometric series is written as $\dfrac{r^{n+1} - 1}{r - 1}$ because in discrete mathematics r is commonly greater than one, making this version of the formula the more convenient.

The concluding section of the chapter applies the technique of mathematical induction to proving algorithm correctness. It is intended to be an introduction to the subject and contains references directing students to sources that treat it at length.

Comments on Exercises

Exercise Set 4.1: #8 and 9: These exercises are designed for students already familiar with logarithms. In classes where such familiarity cannot be assumed, the exercises can simply be skipped or students can be referred to the definitions given in Section 7.1. #55: This exercise is particularly good preparation for the combinatorial manipulations in Chapter 6.

Exercise Set 4.2:. #19-25: These exercises develop skills needed in Section 8.2. #29: The mistake made in this proof fragment is surprisingly common.

Exercise Set 4.3 #21-24 and Exercise Set 4.4 #1-8: The skill developed in doing these exercises is used again in Sections 8.2 and 8.3.

Section 4.1

2. $b_0 = 1 + 2^0 = 2$, $b_1 = 1 + 2^1 = 3$, $b_2 = 1 + 2^2 = 5$, $b_3 = 1 + 2^3 = 9$

4. $d_1 = 1 - (1/10)^1 = 9/10$, $d_2 = 1 - (1/10)^2 = 99/100$, $d_3 = 1 - (1/10)^3 = 999/1000$, $d_4 = 1 - (1/10)^4 = 9999/10000$

6. $f_0 = \lfloor 0/3 \rfloor \cdot 3 = 0 \cdot 3 = 0$, $f_1 = \lfloor 1/3 \rfloor \cdot 3 = 0 \cdot 3 = 0$, $f_2 = \lfloor 2/3 \rfloor \cdot 3 = 0 \cdot 3 = 0$, $f_4 = \lfloor 4/3 \rfloor \cdot 3 = 1 \cdot 3 = 3$

7. $a_0 = 2 \cdot 0 + 1 = 1$, $a_1 = 2 \cdot 1 + 1 = 5$, $a_2 = 2 \cdot 2 + 1 = 5$, $a_3 = 2 \cdot 3 + 1 = 5$
   $b_0 = (0-1)^3 + 0 + 2 = 1$, $b_1 = (1-1)^3 + 1 + 2 = 3$, $b_2 = (2-1)^3 + 2 + 2 = 5$,
   $b_3 = (3-1)^3 + 3 + 2 = 13$
   So $a_0 = b_0$, $a_1 = b_1$, and $a_2 = b_2$, but $a_3 \neq b_3$.

9.

h_1	=	$1 \cdot \lfloor \log_2 1 \rfloor$	=	$1 \cdot 0$
h_2	=	$2 \cdot \lfloor \log_2 2 \rfloor$	=	$2 \cdot 1$
h_3	=	$3 \cdot \lfloor \log_2 3 \rfloor$	=	$3 \cdot 1$
h_4	=	$4 \cdot \lfloor \log_2 4 \rfloor$	=	$4 \cdot 2$
h_5	=	$5 \cdot \lfloor \log_2 5 \rfloor$	=	$5 \cdot 2$
h_6	=	$6 \cdot \lfloor \log_2 6 \rfloor$	=	$6 \cdot 2$
h_7	=	$7 \cdot \lfloor \log_2 7 \rfloor$	=	$7 \cdot 2$
h_8	=	$8 \cdot \lfloor \log_2 8 \rfloor$	=	$8 \cdot 3$
h_9	=	$9 \cdot \lfloor \log_2 9 \rfloor$	=	$9 \cdot 3$
h_{10}	=	$10 \cdot \lfloor \log_2 10 \rfloor$	=	$10 \cdot 3$
h_{11}	=	$11 \cdot \lfloor \log_2 11 \rfloor$	=	$11 \cdot 3$
h_{12}	=	$12 \cdot \lfloor \log_2 12 \rfloor$	=	$12 \cdot 3$
h_{13}	=	$13 \cdot \lfloor \log_2 13 \rfloor$	=	$13 \cdot 3$
h_{14}	=	$14 \cdot \lfloor \log_2 14 \rfloor$	=	$14 \cdot 3$
h_{15}	=	$15 \cdot \lfloor \log_2 15 \rfloor$	=	$15 \cdot 3$

When n is an integral power of 2, h_n is n times the exponent of that power. For instance, $8 = 2^3$ and $h_8 = 8 \cdot 3$. If m and n are integers and $2^m \leq 2^n < 2^{m+1}$, then $h_n = n \cdot m$.

Exercises 10-16 have more than one correct answer.

13. $a_n = \dfrac{1}{n} - \dfrac{1}{n+1}$ for all integers $n \geq 1$

14. $a_n = \dfrac{n}{(n+1)^2}$ for all integers $n \geq 1$

15. $a_n = (-1)^n \dfrac{n-1}{n}$ for all integers $n \geq 2$

16. $a_n = n(n+1)$ for all integers $n \geq 1$

17. $a_n = \lfloor n/2 \rfloor$

18. e. $\prod_{k=2}^{2} a_k = -2$

21. $\sum_{m=0}^{4} \frac{1}{2^m} = \frac{1}{2^0} + \frac{1}{2^1} + \frac{1}{2^2} + \frac{1}{2^3} + \frac{1}{2^4} = \frac{31}{16}$

22. $\prod_{j=1}^{5}(-1)^j = (-1)^1 \cdot (-1)^2 \cdot (-1)^3 \cdot (-1)^4 \cdot (-1)^5 = -1$

23. $\sum_{k=-1}^{1}(k^3 + 2) = ((-1)^3 + 2) + (0^3 + 2) + (1^3 + 2) = 6$

25. $\prod_{i=2}^{5} \frac{(i-1)i}{(i+1)(i+2)} = \frac{1 \cdot 2}{3 \cdot 4} \cdot \frac{2 \cdot 3}{4 \cdot 5} \cdot \frac{3 \cdot 4}{5 \cdot 6} \cdot \frac{4 \cdot 5}{6 \cdot 7} = \frac{1}{105}$

27. $1 \cdot 2 + 2 \cdot 3 + 3 \cdot 4 + \ldots + n \cdot (n+1)$

28. $\frac{1}{0!} + \frac{1}{1!} + \frac{1}{2!} + \ldots + \frac{1}{n!}$

Exercises 29-38 have more than one correct answer.

30. $\sum_{k=1}^{4}(k^3 - 1)$

31. $\prod_{i=2}^{4}(i^2 + 1)$

34. $\prod_{j=1}^{4}(1 - r^j)$

36. $\sum_{k=1}^{n} \frac{k}{(k+1)!}$

38. $\sum_{k=0}^{n-1} \frac{n-k}{(k+1)!}$

40. When $k = 1$, $i = 1 + 1 = 2$. When $k = n$, $i = n + 1$. Since $i = k + 1$, then $k = i - 1$. So $\frac{k^2}{k+1} = \frac{(i-1)^2}{(i-1)+1} = \frac{(i-1)^2}{i}$. Therefore, $\prod_{k=1}^{n}(\frac{k^2}{k+1}) = \prod_{i=2}^{n+1}(\frac{(i-1)^2}{i})$.

43. When $i = 1$, $j = 1 - 1 = 0$. When $i = n - 1$, $j = n - 2$. Since $j = i - 1$, then $i = j + 1$. So $\frac{i}{(n-i)^2} = \frac{j+1}{(n-(j+1))^2} = \frac{j+1}{(n-j-1)^2}$. Therefore, $\sum_{i=1}^{n-1}(\frac{i}{(n-i)^2}) = \sum_{j=0}^{n-2}(\frac{j+1}{(n-j-1)^2})$.

44. When $i = n$, $j = n - 1$. When $i = 2n$, $j = 2n - 1$. Since $j = i - 1$, then $i = j + 1$. So $\frac{n-i+1}{i} = \frac{n-(j+1)+1}{j+1} = \frac{n-j}{j+1}$. Therefore, $\prod_{i=n}^{2n}(\frac{n-i+1}{i}) = \prod_{j=n-1}^{2n-1}(\frac{n-j}{j+1})$.

46. $\left(\prod_{k=1}^{n}(\frac{k}{k+1})\right)\left(\prod_{k=1}^{n}(\frac{k+1}{k+2})\right) = \prod_{k=1}^{n}(\frac{k}{k+1})(\frac{k+1}{k+2}) = \prod_{k=1}^{n}(\frac{k}{k+2})$

47. 1. For $m = 1$ and $n = 1$, the expanded form of the equation is

$$(a_1 + a_2 + a_3 + a_4) + (b_1 + b_2 + b_3 + b_4) = (a_1 + b_1) + (a_2 + b_2) + (a_3 + b_3) + (a_4 + b_4).$$

The two sides of this equation are equal by repeated application of the associative and commutative laws of addition.

2. For $m = 1$ and $n = 1$, the expanded form of the equation is

$$c(a_1 + a_2 + a_3 + a_4) = ca_1 + ca_2 + ca_3 + ca_4.$$

The two sides of this equation are equal by repeated application of the associative law of addition and the (ordinary) distributive law.

3. For $m = 1$ and $n = 1$, the expanded form of the equation is

$$(a_1 \cdot a_2 \cdot a_3 \cdot a_4) \cdot (b_1 \cdot b_2 \cdot b_3 \cdot b_4) = (a_1 \cdot b_1) \cdot (a_2 \cdot b_2) \cdot (a_3 \cdot b_3) \cdot (a_4 \cdot b_4).$$

The two sides of this equation are equal by repeated application of the associative and commutative laws of multiplication.

49. $\dfrac{5!}{7!} = \dfrac{5!}{7 \cdot 6 \cdot 5!} = \dfrac{1}{42}$

50. $\dfrac{6!}{0!} = \dfrac{6!}{1} = 720$

53. $\dfrac{n!}{(n-2)!} = \dfrac{n \cdot (n-1) \cdot (n-2)!}{(n-2)!} = n(n-1)$

56.

$$\dfrac{n!}{(n-k-1)!} = \dfrac{n!}{(n-(k+1))!}$$
$$= \dfrac{n \cdot (n-1) \cdot \ldots \cdot (n-k) \cdot (n-(k+1))(n-(k+2)) \cdot \ldots \cdot 3 \cdot 2 \cdot 1}{(n-(k+1))(n-(k+2)) \cdot \ldots \cdot 3 \cdot 2 \cdot 1}$$
$$= n \cdot (n-1) \cdot \ldots \cdot (n-k)$$

57. *b. Proof:* Let n and k be integers with $n \geq 2$ and $2 \leq k \leq n$. Now $n!$ is the product of all the integers from 1 to n, and so since $n \geq 2$ and $2 \leq k \leq n$, k is a factor of $n!$. That is $n! = k \cdot r$ for some integer r. By substitution $n! + k = k \cdot r + k = k(r+1)$. But $r+1$ is an integer because r is. By definition of divisibility, therefore, $n! + k$ is divisible by k.

c. Yes. If m is any integer that is greater than or equal to 2, then none of the terms of the following sequence of integers is prime: $m! + 2, m! + 3, m! + 4, \ldots, m! + m$. The reason is that each has the form $m! + k$ for an integer k with $2 \leq k \leq m$, and for each such k, by part (*b*) $m! + k$ is divisible by k.

58. *b.* $m + 1$, $sum + a[j-1]$

60.

```
      0       R. 1
    2|1       R. 1
    2|3       R. 0
    2|6       R. 0
    2|12      R. 0
    2|24      R. 1
    2|49      R. 0
    2|98      R. 0
    2|196
```
Hence $196_{10} = 11000100_2$.

61.

```
       0       R. 1
     2⌐1       R. 1
     2⌐3       R. 0
     2⌐6       R. 0
    2⌐12       R. 1
    2⌐25       R. 0
    2⌐50       R. 1
   2⌐101              Hence $101_{10} = 1100101_2$.
```

63.

a	26					
i	0	1	2	3	4	5
q	26	13	6	3	1	0
$r[0]$	0					
$r[1]$		1				
$r[2]$			0			
$r[3]$				1		
$r[4]$					1	

64.

a	37						
i	0	1	2	3	4	5	6
q	37	18	9	4	2	1	0
$r[0]$	1						
$r[1]$		0					
$r[2]$			1				
$r[3]$				0			
$r[4]$					0		
$r[5]$						1	

65. Let a nonnegative integer a be given. Divide a by 16 using the quotient-remainder theorem to obtain a quotient $q[0]$ and a remainder $r[0]$. If the quotient is nonzero, divide by 16 again to obtain a quotient $q[1]$ and a remainder $r[1]$. Continue this process until a quotient of 0 is obtained. The remainders calculated in this way are the hexadecimal digits of a: $a_{10} = (r[k]r[k-1]\ldots r[2]r[1]r[0])_{16}$.

67.

```
       0       R. 2  = $2_{16}$
    16⌐2       R. 10 = $A_{16}$
    16⌐42      R. 12 = $C_{16}$
   16⌐684                Hence $684_{10} = 2AC_{16}$.
```

68.

```
       0       R. 10 = $A_{16}$
    16⌐10      R. 3  = $3_{16}$
    16⌐163     R. 14 = $E_{16}$
   16⌐2622               Hence $2622_{10} = A3E_{16}$.
```

69.

Algorithm Decimal to Hexadecimal Conversion
Using Repeated Division by 16

[In this algorithm the input is a nonnegative integer a. The aim of the algorithm is to produce a sequence of binary digits r[0], r[1], r[2], ...r[i − 1] so that the hexadecimal representation of a is $(r[i-1]r[i-2]\ldots r[2]r[1]r[0])_{16}$.]

Input: a *[a nonnegative integer]*

Algorithm Body:

$q := a,\ i := 0$

while ($i = 0$ or $q = 0$)

$r[i] := q\ mod\ 16$

$q := q\ div\ 16$

$i := i + 1$

end while

[After execution of this step, the values of r[0], r[1], r[2],..., r[i − 1] are all integers from 0 to 15 inclusive, and $a_{10} = (r[i-1]r[i-2]\ldots r[2]r[1]r[0])_{16}$.]

Output: $r[0], r[1], r[2], \ldots, r[i-1]$ *[a sequence of integers]*

end Algorithm

Section 4.2

2. Let $P(n)$ be the property "A postage of $n\!\!\!/$ can be obtained using $3\!\!\!/$ and $5\!\!\!/$ stamps."

   The property is true for n = 8 : A postage of $8\!\!\!/$ can be obtained using one $3\!\!\!/$ stamp and one $5\!\!\!/$ stamp.

   If the property is true for n = k then it is true for n = k + 1 : Suppose that for some integer $k \geq 8$, a postage of k cents can be obtained using $3\!\!\!/$ and $5\!\!\!/$ stamps. We must show that a postage of $k + 1$ cents can be obtained using $3\!\!\!/$ and $5\!\!\!/$ stamps. But if there is a $5\!\!\!/$ stamp among those used to make up the k cents of postage, replace it by two $3\!\!\!/$ stamps; the result will be $(k+1)\!\!\!/$ of postage. And if no $5\!\!\!/$ stamp is used to make up the k cents of postage, then at least three $3\!\!\!/$ stamps must be used because $k \geq 8$. Remove three $3\!\!\!/$ stamps and replace them by two $5\!\!\!/$ stamps; the result will be $(k+1)\!\!\!/$ of postage. Thus in either case $(k+1)\!\!\!/$ of postage can be obtained using $3\!\!\!/$ and $5\!\!\!/$ stamps *[as was to be shown]*.

4. a. $P(2): \sum_{i=1}^{2-1} i(i+1) = \dfrac{2(2-1)(2+1)}{3}$

   $P(2)$ is true because the left-hand side equals $1(1+1) = 2$ and the right-hand side equals $\dfrac{2 \cdot 1 \cdot 3}{3} = 2$ also.

   b. $P(k): \sum_{i=1}^{k-1} i(i+1) = \dfrac{k(k-1)(k+1)}{3}$

   c. $P(k+1): \sum_{i=1}^{(k+1)-1} i(i+1) = \dfrac{(k+1)((k+1)-1)((k+1)+1)}{3}$

Or: $\sum_{i=1}^{k} i(i+1) = \dfrac{(k+1)k(k+2)}{3}$

d. Must show: If k is any integer with $k \geq 2$ and $\sum_{i=1}^{k-1} i(i+1) = \dfrac{k(k-1)(k+1)}{3}$, then $\sum_{i=1}^{k} i(i+1) = \dfrac{(k+1)k(k+2)}{3}$.

7. *Proof (by mathematical induction)*:

   The formula is true for n = 1: The formula holds for $n = 1$ because for $n = 1$ the left-hand side is 1 and the right-hand side is $1 \cdot (2 \cdot 1 - 1) = 1$ also.

   If the formula is true for n = k then it is true for n = k + 1: Suppose $1 + 5 + 9 + \ldots + (4k - 3) = k(2k - 1)$ for some integer $k \geq 1$. We must show that $1 + 5 + 9 + \ldots + (4(k+1) - 3) = (k+1)[2(k+1) - 1]$, or, equivalently, $1 + 5 + 9 + \ldots + (4(k+1) - 3) = (k+1)(2k+1)$. But by the laws of algebra and substitution from the inductive hypothesis, $1 + 5 + 9 + \ldots + (4(k+1) - 3) = 1 + 5 + 9 + \cdots + (4k-3) + (4(k+1) - 3) = k(2k-1) + (4(k+1) - 3) = 2k^2 - k + 4k + 4 - 3 = 2k^2 + 3k + 1 = (k+1)(2k+1)$ *[as was to be shown]*.

8. *Proof (by mathematical induction)*:

   The formula is true for n = 0: The formula holds for $n = 0$ because for $n = 0$ the left-hand side is $2^0 = 1$ and the right-hand side is $2^{0+1} - 1 = 2 - 1 = 1$ also.

   If the formula is true for n = k then it is true for n = k + 1: Suppose $1 + 2 + 2^2 + \ldots + 2^k = 2^{k+1} - 1$ for some integer $k \geq 0$. We must show that $1 + 2 + 2^2 + \ldots + 2^{k+1} = 2^{(k+1)+1} - 1$, or, equivalently, $1 + 2 + 2^2 + \ldots + 2^{k+1} = 2^{k+2} - 1$. But by the laws of algebra and substitution from the inductive hypothesis, $1 + 2 + 2^2 + \ldots + 2^{k+1} = 1 + 2 + 2^2 + \ldots + 2^k + 2^{k+1} = 2^{k+1} - 1 + 2^{k+1} = 2 \cdot 2^{k+1} - 1 = 2^{k+2} - 1$ *[as was to be shown]*.

10. *Proof (by mathematical induction)*:

    The formula is true for n = 1: The formula holds for $n = 1$ because $1^3 = \left(\dfrac{1(1+1)}{2}\right)^2$.

    If the formula is true for n = k then it is true for n = k + 1: Suppose $1^3 + 2^3 + \ldots + k^3 = \left(\dfrac{k(k+1)}{2}\right)^2$ for some integer $k \geq 1$. We must show that $1^3 + 2^3 + \ldots + (k+1)^3 = \left(\dfrac{(k+1)((k+1)+1)}{2}\right)^2$, or, equivalently, $1^3 + 2^3 + \ldots + (k+1)^3 = \left(\dfrac{(k+1)(k+2)}{2}\right)^2$. But by the laws of algebra and substitution from the inductive hypothesis, $1^3 + 2^3 + \ldots + (k+1)^3 = 1^3 + 2^3 + \ldots + k^3 + (k+1)^3 = \left(\dfrac{k(k+1)}{2}\right)^2 + (k+1)^3 = \dfrac{k^2(k+1)^2}{4} + \dfrac{4(k+1)^3}{4} = \dfrac{(k+1)^2(k^2 + 4k + 4)}{4} = \left(\dfrac{(k+1)(k+2)}{2}\right)^2$ *[as was to be shown]*.

11. *Proof (by mathematical induction)*:

    The formula is true for n = 1: The formula holds for $n = 1$ because $\dfrac{1}{1 \cdot 2} = \dfrac{1}{1+1}$.

    If the formula is true for n = k then it is true for n = k + 1: Suppose $\dfrac{1}{1 \cdot 2} + \dfrac{1}{2 \cdot 3} + \ldots + \dfrac{1}{k \cdot (k+1)} = \dfrac{k}{k+1}$ for some integer $k \geq 1$. We must show that $\dfrac{1}{1 \cdot 2} + \dfrac{1}{2 \cdot 3} + \ldots + \dfrac{1}{(k+1)((k+1)+1)} = \dfrac{k+1}{(k+1)+1}$, or, equivalently, $\dfrac{1}{1 \cdot 2} + \dfrac{1}{2 \cdot 3} + \ldots + \dfrac{1}{(k+1)(k+2)} = \dfrac{k+1}{k+2}$.

But by the laws of algebra and substitution from the inductive hypothesis, $\frac{1}{1\cdot 2}+\frac{1}{2\cdot 3}+\ldots+\frac{1}{k\cdot(k+1)}+\frac{1}{(k+1)(k+2)}=\frac{k}{k+1}+\frac{1}{(k+1)(k+2)}=\frac{k(k+2)}{(k+1)(k+2)}+\frac{1}{(k+1)(k+2)}=\frac{k^2+2k+1}{(k+1)(k+2)}=\frac{(k+1)^2}{(k+1)(k+2)}=\frac{k+1}{k+2}$ *[as was to be shown]*.

13. *Proof (by mathematical induction):*

    The formula is true for n = 0: The formula holds for $n=0$ because $\sum_{i=1}^{0+1} i\cdot 2^i = 1\cdot 2^1 = 2$ and $0\cdot 2^{0+2}+2=2$ also.

    If the formula is true for n = k then it is true for n = k + 1: Suppose $\sum_{i=1}^{k+1} i\cdot 2^i = k\cdot 2^{k+2}+2$ for some integer $k\geq 0$. We must show that $\sum_{i=1}^{(k+1)+1} i\cdot 2^i = (k+1)\cdot 2^{(k+1)+2}+2$, or, equivalently, $\sum_{i=1}^{k+2} i\cdot 2^i = (k+1)\cdot 2^{k+3}+2$. But by the laws of algebra and substitution from the inductive hypothesis, $\sum_{i=1}^{k+2} i\cdot 2^i = \sum_{i=1}^{k+1} i\cdot 2^i + (k+2)2^{k+2} = (k\cdot 2^{k+2}+2)+(k+2)2^{k+2} = (k+(k+2))2^{k+2}+2 = (2k+2)2^{k+2}+2 = (k+1)2^{k+3}+2$ *[as was to be shown]*.

14. *Proof (by mathematical induction):*

    The formula is true for n = 1: The formula holds for $n=1$ because $\sum_{i=1}^{1} i(i!) = 1\cdot(1!) = 1$ and $(1+1)!-1 = 2!-1 = 2-1 = 1$ also.

    If the formula is true for n = k then it is true for n = k + 1: Suppose $\sum_{i=1}^{k} i(i!) = (k+1)!-1$ for some integer $k\geq 1$. We must show that $\sum_{i=1}^{k+1} i(i!) = ((k+1)+1)!-1$, or, equivalently, $\sum_{i=1}^{k+1} i(i!) = (k+2)!-1$. But by the laws of algebra and substitution from the inductive hypothesis, $\sum_{i=1}^{k+1} i(i!) = \sum_{i=1}^{k} i(i!)+(k+1)((k+1)!) = [(k+1)!-1]+(k+1)((k+1)!) = ((k+1)!)(1+(k+1))-1 = (k+1)!(k+2)-1 = (k+2)!-1$ *[as was to be shown]*.

15. *Proof (by mathematical induction):*

    The formula is true for n = 2: The formula holds for $n=2$ because $1-\frac{1}{2^2}=\frac{3}{4}$ and $\frac{2+1}{2\cdot 2}=\frac{3}{4}$ also.

    If the formula is true for n = k then it is true for n = k + 1: Suppose

    $$\left(1-\frac{1}{2^2}\right)\cdot\left(1-\frac{1}{3^2}\right)\cdot\ldots\cdot\left(1-\frac{1}{k^2}\right) = \frac{k+1}{2k} \qquad \text{for some integer } k\geq 2.$$

    We must show that

    $$\left(1-\frac{1}{2^2}\right)\cdot\left(1-\frac{1}{3^2}\right)\cdot\ldots\cdot\left(1-\frac{1}{(k+1)^2}\right) = \frac{(k+1)+1}{2(k+1)},$$

    or, equivalently,

    $$\left(1-\frac{1}{2^2}\right)\cdot\left(1-\frac{1}{3^2}\right)\cdot\ldots\cdot\left(1-\frac{1}{(k+1)^2}\right) = \frac{k+2}{2(k+1)}.$$

    But by the laws of algebra and substitution from the inductive hypothesis,

    $$\left(1-\frac{1}{2^2}\right)\cdot\left(1-\frac{1}{3^2}\right)\cdot\ldots\cdot\left(1-\frac{1}{(k+1)^2}\right)$$

    $$= \left(1-\frac{1}{2^2}\right)\cdot\left(1-\frac{1}{3^2}\right)\cdot\ldots\cdot\left(1-\frac{1}{k^2}\right)\cdot\left(1-\frac{1}{(k+1)^2}\right)$$

    $$= \frac{k+1}{2k}\cdot\left(1-\frac{1}{(k+1)^2}\right)$$

$$= \frac{k+1}{2k} - \frac{1}{2k(k+1)}$$
$$= \frac{(k+1)^2 - 1}{2k(k+1)}$$
$$= \frac{k^2 + 2k}{2k(k+1)}$$
$$= \frac{k+2}{2(k+1)}$$

[as was to be shown].

16. *Proof (by mathematical induction)*:

   The formula is true for n = 0 : The formula holds for $n = 0$ because $\prod_{i=0}^{0}\left(\frac{1}{2i+1} \cdot \frac{1}{2i+2}\right) = \frac{1}{1} \cdot \frac{1}{2} = \frac{1}{2}$ and $\frac{1}{(2 \cdot 0 + 2)!} = \frac{1}{2}$ also.

   If the formula is true for n = k then it is true for n = k + 1 : Suppose $\prod_{i=0}^{k}\left(\frac{1}{2i+1} \cdot \frac{1}{2i+2}\right) = \frac{1}{(2k+2)!}$ for some integer $k \geq 0$. We must show that $\prod_{i=0}^{k+1}\left(\frac{1}{2i+1} \cdot \frac{1}{2i+2}\right) = \frac{1}{(2 \cdot (k+1) + 2)!}$, or, equivalently, $\prod_{i=0}^{k+1}\left(\frac{1}{2i+1} \cdot \frac{1}{2i+2}\right) = \frac{1}{((2k+4)!}$. But by the laws of algebra and substitution from the inductive hypothesis,

$$\prod_{i=0}^{k+1}\left(\frac{1}{2i+1} \cdot \frac{1}{2i+2}\right) = \prod_{i=0}^{k}\left(\frac{1}{2i+1} \cdot \frac{1}{2i+2}\right) \cdot \left(\frac{1}{2(k+1)+1} \cdot \frac{1}{2(k+1)+2}\right)$$
$$= \left(\frac{1}{(2k+2)!}\right) \cdot \left(\frac{1}{2(k+1)+1} \cdot \frac{1}{2(k+1)+2}\right)$$
$$= \frac{1}{(2k+2)!} \cdot \frac{1}{2k+3} \cdot \frac{1}{2k+4}$$
$$= \frac{1}{(2k+4)!}$$

[as was to be shown].

17. *Proof (by mathematical induction)*:

   The formula is true for n = 2: The formula holds for $n = 2$ because $c(a_1 + a_2) = ca_1 + ca_2$ by the distributive law for the real numbers. (See Appendix A.)

   If the formula is true for n = k then it is true for n = k + 1 : Suppose $c(a_1+a_2+\ldots+a_k) = ca_1 + ca_2 + \ldots + ca_k$ for some integer $k \geq 2$. We must show that $c(a_1 + a_2 + \ldots + a_{k+1}) = ca_1 + ca_2 + \ldots + ca_{k+1}$. But by the associative law of addition, the ordinary distributive law for the real numbers, and substitution from the inductive hypothesis, $c(a_1+a_2+\ldots+a_k+a_{k+1}) = c[(a_1+a_2+\ldots+a_k)+a_{k+1}] = c(a_1+a_2+\ldots+a_k)+ca_{k+1} = (ca_1+ca_2+\ldots+ca_k)+ca_{k+1} = ca_1 + ca_2 + \ldots + ca_k + ca_{k+1}$. *[as was to be shown]*.

20. $5 + 10 + 15 + 20 + \ldots + 300 = 5(1 + 2 + 3 + \ldots + 60) = 5\left(\frac{60 \cdot 61}{2}\right) = 9150$

22. *b.* $2 + 2^2 + 2^3 + \ldots + 2^{26} = 2(1 + 2 + 2^2 + \ldots + 2^{25}) = 2(2^{26} - 1)$ *[by part (a)]* $= 134,217,726$

23. $3 + 3^2 + 3^3 + \ldots + 3^n = 3(1 + 3 + 3^2 + \ldots + 3^{n-1}) = 3\left(\dfrac{3^{(n-1)+1} - 1}{3 - 1}\right) = \dfrac{3(3^n - 1)}{2}$

25. $1 - 2 + 2^2 - 2^3 + \ldots + (-1)^n 2^n = \ldots + (-2)^n = \dfrac{(-2)^{n+1} - 1}{(-2) - 1} = \dfrac{(-2)^{n+1} - 1}{-3} = \dfrac{1}{3}\left(1 + (-1)^n 2^{n+1}\right)$

26. $(a + md) + (a + (m+1)d) + (a + (m+2)d) + \ldots + (a + (m+n)d)$

$\quad = \underbrace{(a + a + \ldots + a)}_{n+1 \text{ terms}} + (m + (m+1) + (m+2) + \ldots + (m+n))d$

$\quad = a(n+1) + \underbrace{(m + m + \ldots + m)}_{n+1 \text{ terms}} d + (0 + 1 + 2 + \ldots + n)d$

$\quad = a(n+1) + m(n+1)d + \dfrac{n(n+1)}{2}d \qquad\qquad\qquad\qquad\qquad$ by Theorem 4.2.2

$\quad = (a + md)(n+1) + \dfrac{n(n+1)}{2}d$

$\quad = (a + md + \dfrac{n}{2}d)(n+1)$

$\quad = [a + (m + \dfrac{n}{2})d](n+1)$

27. $ar^m + ar^{m+1} + ar^{m+2} + \ldots + ar^{m+n} = ar^m(1 + r + r^2 + \ldots + r^n) = ar^m\left(\dfrac{r^{n+1} - 1}{r - 1}\right)$ by Theorem 4.2.3

28. *a.* $2 + 4 + 8 + \ldots + 2^{40} = 2(1 + 2 + 2^2 + \ldots + 2^{39}) = 2\left(\dfrac{2^{39+1} - 1}{2 - 1}\right) = 2^{41} - 2$

b. 40 generations = 1200 years (at 30 years per generation)

c. Since $2^{41} - 2 \cong 2.2 \times 10^{12} > 10^{10}$, not all ancestors are distinct: some ancestors on different branches of the family tree must be the same.

Section 4.3

2. *Formula:* $(1 + \dfrac{1}{1}) \cdot (1 + \dfrac{1}{2}) \cdot \ldots \cdot (1 + \dfrac{1}{n}) = n + 1$ for all integers $n \geq 1$.

Proof (by mathematical induction):

The formula is true for n = 1: The formula holds for $n = 1$ because $1 + \dfrac{1}{1} = 1 + 1$.

If the formula is true for n = k then it is true for n = k + 1: Suppose $(1 + \dfrac{1}{1}) \cdot (1 + \dfrac{1}{2}) \cdot \ldots \cdot (1 + \dfrac{1}{k}) = k + 1$, for some integer $k \geq 1$. We must show that $(1 + \dfrac{1}{1}) \cdot (1 + \dfrac{1}{2}) \cdot \ldots \cdot (1 + \dfrac{1}{k+1}) = (k+1) + 1$. But by the laws of algebra and substitution from the inductive hypothesis, $(1 + \dfrac{1}{1}) \cdot (1 + \dfrac{1}{2}) \cdot \ldots \cdot (1 + \dfrac{1}{k+1}) = (1 + \dfrac{1}{1}) \cdot (1 + \dfrac{1}{2}) \cdot \ldots \cdot (1 + \dfrac{1}{k}) \cdot (1 + \dfrac{1}{k+1}) = (k+1) \cdot (1 + \dfrac{1}{k+1}) = (k+1) + 1$ *[as was to be shown]*.

4. Formula: $\sum_{i=1}^{n}(-1)^{i-1}i^2 = (-1)^{n-1}(1+2+3+\ldots+n)$ or $\sum_{i=1}^{n}(-1)^{i-1}i^2 = (-1)^{n-1} \cdot \frac{n(n+1)}{2}$ (by Theorem 4.2.1) for all integers $n \geq 1$.

Proof (by mathematical induction):

The formula is true for n = 1: The formula holds for $n = 1$ because $\sum_{i=1}^{1}(-1)^{i-1}i^2 = (-1)^0 \cdot 1^2 = 1$ and $(-1)^{1-1} \cdot \frac{1(1+1)}{2} = 1$ also.

If the formula is true for n = k then it is true for n = k + 1: Suppose

$$\sum_{i=1}^{k}(-1)^{i-1}i^2 = (-1)^{k-1} \cdot \frac{k(k+1)}{2} \quad \text{for some integer } k \geq 1.$$

We must show that

$$\sum_{i=1}^{k+1}(-1)^{i-1}i^2 = (-1)^{(k+1)-1} \cdot \frac{(k+1)((k+1)+1)}{2},$$

or, equivalently,

$$\sum_{i=1}^{k+1}(-1)^{i-1}i^2 = (-1)^k \cdot \frac{(k+1)(k+2)}{2}.$$

But by the laws of algebra and substitution from the inductive hypothesis,

$$\begin{aligned}
\sum_{i=1}^{k+1}(-1)^{i-1}i^2 &= \sum_{i=1}^{k}(-1)^{i-1}i^2 + (-1)^{(k+1)-1}(k+1)^2 \\
&= (-1)^{k-1} \cdot \frac{k(k+1)}{2} + (-1)^k(k+1)^2 \\
&= (-1)^{k-1}(k+1)\left(\frac{k}{2} + (-1)(k+1)\right) \\
&= (-1)^{k-1}(k+1)\left(\frac{k - 2k - 2}{2}\right) \\
&= (-1)^{k-1}(k+1)\left(\frac{-k-2}{2}\right) \\
&= (-1)^{k-1}(k+1)\left(\frac{-(k+2)}{2}\right) \\
&= (-1)^k \cdot \frac{(k+1)(k+2)}{2}
\end{aligned}$$

[as was to be shown].

5.

$$\sum_{k=1}^{1} \frac{k}{(k+1)!} = \frac{1}{2!} = \frac{1}{2}$$

$$\sum_{k=1}^{2} \frac{k}{(k+1)!} = \frac{1}{2} + \frac{1}{3!} = \frac{5}{6}$$

$$\sum_{k=1}^{3} \frac{k}{(k+1)!} = \frac{5}{6} + \frac{1}{4!} = \frac{23}{24}$$

$$\sum_{k=1}^{4} \frac{k}{(k+1)!} = \frac{23}{24} + \frac{1}{5!} = \frac{119}{120}$$

$$\sum_{k=1}^{5} \frac{k}{(k+1)!} = \frac{119}{120} + \frac{1}{6!} = \frac{719}{720}$$

Note that $\frac{1}{2} = \frac{2!-1}{2!}$, $\frac{5}{6} = \frac{3!-1}{3!}$, $\frac{23}{24} = \frac{4!-1}{4!}$, $\frac{119}{120} = \frac{5!-1}{5!}$, and $\frac{719}{720} = \frac{6!-1}{6!}$. So we conjecture that $\sum_{k=1}^{n} \frac{k}{(k+1)!} = \frac{(n+1)!-1}{(n+1)!}$.

Proof by mathematical induction:

The formula holds for n = 1: The calculation above shows that $\sum_{k=1}^{1} \frac{k}{(k+1)!} = \frac{(1+1)!-1}{(1+1)!}$.

If the formula holds for n = r then it holds for n = r+1: Suppose that for some integer $r \geq 1$, $\sum_{k=1}^{r} \frac{k}{(k+1)!} = \frac{(r+1)!-1}{(r+1)!}$. We must show that $\sum_{k=1}^{r+1} \frac{k}{(k+1)!} = \frac{[(r+1)+1]!-1}{[(r+1)+1]!}$, or, equivalently, $\sum_{k=1}^{r+1} \frac{k}{(k+1)!} = \frac{(r+2)!-1}{(r+2)!}$. But by substitution from the inductive hypothesis and by the laws of algebra, $\sum_{k=1}^{r+1} \frac{k}{(k+1)!} = \sum_{k=1}^{r} \frac{k}{(k+1)!} + \frac{r+1}{[(r+1)+1]!} = \frac{(r+1)!-1}{(r+1)!} + \frac{r+1}{(r+2)!} = \frac{(r+1)!-1}{(r+1)!} \cdot \frac{(r+2)}{(r+2)} + \frac{r+1}{(r+2)!} = \frac{(r+2)!-(r+2)}{(r+2)!} + \frac{r+1}{(r+2)!} = \frac{(r+2)!-1}{(r+2)!}$ [as was to be shown].

7. a. $P(5): 5^2 < 2^5$

   $P(5)$ is true because $25 < 32$.

   b. $P(k): k^2 < 2^k$

   c. $P(k+1): (k+1)^2 < 2^{k+1}$

   d. *Must show*: If k is any integer with $k \geq 5$ and $k^2 < 2^k$, then $(k+1)^2 < 2^{k+1}$.

9. *Proof (by mathematical induction):*

   The property holds for n = 1: The property holds for $n = 1$ because $2^{3 \cdot 1} - 1 = 7$ and 7 is divisible by 7.

   If the property holds for n = k then it holds for n = k + 1: Suppose $2^{3k} - 1$ is divisible by 7 for some integer $k \geq 1$. We must show that $2^{3(k+1)} - 1$ is divisible by 7. By definition of divisibility $2^{3k} - 1 = 7r$ for some integer r. Then by the laws of algebra and the inductive hypothesis, $2^{3(k+1)} - 1 = 2^{3k+3} - 1 = 2^3 2^{3k} - 1 = (7+1)2^{3k} - 1 = 7 \cdot 2^{3k} + (2^{3k} - 1) = 7 \cdot 2^{3k} + 7r = 7(2^{3k} + r)$, which is divisible by 7 because $2^{3k} + r$ is an integer. Therefore, $2^{3(k+1)} - 1$ is divisible by 7 [as was to be shown].

10. *Proof (by mathematical induction)*:

    The property holds for n = 0: The property holds for $n = 0$ because $0^3 - 7 \cdot 0 + 3 = 3$ and 3 is divisible by 3.

    If the property holds for n = k then it holds for n = k + 1: Suppose $k^3 - 7k + 3$ is divisible by 3 for some integer $k \geq 0$. We must show that $(k+1)^3 - 7(k+1) + 3$ is divisible by 3. By definition of divisibility $k^3 - 7k + 3 = 3r$ for some integer r. Then by the laws of algebra and the inductive hypothesis, $(k+1)^3 - 7(k+1) + 3 = k^3 + 3k^2 + 3k + 1 - 7k - 7 + 3 = (k^3 - 7k + 3) + (3k^2 + 3k - 6) = 3r + 3(k^2 + k - 2) = 3(r + k^2 + k - 2)$, which is divisible by 3 because $r + k^2 + k - 2$ is an integer. Therefore, $(k+1)^3 - 7(k+1) + 3$ is divisible by 3 *[as was to be shown]*.

11. *Proof (by mathematical induction)*:

    The property holds for n = 0: The property holds for $n = 0$ because $3^{2 \cdot 0} - 1 = 1 - 1 = 0$ and 0 is divisible by 8.

    If the property holds for n = k then it holds for n = k + 1: Suppose $3^{2k} - 1$ is divisible by 8 for some integer $k \geq 0$. We must show that $3^{2(k+1)} - 1$ is divisible by 8. By definition of divisibility $3^{2k} - 1 = 8r$ for some integer r. Then by the laws of algebra and the inductive hypothesis, $3^{2(k+1)} - 1 = 3^{2k} 3^2 - 1 = 9 \cdot 3^{2k} - 1 = (8+1)3^{2k} - 1 = 8 \cdot 3^{2k} + (3^{2k} - 1) = 8 \cdot 3^{2k} + 8r = 8(3^{2k} + r)$, which is divisible by 3 because $3^{2k} + r$ is an integer. Therefore, $3^{2(k+1)} - 1$ is divisible by 8 *[as was to be shown]*.

13. *Proof:* Suppose x and y are any integers with $x \neq y$. We show by mathematical induction on n that for all integers $n \geq 1$, $x^n - y^n$ is divisible by $x - y$.

    The property holds for n = 1: The property holds for $n = 1$ because $x^1 - y^1 = x - y$ and $x - y$ is divisible by $x - y$ as long as $x \neq y$.

    If the property holds for n = k then it holds for n = k + 1: Suppose $x^k - y^k$ is divisible by $x - y$ for some integer $k \geq 1$. We must show that $x^{k+1} - y^{k+1}$ is divisible by $x - y$. By definition of divisibility $x^k - y^k = (x - y)r$ for some integer r. Then by the laws of algebra and the inductive hypothesis, $x^{k+1} - y^{k+1} = x^{k+1} - xy^k + xy^k - y^{k+1} = x(x^k - y^k) + y^k(x - y) = (x - y)r + y^k(x - y) = (x - y)(r + y^k)$, which is divisible by $x - y$ because $r + y^k$ is an integer. Therefore, $x^{k+1} - y^{k+1}$ is divisible by $x - y$ *[as was to be shown]*.

14. *Proof (by mathematical induction)*:

    The property holds for n = 2: The property holds for $n = 2$ because $2^3 - 2 = 6$ and 6 is divisible by 6.

    If the property holds for n = k then it holds for n = k + 1: Suppose $k^3 - k$ is divisible by 6 for some integer $k \geq 2$. We must show that $(k+1)^3 - (k+1)$ is divisible by 6. By definition of divisibility $k^3 - k = 6r$ for some integer r. Then by the laws of algebra and the inductive hypothesis, $(k+1)^3 - (k+1) = k^3 + 3k^2 + 3k + 1 - k - 1 = (k^3 - k) + 3(k^2 + k) = 6r + 3(k(k+1))$. Now $k(k+1)$ is a product of two consecutive integers. By the parity theorem one of these is even, and so *[by exercise 20 of Section 3.1]* the product $k(k+1)$ is even. Hence $k(k+1) = 2s$ for some integer s. Thus $6r + 3(k(k+1)) = 6r + 3(2s) = 6(r+s)$, and so by substitution, then, $(k+1)^3 - (k+1) = 6(r+s)$, which is divisible by 6 because $r + s$ is an integer. Therefore, $(k+1)^3 - (k+1)$ is divisible by 6 *[as was to be shown]*.

15. *Proof (by mathematical induction)*:

    The property holds for n = 1: The property holds for $n = 1$ because $1(1^2 + 5) = 6$ and 6 is divisible by 6.

    If the property holds for n = k then it holds for n = k + 1: Suppose $k(k^2 + 5)$ is divisible by 6 for some integer $k \geq 1$. We must show that $(k+1)((k+1)^2 + 5)$ is divisible by 6. By definition of divisibility $k(k^2 + 5) = 6r$ for some integer r. Then by the laws of algebra and the inductive hypothesis, $(k+1)((k+1)^2 + 5) = (k+1)(k^2 + 2k + 1 + 5) = k(k^2 + 5) + (k(2k + 1) + k^2 + 2k + 1 + 5) = k(k^2 + 5) + (3k^2 + 3k + 6) = 6r + 3(k^2 + k) + 6$. Now $k(k+1)$ is a product of two consecutive integers. By the parity theorem one of these is even, and so *[by exercise 20 of Section 3.1]* the product $k(k+1)$ is even. Hence $k(k+1) = 2s$ for some integer s. Thus $6r + 3(k^2 + k) + 6 = 6r + 3(2s) + 6 = 6(r + s + 1)$. By substitution, then, $(k+1)((k+1)^2 + 5) = 6(r + s + 1)$, which is divisible by 6 because $r + s + 1$ is an integer. Therefore, $(k+1)((k+1)^2 + 5)$ is divisible by 6 *[as was to be shown]*.

17. *Proof (by mathematical induction)*:

    The inequality holds for n = 0 : The inequality holds for $n = 0$ because $2^0 = 1$ and $(0+2)! = 2$ and $1 < 2$.

    If the inequality holds for n = k then it holds for n = k + 1 : Suppose $2^k < (k+2)!$ for some integer $k \geq 0$. We must show that $2^{k+1} < ((k+1)+2)!$. But by the laws of algebra and substitution from the inductive hypothesis, $2^{k+1} = 2^k \cdot 2 < (k+2)! \cdot 2 < (k+2)! \cdot (k+3)$ *[2 < k + 3 because k ≥ 0]*. Thus $2^{k+1} < (k+2)! \cdot (k+3) = (k+3)!$ and so $2^{k+1} < ((k+1)+2)!$ *[as was to be shown]*.

18. *Proof (by mathematical induction)*:

    The inequality holds for n = 2 : To show that the inequality holds for $n = 2$ we must show that $\sqrt{2} < \dfrac{1}{\sqrt{1}} + \dfrac{1}{\sqrt{2}}$. But this inequality is true if, and only if, $2 < \sqrt{2} + 1$ (by multiplying both sides by $\sqrt{2}$). And this is true if, and only if, $1 < \sqrt{2}$ (by subtracting 1 from both sides). But $1 < \sqrt{2}$, and so the inequality holds for $n = 2$.

    If the inequality holds for n = k then it holds for n = k + 1 : Suppose $\sqrt{k} < \dfrac{1}{\sqrt{1}} + \dfrac{1}{\sqrt{2}} + \dfrac{1}{\sqrt{3}} + \ldots + \dfrac{1}{\sqrt{k}}$ for some integer $k \geq 2$. We must show that $\sqrt{k+1} < \dfrac{1}{\sqrt{1}} + \dfrac{1}{\sqrt{2}} + \dfrac{1}{\sqrt{3}} + \ldots + \dfrac{1}{\sqrt{k+1}}$. But for each integer $k \geq 2$, $\sqrt{k} < \sqrt{k+1}$, and so (multiplying both sides by $\sqrt{k}$) $k < \sqrt{k} \cdot \sqrt{k+1}$. Adding 1 to both sides gives $k + 1 < \sqrt{k} \cdot \sqrt{k+1} + 1$, and dividing both sides by $\sqrt{k+1}$ gives $\sqrt{k+1} = \sqrt{k} + \dfrac{1}{\sqrt{k+1}}$. By substitution from the inductive hypothesis, then, $\sqrt{k+1} < \dfrac{1}{\sqrt{1}} + \dfrac{1}{\sqrt{2}} + \dfrac{1}{\sqrt{3}} + \ldots + \dfrac{1}{\sqrt{k}} + \dfrac{1}{\sqrt{k+1}}$ *[as was to be shown]*.

19. Suppose x is a *[particular but arbitrarily chosen]* real number that is greater than -1. We show by mathematical induction that $1 + nx \leq (1+x)^n$ for all integers $n \geq 2$.

    The inequality holds for n = 2 : To show that the inequality holds for $n = 2$ we must show that $1 + 2x \leq (1+x)^2$. But $(1+x)^2 = 1 + 2x + x^2$ and $1 + 2x \leq 1 + 2x + x^2$ because $x^2 \geq 0$ for all real numbers x. Hence the inequality holds for $n = 2$.

    If the inequality holds for n = k then it holds for n = k + 1 : Suppose $1 + kx \leq (1+x)^k$ for some integer $k \geq 2$. We must show that $1 + (k+1)x \leq (1+x)^{k+1}$. But by the laws of algebra and the inductive hypothesis, $(1+x)^{k+1} = (1+x)^k(1+x) \geq (1+kx)(1+x)$ provided $1 + x > 0$ (which is true because $x > -1$). Then $(1+kx)(1+x) = 1 + kx + x + kx^2 = 1 + (k+1)x + kx^2 \geq 1 + (k+1)x$ because $kx^2 \geq 0$ for all x. Thus $1 + (k+1)x \leq (1+x)^{k+1}$ *[as was to be shown]*.

20. a. *Proof (by mathematical induction)*:

    The inequality holds for n = 2 : The inequality holds for $n = 2$ because $2^3 = 8$ and $2 \cdot 2 + 1 = 5$ and $8 > 5$.

    If the inequality holds for n = k then it holds for n = k + 1 : Suppose $k^3 > 2k + 1$ for some integer $k \geq 2$. We must show that $(k+1)^3 > 2(k+1) + 1$. But by the laws of algebra and the inductive hypothesis, $(k+1)^3 = k^3 + 3k^2 + 3k + 1 > (2k+1) + 3k^2 + 3k + 1 =$

$(2k+3) + (3k^2 + 3k - 2)$. Since $k \geq 2$, $3k(k+1) \geq 2$, and so $3k^2 + 3k - 2 \geq 0$. Hence, $(2k+3) + (3k^2 + 3k - 2) \geq 2k+3 = 2(k+1) + 1$, and thus $(k+1)^3 > 2(k+1) + 1$ [as was to be shown].

b. *Proof (by mathematical induction)*:

The inequality holds for n = 4: The inequality holds for $n = 4$ because $4! = 24$ and $4^2 = 16$ and $24 > 16$.

If the inequality holds for n = k then it holds for n = k+1: Suppose $k! > k^2$ for some integer $k \geq 4$. We must show that $(k+1)! > (k+1)^2$. But by the laws of algebra and the inductive hypothesis, $(k+1)! = (k+1)k! > (k+1)k^2 = k^3 + k^2$. By part (a) $k^3 > 2k+1$. Hence $k^3 + k^2 > k^2 + 2k + 1 = (k+1)^2$. Thus $(k+1)! > (k+1)^2$ [as was to be shown].

22. *Proof by mathematical induction*:

The formula holds for n = 0: We must show that $b_0 = 5 + 4 \cdot 0$. But $5 + 4 \cdot 0 = 5$ and $b_0 = 5$ by definition of $b_0, b_1, b_2, \ldots$. So the formula holds for $n = 0$.

If the formula holds for n = k then it holds for n = k+1: Suppose that for some integer $k \geq 0$, $b_k = 5 + 4k$. We must show that $b_{k+1} = 5 + 4(k+1)$. But by definition of $b_0, b_1, b_2, \ldots$, substitution from the inductive hypothesis, and the laws of algebra, $b_{k+1} = 4 + b_k = 4 + (5 + 4k) = 5 + 4(k+1)$ [as was to be shown].

23. *Proof by mathematical induction*:

The formula holds for n = 0: We must show that $c_0 = 3^{2^0}$. But $3^{2^0} = 3^1 = 1$ and $c_0 = 3$ by definition of $c_0, c_1, c_2, \ldots$. So the formula holds for $n = 0$.

If the formula holds for n = k then it holds for n = k+1: Suppose that for some integer $k \geq 0$, $c_k = 3^{2^k}$. We must show that $c_{k+1} = 3^{2^{k+1}}$. But by definition of $c_0, c_1, c_2, \ldots$, substitution from the inductive hypothesis, and the laws of algebra, $c_{k+1} = (c_k)^2 = (3^{2^k})^2 = 3^{2^k \cdot 2} = 3^{2^{k+1}}$ [as was to be shown].

24. *Proof by mathematical induction*:

The formula holds for n = 1: We must show that $d_1 = \dfrac{2}{1!}$. But $\dfrac{2}{1!} = 2$ and $d_1 = 2$ by definition of $d_1, d_2, d_3, \ldots$. So the formula holds for $n = 1$.

If the formula holds for n = k then it holds for n = k+1: Suppose that for some integer $k \geq 1$, $d_k = \dfrac{2}{k!}$. We must show that $d_{k+1} = \dfrac{2}{(k+1)!}$. But by definition of $d_1, d_2, d_3, \ldots$, substitution from the inductive hypothesis, and the laws of algebra, $d_{k+1} = \dfrac{d_k}{k+1} = \dfrac{2/k!}{k+1} = \dfrac{2}{(k+1)k!} = \dfrac{2}{(k+1)!}$ [as was to be shown].

25. *Proof by mathematical induction*:

The formula holds for n = 1: For $n = 1$ the formula simply asserts that $\dfrac{1}{3} = \dfrac{1}{3}$, which is true.

If the formula holds for n = k then it holds for n = k+1: Suppose that for some integer $k \geq 1$, $\dfrac{1}{3} = \dfrac{1 + 3 + 5 + \ldots + (2k-1)}{(2k+1) + (2k+3) + \ldots + (4k-1)}$. We must show that

$$\dfrac{1}{3} = \dfrac{1 + 3 + 5 + \ldots + [2(k+1) - 1]}{[2(k+1) + 1] + [2(k+1) + 3] + \ldots + [4(k+1) - 1]},$$

or, equivalently,

$$\dfrac{1}{3} = \dfrac{1 + 3 + 5 + \ldots + (2k+1)}{(2k+3) + (2k+5) + \ldots + (4k+3)}.$$

But by the inductive hypothesis,

$$\frac{1}{3} = \frac{1+3+5+\ldots+(2k-1)}{(2k+1)+(2k+3)+\ldots+(4k-1)},$$

and so $(2k+1)+(2k+3)+\ldots+(4k-1) = 3[1+3+5+\ldots+(2k-1)]$, or equivalently, $(2k+3)+\ldots+(4k-1) = 3[1+3+5+\ldots+(2k-1)]-(2k+1)$. Hence,

$$\frac{1+3+5+\ldots+(2k+1)}{(2k+3)+(2k+5)+\ldots+(4k+3)}$$

$$= \frac{1+3+5+\ldots+(2k-1)+(2k+1)}{(2k+3)+(2k+5)+\ldots+(4k-1)+(4k+1)+(4k+3)} \quad \text{by making more terms explicit}$$

$$= \frac{1+3+5+\ldots+(2k-1)+(2k+1)}{[3(1+3+5+\ldots+(2k-1))-(2k+1)]+(4k+1)+(4k+3)} \quad \text{by inductive hypothesis}$$

$$= \frac{1+3+5+\ldots+(2k-1)+(2k+1)}{3[1+3+5+\ldots+(2k-1)]+(6k+3)}$$

$$= \frac{1+3+5+\ldots+(2k-1)+(2k+1)}{3[1+3+5+\ldots+(2k-1)]+3(2k+1)}$$

$$= \frac{1+3+5+\ldots+(2k-1)+(2k+1)}{3[1+3+5+\ldots+(2k-1)+(2k+1)]}$$

$$= \frac{1}{3} \quad \text{by basic algebra}$$

[as was to be shown].

26. Let $P(n)$ be the property "if n people come to the meeting and each shakes hands with all the others present, then $\frac{k(k-1)}{2}$ handshakes occur." We show by mathematical induction that this property holds for all integers $n \geq 2$.

Proof (by mathematical induction):

The property holds for n = 2: The property holds for $n = 2$ because on the one hand if two people come to the meeting and each shakes hands with each of the others present, then just one handshake occurs, and on the other hand $\frac{2 \cdot (2-1)}{2}$ equals 1 also.

If the property holds for n = k then it holds for n = k + 1: Let k be an integer with $k \geq 2$, and suppose that if k people come to the meeting and each shakes hands with all the others present, then $\frac{k(k-1)}{2}$ handshakes occur. We must show that if $k+1$ people come to the meeting and each shakes hands with all the others present, then $\frac{(k+1)((k+1)-1)}{2} = \frac{(k+1)k}{2}$ handshakes occur. But if $k+1$ people come to the meeting, then after the kth person has arrived and shaked hands all around, $\frac{k(k-1)}{2}$ handshakes have occurred (by inductive hypothesis). When the $(k+1)$st person arrives and shakes hands with all k others, k additional handshakes occur. Thus the total number of handshakes is

$$\frac{k(k-1)}{2} + k = \frac{k^2-k+2k}{2} = \frac{k^2+k}{2} = \frac{(k+1)k}{2}$$

[as was to be shown].

29. *Proof by mathematical induction*:

The formula holds for n = 1: For $n = 1$ the left-hand side of the equation is $\begin{bmatrix} 2 & 1 \\ 0 & 2 \end{bmatrix}^1 = \begin{bmatrix} 2 & 1 \\ 0 & 2 \end{bmatrix}$, and the right-hand side is $\begin{bmatrix} 2^1 & 1 \cdot 2^{1-1} \\ 0 & 2^1 \end{bmatrix} = \begin{bmatrix} 2 & 1 \\ 0 & 2 \end{bmatrix}$ also. So the formula holds for $n = 1$.

If the formula holds for n = k then it holds for n = k+1: Suppose that for some integer $k \geq 1$, $\begin{bmatrix} 2 & 1 \\ 0 & 2 \end{bmatrix}^k = \begin{bmatrix} 2^k & k \cdot 2^{k-1} \\ 0 & 2^k \end{bmatrix}$. We must show that

$$\begin{bmatrix} 2 & 1 \\ 0 & 2 \end{bmatrix}^{k+1} = \begin{bmatrix} 2^{k+1} & (k+1) \cdot 2^{(k+1)-1} \\ 0 & 2^{(k+1)} \end{bmatrix},$$

or, equivalently,

$$\begin{bmatrix} 2 & 1 \\ 0 & 2 \end{bmatrix}^{k+1} = \begin{bmatrix} 2^{k+1} & (k+1) \cdot 2^k \\ 0 & 2^{k+1} \end{bmatrix}.$$

But by definition of matrix multiplication, substitution from the inductive hypothesis, and the laws of algebra,

$$\begin{bmatrix} 2 & 1 \\ 0 & 2 \end{bmatrix}^{k+1} = \begin{bmatrix} 2 & 1 \\ 0 & 2 \end{bmatrix} \begin{bmatrix} 2 & 1 \\ 0 & 2 \end{bmatrix}^k = \begin{bmatrix} 2 & 1 \\ 0 & 2 \end{bmatrix} \begin{bmatrix} 2^k & k \cdot 2^{k-1} \\ 0 & 2^k \end{bmatrix}$$

$$= \begin{bmatrix} 2 \cdot 2^k + 1 \cdot 0 & 2 \cdot k \cdot 2^{k-1} + 1 \cdot 2^k \\ 0 \cdot 2^k + 2 \cdot 0 & 0 \cdot k \cdot 2^{k-1} + 2 \cdot 2^k \end{bmatrix} = \begin{bmatrix} 2^{k+1} & (k+1) \cdot 2^k \\ 0 & 2^{k+1} \end{bmatrix}$$

[as was to be shown].

30. Let $P(n)$ be the property that "in any round-robin tournament involving n teams, it is possible to label the teams $T_1, T_2, T_3, \ldots, T_n$ so that T_i beats T_{i+1} for all $i = 1, 2, 3, \ldots, n-1$." We will show by mathematical induction that this property is true for all integers $n \geq 2$.

The property holds for n = 2: Consider any round-robin tournament involving two teams. By definition of round-robin tournament, these teams play each other exactly once. Let T_1 be the winner and T_2 the loser of this game. Then T_1 beats T_2, and so the labeling is as required.

If the property holds for n = k then it holds for n = k+1: Let k be an integer with $k \geq 2$ and suppose that in any round-robin tournament involving k teams it is possible to label the teams in the way described. *[We must show that in any round-robin tournament involving $k+1$ teams it is possible to label the teams in the way described.]* Consider any round-robin tournament with $k+1$ teams. Pick one and call it T'. Temporarily remove T' and consider the remaining k teams. Since each of these teams plays each other team exactly once, the games played by these k teams form a round-robin tournament. It follows by inductive hypothesis that these k teams may be labeled $T_1, T_2, T_3, \ldots, T_k$ where T_i beats T_{i+1} for all $i = 1, 2, 3, \ldots, k-1$.

Case 1 (T' beats T_1): In this case, relabel each T_i to be T_{i+1}, and let $T_1 = T'$. Then T_1 beats the newly labeled T_2 (because T' beats the old T_1), and T_i beats T_{i+1} for all $i = 2, 3, \ldots k$ (by inductive hypothesis).

Case 2 (T' loses to $T_1, T_2, T_3, \ldots, T_m$ and beats T_{m+1} where $1 \leq m \leq k-1$): In this case, relabel teams $T_{m+1}, T_{m+2}, \ldots, T_k$ to be $T_{m+2}, T_{m+3}, \ldots, T_{k+1}$ and let $T_{m+1} = T'$. Then for each i with $1 \leq i \leq m-1$, T_i beats T_{i+1} (by inductive hypothesis), T_m beats T_{m+1} (because T_m beats T'), T_{m+1} beats T_{m+2} (because T' beats the old T_{m+1}), and for each i with $m+2 \leq i \leq k$, T_i beats T_{i+1} (by inductive hypothesis)..

Case 3 (T' loses to T_i for all $i = 1, 2, \ldots, k$): In this case, let $T_{k+1} = T'$. Then for all $i = 1, 2, \ldots, k-1$, T_i beats T_{i+1} (by inductive hypothesis) and T_k beats T_{k+1} (because T_k beats T').

Thus in all three cases the teams may be relabeled in the way specified.

31. *Proof by contradiction*: Suppose not. Suppose it is impossible to find three successive integers on the rim of the disk whose sum is at least 45. *[We must derive a contradiction.]* Then there is some ordering of the integers from 1 to 30, say $x_1, x_2, \ldots, x_{30}$ such that

$$x_1 + x_2 + x_3 < 45$$
$$x_2 + x_3 + x_4 < 45$$
$$x_3 + x_4 + x_5 < 45$$
$$\vdots$$
$$x_{29} + x_{30} + x_1 < 45$$
$$x_{30} + x_1 + x_2 < 45.$$

Adding all these inequalities gives $3\sum_{i=1}^{30} x_i < 30 \cdot 45 = 1350$. But $\sum_{i=1}^{30} x_i = \sum_{i=1}^{30} i$ because the sequence $x_1, x_2, \ldots, x_{30}$ is a rearrangement of the integers from 1 to 30. Hence $3\sum_{i=1}^{30} i < 1350$, and so by the formula for the sum of the first n integers, $3 \cdot \dfrac{30 \cdot 31}{2} < 1350$, or, equivalently, $1395 < 1350$. But $1395 \geq 1350$, and so we have arrived at a contradiction. *[Hence the supposition is false and the given statement is true.]*

32. Sara shook hands with n people.

    Proof (by mathematical induction): Let $P(n)$ be the property that "if n couples attended the party and each person other than John shook hands with a different number of people and no one shook their own or their spouse's hand, then Sara shook hands with n people."

    The property is true for n = 0 : If no couples attended the party, then Sara shook 0 people's hands.

    If the property is true for n = k then it is true for n = k + 1 : Suppose that if k couples came to the party and each person other than John shook hands with a different number of people and no one shook their own or their spouse's hand, then Sara shook hands with k people. *[This is the inductive hypothesis.]* Suppose $k + 1$ couples came to the party and each person other than John shook hands with a different number of people and no one shook teir own or their spouses hand. We must show that Sara shook hands with $k + 1$ people. But since $k + 1$ couples came to the party, then the maximum number of people anyone could shake hands with was $2k$ *[because no one shakes his or her own hand or his or her spouse's hand]*, and there are $2k + 1$ integers from 0 to $2k$ inclusive. Since $2k + 1$ people (other than John) were present and each shook a different number of hands, then for each integer r from 0 to $2k$ inclusive there was exactly one person who shook r hands. In particular, there was a person A who shook $2k$ hands. Let B be A's spouse. Now since A shook $2k$ hands, A shook hands with every person at the party except A and B. Consequently, every person at the party except, possibly, B shook at least one person's hand. But someone shook 0 hands. Thus B shook 0 hands. Now imagine the party without A and B. The number of hands each person would have shaken would be one less than the number he or she actually shook (because the handshake with A would not occur). Thus every person (other than John) would still have shaken a different number of hands from every other person, and so by inductive hypothesis, Sara would shake k hands. In the actual situation in which A and B are present, Sara shakes A's hand as well as the k others, and so Sara shakes a total of $k + 1$ hands *[as was to be shown]*.

Section 4.4

2. *Proof:* Let $P(n)$ be the property "b_n is divisible by 3." We prove by strong mathematical induction that this property is true for all integers $n \geq 1$.

   The property is true for n = 1 and n = 2: $b_1 = 3$ and $b_2 = 6$ and both 3 and 6 are divisible by 3.

   If the property is is true for all integers i with $1 \leq i < k$ then it is true for k: Let $k > 2$ be an integer and suppose b_i is divisible by 3 for all integers i with $1 \leq i < k$. *[This is the inductive hypothesis.]* We must show that b_k is divisible by 3. But by definition of $b_1, b_2, b_3, \ldots$, $b_k = b_{k-2} + b_{k-1}$. Since $k > 2$, $0 \leq k-2 < k$ and $0 \leq k-1 < k$, and so by inductive hypothesis, both b_{k-2} and b_{k-1} are divisible by 3. Thus *[by exercise 14 of Section 3.3]* the sum of b_{k-2} and b_{k-1}, which equals b_k, is also divisible by 3 *[as was to be shown]*.

3. Let $P(n)$ be the property "c_n is even." We prove by strong mathematical induction that this property is true for all integers $n \geq 0$.

   The property is true for n = 0, 1, and 2: $c_0 = 2$, $c_1 = 4$, and $c_2 = 6$ and 2, 4, and 6 are all even.

   If the property is true for all integers i with $1 \leq i < k$ then it is true for k: Let $k > 2$ be an integer and suppose c_i is even for all integers i with $1 \leq i < k$. *[This is the inductive hypothesis.]* We must show that c_k is even. But by definition of $c_0, c_1, c_2, \ldots$, $c_k = 5c_{k-3}$. Since $k > 2$, $0 \leq k-3 < k$, and so by inductive hypothesis, c_{k-3} is even. Thus *[by exercise 20 of Section 3.1]* $5c_{k-3}$, which equals c_k, is also even *[as was to be shown]*.

5. Let $P(n)$ be the inequality $e_n \leq 3^n$. We prove by strong mathematical induction that this inequality is true for all integers $n \geq 0$.

   The inequality is true for n = 0, 1, and 2: $e_0 = 1$, $e_1 = 2$, and $e_2 = 3$ and $1 \leq 3^0$, $2 \leq 3^1$, and $3 \leq 3^2$.

   If the inequality is true for all integers i with $1 \leq i < k$ then it is true for k: Let $k > 2$ be an integer and suppose $e_i \leq 3^i$ for all integers i with $1 \leq i < k$. *[This is the inductive hypothesis.]* We must show that $e_k \leq 3^k$. But by definition of $e_0, e_1, e_2, \ldots$, $e_k = e_{k-1} + e_{k-2} + e_{k-3}$. Since $k > 2$, $0 \leq k-3 < k$, and so by inductive hypothesis, $e_{k-1} \leq 3^{k-1}$, $e_{k-2} \leq 3^{k-2}$, and $e_{k-3} \leq 3^{k-3}$. Adding the inequalities and using the laws of basic algebra gives $e_{k-1} + e_{k-2} + e_{k-3} \leq 3^{k-1} + 3^{k-2} + 3^{k-3} \leq 3^{k-1} + 3^{k-1} + 3^{k-1} = 3 \cdot 3^{k-1} = 3^k$. But $e_{k-1} + e_{k-2} + e_{k-3} = e_k$, and so $e_k \leq 3^k$ *[as was to be shown]*.

6. Let $P(n)$ be the inequality $f_n \leq n$. We prove by strong mathematical induction that this inequality is true for all integers $n \geq 1$.

   The inequality is true for n = 1: $f_1 = 1$ and $1 \leq 1$.

   If the inequality is true for all integers i with $1 \leq i < k$ then it is true for k: Let $k > 1$ be an integer and suppose $f_i \leq i$ for all integers i with $1 \leq i < k$. *[This is the inductive hypothesis.]* We must show that $f_k \leq k$. But by definition of $f_1, f_2, f_3, \ldots$, $f_k = 2 \cdot f_{\lfloor k/2 \rfloor}$. Since $k > 1$, $1 \leq \lfloor k/2 \rfloor < k$, and so by inductive hypothesis, $f_{\lfloor k/2 \rfloor} \leq \lfloor k/2 \rfloor$. Thus

   $$f_k = 2 \cdot f_{\lfloor k/2 \rfloor} \leq 2 \cdot \lfloor k/2 \rfloor = \begin{cases} 2 \cdot ((k-1)/2) & \text{if } k \text{ is odd} \\ 2 \cdot (k/2) & \text{if } k \text{ is even} \end{cases} = \begin{cases} k-1 & \text{if } k \text{ is odd} \\ k & \text{if } k \text{ is even} \end{cases} \leq k,$$

   and so $f_k \leq k$ *[as was to be shown]*.

8. a. Let $P(n)$ be the inequality $h_n \leq 3^n$. We prove by strong mathematical induction that this inequality is true for all integers $n \geq 0$.

   The inequality is true for n = 0, 1, and 2: Note that $h_0 = 1 \leq 3^0$, $h_1 = 2 \leq 3^1$, and $h_2 = 3 \leq 3^2$. So the inequality is true for $n = 0, 1$, and 2.

If the inequality is true for all integers i with i ≤ k then it is true for k: Let k be an integer with $k > 2$, and suppose the inequality holds for all integers i with $0 \leq i < k$. We must show that the inequality is true for k. But

$$\begin{aligned} h_k &= h_{k-1} + h_{k-2} + h_{k-3} \\ \Rightarrow h_k &\leq 3^{k-1} + 3^{k-2} + 3^{k-3} \quad \text{by inductive hypothesis} \\ \Rightarrow h_k &\leq 3^{k-3}(3^2 + 3 + 1) \end{aligned}$$

But $3^2 + 3 + 1 = 13 \leq 27 = 3^3$. Hence by the transitive properties of = and <,

$$h_k \leq 3^{k-3} \cdot 3^3 = 3^k$$

[as was to be shown].

b. Let s be any real number such that $s^3 \geq s^2 + s + 1$, and let $P(n)$ be the inequality $h_n \leq s^n$. We prove by strong mathematical induction that this inequality is true for all integers $n \geq 2$.

The inequality is true for n = 2, 3, and 4: Because $s \geq 1.83$, $h_2 = 3 \leq 3.3489 = 1.83^2 < s^2$, $h_3 = h_0 + h_1 + h_2 = 1 + 2 + 3 = 6 \leq 6.128487 = 1.83^3$, and $h_4 = h_1 + h_2 + h_3 = 2 + 3 + 11 \leq 11.21513121 = 1.83^4$. So the inequality is true for $n = 2, 3,$ and 4.

If the inequality is true for all integers i with i ≤ k then it is true for k: Let k be an integer with $k > 4$, and suppose the inequality is true for all integers i with $0 \leq i < k$. We must show that the inequality is true for k. But

$$\begin{aligned} h_k &= h_{k-1} + h_{k-2} + h_{k-3} \\ \Rightarrow h_k &\leq s^{k-1} + s^{k-2} + s^{k-3} \quad \text{by inductive hypothesis} \\ \Rightarrow h_k &\leq s^{k-3}(s^2 + s + 1) \\ \Rightarrow h_k &\leq s^{k-3} \cdot s^3 = s^k \quad \text{since by hypothesis about } s, s^2 + s + 1 \leq s^3 \end{aligned}$$

[as was to be shown].

9. Let $P(n)$ be the property "A jigsaw puzzle consisting of n pieces takes $n - 1$ steps to put together." We prove by strong mathematical induction that this property is true for all integers $n \geq 1$.

   The property is true for n = 1: A jigsaw puzzle consisting of just one piece does not take any steps to put together. Hence it is correct to say that it takes zero steps to put together.

   If the property is true for all integers i with 1 ≤ i < k then it is true for k: Let $k > 1$ be an integer and suppose that for all integers i with $1 \leq i < k$, a jigsaw puzzle consisting of i pieces takes $i - 1$ steps to put together. *[This is the inductive hypothesis.]* We must show that jigsaw puzzle consisting of k pieces takes $k - 1$ steps to put together. Consider assembling a jigsaw puzzle consisting of k pieces. The last step involves fitting together two blocks. Suppose one of the blocks consists of r pieces and the other consists of s pieces. Then $r + s = k$, and $1 \leq r < k$ and $1 \leq s < k$. So by inductive hypothesis the number of steps required to assemble the blocks are $r - 1$ and $s - 1$ respectively. Then the total number of steps required to assemble the puzzle is $(r - 1) + (s - 1) + 1 = (r + s) - 1 = k - 1$ *[as was to be shown]*.

10. The inductive step fails when $k = 1$. The reason is that to go from $k = 1$ to $k = 2$, one evaluates $\dfrac{r^{k-1} \cdot r^{k-1}}{r^{k-2}}$ for $k = 1$. But the inductive hypothesis only says that $r^i = 1$ for all i with $0 \leq i < k$ and when $k = 1$ then $k - 2 = -1 < 0$. Therefore it cannot be deduced from the inductive hypothesis that $r^{k-2} = 1$.

12. *Proof*: Suppose a and b are any integers that are not both zero. Let S be the set of all positive integers of the form $ua + vb$ for some integers u and v.

    We first show that $S \neq \emptyset$. Since a and b are not both zero, one of them is nonzero. Without loss of generality, we may assume that $a \neq 0$. In case $a > 0$, let $u = 1$ and $v = 0$, and in case $a < 0$, let $u = -1$ and $v = 0$. In the first case, $ua + vb = a \in S$, and in the second case $ua + vb = -a \in S$. So in either case, $S \neq \emptyset$. Hence by the well-ordering principle for the integers, S has a least element, say m.

We next show that $m \mid a$ and $m \mid b$. By the quotient-remainder theorem, $a = mq + r$ where $0 \leq r < m$. Then $r = a - mq = a - (ua + vb)q = (1 - uq)a - (vq)b$. Thus by definition of S, $r \in S$ (if r is positive) or else $r = 0$. But if $r \in S$, then $m \leq r$ (since m is the *least* element of S) which is impossible because $r < m$. Hence $r = 0$, and so $m \mid a$. A similar argument shows that $m \mid b$. Hence m is a common divisor of a and b.

Finally we show that m is the *greatest* common divisor of a and b. Suppose c is any integer such that $c \mid a$ and $c \mid b$. Then $c \mid ua$ and $c \mid vb$ (by exercise 21 of Section 3.3), and so $c \mid (ua+vb)$ (by exercise 14 of Section 3.3). Hence, if $c \mid a$ and $c \mid b$ then $c \mid m$. But since $c \mid m$ and $m > 0$, then $c \leq m$. *[If c is positive, this is true by Example 3.3.3. If c is negative, it is true because each negative integer is less than each positive integer.]* Therefore, $m = \gcd(a, b)$.

13. No.

    Counterexample 1: Let $P(n)$ be "$n + 24$ is composite." Then $P(0)$, $P(1)$, and $P(2)$ are all true (because 24, 25, and 26 are composite), and for all integers $k \geq 0$, if $P(k)$ is true then $P(3k)$ is true (because if $k + 24$ is composite then $3k + 24$ is also composite — in fact, $3k + 24$ is composite for *any* integer $k \geq 0$). But it is false that $n + 24$ is composite for all integers $n \geq 0$ (because, for instance, $5 + 24 = 29$ is prime).

    Counterexample 2: Let $P(n)$ be "$n = 0$, or $n = 1$, or $n = 2$, or $n = 3r$ for some integer r." Then $P(0)$, $P(1)$, and $P(2)$ are all true (because an *or* statement is true if any component is true), and for all integers $k \geq 0$, if $P(k)$ is true then $P(3k)$ is true (also by definition of the truth value of an *or* statement and by definition of $P(n)$). But $P(n)$ is not true for all integers $n \geq 0$ (because, for instance, $P(4)$ is not true).

    (There are many other counterexamples besides these two.)

14. *Proof*: Suppose the principle of ordinary mathematical induction is true. Let $P(n)$ be a predicate that is defined for integers n, and suppose the following two statements are true:

    1. $P(a)$, $P(a+1)$, $P(a+2)$, ..., $P(b)$ are all true.
    2. For any integer $k > b$, if $P(i)$ is true for all integers i with $a \leq i < k$, then $P(k)$ is true.

    Let $Q(n)$ be the property "$P(j)$ is true for all integers j with $a \leq j \leq n$."

    The property is true for n = b: For $n = b$, the property is "$P(j)$ is true for all integers j with $a \leq j \leq b$." But this is true by (1) above.

    If the property is true for n = k then it is true for n = k + 1: Let k be an integer with $k \geq b$ and suppose $Q(k)$ is true. By definition of Q this means that $P(j)$ is true for all integers j with $a \leq j \leq k$. It follows from (2) above that $P(k+1)$ is also true. Hence $P(j)$ is true for all integers j with $a \leq j \leq k+1$. By definition of Q this means that $Q(k+1)$ is true.

    It follows from the above by the principle of ordinary mathematical induction that $Q(n)$ is true for all integers $n \geq b$. From this and from (1) above, we conclude that $P(n)$ is true for all integers $n \geq a$.

15. *Example 1 (k is even)*: Let $k = 42$. Then $k/2 = 21 = 16+4+1 = 1 \cdot 2^4 + 0 \cdot 2^3 + 1 \cdot 2^2 + 0 \cdot 2 + 1$. To obtain the binary representation for k, multiply the representation for $k/2$ by 2 and add $0 \ (= 0 \cdot 1)$: $k = 2 \cdot (1 \cdot 2^4 + 0 \cdot 2^3 + 1 \cdot 2^2 + 0 \cdot 2 + 1) + 0 \cdot 1 = 1 \cdot 2^5 + 0 \cdot 2^4 + 1 \cdot 2^3 + 0 \cdot 2^2 + 1 \cdot 2 + 0 \cdot 1$. Thus $k_2 = 42_2 = 101010_2$.

    Example 2 (k is odd): Let $k = 43$. Then $(k-1)/2 = 21 = 16+4+1 = 1 \cdot 2^4 + 0 \cdot 2^3 + 1 \cdot 2^2 + 0 \cdot 2 + 1$. To obtain the binary representation for k, multiply the representation for $(k-1)/2$ by 2 and add $1 \ (= 1 \cdot 1)$: $k = 2 \cdot (1 \cdot 2^4 + 0 \cdot 2^3 + 1 \cdot 2^2 + 0 \cdot 2 + 1) + 1 \cdot 1 = 1 \cdot 2^5 + 0 \cdot 2^4 + 1 \cdot 2^3 + 0 \cdot 2^2 + 1 \cdot 2 + 1 \cdot 1$. Thus $k_2 = 43_2 = 101011_2$.

16. *Sketch of Proof*: Consider the property "n can be written in the form

    $$n = c_r \cdot 3^r + c_{r-1} 3^{r-1} + \ldots + c_2 \cdot 3^2 + c_1 \cdot 3 + c_0$$

where r is a nonnegative integer, $c_r = 1$ or 2 and $c_j = 0, 1,$ or 2 for all $j = 0, 1, 2, \ldots, r-1$."

We will show that the property holds for all integers $n \geq 1$.

The property holds for n = 1: Observe that $1 = c_r 3^r$ for $r = 0$ and $c_r = 1$. Thus 1 can be written in the required form.

If the property holds for all i with $1 \leq i < k$ then it holds for k: Let k be an integer with $k > 1$, and suppose that for all integers i with $1 \leq i < k$, i can be written in the required form:
$$i = c_r \cdot 3^r + c_{r-1} 3^{r-1} + \ldots + c_2 \cdot 3^2 + c_1 \cdot 3 + c_0$$
where r is a nonnegative integer, $c_r = 1$ or 2 and $c_j = 0, 1,$ or 2 for all $j = 0, 1, 2, \ldots, r-1$.

We must show that k can be written in the required form. By the quotient-remainder theorem, k can be written as $3m$, $3m+1$, or $3m+2$ for some integer m. In each case m satisfies $0 \leq m < k$, and so m can be written in the required form:
$$m = c_r \cdot 3^r + c_{r-1} 3^{r-1} + \ldots + c_2 \cdot 3^2 + c_1 \cdot 3 + c_0$$
where r is a nonnegative integer, $c_r = 1$ or 2 and $c_j = 0, 1,$ or 2 for all $j = 0, 1, 2, \ldots, r-1$. Then

$$\begin{aligned}
3m &= c_r \cdot 3^{r+1} + c_{r-1} 3^r + \ldots + c_2 \cdot 3^3 + c_1 \cdot 3^2 + c_0 \cdot 3 + 0 \\
3m + 1 &= c_r \cdot 3^{r+1} + c_{r-1} 3^r + \ldots + c_2 \cdot 3^3 + c_1 \cdot 3^2 + c_0 \cdot 3 + 1 \\
3m + 2 &= c_r \cdot 3^{r+1} + c_{r-1} 3^r + \ldots + c_2 \cdot 3^3 + c_1 \cdot 3^2 + c_0 \cdot 3 + 2
\end{aligned}$$

which all have the required form. Hence k can be written in the required form.

17. *Theorem*: Given any nonnegative integer n and any positive integer d, there exist integers q and r such that $n = d \cdot q + r$ and $0 \leq r < d$.

Proof: Let a nonnegative integer d be given, and for each integer n let $P(n)$ be the property "∃ integers q and r such that $n = dq + r$ and $0 \leq r < d$."

The property is true for n = 0: We must show that there exist nonnegative integers q and r such that $0 = d \cdot q + r$ and $0 \leq r < d$. Let $q = r = 0$. Then $0 = d \cdot 0 + 0$ and $0 \leq 0 < d$. Hence the theorem is true for $n = 0$.

If the property is true for n = k then it is true for n = k + 1: Let $k \geq 0$ be given, and suppose that there exist integers q' and r' such that $k = d \cdot q' + r'$ and $0 \leq r' < d$. *[We must show that there exist nonnegative integers q and r such that $k + 1 = d \cdot q + r$ and $0 \leq r < d$.]* Then $k + 1 = (d \cdot q' + r') + 1$. Note that since r' is an integer and $0 \leq r' < d$, then either $r' < d - 1$ or $r' = d - 1$.

Case 1 ($r' < d - 1$): In this case, $k + 1 = (d \cdot q' + r') + 1 = d \cdot q' + (r' + 1)$. Let $q = q'$ and $r = r' + 1$. Then by substitution, $k + 1 = d \cdot q + r$. Since $r' < d - 1$ then $r = r' + 1 < d$, and since $r = r' + 1$ and $r' \geq 0$, then $r \geq 0$. Hence $0 \leq r < d$.

Case 2 ($r' = d - 1$): In this case, $k + 1 = (d \cdot q' + r') + 1 = d \cdot q' + (r' + 1) = d \cdot q' + ((d-1) + 1) = d \cdot q' + d = d \cdot (q' + 1)$. Let $q = q' + 1$ and $r = 0$. Then by substitution $k + 1 = d \cdot q + r$, and since $r = 0$ and $d > 0$, $0 \leq r < d$.

Thus in either case there exist nonnegative integers q and r such that $k + 1 = d \cdot q + r$ and $0 \leq r < d$ *[as was to be shown]*.

19. *Proof*: Suppose the well-ordering principle is true. Let $P(n)$ be a predicate that is defined for integers n and satisfies

  1. $P(a)$ is true;

  2. For all integers $k \geq a$, if $P(k)$ is true then $P(k+1)$ is true.

Let S be the set of all integers greater than or equal to a for which $P(n)$ is false. Suppose S has one or more elements. *[We must derive a contradiction.]* By the well-ordering principle, S has a least element, b, and by definition of S, $P(b)$ is false. Now $P(b-1)$ is true because $b-1 < b$ and b is the *least* element for which $P(n)$ is false. Also $b - 1 \geq a$ because S consists entirely of integers greater than or equal to a and $b \neq a$ because $P(a)$ is true and $P(b)$ is false. Thus $b - 1 \geq a$ and $P(b-1)$ is true, and so by (2) above, $P((b-1)+1)$, which equals $P(b)$, is true. Hence $P(b)$ is both true and false, which is a contradiction. This contradiction shows that the supposition is false, and so S has no elements. But that means that $P(n)$ is true for all integers $n \geq a$.

Section 4.5

2. *Proof*: Suppose the condition $m + n$ is odd is true before entry to the loop. Then $m_{\text{old}} + n_{\text{old}}$ is odd. After execution of the loop, $m_{\text{new}} = m_{\text{old}} + 4$ and $n_{\text{new}} = n_{\text{old}} - 2$. So $m_{\text{new}} + n_{\text{new}} = (m_{\text{old}} + 4) + (n_{\text{old}} - 2) = m_{\text{old}} + n_{\text{old}} + 2$. But $m_{\text{old}} + n_{\text{old}}$ is odd and 2 is even, and (by exercise 26 of Section 3.1) the sum of an odd and an even integer is odd. Hence $m_{\text{new}} + n_{\text{new}}$ is odd.

4. *Proof*: Suppose the condition $2^n < (n+2)!$ is true before entry to the loop. Then $2^{n_{\text{old}}} < (n_{\text{old}} + 2)!$. After execution of the loop, $n_{\text{new}} = n_{\text{old}} + 1$. So $2^{n_{\text{new}}} = 2^{n_{\text{old}}+1} = 2 \cdot 2^{n_{\text{old}}} < 2 \cdot (n_{\text{old}} + 2)! < (n_{\text{old}} + 3) \cdot (n_{\text{old}} + 2)!$ *[because $n_{\text{old}} \geq 0$]* $= (n_{\text{old}} + 3)! = (n_{\text{new}} + 2)!$. Hence $2^{n_{\text{new}}} < (n_{\text{new}} + 2)!$.

5. *Proof*: Suppose the condition $2n+1 \leq 2^n$ is true before entry to the loop. Then $2n_{\text{old}}+1 \leq 2^{n_{\text{old}}}$. After execution of the loop, $n_{\text{new}} = n_{\text{old}} + 1$. So $2n_{\text{new}} + 1 = 2(n_{\text{old}}+1)+1 = (2n_{\text{old}}+1)+2 \leq 2^{n_{\text{old}}}+2$. But since $n_{\text{old}} \geq 3$, $2 \leq 2^{n_{\text{old}}}$, and so $2^{n_{\text{old}}}+2 \leq 2^{n_{\text{old}}}+2^{n_{\text{old}}} = 2 \cdot 2^{n_{\text{old}}} = 2^{n_{\text{old}}+1} = 2^{n_{\text{new}}}$. Putting the inequalities together gives $2n_{\text{new}} + 1 \leq 2^{n_{\text{old}}} + 2 \leq 2^{n_{\text{new}}}$ *[as was to be shown]*.

7. *Proof*:

I. Basis Property: $I(0)$ is the statement "*largest* = the maximum value of $A[1]$ and $i = 1$." According to the pre-condition this statement is true.

II. Inductive Property: Suppose k is a nonnegative integer such that $G \wedge I(k)$ is true before an iteration of the loop. Then as execution reaches the top of the loop, $i \neq m$, *largest* = the maximum value of $A[1], A[2], \ldots, A[k+1]$, and $i = k+1$. Since $i \neq m$, the guard is passed and statement 1 is executed. Now before execution of statement 1, $i_{\text{old}} = k+1$. So after execution of statement 1, $i_{\text{new}} = i_{\text{old}}+1 = (k+1)+1 = k+2$. Also before statement 2 is executed, *largest*$_{\text{old}}$ = the maximum value of $A[1], A[2], \ldots, A[k+1]$. Statement 2 checks whether $A[i_{\text{new}}] = A[k+2] > $ *largest*$_{\text{old}}$. If the condition is true, then *largest*$_{\text{new}}$ is set equal to $A[k+2]$ which is the maximum value of $A[1], A[2], \ldots, A[k+1], A[k+2]$. If the condition is false then $A[k+2] \leq$ *largest*$_{\text{old}}$, and so *largest*$_{\text{old}}$ is the maximum value of $A[1], A[2], \ldots, A[k+1], A[k+2]$. Now in this case since the condition is false, the variable *largest* retains its previous value, and so *largest*$_{\text{new}} = $ *largest*$_{\text{old}}$. Thus in either case *largest*$_{\text{new}}$ is the maximum value of $A[1], A[2], \ldots, A[k+1], A[k+2]$. Hence $I(k+1)$ is true.

III. Eventual Falsity of Guard: The guard G is the condition $i \neq m$. By I and II, it is known that for all integers $n \geq 1$, after n iterations of the loop $I(n)$ is true. Hence after m iterations of the loop $I(m)$ is true, which implies that $i = m$ and G is false.

IV. Correctness of the Post-Condition: Suppose that N is the least number of iterations after which G is false and $I(N)$ is true. Then (since G is false) $i = m$ and (since $I(N)$ is true) *largest* = the maximum value of $A[1], A[2], \ldots, A[N+1]$ and $i = N+1$. Putting these together gives $m = N+1$ and so *largest* = the maximum value of $A[1], A[2], \ldots, A[m]$, which is the post-condition.

9. *Proof*:

   I. Basis Property: I(0) is the statement "both a and A are even integers or both are odd integers and $a \geq -1$." According to the pre-condition this statement is true.

   II. Inductive Property: Suppose k is a nonnegative integer such that $G \wedge I(k)$ is true before an iteration of the loop. Then as execution reaches the top of the loop, $a_{\text{old}} > 0$ and a_{old} and A are both even integers or both are odd integers, and $a_{\text{old}} \geq -1$. Execution of statement 1 sets a_{new} equal to $a_{\text{old}} - 2$. Hence a_{new} has the same parity as a_{old} which is the same as A. Also since $a_{\text{old}} > 0$, then $a_{\text{new}} = a_{\text{old}} - 2 > 0 - 2 = -2$. But a_{new} is an integer. So since $a_{\text{new}} > -2$, $a_{\text{new}} \geq -1$. Hence after the loop iteration, $I(k+1)$ is true.

   III. Eventual Falsity of Guard: The guard G is the condition $a > 0$. After each iteration of the loop, $a_{\text{new}} = a_{\text{old}} - 2 < a_{\text{old}}$, and so successive iterations of the loop give a strictly decreasing sequence of integer values of a which eventually becomes less than or equal to zero, at which point G becomes false.

   IV. Correctness of the Post-Condition: Suppose that N is the least number of iterations after which G is false and $I(N)$ is true. Then (since G is false) $a \leq 0$ and (since $I(N)$ is true) both a and A are even integers or both are odd integers, and $a \geq -1$. Putting the inequalities together gives $-1 \leq a \leq 0$, and so since a is an integer, $a = -1$ or $a = 0$. Since a and A have the same parity, then, $a = 0$ if A is even and $a = -1$ if A is odd. This is the post-condition.

10. **I. Basis Property**: I(0) is the statement "(1) a and b are nonnegative integers with $\gcd(a, b) = \gcd(A, B)$, (2) at most one of a and b equals 0, (3) $0 \leq a + b \leq A + B - 0$." According to the pre-condition, A and B are positive integers and $a = A$ and $b = B$, and so I(0) is true.

    II. Inductive Property: Suppose k is a nonnegative integer such that $G \wedge I(k)$ is true before an iteration of the loop. Then as execution reaches the top of the loop, $a_{\text{old}} \neq 0$, $b_{\text{old}} \neq 0$ and

    (1) a_{old} and b_{old} are nonnegative integers with $\gcd(a_{\text{old}}, b_{\text{old}}) = \gcd(A, B)$,
    (2) at most one of a_{old} and b_{old} equals 0, and
    (3) $0 \leq a_{\text{old}} + b_{\text{old}} \leq A + B - k$.

    After execution of the statement "**if** $a \geq b$ **then** $a := a - b$ **else** $b := b - a$", we have the following:

    (1) *Case 1* ($a_{\text{old}} \geq b_{\text{old}}$): In this case $a_{\text{new}} = a_{\text{old}} - b_{\text{old}} \geq 0$ and $b_{\text{new}} = b_{\text{old}} \geq 0$, and so both a_{new} and b_{new} are nonnegative. Also $\gcd(a_{\text{new}}, b_{\text{new}}) = \gcd(a_{\text{old}} - b_{\text{old}}, b_{\text{old}}) = \gcd(a_{\text{old}}, b_{\text{old}})$ by Lemma 3.8.3. But $\gcd(a_{\text{old}}, b_{\text{old}}) = \gcd(A, B)$ by (1) above. Hence $\gcd(a_{\text{new}}, b_{\text{new}}) = \gcd(A, B)$.

    Case 2 ($a_{\text{old}} < b_{\text{old}}$): In this case $b_{\text{new}} = b_{\text{old}} - a_{\text{old}} \geq 0$ and $a_{\text{new}} = a_{\text{old}} \geq 0$, and so both a_{new} and b_{new} are nonnegative. Also $\gcd(a_{\text{new}}, b_{\text{new}}) = \gcd(a_{\text{old}}, b_{\text{old}} - a_{\text{old}}) = \gcd(a_{\text{old}}, b_{\text{old}})$ by Lemma 3.8.3. But $\gcd(a_{\text{old}}, b_{\text{old}}) = \gcd(A, B)$ by (1) above. Hence $\gcd(a_{\text{new}}, b_{\text{new}}) = \gcd(A, B)$.

    (2) Because G is true $a_{\text{old}} \neq 0$ and $b_{\text{old}} \neq 0$, and by (1) above both a_{old} and b_{old} are nonnegative. Hence $a_{\text{old}} > 0$ and $b_{\text{old}} > 0$. Since either $a_{\text{new}} = a_{\text{old}}$ or $b_{\text{new}} = b_{\text{old}}$, at most one of a_{new} or b_{new} equals zero.

    (3) Observe that

    $$a_{\text{new}} + b_{\text{new}} = \begin{cases} a_{\text{old}} - b_{\text{old}} + b_{\text{old}} & \text{if } a_{\text{old}} \geq b_{\text{old}} \\ a_{\text{old}} + b_{\text{old}} - a_{\text{old}} & \text{if } b_{\text{old}} > a_{\text{old}} \end{cases} = \begin{cases} a_{\text{old}} & \text{if } a_{\text{old}} \geq b_{\text{old}} \\ b_{\text{old}} & \text{if } b_{\text{old}} > a_{\text{old}} \end{cases}$$

    But since $a_{\text{old}} \neq 0$ and $b_{\text{old}} \neq 0$ and a_{old} and b_{old} are nonnegative integers, then $a_{\text{old}} \geq 1$ and $b_{\text{old}} \geq 1$. Hence $a_{\text{old}} - 1 \geq 0$ and $b_{\text{old}} - 1 \geq 0$, and so $a_{\text{old}} \leq a_{\text{old}} + b_{\text{old}} - 1$ and $b_{\text{old}} \leq b_{\text{old}} + a_{\text{old}} - 1$. But by (3) above $a_{\text{old}} + b_{\text{old}} \leq A + B - k$. It follows that $a_{\text{new}} + b_{\text{new}} \leq a_{\text{old}} + b_{\text{old}} - 1 \leq A + B - k - 1 = A + B - (k+1)$.

    III. Eventual Falsity of Guard: The guard G is the condition $a \neq 0$ and $b \neq 0$. Suppose G is never false. *[We will derive a contradiction.]* Then the loop would iterate indefinitely. But by II above, for all integers n, after n iterations of the loop, $a \geq 0$, $b \geq 0$, at most one of a and b equals 0, and $0 \leq a + b \leq A + B - n$. Hence after $A + B$ iterations $0 \leq a + b \leq A + B - (A + B) = 0$. Since $a \geq 0$ and $b \geq 0$, then, $a = b = 0$. But this contradicts the fact that at most one of a and b equals 0. Hence the supposition is false, and so after some iteration G is false.

IV. Correctness of the Post-Condition: Suppose that N is the least number of iterations after which G is false and $I(N)$ is true. Then (since G is false) $a = 0$ or $b = 0$, and (since $I(N)$ is true) both a and b are nonnegative, at most one of a and b equals 0, and $\gcd(a, b) = \gcd(A, B)$. Thus $\gcd(a, b) = \begin{cases} \gcd(0, b) & \text{if } a = 0 \\ \gcd(a, 0) & \text{if } b = 0 \end{cases}$. By Lemma 3.8.1, $\gcd(0, b) = b$ and $\gcd(a, 0) = a$. Hence one of a or b is zero and the other is nonzero, and $\gcd(A, B)$ equals whichever of a or b is nonzero.

Alternative Solution: Omit part (3) of the loop invariant, and change the proof in III to the following: The guard G is the condition $a \neq 0$ and $b \neq 0$. Observe that after each iteration of the loop

$$a_{\text{new}} + b_{\text{new}} = \begin{cases} a_{\text{old}} - b_{\text{old}} + b_{\text{old}} & \text{if } a_{\text{old}} \geq b_{\text{old}} \\ a_{\text{old}} + b_{\text{old}} - a_{\text{old}} & \text{if } b_{\text{old}} > a_{\text{old}} \end{cases} = \begin{cases} a_{\text{old}} & \text{if } a_{\text{old}} \geq b_{\text{old}} \\ b_{\text{old}} & \text{if } b_{\text{old}} > a_{\text{old}} \end{cases} < a_{\text{old}} + b_{\text{old}}.$$

Therefore, the values of $a + b$ form a strictly decreasing sequence of nonnegative integers (since a and b are nonnegative). By the well-ordering principle, this sequence has a least value. Let N be the least integer for which this value is attained, and let $a_N + b_N$ be this least value. Suppose G is true after the Nth iteration of the loop. *[We will derive a contradiction.]* Then the loop iterates another time, which results in new values a_{N+1} and b_{N+1} for a and b. But by the argument above, $a_{N+1} + b_{N+1} < a_N + b_N$. This contradicts the fact that $a_N + b_N$ is the least value of the sum $a + b$. Hence the supposition is false, and so G is false after the Nth iteration of the loop.

11. a. Suppose the following condition is satisfied before entry to the loop: "there exist integers u, v, s, and t such that $a = uA + vB$ and $b = sA + tB$." Then

$$a_{\text{old}} = u_{\text{old}} A + v_{\text{old}} B \quad \text{and} \quad b_{\text{old}} = s_{\text{old}} A + t_{\text{old}} B,$$

for some integers u_{old}, v_{old}, s_{old}, and t_{old}. Observe that $b_{\text{new}} = r_{\text{new}} = a_{\text{old}} \bmod b_{\text{old}}$. So by the quotient-remainder theorem, there exists a unique integer q_{new} with $a_{\text{old}} = b_{\text{old}} \cdot q_{\text{new}} + r_{\text{new}} = b_{\text{old}} \cdot q_{\text{new}} + b_{\text{new}}$. Solving for b_{new} gives

$$b_{\text{new}} = a_{\text{old}} - b_{\text{old}} \cdot q_{\text{new}} = (u_{\text{old}} A + v_{\text{old}} B) - (s_{\text{old}} A + t_{\text{old}} B) q_{\text{new}} = (u_{\text{old}} - s_{\text{old}} q_{\text{new}}) A + (v_{\text{old}} - t_{\text{old}} q_{\text{new}}) B.$$

Therefore, let $s_{\text{new}} = u_{\text{old}} - s_{\text{old}} q_{\text{new}}$ and $t_{\text{new}} = v_{\text{old}} - t_{\text{old}} q_{\text{new}}$. Also since $a_{\text{new}} = b_{\text{old}} = s_{\text{old}} A + t_{\text{old}} B$, let $u_{\text{new}} = s_{\text{old}}$ and $v_{\text{new}} = t_{\text{old}}$. Hence $a_{\text{new}} = u_{\text{new}} \cdot A + v_{\text{new}} \cdot B$ and $b_{\text{new}} = s_{\text{new}} \cdot A + t_{\text{new}} \cdot B$, and so the condition is true after exit from the loop.

b. Initially $a = A$ and $b = B$. Let $u = 1$, $v = 0$, $s = 0$, and $t = 1$. Then before the first iteration of the loop, $a = uA + vB$ and $b = sA + tB$ as was to be shown.

c. By part (b) there exist integers u, v, s, and t such that before the first iteration of the loop, $a = uA + vB$ and $b = sA + tB$. So by part (a), after each subsequent iteration of the loop, there exist integers u, v, s, and t such that $a = uA + vB$ and $b = sA + tB$. Now after the final iteration of the **while** loop in the Euclidean algorithm, the variable *gcd* is given the current value of a. But by the correctness proof for the Euclidean algorithm, $gcd = \gcd(A, B)$. Hence there exist integers u and v such that $\gcd(A, B) = uA + vB$.

d. The method discussed in part (a) gives the following formulas for u, v, s, and t:

$$u_{\text{new}} = s_{\text{old}}, \quad v_{\text{new}} = t_{\text{old}}, \quad s_{\text{new}} = u_{\text{old}} - s_{\text{old}} q_{\text{new}}, \quad \text{and} \quad t_{\text{new}} = v_{\text{old}} - t_{\text{old}} q_{\text{new}},$$

where in each iteration q_{new} is the quotient obtained by dividing a_{old} by b_{old}. The trace table below shows the values of a, b, r, q, gcd, and u, v, s, and t for the iterations of the **while** loop from the Euclidean algorithm. By part (b) the initial values of u, v, s, and t are $u = 1$, $v = 0$, $s = 0$, and $t = 1$.

r		18	12	6	0
q		2	8	1	1
a	330	156	18	12	6
b	156	18	12	6	0
gcd					6
u	1	0	1	-8	9
v	0	1	-2	17	-19
s	0	1	-8	9	-17
t	1	-2	17	-19	36

Since the final values of gcd, u, and v are 6, 9 and -19 and since $A = 330$ and $B = 156$, we have $\gcd(330, 156) = 6 = 330u + 156v = 330 \cdot 9 + 156 \cdot (-19)$, which is true.

Chapter 5 Set Theory

The first section of this chapter introduces the structures of set and ordered set and illustrates them with a variety of examples. The aim of this section is to provide a solid basis of experience for deriving the set properties discussed in the remainder of the chapter. In addition, this section introduces the notation of formal languages, which are discussed at length in Sections 7.2 and 10.4 and in passing in Sections 7.1, 8.1, 8.4, 10.1, 10.2, 10.3, and 10.5.

Students of computer science may be motivated to take seriously the formal structures and notation of set theory if they are shown a relation between them and the formal structures and notation of computer science. In programming, for instance, it is important to distinguish among different kinds of data structures and to respect the notations that are used to refer to them. Similarly, in set theory it is important to distinguish between, say, $\{1,2,3\}$ and $(1,2,3)$ or between $((u,v),w)$ and $(u,(v,w))$.

The reason for delaying discussion of set theory to this chapter is that considerable sophistication is needed to understand the derivations of set properties. Even after having studied the first four chapters of the book, many students have difficulty constructing simple element proofs. One reason for the difficulty is a tendency for students to interpret "if $x \in A$ then $x \in B$" as "$x \in A$ and $x \in B$". I have found this tendency to be quite strong. For example, even when I tell students that part of a test will be on definitions, specifically warn them against this error, and repeatedly emphasize the dynamic if-then nature of the definition of subset, there are always students who write the definition as an *and* statement anyway. Part of the reason for the misunderstanding may be that if it is true for a particular x that "if $x \in A$ then $x \in B$" and if $x \in A$, then the statement "$x \in A$ and $x \in B$" is also true. In any case, you may be taken aback by the confusion some of your students manifest in tackling the proofs of Section 5.2. However, many students who have difficulty at the outset catch on to the idea of element proof eventually, particularly if given feedback on their work through discussion of student presentations of proofs at the board or if allowed to resubmit some homework problems for a better grade. And at this stage of the course, most students actually enjoy the "algebraic" derivation of set theory properties, often negotiating to use this method rather than the element method on tests.

It is possible to cover the material of this chapter lightly and still give students a bare introduction to the idea of element proof. One way to do so is to start with Section 5.1, making sure to assign exercises 12 and 13. Then you would simply state the various set properties from Sections 5.2 and 5.3, perhaps giving a proof or two using element arguments but not asking students to write such proofs themselves. Instead you would assign the "algebraic" proofs of exercises 28–36 from Section 5.2 and exercises 21–25 of Section 5.3.

Some of the set theory properties developed in Sections 5.2 and 5.3 are used to derive counting principles in Chapter 6. However, you can cover Chapter 6 before Chapter 5 by referring to those properties in informal terms. For instance, if you are on the quarter system, you might consider covering Chapters 1-4 and 6 during the first quarter and leaving Chapter 5 to start the second quarter. This allows time for the more abstract ideas of Chapters 3 and 4 to settle in before the final examination, while ending the course with material that is concrete and of obviously practical use. Then the second quarter can start with sets and move on to other discrete structures such as functions, relations, and graphs.

Comments on Exercises

Exercise Set 5.1: #12 and (to a lesser extent) #13 are useful for introducing students to the idea of element argument as preparation for Section 5.2. #12 is also preparation for understanding the equivalence of various representations of congruence classes, which are discussed in Section 10.3.
Exercise Set 5.3: The discussion of partitions in Section 5.3 is part of the groundwork for the discussion of equivalence relations in Section 10.3. #36 and #38 are especially helpful as background for understanding congruence slasses. #46 is a warm-up exercise for the proof of the theorem that

Section 5.1

2. No. $\{4\}$ is the set whose only element is 4. This does not equal the number 4.

3. $A = D$

4. b. $T = \{0, 2\}$

5. b. Yes. Every element in C is in A.

   c. Yes. Every element in C is in C.

6. c. No. d. Yes e. Yes. g. Yes h. No j. Yes

7. a. $A \cup B = \{a, b, c, d, f, g\}$ b. $A \cap B = \{b, c\}$ d. $B - A = \{a\}$

9.
   a. $A \cup B = \{x \in \mathbf{R} \mid -2 \leq x < 3\}$ b. $A \cap B = \{x \in \mathbf{R} \mid -1 < x \leq 1\}$
   c. $A^c = \{x \in \mathbf{R} \mid x < -2 \text{ or } x > 1\}$ d. $B^c = \{x \in \mathbf{R} \mid x \leq -1 \text{ or } x \geq 3\}$
   e. $A^c \cap B^c = \{x \in \mathbf{R} \mid x < -2 \text{ or } x \geq 3\}$ f. $A^c \cup B^c = \{x \in \mathbf{R} \mid x \leq -1 \text{ or } x > 1\}$
   g. $(A \cap B)^c = \{x \in \mathbf{R} \mid x \leq -1 \text{ or } x > 1\}$ h. $(A \cup B)^c = \{x \in \mathbf{R} \mid x < -2 \text{ or } x \geq 3\}$

10. a. True: every positive integer is a rational number.

    c. False: there are many rational numbers that are not integers. For instance, $1/2 \in \mathbf{Q}$ but $1/2 \notin \mathbf{Z}$.

    e. True: every rational number is real. So the set of all numbers that are both rational and real is the same as the set of all numbers that are rational.

    f. True: every integer is a rational number, and so the set of all numbers that are integers or rational numbers is the same as the set of all rational numbers. Thus $\mathbf{Q} \cup \mathbf{Z} = \mathbf{Q}$.

    g. True: every positive integer is a real number, and so the set of all numbers that are both positive integers and real numbers is the same as the set of all positive integers.

    h. False: every integer is a rational number, and so the set of all numbers that are integers or rational numbers is the same as the set of all rational numbers. However there are many rational numbers that are not integers, and so $\mathbf{Z} \cup \mathbf{Q} = \mathbf{Q} \neq \mathbf{Z}$.

11. b. *Negation*: $\exists$ a set S such that $S \subseteq \mathbf{Q}^+$ and $S \not\subseteq \mathbf{Q}^-$. The negation is true. For example let $S = \{1/2\}$. Then $S \subseteq \mathbf{Q}^+$ and $S \not\subseteq \mathbf{Q}^-$.

12. c. No. $8 \in D$ because $8 = 3 \cdot 3 - 1$. But $8 \notin A$. For if 8 were in A, then $8 = 2i - 1$ for some integer i. Solving for i would give $2i = 9$, or $i = 9/2$, which is not an integer. Hence $8 \notin A$.

    d. Yes.

    Suppose $n \in B$. By definition of B, $n = 3j + 2$ for some integer j. But then $n = 3j + 2 = 3j + 3 - 1 = 3(j+1) - 1$. Let $s = j + 1$. Then s is an integer and $n = 3s - 1$. So by definition of D, $n \in D$. Hence any element of B is in D, or, symbolically, $B \subseteq D$.

    Conversely, suppose $q \in D$. By definition of D, $q = 3s - 1$ for some integer s. But then $q = 3s - 1 = 3s - 3 + 2 = 3(s-1) + 2$ for some integer s. Let $j = s - 1$. Then j is an integer and $q = 3j + 2$. So by definition of B, $q \in B$. Hence any element of D is in B, or, symbolically, $D \subseteq B$.

    Since $B \subseteq D$ and $D \subseteq B$, by definition of set equality $B = D$.

13. c. Yes. Every integer that is divisible by 6 is also divisible by 3.

    d. $R \cap S = \{u \in \mathbf{Z} \mid u \text{ is divisible by 2 and } u \text{ is divisible by 3}\}$
    $= \{u \in \mathbf{Z} \mid u \text{ is divisible by 6}\} = T$

14. b. $A \cap (B \cup C) = \{a,b,c\} \cap \{b,c,d,e\} = \{b,c\}$, $(A \cap B) \cup C = \{b,c\} \cup \{b,c,e\} = \{b,c,e\}$, and $(A \cap B) \cup (A \cap C) = \{b,c\} \cup \{b,c\} = \{b,c\}$. Hence $A \cap (B \cup C) = (A \cap B) \cup (A \cap C)$.

  c. $(A - B) - C = \{a\} - \{b,c,e\} = \{a\}$ and $A - (B - C) = \{a,b,c\} - \{d\} = \{a,b,c\}$. These sets are not equal.

15.

    b.

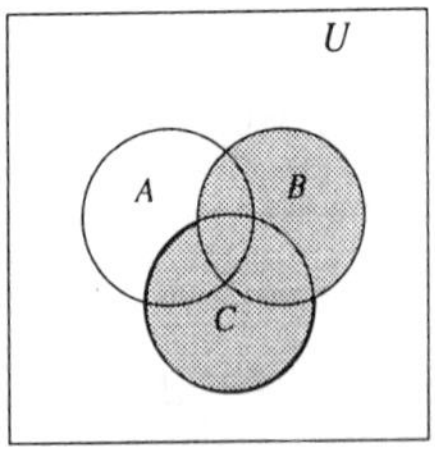

    c.

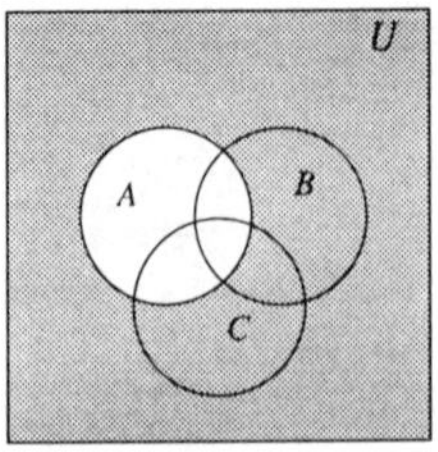

    d.

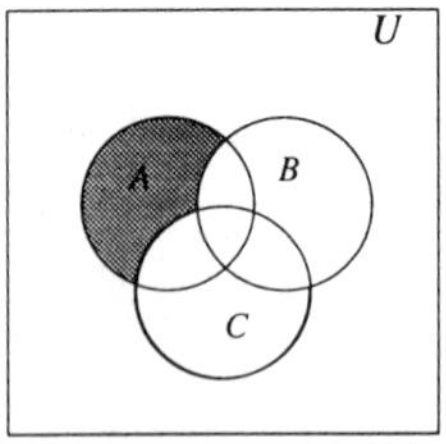

    e. 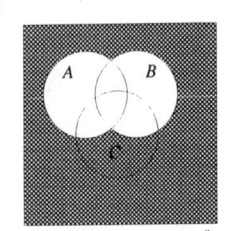 Shaded region is $(A \cup B)^c$

    f. Region shaded ||| is A^c
       Region shaded === is B^c
       Cross-hatched region is $A^c \cap B^c$

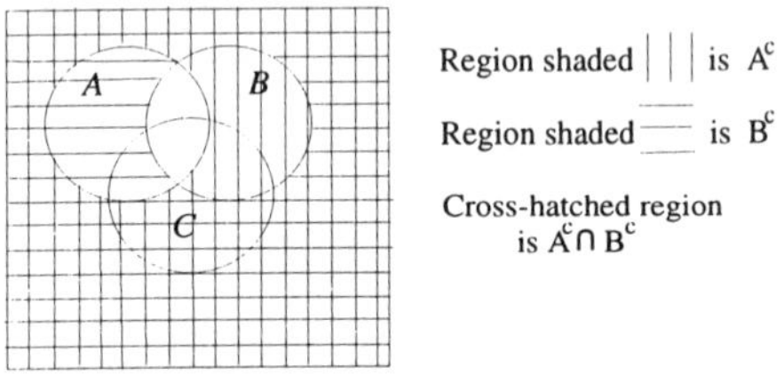

16. b.

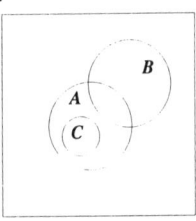

17. b. $B \times A = \{(a,x),(a,y),(a,z),(a,w),(b,x),(b,y),(b,z),(b,w)\}$

  c. $A \times A = \{(x,x),(x,y),(x,z),(x,w),(y,x),(y,y),(y,z),(y,w),(z,x),(z,y),(z,z),(z,w),(w,x),(w,y),(w,z),(w,w)\}$

  d. $B \times B = \{(a,a),(a,b),(b,a),(b,b)\}$

18. b. $(A \times B) \times C = \{((1,u),m),((1,u),n),((1,v),m),((1,v),n),((2,u),m),((2,u),n),((2,v),m),((2,v),n),((3,u),m),((3,u),n),((3,v),m),((3,v),n)\}$

  c. $A \times B \times C = \{(1,u,m),(1,u,n),(1,v,m),(1,v,n),(2,u,m),(2,u,n),(2,v,m),(2,v,n),(3,u,m),(3,u,n),(3,v,m),(3,v,n)\}$

19. b. $L_2 = \{x, xx, xy, xxx, xxy, xyx, xyy\}$

  d. $\Sigma^4 = \{xxxx, xxxy, xxyx, xxyy, xyxx, xyxy, xyyx, xyyy, yxxx, yxxy, yxyx, yxyy, yyxx, yyxy, yyyx, yyyy\}$

  e. A is the set of all strings over Σ that have length 1 or 2. B is the set of all strings over Σ that have length 3 or 4. $A \cup B$ is the set of all strings over Σ that have length from 1 to 4 inclusive.

20. $L = \{11*, 11/, 12*, 12/, 21*, 21/, 22*, 22/\}$

22.

i	1				2			3				4
j	1	2	3	4	1	2	3	1	2	3	4	5
found	no	yes		no		yes		no				
answer	$A \subseteq B$											$A \not\subseteq B$

23.

Algorithm   Testing Whether x ∈ A

[This algorithm checks whether an element x is in a set A, which is represented as a one-dimensional array $a[1], a[2], \ldots, a[n]$. Initially answer is set equal to "$x \notin A$." Then for successive integers i from 1 to n, x is compared to $a[i]$. If at any stage $x=a[i]$, the value of answer is changed to "$x \in A$" and iteration of the loop ceases.]

Input: $a[1], a[2], \ldots, a[n]$ *[a one-dimensional array]*, x *[an element of the same data type as the elements of the array]*

Algorithm Body:

$i := 1$, $answer := $ "$x \notin A$"

while ($i \leq n$ and $answer = $ "$x \notin A$")

   if $x = a[i]$ **then** $answer := $ "$x \in A$"

   $i := i + 1$

end while

Output: *answer [a string]*

end Algorithm

Section 5.2

1. *c.* (1) A (2) $B \cap C$

4. *a.* $A \cup B \subseteq B$ *b.* $A \cup B$ *c.* $x \in B$ *d.* A *e.* or   *f.* B *g.* A *h.* B *i.* B

7. *Proof:*

   Suppose A and B are sets.

   $(A \cap B)^c \subseteq A^c \cup B^c$: Suppose $x \in (A \cap B)^c$. By definition of complement, $x \notin A \cap B$, which means that it is false that (x is in A and x is in B). By De Morgan's laws of logic, this implies that x is not in A or x is not in B, which can be written $x \notin A$ or $x \notin B$. Hence $x \in A^c$ or $x \in B^c$ by definition of complement. It follows by definition of union that $x \in A^c \cup B^c$. *[Thus $(A \cap B)^c \subseteq A^c \cup B^c$ by definition of subset.]*

   $A^c \cup B^c \subseteq (A \cap B)^c$: Suppose $x \in A^c \cup B^c$. Then by definition of union $x \in A^c$ or $x \in B^c$. By definition of complement $x \notin A$ or $x \notin B$. In other words, x is not in A or x is not in B. By De Morgan's laws of logic this implies that it is false that (x is in A and x is in B), which can be written $x \notin A \cap B$ by definition of intersection. Hence by definition of complement, $x \in (A \cap B)^c$. *[Thus $A^c \cup B^c \subseteq (A \cap B)^c$ by definition of subset.]*

   [Since both set containments have been proved, $(A \cap B)^c = A^c \cup B^c$ by definition of set equality.]

9. *Proof:* Suppose A and B are arbitrarily chosen sets.

   $(A - B) \cup (A \cap B) \subseteq A$: Suppose $x \in (A - B) \cup (A \cap B)$. By definition of union, $x \in (A - B)$ or $x \in A \cap B$.

   Case 1 ($x \in A - B$): Then by definition of set difference $x \in A$ and $x \notin B$. In particular, $x \in A$.

Case 2 ($x \in A \cap B$): Then by definition of intersection $x \in A$ and $x \in B$. In particular, $x \in A$.

Hence in either case, $x \in A$, and so by definition of subset, $(A - B) \cup (A \cap B) \subseteq A$.

A $\subseteq$ (A − B) $\cup$ (A $\cap$ B): Suppose $x \in A$. Either $x \in B$ or $x \notin B$.

Case 1 ($x \in B$): Then since $x \in A$ also, by definition of intersection $x \in A \cap B$, and so by the inclusion in union property, $x \in (A - B) \cup (A \cap B)$.

Case 2 ($x \notin B$): Then since $x \in A$, by definition of set difference $x \in A - B$, and so by the inclusion in union property, $x \in (A - B) \cup (A \cap B)$.

Hence in either case, $x \in (A-B) \cup (A \cap B)$, and so by definition of subset, $A \subseteq (A-B) \cup (A \cap B)$.

Since both set containments have been proved, $(A - B) \cup (A \cap B) = A$ by definition of set equality.

10. False. *Counterexample*: Let $A = \{1, 2, 3\}$, $B = \{2, 3\}$, and $C = \{3\}$. Then $B - C = \{2\}$, and so $A - (B - C) = \{1, 2, 3\} - \{2\} = \{1, 3\}$. On the other hand $A - B = \{1, 2, 3\} - \{2, 3\} = \{1\}$, and so $(A - B) - C = \{1\} - \{3\} = \{1\}$. Hence $A - (B - C) \neq (A - B) - C$.

11. True. *Proof*: Let A, B, and C be any sets.

    (A − B) $\cap$ (C − B) $\subseteq$ (A $\cap$ C) − B: Suppose $x \in (A-B) \cap (C-B)$. By definition of intersection, $x \in A - B$ and $x \in C - B$, and so by definition of set difference, $x \in A$ and $x \notin B$ and $x \in C$ and $x \notin B$. Thus $x \in A$ and $x \in C$ and $x \notin B$. So by definition of intersection, $x \in A \cap C$ and $x \notin B$, and by definition of set difference $x \in (A \cap C) - B$. *[Thus $(A-B) \cap (C-B) \subseteq (A \cap C) - B$ by definition of subset.]*

    (A $\cap$ C) − B $\subseteq$ (A − B) $\cap$ (C − B): Suppose $x \in (A \cap C) - B$. By definition of set difference, $x \in (A \cap C)$ and $x \notin B$, and by definition of intersection, $x \in A$ and $x \in C$ and $x \notin B$. Thus it is true that $x \in A$ and $x \notin B$ and $x \in C$ and $x \notin B$, and so by definition of set difference, $x \in A - B$ and $x \in C - B$, and by definition of intersection, $x \in (A - B) \cap (C - B)$. *[Thus $(A \cap C) - B \subseteq (A - B) \cap (C - B)$ by definition of subset.]*

    [Since both subset containments have been proved, $(A - B) \cap (C - B) = (A \cap C) - B$ by definition of set equality.]

12. False. *Counterexample*: Let $A = \{1, 2, 3\}$, $B = \{2\}$, and $C = \{1\}$. Then $A - B = \{1, 2, 3\} - \{2\} = \{1, 3\}$ and $C - B = \{1\} - \{2\} = \{1\}$, and so $(A - B) \cap (C - B) = \{1\} \cap \{1\} = \{1\}$. On the other hand, $B \cup C = \{1, 2\}$, and so $A - (B \cup C) = \{1, 2, 3\} - \{1, 2\} = \{3\}$. Hence $(A - B) \cap (C - B) \neq A - (B \cup C)$.

13. True. *Proof*: Suppose A, B, and C are sets and $A \subseteq B$. Let $x \in A \cap C$. By definition of intersection, $x \in A$ and $x \in C$. But since $A \subseteq B$ and $x \in A$, then $x \in B$. Hence $x \in B$ and $x \in C$, and so by definition of intersection $x \in B \cap C$. *[Thus $A \cap C \subseteq B \cap C$ by definition of subset.]*

14. True. *Proof*: Suppose A, B, and C are sets and $A \subseteq B$. Let $x \in A \cup C$. By definition of union, $x \in A$ or $x \in C$. But since $A \subseteq B$ and $x \in A$, then $x \in B$. Hence $x \in B$ or $x \in C$, and so by definition of union $x \in B \cup C$. *[Thus $A \cup C \subseteq B \cup C$ by definition of subset.]*

16. False. *Counterexample*: Let $A = \{1\}$, $B = \{2\}$, and $C = \{1, 2\}$. Then $A \cup C = \{1, 2\} = B \cup C$ but $A \neq B$.

17. False. *Counterexample*: Let $A = \{1, 3\}$, $B = \{1, 2, 3\}$, and $C = \{2, 3\}$. Then $A \cap C = \{3\}$ and $B \cap C = \{2, 3\}$ and so $A \cap C \subseteq B \cap C$. Also $A \cup C = \{1, 2, 3\} = B \cup C$, and so $A \cup C \subseteq B \cup C$. But $A \neq B$.

18. False. One counterexample is given in exercise 14 of Section 5.1. Here is another: Let $A = \{1\}$, $B = \{2\}$, and $C = \{2, 3\}$. Then $(A \cup B) \cap C = \{1, 2\} \cap \{2, 3\} = \{2\}$ whereas $A \cup (B \cap C) = \{1\} \cup \{2\} = \{1, 2\}$.

19. True. *Proof:* Suppose A and B are sets and $A \subseteq B$. Let $x \in B^c$. By definition of complement, $x \notin B$. Then $x \notin A$ because if $x \in A$ then $x \in B$ (since $A \subseteq B$), and $x \notin B$. Hence by definition of complement $x \in A^c$. *[Thus $B^c \subseteq A^c$ by definition of subset.]*

20. False. *Counterexample:* Let $A = \{1\}$, $B = \{2\}$, and $C = \{1,3\}$. Then $A \not\subseteq B$ and $B \not\subseteq C$ but $A \subseteq C$.

21. True. *Proof:* Suppose A, B, and C are sets and $A \subseteq B$ and $A \subseteq C$. Let $x \in A$. Since $x \in A$ and $A \subseteq B$, then $x \in B$ (by definition of subset). Similarly, since $x \in A$ and $A \subseteq C$, then $x \in C$. Hence $x \in B$ and $x \in C$, and so by definition of intersection, $x \in B \cap C$. *[By definition of subset, therefore, $A \subseteq B \cap C$.]*

22. True. *Proof:* Suppose A, B, and C are sets and $A \subseteq C$ and $B \subseteq C$. Let $x \in A \cup B$. By definition of union, $x \in A$ or $x \in B$. But if $x \in A$ then $x \in C$ (because $A \subseteq C$), and if $x \in B$ then $x \in C$ (because $B \subseteq C$). Hence in either case $x \in C$. *[So by definition of subset, $A \cup B \subseteq C$.]*

23. *Completion of Proof:*

    $(\mathbf{A} \times \mathbf{B}) \cup (\mathbf{A} \times \mathbf{C}) \subseteq \mathbf{A} \times (\mathbf{B} \cup \mathbf{C})$: Suppose $(x,y) \in (A \times B) \cup (A \times C)$. Then $(x,y) \in A \times B$ or $(x,y) \in A \times C$.

    Case 1 $((x,y) \in A \times B)$: In this case $x \in A$ and $y \in B$. By definition of union, since $y \in B$, then $y \in B \cup C$. Hence $x \in A$ and $y \in B \cup C$, and so by definition of Cartesian product, $(x,y) \in A \times (B \cup C)$.

    Case 2 $((x,y) \in A \times C)$: In this case $x \in A$ and $y \in C$. By definition of union, since $y \in C$, then $y \in B \cup C$. Hence $x \in A$ and $y \in B \cup C$, and so by definition of Cartesian product, $(x,y) \in A \times (B \cup C)$.

    Thus in either case $(x,y) \in A \times (B \cup C)$. *[Hence by definition of subset, $(A \times B) \cup (A \times C) \subseteq A \times (B \cup C)$.]*

    [Since both subset relations have been proved (one above and the other in Appendix B), $A \times (B \cup C) = (A \times B) \cup (A \times C)$ by definition of set equality.]

24. True. *Proof:* Suppose A, B, and C are sets.

    $\mathbf{A} \times (\mathbf{B} \cap \mathbf{C}) \subseteq (\mathbf{A} \times \mathbf{B}) \cap (\mathbf{A} \times \mathbf{C})$: Suppose $(x,y) \in A \times (B \cap C)$. By definition of Cartesian product, $x \in A$ and $y \in B \cap C$. By definition of intersection, $y \in B$ and $y \in C$. It follows that both statements "$x \in A$ and $y \in B$" and "$x \in A$ and $y \in C$" are true. Hence by definition of Cartesian product, $x \in A \times B$ and $x \in A \times C$, and so by definition of intersection, $x \in (A \times B) \cap (A \times C)$. *[Thus $A \times (B \cap C) \subseteq (A \times B) \cap (A \times C)$ by definition of subset.]*

    $(\mathbf{A} \times \mathbf{B}) \cap (\mathbf{A} \times \mathbf{C}) \subseteq \mathbf{A} \times (\mathbf{B} \cap \mathbf{C})$: Suppose $x \in (A \times B) \cap (A \times C)$. By definition of intersection, $x \in A \times B$ and $x \in A \times C$, and so by definition of Cartesian product $x \in A$ and $y \in B$ and also $x \in A$ and $y \in C$. Consequently, the statement "$x \in A$ and both $y \in B$ and $y \in C$" is true. It follows by definition of intersection that $x \in A$ and $y \in B \cap C$, and so by definition of Cartesian product, $(x,y) \in A \times (B \cap C)$. *[Thus $(A \times B) \cap (A \times C) \subseteq A \times (B \cap C)$ by definition of subset.]*

    [Since both subset containments have been proved, $A \times (B \cap C) = (A \times B) \cap (A \times C)$ by definition of set equality.]

25. b

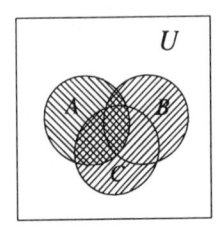

darkly shaded region is $A \cap (B \cup C)$

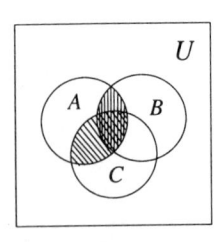

entire shaded region is $(A \cap B) \cup (A \cap C)$

c

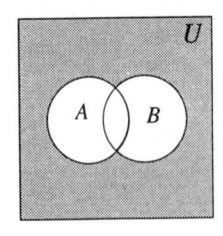

shaded region is $(A \cup B)^c$

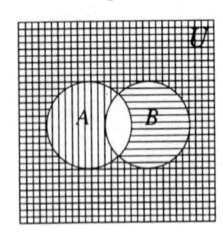

cross-hatched region is $A^c \cap B^c$

d

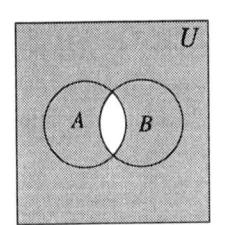

shaded region is $(A \cap B)^c$

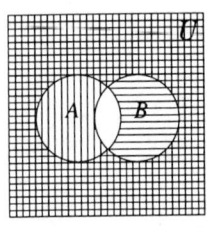

entire shaded region is $A^c \cup B^c$

27. *Proof (by mathematical induction)*:

The formula holds for n = 1: For $n = 1$ the formula is $A_1 - B = A_1 - B$, which is clearly true.

If the formula holds for n = k, then it holds for n = k+1: Let k be an integer with $k \geq 1$, and suppose $A_1, A_2, \ldots, A_k, A_{k+1}$, and B are any sets such that

$$(A_1 - B) \cap (A_2 - B) \cap \cdots \cap (A_k - B) = (A_1 \cap A_2 \cap \cdots \cap A_k) - B.$$

We must show that

$$(A_1 - B) \cap (A_2 - B) \cap \cdots \cap (A_{k+1} - B) = (A_1 \cap A_2 \cap \cdots \cap A_{k+1}) - B.$$

But

$(A_1 - B) \cap (A_2 - B) \cap \cdots \cap (A_{k+1} - B)$

$\quad = [(A_1 - B) \cap (A_2 - B) \cap \cdots \cap (A_k - B)] \cap (A_{k+1} - B)$ by assumption
$\quad = [(A_1 \cap A_2 \cap \cdots \cap A_k) - B] \cap (A_{k+1} - B)$ by inductive hypothesis
$\quad = [(A_1 \cap A_2 \cap \cdots \cap A_k) \cap A_{k+1}] - B$ by exercise 11
$\quad = (A_1 \cap A_2 \cap \cdots \cap A_{k+1}) - B$ by assumption

[*as was to be shown*].

30. *Proof*: Let sets A, B, and C be given. Then

$$\begin{aligned}(A \cap B) \cup C &= C \cup (A \cap B) && \text{by the commutative law for } \cup \\ &= (C \cup A) \cap (C \cup B) && \text{by the distributive law} \\ &= (A \cup C) \cap (B \cup C) && \text{by the commutative law for } \cup.\end{aligned}$$

31. *Proof:* Let sets A, B, and C be given. Then

$$\begin{aligned}(A - B) - C &= (A \cap B^c) \cap C^c && \text{by the alternate representation for set difference law} \\ &= A \cap (B^c \cap C^c) && \text{by the associative law for } \cap \\ &= A \cap (C^c \cap B^c) && \text{by the commutative law for } \cap \\ &= (A \cap C^c) \cap B^c && \text{by the associative law for } \cap \\ &= (A - C) - B && \text{by the alternate representation for set difference law.}\end{aligned}$$

33. *Proof:* Let A and B be any sets. Then

$(A - B) \cup (B - A)$

$$\begin{aligned}&= (A \cap B^c) \cup (B \cap A^c) && \text{by the alternate representation} \\ & && \text{for set difference law} \\ &= [(A \cap B^c) \cup B] \cap [(A \cap B^c) \cup A^c)] && \text{by the distributive law} \\ &= [B \cup (A \cap B^c)] \cap [A^c \cup (A \cap B^c)] && \text{by the commutative law for } \cup \\ &= [(B \cup A) \cap (B \cup B^c)] \cap [(A^c \cup A) \cap (A^c \cup B^c)] && \text{by the distributive law} \\ &= [(A \cup B) \cap (B \cup B^c)] \cap [(A \cup A^c) \cap (A^c \cup B^c)] && \text{by the commutative law for } \cup \\ &= [(A \cup B) \cap U] \cap [U \cap (A^c \cup B^c)] && \text{because the union of a set} \\ & && \text{and its complement is } U \\ &= [(A \cup B) \cap U] \cap [(A^c \cup B^c) \cap U] && \text{by the commutative law for } \cap \\ &= (A \cup B) \cap (A^c \cup B^c) && \text{by the intersection with } \cap \text{ law} \\ &= (A \cup B) \cap (A \cap B)^c && \text{by De Morgan's law} \\ &= (A \cup B) - (A \cap B) && \text{by the alternate representation} \\ & && \text{for set difference law.}\end{aligned}$$

34. *Proof:* Let sets A, B, and C be given. Then

$$\begin{aligned}(A - B) - C &= (A \cap B^c) \cap C^c && \text{by the alternate representation for set difference law} \\ &= A \cap (B^c \cap C^c) && \text{by the associative law for } \cap \\ &= A \cap (B \cup C)^c && \text{by De Morgan's law} \\ &= A - (B \cup C) && \text{by the alternate representation for set difference law.}\end{aligned}$$

36. *Proof:* Let sets A and B be given. Then

$$\begin{aligned}B^c \cup (B^c - A)^c &= (B^c \cup (B^c \cap A^c))^c && \text{by the alternate representation} \\ & && \text{for set difference law} \\ &= (B^c)^c \cap (B^c \cap A^c)^c && \text{by De Morgan's law} \\ &= B \cap (B^c \cap A^c)^c && \text{by the double complement law} \\ &= B \cap ((B^c)^c \cup (A^c)^c) && \text{by De Morgan's law} \\ &= B \cap (B \cup A) && \text{by the double complement law} \\ &= B && \text{by the absorption law.}\end{aligned}$$

37. The second error in this proof occurs in the last sentence. Just because there is an element in A that is in B and an element in B that is in C, it does not follow that there is an element in A that is in C. For instance, suppose $A = \{1, 2\}$, $B = \{2, 3\}$, and $C = \{3, 4\}$. Then there is an element in A that is in B (namely 2) and there is an element in B that is in C (namely 3), but there is no element in A that is in C.

38. Let $A = \{1, 2\}$, $B = \{2, 3\}$, and $x = 3$. Then since $3 \notin \{1, 2\}$, the statement "$x \notin A$ or $x \notin B$" is true. But since $A \cup B = \{1, 2, 3\}$ and $3 \in \{1, 2, 3\}$, the statement "$x \notin A \cup B$" is false.

39. A correct proof of the given statement must show that if $x \in (A - B) \cup (A \cap B)$ then $x \in A$. This incorrect proof uses the assumption that $x \in A$ as a basis for concluding that $x \in A$. In other words, this incorrect proof begs the question.

Section 5.3

1. *a.* No, because $\emptyset$ does not have any elements.

   c. Yes, because $\{\emptyset\}$ is the set that contains the one element $\emptyset$.

4. *a.* For any subset A of a universal set U, $A \cap A^c = \emptyset$.

   Proof: Let A be a subset of a universal set U. Suppose $A \cap A^c \neq \emptyset$, that is, suppose there is an element x such that $x \in A \cap A^c$. Then by definition of intersection, $x \in A$ and $x \in A^c$, and so by definition of complement, $x \in A$ and $x \notin A$. This is a contradiction. *[Hence the supposition is false, and we conclude that $A \cap A^c = \emptyset$.]*

   b. For any subset A of a universal set U, $A \cup A^c = U$.

   Proof: Suppose A is a subset of a universal set U.

   $A \cup A^c \subseteq U$: Let $x \in A \cup A^c$. By definition of union, $x \in A$ or $x \in A^c$. But A is a subset of U by hypothesis and A^c is also a subset of U by definition of complement. Thus by definition of subset, $x \in U$ regardless of whether $x \in A$ or $x \in A^c$ *[and hence $A \cup A^c \subseteq U$]*.

   $U \subseteq A \cup A^c$: Let $x \in U$. It is certainly true that $x \in A$ or $x \notin A$ (this is a tautology). But by definition of complement, if $x \notin A$ then $x \in A^c$. Thus $x \in A$ or $x \in A^c$, and so by definition of union $x \in A \cup A^c$ and hence $U \subseteq A \cup A^c$*]*.

   [Since both set containments $U \subseteq A \cup A^c$ and $A \cup A^c \subseteq U$ have been proved, $U = A \cup A^c$ by definition of set equality.]

5. *a.* If U is a universal set, then $U^c = \emptyset$.

   Proof: Let U be a universal set. Suppose $U^c \neq \emptyset$, that is, suppose there were an element x in U^c. Then by definition of complement $x \notin U$. But, by definition, a universal set contains all elements under discussion, and thus it is impossible that $x \notin U$. *[Hence the supposition is false, and so $U^c = \emptyset$.]*

   b. If U is a universal set, then $\emptyset^c = U$.

   Proof: Let U be a universal set.

   $\emptyset^c \subseteq U$: Let $x \in \emptyset^c$. By definition of complement, $x \in U$ and $x \notin \emptyset$. In particular, $x \in U$. *[Hence $\emptyset^c \subseteq U$ by definition of subset.]*

   $U \subseteq \emptyset^c$: Let $x \in U$. By definition of $\emptyset$, $x \notin \emptyset$ because *no* element is in $\emptyset$. Thus $x \in U$ and $x \notin \emptyset$, and so by definition of complement $x \in \emptyset^c$. *[Hence $U \subseteq \emptyset^c$ by definition of subset.]*

   [Since both subset containments $\emptyset^c \subseteq U$ and $U \subseteq \emptyset^c$ have been proved, $\emptyset^c = U$ by definition of set equality.]

6. *b.*

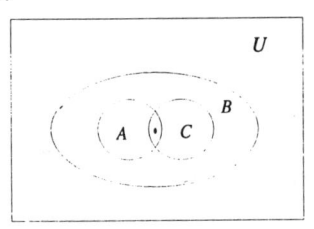

c.

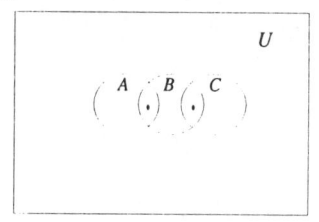

8. This is true. *Illustration*:

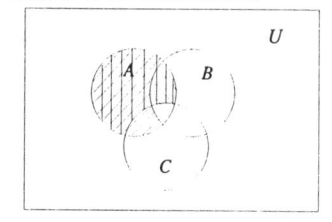

A - B: ☐
A - C: ▦
B - C: ☐

Proof: Let A, B, and C be any sets, and suppose that $(A-B) \cap (B-C) \cap (A-C) \neq \emptyset$. Then there is an element x such that $x \in (A-B) \cap (B-C) \cap (A-C)$. By definition of intersection, $x \in A-B$ and $x \in B-C$ and $x \in A-C$, and so by definition of set difference, $x \in A$ and $x \notin B$ and $x \in B$ and $x \notin C$ and $x \in A$ and $x \notin C$. In particular, $x \notin B$ and $x \in B$, which is a contradiction. Hence the supposition is false. That is, $(A-B) \cap (B-C) \cap (A-C) = \emptyset$.

11. This is true. *Illustration*:

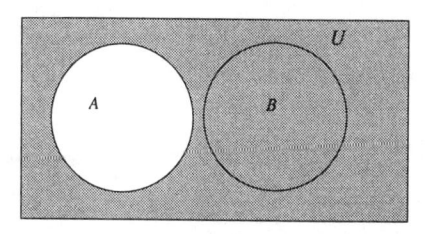

Shaded region is A^c.

Proof: Let A and B be sets with $B \subseteq A^c$. Suppose $A \cap B \neq \emptyset$, that is, suppose there were an element x in $A \cap B$. Then by definition of intersection, $x \in A$ and $x \in B$. But $B \subseteq A^c$, and so by definition of subset, $x \in A^c$. By definition of complement this means that $x \notin A$. Hence $x \in A$ and $x \notin A$, which is a contradiction. *[Thus the supposition is false, and we conclude that $A \cap B = \emptyset$.]*

12. This is true. *Illustration*:

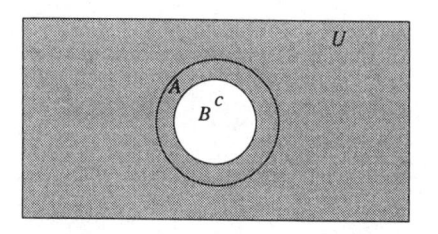

Shaded region is B.

Proof: Let A and B be sets with $A^c \subseteq B$. To prove that $A \cup B \subseteq U$, suppose $x \in A \cup B$. Then $x \in A$ or $x \in B$ by definition of union. But both sets A and B are subsets of the universal set U, and so regardless of whether $x \in A$ or $x \in B$, $x \in U$ by definition of subset. *[Therefore, $A \cup B \subseteq U$.]* To prove that $U \subseteq A \cup B$, suppose $x \in U$. It is certainly true that $x \in A$ or

$x \notin A$ (a tautology), and so by definition of complement $x \in A$ or $x \in A^c$. In case $x \in A$, then $x \in A \cup B$ by definition of union. In case $x \in A^c$, then $x \in B$ because $A^c \subseteq B$ by hypothesis, and so $x \in A \cup B$ by definition of union. Thus in either case $x \in A \cup B$. *[Therefore, $U \subseteq A \cup B$.]* *[Since both set containments $A \cup B \subseteq U$ and $U \subseteq A \cup B$ have been proved, $A \cup B = U$ by definition of set equality.]*

13. This is true. *Illustration*:

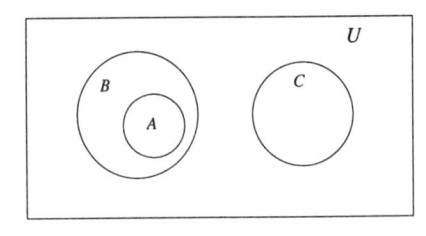

Proof: Let A, B, and C be any sets such that $A \subseteq B$ and $B \cap C = \emptyset$. Suppose $A \cap C \neq \emptyset$, that is, suppose there were an element x in $A \cap C$. By definition of intersection, $x \in A$ and $x \in C$. By hypothesis $A \subseteq B$, and so since $x \in A$, $x \in B$ also. Hence $x \in B \cap C$, which implies that $B \cap C \neq \emptyset$. But $B \cap C = \emptyset$ by hypothesis. This is a contradiction. *[Thus the supposition that $A \cap C \neq \emptyset$ is false, or equivalently, $A \cap C = \emptyset$.]*

14. This is false. *Illustration*:

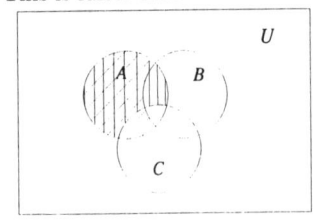

    A – B:
    A – C:
    (A – B) ∩ (A – C):

Counterexample: Let $A = \{1, 2, 3\}$, $B = \{2\}$, and $C = \{3\}$. Then $B \cap C = \emptyset \subseteq A$, and $A - B = \{1, 3\}$ and $A - C = \{1, 2\}$, and so $(A - B) \cap (A - C) = \{1\} \neq \emptyset$.

15. This is true. *Illustration*:

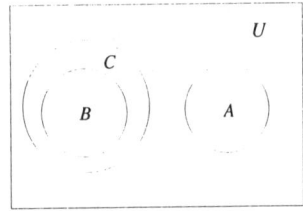

Proof: Let A, B, and C be any sets such that $B \subseteq C$ and $A \cap C = \emptyset$. Suppose $A \cap B \neq \emptyset$. Then there is an element x such that $x \in A \cap B$. By definition of intersection, $x \in A$ and $x \in B$. Since $B \subseteq C$, then, $x \in C$. So $x \in A$ and $x \in C$. But then $x \in A \cap C$ by definition of intersection. However, this is impossible because $A \cap C = \emptyset$. Hence the supposition is false, and so $A \cap B = \emptyset$.

16. This is true. *Illustration*:

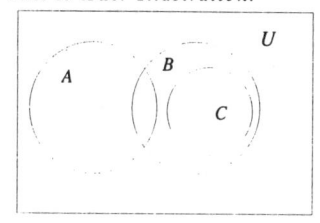

Proof: Let A, B, and C be any sets such that $C \subseteq B - A$. Suppose $A \cap C \neq \emptyset$. Then there is an element x such that $x \in A \cap C$. By definition of intersection, $x \in A$ and $x \in C$. Since $C \subseteq B - A$, then, $x \in B$ and $x \notin A$. So $x \in A$ and $x \notin A$, which is a contradiction. Hence the supposition is false, and thus $A \cap C = \emptyset$.

17. This is true. *Illustration*:

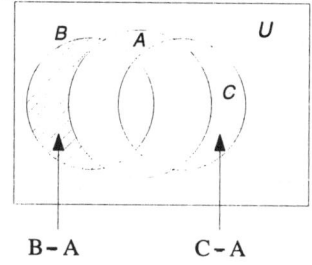

Proof: Let A, B, and C be any sets such that $B \cap C \subseteq A$. Suppose $(C - A) \cap (B - A) \neq \emptyset$. Then there is an element x such that $x \in (C - A) \cap (B - A)$. By definition of intersection, $x \in C - A$ and $x \in B - A$, and so by definition of set difference, $x \in C$ and $x \notin A$ and $x \in B$ and $x \notin A$. Since $x \in C$ and $x \in B$, $x \in B \cap C$. But $B \cap C \subseteq A$, and so $x \in A$. Thus $x \notin A$ and $x \in A$, which is a contradiction. Hence the supposition is false, and so $(C - A) \cap (B - A) = \emptyset$.

18. This is false. *Counterexample*: Let $A = \{1\}$ and $B = \{2\}$. Then $A \cap B = \emptyset$ but $A \times B = \{(1,2)\} \neq \emptyset$.

19. This is true. *Proof:* Let A be a set. Suppose $A \times \emptyset \neq \emptyset$. Then there would be an element (x, y) in $A \times \emptyset$. By definition of Cartesian product, $x \in A$ and $y \in \emptyset$. But there are no elements y such that $y \in \emptyset$. Hence there are no elements (x, y) such that $x \in A$ and $y \in \emptyset$. Consequently, $(x, y) \notin A \times \emptyset$. [Thus the supposition is false, and so $A \times \emptyset = \emptyset$.]

20. *b. Negation*: $\forall$ sets S, $\exists$ a set T such that $S \cup T \neq \emptyset$. The negation is true. Given any set S, let $T = \{1\}$. Then $S \cup T = S \cup \{1\}$. Since $1 \in S \cup \{1\}$, $S \cup \{1\} \neq \emptyset$.

22. *Proof:* Let A and B be sets. Then

$$\begin{aligned}
A - (A - B) &= A \cap (A \cap B^c) && \text{by the alternate representation for set difference law} \\
&= (A \cap A) \cap B^c && \text{by the associative law for } \cap \\
&= A \cap B^c && \text{by the idempotent law for } \cap \\
&= A - B && \text{by the alternate representation for set difference law.}
\end{aligned}$$

23. *Proof:* Let A and B be sets. Then

$$\begin{aligned}
A - (A \cap B) &= A \cap (A \cap B)^c && \text{by the alternate representation} \\
&&& \text{for set difference law} \\
&= A \cap (A^c \cup B^c) && \text{by De Morgan's law} \\
&= (A \cap A^c) \cup (A \cap B^c) && \text{by the distributive law} \\
&= \emptyset \cup (A \cap B^c) && \text{by the intersection with the complement law} \\
&= (A \cap B^c) \cup \emptyset && \text{by the commutative law for } \cup \\
&= A \cap B^c && \text{by the union with } \emptyset \text{ law} \\
&= A - B && \text{by the alternate representation} \\
&&& \text{for set difference law.}
\end{aligned}$$

24. *Proof:* Let A and B be sets. Then

$$\begin{aligned}
(A \cup B) - B &= (A \cup B) \cap B^c && \text{by the alternate representation} \\
&&& \text{for set difference law} \\
&= B^c \cap (A \cup B) && \text{by the commutative law for } \cap \\
&= (B^c \cap A) \cup (B^c \cap B) && \text{by the distributive law} \\
&= (A \cap B^c) \cup (B \cap B^c) && \text{by the commutative law for } \cap \\
&= (A \cap B^c) \cup \emptyset && \text{by the intersection with the complement law} \\
&= A \cap B^c && \text{by the union with the complement law} \\
&= A - B && \text{by the alternate representation} \\
&&& \text{for set difference law.}
\end{aligned}$$

25. *Proof:* Let A and B be sets. Then

$$\begin{aligned}
(A - B) \cup (A \cap B) &= (A \cap B^c) \cup (A \cap B) && \text{by the alternate representation} \\
&&& \text{for set difference law} \\
&= A \cap (B^c \cup B) && \text{by the distributive law} \\
&= A \cap (B \cup B^c) && \text{by the commutative law for } \cup \\
&= A \cap U && \text{by the union with the complement law} \\
&= A && \text{by the intersection with } U \text{ law.}
\end{aligned}$$

26. $A \cap ((B \cup A^c) \cap B^c)$

$$\begin{aligned}
&= A \cap (B^c \cap (B \cup A^c)) && \text{by the commutative law for } \cap \\
&= A \cap ((B^c \cap B) \cup (B^c \cap A^c)) && \text{by the distributive law} \\
&= A \cap ((B \cap B^c) \cup (B^c \cap A^c)) && \text{by the commutative law for } \cap \\
&= A \cap (\emptyset \cup (B^c \cap A^c)) && \text{by the intersection with the complement law} \\
&= A \cap ((B^c \cap A^c) \cup \emptyset) && \text{by the commutative law for } \cup \\
&= A \cap (B^c \cap A^c) && \text{by the union with } \emptyset \text{ law} \\
&= A \cap (A^c \cap B^c) && \text{by the commutative law for } \cap \\
&= (A \cap A^c) \cap B^c && \text{by the associative law for } \cap \\
&= \emptyset \cap B^c && \text{by the intersection with the complement law} \\
&= B^c \cap \emptyset && \text{by the commutative law for } \cap \\
&= \emptyset && \text{by the intersection with } \emptyset \text{ law.}
\end{aligned}$$

Alternate derivation:

$A \cap ((B \cup A^c) \cap B^c)$

$$\begin{aligned}
&= A \cap (B^c \cap (B \cup A^c)) && \text{by the commutative law for } \cap \\
&= (A \cap B^c) \cap (B \cup A^c) && \text{by the associative law for } \cap \\
&= (A \cap B^c) \cap (A^c \cup B) && \text{by the commutative law for } \cup \\
&= (A \cap B^c) \cap (A^c \cup (B^c)^c) && \text{by the double complement law} \\
&= (A \cap B^c) \cap (A \cap B^c)^c && \text{by De Morgan's law} \\
&= \emptyset && \text{by the intersection with the complement law.}
\end{aligned}$$

27. $(A - (A \cap B)) \cap (B - (A \cap B))$

$\begin{aligned}
&= (A \cap (A \cap B)^c) \cap (B \cap (A \cap B)^c) &&\text{by the alternate representation} \\
&&&\text{for set difference law} \\
&= A \cap ((A \cap B)^c \cap (B \cap (A \cap B)^c)) &&\text{by the associative law for } \cap \\
&= A \cap (((A \cap B)^c \cap B) \cap (A \cap B)^c) &&\text{by the associative law for } \cap \\
&= A \cap ((B \cap (A \cap B)^c) \cap (A \cap B)^c) &&\text{by the commutative law for } \cap \\
&= A \cap (B \cap ((A \cap B)^c \cap (A \cap B)^c)) &&\text{by the associative law for } \cap \\
&= A \cap (B \cap (A \cap B)^c) &&\text{by the idempotent law for } \cap \\
&= (A \cap B) \cap (A \cap B)^c &&\text{by the associative law for } \cap \\
&= \emptyset &&\text{by the intersection with the complement law.}
\end{aligned}$

28. $((A \cap (B \cup C)) \cap (A - B)) \cap (B \cup C^c)$

$\begin{aligned}
&= ((A \cap (B \cup C)) \cap (A \cap B^c)) \cap (B \cup C^c) &&\text{by the alternate representation} \\
&&&\text{for set difference law} \\
&= ((A \cap B^c) \cap (A \cap (B \cup C))) \cap (B \cup C^c) &&\text{by the commutative law for } \cap \\
&= (((A \cap B^c) \cap A) \cap (B \cup C)) \cap (B \cup C^c) &&\text{by the commutative law for } \cap \\
&= ((A \cap (A \cap B^c)) \cap (B \cup C)) \cap (B \cup C^c) &&\text{by the commutative law for } \cap \\
&= (((A \cap A) \cap B^c) \cap (B \cup C)) \cap (B \cup C^c) &&\text{by the associative law for } \cap \\
&= ((A \cap B^c) \cap (B \cup C)) \cap (B \cup C^c) &&\text{by the idempotent law for } \cap \\
&= (A \cap B^c) \cap ((B \cup C) \cap (B \cup C^c)) &&\text{by the associative law for } \cap \\
&= (A \cap B^c) \cap (B \cup (C \cap C^c)) &&\text{by the distributive law} \\
&= (A \cap B^c) \cap (B \cup \emptyset) &&\text{by the intersection with} \\
&&&\text{the complement law} \\
&= (A \cap B^c) \cap B &&\text{by the union with } \emptyset \text{ law} \\
&= A \cap (B^c \cap B) &&\text{by the associative law for } \cap \\
&= A \cap (B \cap B^c) &&\text{by the commutative law for } \cap \\
&= A \cap \emptyset &&\text{by the intersection with} \\
&&&\text{the complement law} \\
&= \emptyset &&\text{by the intersection with } \emptyset \text{ law.}
\end{aligned}$

29. *Proof:* Let A and B be any sets. Then

$\begin{aligned}
(A - B) \cap B &= (A \cap B^c) \cap B &&\text{by the alternate representation} \\
&&&\text{for set difference law} \\
&= A \cap (B^c \cap B) &&\text{by the associative law for } \cap \\
&= A \cap (B \cap B^c) &&\text{by the commutative law for } \cap \\
&= A \cap \emptyset &&\text{by the intersection with the complement law} \\
&= \emptyset &&\text{by the intersection with } \emptyset \text{ law.}
\end{aligned}$

30. *Proof:* Let A be any set. By Theorem 5.2.1(1)(b), $A \cap \emptyset \subseteq \emptyset$, and by Theorem 5.3.1, $\emptyset \subseteq A \cap \emptyset$. Hence by definition of set equality, $A \cap \emptyset = \emptyset$.

31. *Proof:* Let sets A and B be given. By the union with the universal set law, $B \cup U = U$. Thus by the commutative law, $U \cup B = U$. Take the intersection of both sides with A to obtain $A \cap (U \cup B) = A \cap U$. But the left-hand side of this equation is $A \cap (U \cup B) = (A \cap U) \cup (A \cap B) = A \cup (A \cap B)$ by the distributive law and the intersection with the universal set law. The right-hand side of the equation equals A by the intersection with the universal set law. Hence $A \cup (A \cap B) = A$ as was to be shown.

32. *Proof:* Let sets A and B be given. By the intersection with the empty set law, $B \cup \emptyset = \emptyset$. Thus by the commutative law, $\emptyset \cap B = \emptyset$. Take the union of both sides with A to obtain $A \cup (\emptyset \cap B) = A \cup \emptyset$. But the left-hand side of this equation is $A \cup (\emptyset \cap B) = (A \cup \emptyset) \cap (A \cup B) = A \cap (A \cup B)$ by the distributive law and the union with the empty set law. The right-hand side of the equation equals A by the union with the empty set law. Hence $A \cap (A \cup B) = A$ as was to be shown.

33. When you use element arguments to solve parts (b) and (f) of this problem, you will notice that a certain line of reasoning occurs over and over again To avoid unnecessary repetition, we state the result as a lemma.

Lemma: For any sets A and B and any element x,

(1) $x \in A \oplus B \Leftrightarrow (x \in A$ and $x \notin B)$ or $(x \notin A$ and $x \in B)$

(2) $x \notin A \oplus B \Leftrightarrow (x \notin A$ and $x \notin B)$ or $(x \in A$ and $x \in B)$.

Proof:

(1) Suppose A and B are any sets and x is any element. Then

$$\begin{aligned} x \in A \oplus B &\Leftrightarrow x \in (A-B) \cup (B-A) & \text{by definition of } \oplus \\ &\Leftrightarrow x \in A-B \text{ or } x \in B-A & \text{by definition of } \cup \\ &\Leftrightarrow (x \in A \text{ and } x \notin B) \text{ or } (x \in B \text{ and } x \notin A) & \text{by definition of } \cap. \end{aligned}$$

(2) Suppose A and B are any sets and x is any element. Observe that there are only four possibilities for the relation of x to A and B: $x \in A$ and $x \notin B$, $x \notin A$ and $x \in B$, $x \in A$ and $x \in B$, and $x \notin A$ and $x \notin B$. In the first two cases $x \in A \oplus B$, and thus in the second two cases $x \notin A \oplus B$. Hence $x \notin A \oplus B \Leftrightarrow (x \in A$ and $x \in B)$ or $(x \notin A$ and $x \notin B)$.

More formally, observe that

$$\begin{aligned} (A \oplus B)^c &= [(A-B) \cup (B-A)]^c & \text{by definition of } \oplus \\ &= (A-B)^c \cap (B-A)^c) & \text{by De Morgan's law} \\ &= (A \cap B^c)^c \cap (B \cap A^c)^c) & \text{by the alternate representation} \\ & & \text{for set difference law} \\ &= [A^c \cup (B^c)^c] \cap [B^c \cup (A^c)^c] & \text{by De Morgan's law} \\ &= [A^c \cup B] \cap [B^c \cup A] & \text{by the double negative law} \\ &= [(A^c \cup B) \cap B^c] \cup [(A^c \cup B) \cap A] & \text{by the distributive law} \\ &= [B^c \cap (A^c \cup B)] \cup [A \cap (A^c \cup B)] & \text{by the commutative law for } \cap \\ &= [(B^c \cap A^c) \cup (B^c \cap B)] \cup [(A \cap A^c) \cup (A \cap B)] & \text{by the distributive law} \\ &= [(B^c \cap A^c) \cup (B \cap B^c)] \cup [(A \cap A^c) \cup (A \cap B)] & \text{by the commutative law for } \cap \\ &= [(B^c \cap A^c) \cup \emptyset] \cup [\emptyset \cup (A \cap B)] & \text{by the intersection with the} \\ & & \text{complement law} \\ &= [(B^c \cap A^c) \cup \emptyset] \cup [(A \cap B) \cup \emptyset] & \text{by the commutative law for } \cup \\ &= (B^c \cap A^c) \cup [(A \cap B)] & \text{by the union with } \emptyset \text{ law.} \end{aligned}$$

Thus for any element x, $x \notin A \oplus B \Leftrightarrow x \in (B^c \cap A^c) \cup [(A \cap B)]$. By definition of intersection, union, and complement (as in (1) above), this implies that $(x \in A$ and $x \in B)$ or $(x \notin A$ and $x \notin B)$.

b. *Proof*: Let A, B, and C be any sets. Then

$$\begin{aligned} x \in (A \oplus B) \oplus C &\Leftrightarrow (x \in A \oplus B \text{ and } x \notin C) \text{ or } (x \in C \text{ and } x \notin A \oplus B) \\ & \hspace{4cm} \text{by the lemma} \\ &\Leftrightarrow [(x \in A \text{ and } x \notin B) \text{ or } (x \in B \text{ and } x \notin A) \text{ and } x \notin C] \text{ or} \\ & \quad (x \in C \text{ and } [(x \in A \text{ and } x \in B) \text{ or } (x \notin A \text{ and } x \notin B)]) \\ & \hspace{4cm} \text{by the lemma} \\ &\Leftrightarrow [(x \in A \text{ and } x \notin B \text{ and } x \notin C) \text{ or } (x \in B \text{ and } x \notin A \text{ and } x \notin C)] \text{ or} \\ & \quad [(x \in C \text{ and } x \notin A \text{ and } x \notin B) \text{ or } (x \in C \text{ and } x \in A \text{ and } x \in B)] \\ & \hspace{4cm} \text{by the distributive law of logic} \\ &\Leftrightarrow x \text{ is in exactly one of the sets } A, B, \text{ and } C, \text{ or} \\ & \quad x \text{ is in all three of the sets } A, B, \text{ and } C. \end{aligned}$$

On the other hand,

$$\begin{aligned} x \in A \oplus (B \oplus C) &\Leftrightarrow (x \in A \text{ and } x \notin B \oplus C) \text{ or } (x \in B \oplus C \text{ and } x \notin A) \\ & \hspace{4cm} \text{by the lemma} \\ &\Leftrightarrow (x \in B \oplus C \text{ and } x \notin A) \text{ or } (x \in A \text{ and } x \notin B \oplus C) \\ & \hspace{4cm} \text{by the commutative law for } and. \end{aligned}$$

By exactly the same sequence of steps as in the first part of this proof but with B in place of A, C in place of B, and A in place of C, we deduce that

$x \in A \oplus (B \oplus C) \Leftrightarrow$ x is in exactly one of the sets A, B, and C, or
x is in all three of the sets A, B, and C.

So $x \in (A \oplus B) \oplus C \Leftrightarrow x \in A \oplus (B \oplus C)$, and hence $(A \oplus B) \oplus C = A \oplus (B \oplus C)$.

c. *Proof:* Let A be any set. Then

$$\begin{aligned} A \oplus \emptyset &= (A - \emptyset) \cup (\emptyset - A) & &\text{by definition of } \oplus \\ &= (A \cap \emptyset^c) \cup (\emptyset \cap A^c) & &\text{by the alternate representation for set difference law} \\ &= (A \cap U) \cup (A^c \cap \emptyset) & &\text{by the complement of } U \text{ law and} \\ & & &\text{the commutative law for } \cap \\ &= A \cup \emptyset & &\text{by the intersection with U and } \emptyset \text{ laws} \\ &= A & &\text{by the union with } \emptyset \text{ law.} \end{aligned}$$

d. *Proof:* Let A be any set. Then

$$\begin{aligned} A \oplus A^c &= (A - A^c) \cup (A^c - A) & &\text{by definition of } \oplus \\ &= (A \cap (A^c)^c) \cup (A^c \cap A^c) & &\text{by the alternate representation for set difference law} \\ &= (A \cap A) \cup A^c & &\text{by the double negative law and} \\ & & &\text{the idempotent law for } \cap \\ &= A \cup A^c & &\text{by the the idempotent law for } \cap \\ &= U & &\text{by the union with the complement law.} \end{aligned}$$

e. *Proof:* Let A be any set. Then

$$\begin{aligned} A \oplus A &= (A - A) \cup (A - A) & &\text{by definition of } \oplus \\ &= A - A & &\text{by the the idempotent law for } \cup \\ &= (A \cap A^c) & &\text{by the alternate representation for set difference law} \\ &= \emptyset & &\text{by the intersection with the complement law.} \end{aligned}$$

f. Let A, B, and C be any sets with $A \oplus C = B \oplus C$.

A ⊆ B: Suppose $x \in A$. Either $x \in C$ or $x \notin C$. If $x \in C$, then $x \in A$ and $x \in C$ and so by the lemma, $x \notin A \oplus C$. But $A \oplus C = B \oplus C$. Thus $x \notin B \oplus C$ either. So, again by the lemma, since $x \in C$ and $x \notin B \oplus C$, then $x \in B$. On the other hand, if $x \notin C$, then by the lemma, since $x \in A$, $x \in A \oplus C$. But $A \oplus C = B \oplus C$. So, again by the lemma, since $x \notin C$ and $x \in B \oplus C$, then $x \in B$. Hence in either case, $x \in B$ *as was to be shown.*

B ⊆ A: The proof is exactly the same as for $A \subseteq B$ with the letters A and B reversed.

Since $A \subseteq B$ and $B \subseteq A$, by definition of set equality $A = B$.

34. a. Yes because by definition of set difference, no element in $A - B$ can be in B whereas all elements in $B - C$ are in B. Hence there can be no elements simultaneously in the sets $A - B$ and $B - C$, and so these sets are disjoint.

    b. No. For example, let $A = \{1, 2\} = C$ and $B = \{2\}$. Then $A - B = C - B = \{1\} \neq \emptyset$.

    c. Yes because by definition of set difference, no element in $A - (B \cup C)$ can be in $B \cup C$, which implies (by definition of union) that no element in $A - (B \cup C)$ can be in B, whereas all elements in $B - (A \cup C)$ are in B. Hence there can be no elements simultaneously in the sets $A - (B \cup C)$ and $B - (A \cup C)$, and so these sets are disjoint.

    d. No. For example, let $A = \{1, 2\}$, $B = \{1, 3\}$, and $C = \{2, 3\}$. Then $A - (B \cap C) = \{1, 2\} - \{3\} = \{1, 2\}$ and $B - (A \cap C) = \{1, 3\} - \{2\} = \{1, 3\}$ and $\{1, 2\} \cap \{1, 3\} \neq \emptyset$.

35. b. Yes. Every element in $\{p, q, u, v, w, x, y, z\}$ is in one of the sets of the partition and no element is in more than one set of the partition.

    c. No. The number 4 is in both sets $\{5, 4\}$ and $\{1, 3, 4\}$.

    e. Yes. Every element in $\{1, 2, 3, 4, 5, 6, 7, 8\}$ is in one of the sets of the partition and no element is in more than one set of the partition.

37. Yes. Every real number x satisfies exactly one of the conditions: $x > 0$ or $x = 0$ or $x < 0$. (See property T16 of Appendix A.)

38. Yes. By the quotient-remainder theorem, every integer can be represented in exactly one of the following forms: $4k$ or $4k+1$ or $4k+2$ or $4k+3$ for some integer k.

39. Yes. Every string in A has length no greater than m. So $A \subseteq \Sigma^0 \cup \Sigma^1 \cup \ldots \cup \Sigma^m$. Conversely, every string in $\Sigma^0 \cup \Sigma^1 \cup \ldots \cup \Sigma^m$ is in Σ^* and has length no greater than m. So $\Sigma^0 \cup \Sigma^1 \cup \ldots \cup \Sigma^m \subseteq A$. Thus $A = \Sigma^0 \cup \Sigma^1 \cup \ldots \cup \Sigma^m$. The sets $\Sigma^0, \Sigma^1, \ldots, \Sigma^m$ are mutually disjoint because each string has one and only one length and so no string is in two distinct sets.

41. b. $X \times Y = \{(a,x), (a,y), (b,x), (b,y)\}$

$\mathcal{P}(X \times Y) = \{\emptyset, \{(a,x)\}, \{(a,y)\}, \{(b,x)\}, \{(b,y)\}, \{(a,x),(a,y)\}, \{(a,x),(b,x)\},$
$\{(a,x),(b,y)\}, \{(a,y),(b,x)\}, \{(a,y),(b,y)\}, \{(b,x),(b,y)\},$
$\{(a,x),(a,y),(b,x)\}, \{(a,x),(a,y),(b,y)\}, \{(a,y),b,y)\},$
$\{(a,x),(a,y),(b,x),(b,y)\}\}$

42. a. $\mathcal{P}(\emptyset) = \{\emptyset\}$

c. $\mathcal{P}(\mathcal{P}(\mathcal{P}(\emptyset))) = \{\emptyset, \{\emptyset\}, \{\{\emptyset\}\}, \{\emptyset, \{\emptyset\}\}\}$

43. a. False. *Counterexample*: Let $A = \{1\}$ and $B = \{2\}$. Then $\mathcal{P}(A) = \{\emptyset, \{1\}\}$ and $\mathcal{P}(B) = \{\emptyset, \{2\}\}$, and so $\mathcal{P}(A) \cup \mathcal{P}(B) = \{\emptyset, \{1\}, \{2\}\}$. However, $A \cup B = \{1, 2\}$, and so $\mathcal{P}(A \cup B) = \{\emptyset, \{1\}, \{2\}, \{1, 2\}\}$. Then $\{1, 2\} \in \mathcal{P}(A \cup B)$ but $\{1, 2\} \notin \mathcal{P}(A) \cup \mathcal{P}(B)$.

b. True.

Proof: Let A and B be sets.

$\mathcal{P}(A \cap B) \subseteq \mathcal{P}(A) \cap \mathcal{P}(B)$: Let $X \in \mathcal{P}(A \cap B)$. Then X is a subset of $A \cap B$ *[by definition of power set]*, and so every element in X is in both A and B. Thus $X \subseteq A$ and $X \subseteq B$ *[by definition of subset]*, and so $X \in \mathcal{P}(A)$ and $X \in \mathcal{P}(B)$ *[by definition of power set]*. Hence $X \in \mathcal{P}(A) \cap \mathcal{P}(B)$ *[by definition of intersection]*. Consequently, $\mathcal{P}(A \cap B) \subseteq \mathcal{P}(A) \cap \mathcal{P}(B)$ *[by definition of subset]*.

$\mathcal{P}(A) \cap \mathcal{P}(B) \subseteq \mathcal{P}(A \cap B)$: Let $X \in \mathcal{P}(A) \cap \mathcal{P}(B)$. Then $X \in \mathcal{P}(A)$ and $X \in \mathcal{P}(B)$ *[by definition of intersection]*. Hence $X \subseteq A$ and $X \subseteq B$ *[by definition of power set]*. So every element of X is in both A and B, and thus $X \subseteq A \cap B$ *[by definition of subset]*. It follows that $X \in \mathcal{P}(A \cap B)$ *[by definition of power set]*, and so $\mathcal{P}(A) \cap \mathcal{P}(B) \subseteq \mathcal{P}(A \cap B)$ *[by definition of subset]*.

[Since both subset containments have been proved, $\mathcal{P}(A \cap B) = \mathcal{P}(A) \cap \mathcal{P}(B)$ by definition of set equality.]

c. True.

Proof: Let A and B be sets and suppose $X \in \mathcal{P}(A) \cup \mathcal{P}(B)$. Then $X \in \mathcal{P}(A)$ or $X \in \mathcal{P}(B)$ *[by definition of union]*. In case $X \in \mathcal{P}(A)$, then $X \subseteq A$ *[by definition of power set]*, and so $X \subseteq A \cup B$ *[by definition of union]*. In case $X \in \mathcal{P}(B)$, then $X \subseteq B$ *[by definition of power set]*, and so $X \subseteq A \cup B$ *[by definition of union]*. Thus in either case, $X \subseteq A \cup B$, and so $X \in \mathcal{P}(A \cup B)$ *[by definition of power set]*. Hence $\mathcal{P}(A) \cup \mathcal{P}(B) \subseteq \mathcal{P}(A \cup B)$ *[by definition of subset]*.

d. False. The elements of $\mathcal{P}(A \times B)$ are subsets of $A \times B$, whereas the elements of $\mathcal{P}(A) \times \mathcal{P}(B)$ are ordered pairs whose first element is a subset of A and whose second element is a subset of B. To be concrete, let $A = B = \{1\}$. Then $\mathcal{P}(A) = \{\emptyset, \{1\}\}$, $\mathcal{P}(B) = \{\emptyset, \{1\}\}$, and $\mathcal{P}(A) \times \mathcal{P}(B) = \{(\emptyset, \emptyset), (\emptyset, \{1\}), (\{1\}, \emptyset), (\{1\}, \{1\})\}$. On the other hand, $A \times B = \{(1,1)\}$, and so $\mathcal{P}(A \times B) = \{\emptyset, \{(1,1)\}\}$. By inspection $\mathcal{P}(A) \times \mathcal{P}(B) \neq \mathcal{P}(A \times B)$.

44. $S_0 = \emptyset$, $S_1 = \{\{a\}, \{b\}, \{c\}\}$, $S_2 = \{\{a,b\}, \{a,c\}, \{b,c\}\}$, $S_3 = \{\{a,b,c\}\}$. Yes, $\{S_0, S_1, S_2, S_3\}$ is a partition of $\mathcal{P}(S)$ because the sets S_0, S_1, S_2, and S_3 are mutually disjoint and their union is $\mathcal{P}(S)$.

45. No. The sets S_a, S_b, S_c, and $S_\emptyset$ are not mutually disjoint. For example, $\{a,b\} \in S_a$ and $\{a,b\} \in S_b$.

46. b. $S_2 = \{\{w\}, \{t,w\}, \{u,w\}, \{v,w\}, \{t,u,w\}, \{t,v,w\}, \{u,v,w\}, \{t,u,v,w\}\}$

    c. Yes

    d. S_1 and S_2 each have eight elements.

    e. $S_1 \cup S_2$ has sixteen elements.

    f. $S_1 \cup S_2 = \mathcal{P}(A)$.

47. This is true.

    Theorem: For any positive integer $n \geq 2$, let S be the set of all nonempty subsets of $\{2, 3, \ldots, n\}$, and for each $S_i \in S$, let P_i be the product of all the elements in S_i. Then

    $$\sum_{i=1}^{2^{n-1}-1} P_i = \frac{(n+1)!}{2} - 1.$$

    Note: S has $2^{n-1} - 1$ elements, and $\sum_{i=1}^{2^{n-1}-1} P_i$ equals the sum of all products of elements of nonempty subsets of $\{2, 3, \ldots, n\}$.

    Proof (by mathematical induction):

    The formula is true for n = 2: For $n = 2$, $S = \{\{2\}\}$ and there is only one element of S, namely $S_1 = \{2\}$. Then $P_1 = 2$, and so the left-hand side of the equation equals $\sum_{i=1}^{2^1-1} P_i = \sum_{i=1}^{1} P_i = P_1 = 2$. The right-hand side of the equation is $\frac{(2+1)!}{2} - 1 = \frac{3!}{2} - 1 = 3 - 1 = 2$ also. Hence the equation holds for $n = 2$.

    If the formula is true for n = k then it is true for n = k+1: Let k be an integer such that $k \geq 2$, let S be the set of all nonempty subsets of $\{2, 3, \ldots, k\}$, and for each $S_i \in S$, let P_i be the product of all the elements in S_i. Suppose the equation holds for $n = k$. In other words, suppose $\sum_{i=1}^{2^{k-1}-1} P_i = \frac{(k+1)!}{2} - 1$. *[This is the inductive hypothesis.]* Let S' be the set of all nonempty subsets of $\{2, 3, \ldots, k+1\}$, and for each $S'_i \in S'$, let P'_i be the product of all the elements in S'_i. Now any subset of $\{2, 3, \ldots, k+1\}$ either contains $k+1$ or does not contain $k+1$. Any subset of $\{2, 3, \ldots, k+1\}$ that does not contain $k+1$ is a subset of $\{2, 3, \ldots, k\}$, and any subset of $\{2, 3, \ldots, k+1\}$ that contains $k+1$ is the union of a subset of $\{2, 3, \ldots, k\}$ and $\{k+1\}$. Also by Theorem 5.3.6, there are $2^{k-1} - 1$ nonempty subsets of $\{2, 3, \ldots, k\}$ *[because there are $k-1$ elements in the set $\{2, 3, \ldots, k\}$]*, and by the same reasoning there are $2^k - 1$ nonempty subsets of $\{2, 3, \ldots, k+1\}$. Thus

    $$\sum_{i=1}^{2^k-1} P'_i = \sum_{i=1}^{2^{k-1}-1} P_i + \begin{bmatrix} \text{the sum of all products of elements of nonempty} \\ \text{subsets of } \{2, 3, \ldots, k+1\} \text{ that contain } k+1 \end{bmatrix}.$$

    But given any nonempty subset A of $\{2, 3, \ldots, k+1\}$ that contains $k+1$, the product of all the elements of A equals $k+1$ times the product of the elements of $A - \{k+1\}$. Thus the sum of all products of elements of nonempty subsets of $\{2, 3, \ldots, k+1\}$ that contain $k+1$ equals the product of $k+1$ times the sum of all products of elements of nonempty subsets of $\{2, 3, \ldots, k\}$ plus 1 *[for the case $A = \{k+1\}$ for which $A - \{k+1\}$ equals the empty set]*. Hence

    $$\sum_{i=1}^{2^k-1} P'_i = \sum_{i=1}^{2^{k-1}-1} P_i + (k+1) \cdot \left(\sum_{i=1}^{2^{k-1}-1} P_i + 1 \right)$$

$$= (k+2) \cdot \sum_{i=1}^{2^{k-1}-1} P_i + (k+1)$$
$$= (k+2) \cdot (\frac{(k+1)!}{2} - 1) + (k+1) \quad \text{by inductive hypothesis}$$
$$= \frac{(k+2)!}{2} - (k+2) + (k+1)$$
$$= \frac{(k+2)!}{2} - 1.$$

Thus the equation holds for $n = k+1$.

48. c. (i) *Proof:* For all elements x in B,

$x + 1$	$= x + (x + \bar{x})$	by the complement law for $+$
	$= (x + x) + \bar{x}$	by the associative law for $+$
	$= x + \bar{x}$	by part (a)(ii)
	$= 1$	by the complement law for $+$.

(ii) *Proof:* For all elements x in B,

$x \cdot 0$	$= x \cdot (x \cdot \bar{x})$	by the complement law for $\cdot$
	$= (x \cdot x) \cdot \bar{x}$	by the associative law for $\cdot$
	$= x \cdot \bar{x}$	by part (a)(i)
	$= 0$	by the complement law for $\cdot$.

d. (i) *Proof:* For all elements x and y in B,

$(x + y) \cdot x$	$= x \cdot (x + y)$	by the commutative law for $\cdot$
	$= (x \cdot x) + (x \cdot y)$	by the distributive law for $\cdot$ over $+$
	$= x + (x \cdot y)$	by part (a)(i)
	$= (x \cdot 1) + (x \cdot y)$	because 1 is an identity for $\cdot$
	$= x \cdot (1 + y)$	by the distributive law for $\cdot$ over $+$
	$= x \cdot (y + 1)$	by the commutative law for $+$
	$= x \cdot 1$	by part (c)(i)
	$= x$	because 1 is an identity for $\cdot$.

(ii) *Proof:* For all elements x in B,

x	$= (x + y) \cdot x$	by part (d)(i)
	$= x \cdot (x + y)$	by the commutative law for $\cdot$
	$= (x \cdot x) + (x \cdot y)$	by the distributive law for $\cdot$ over $+$
	$= x + (x \cdot y)$	by part (a)(i)
	$= (x \cdot y) + x$	by the commutative law for $+$.

e. *Proof:* Let x, y, and z be any elements in B such that $x + y = x + z$ and $x \cdot y = x \cdot z$. Then

y	$= (y + x) \cdot y$	by part (d)(i)
	$= (x + y) \cdot y$	by the commutative law for $+$
	$= (x + z) \cdot y$	by hypothesis
	$= (x \cdot y) + (z \cdot y)$	by the distributive law for $\cdot$ over $+$
	$= (x \cdot z) + (z \cdot y)$	by hypothesis
	$= (z \cdot x) + (z \cdot y)$	by the commutative law for $\cdot$
	$= z \cdot (x + y)$	by the distributive law for $\cdot$ over $+$
	$= z \cdot (x + z)$	by hypothesis
	$= (z + x) \cdot z$	by the commutative laws for $\cdot$ and $+$
	$= z$	by part (d)(i).

Section 5.4

3. This statement contradicts itself. If it were true, then because it declares itself to be a lie, it would be false. Consequently, it is not true. On the other hand, if it were false, then it would be false that "the sentence in this box is a lie," and so the sentence would be true. Consequently, the sentence is not false. Thus the sentence is neither true nor false, which contradicts the definition of a statement. Hence the sentence is not a statement.

4. Since there are no real numbers with negative squares, this sentence is vacuously true, and hence it is a statement.

6. *a.* Assuming that the sentence "If this sentence is true, then $1+1=3$" is a statement, then the sentence is either true or false. By definition of truth values for an if-then statement, the only way the sentence can be false is for its hypothesis to be true and its conclusion false. But if its hypothesis is true, then the sentence is true and therefore it is not false. So it is impossible for the sentence to be false, and hence it is true. Consequently, what its hypothesis asserts is true, and so (again by definition of truth values for if-then statements) its conclusion must also be true. Therefore, $1+1=3$.

   b. We can deduce that "This sentence is true" is not a statement. For if it were a statement, then since $1+1=3$ is also a statement, the sentence "If this sentence is true, then $1+1=3$" would also be a statement. It would then follow by part (a) that $1+1=3$, which we know to be false. So "This sentence is true" is not a statement.

7. Suppose Nixon says (ii) and the only utterance Jones makes about Watergate is (i). Suppose also that apart from (ii) all of Nixon's other assertions about Watergate are evenly split between true and false.

   Case 1 (Statement (i) is true): In this case, more than half of Nixon's assertions about Watergate are false, and so (since all of Nixon's other assertions about Watergate are evenly split between true and false) statement (ii) must be false (because it is an assertion about Watergate). So at least one of Jones' statements about Watergate is false. But the only statement Jones makes about Watergate is (i). So statement (i) is false.

   Case 2 (Statement (i) is false): In this case, half or more of Nixon's assertions about Watergate are true, and so (since all of Nixon's other assertions about Watergate are evenly split between true and false) statement (ii) must be true. But statement (ii) asserts that everything Jones says about Watergate is true. And so, in particular, statement (i) is true.

   The above arguments show that under the given circumstances, statements (i) and (ii) are contradictory.

9. No. Suppose there were such a book. If such a book did not refer to itself, then it would belong to the set of all books that do not refer to themselves. But it is supposed to refer to all books in this set, and so it would refer to itself. On the other hand, if such a book referred to itself, then it would belong to the set of books to which it refers and this set only contains books that do not refer to themselves. Thus it would not refer to itself. It follows that the assumption that such a book exists leads to a contradiction, and so there is no such book.

10. The answer is neither yes nor no. (In other words, the definition of heterological is inherently contradictory.) For if heterological were heterological, then it would describe itself. But by definition of heterological, this would mean that it would not describe itself. Hence it is impossible for heterological to be heterological. On the other hand, if heterological were not heterological, then it would not describe itself. But by definition of heterological this would mean that it would be a heterological word, and so it would be heterological. Hence it is impossible for heterological to be not heterological. These arguments show that heterological is neither heterological nor not heterological.

11. Suppose n exists. Then n is "the smallest integer not describable in fewer than 12 English words." But the description of n contains only 11 words. So n is describable in fewer than 12 English words, which is a contradiction. (Comment: Since the total number of strings consisting of 11 or fewer English words is finite, the number of such strings that describe integers would appear to be a finite number. Thus, it would seem reasonable that the number of integers described by such strings would be finite, and hence that there would be a largest such integer. But that would imply that there would exist an integer n that would be "the smallest integer not describable in fewer than 12 English words." Curious!)

12. There is no such algorithm.

    Proof: Suppose there were an algorithm, call it CheckPrint, such that if a fixed quantity a, an algorithm X, and a data set D are input to it, then

    CheckPrint(a, X, D) prints

    "yes" if X prints a when it is run with data set D

    "no" if X does not print a when it is run with data set D.

    Let SignalHalt be an algorithm that operates on an algorithm X and a data set D as follows: SignalHalt runs X with D and prints "halts" if X terminates. If X does not terminate, SignalHalt does not terminate either. Observe that

    CheckPrint("halts", SignalHalt, (X, D)) prints

    "yes" if SignalHalt prints "halts" when it is run with data set (X, D)

    "no" if SignalHalt does not print "halts" when it is run with data set (X, D).

    Thus, we may define a new algorithm CheckHalt, whose input is an algorithm X and a data set D, as follows:

    CheckHalt(X, D) prints

    "halts" if CheckPrint("halts", SignalHalt, (X, D)) prints "yes"

    "loops forever" if CheckPrint("halts", SignalHalt, (X, D)) prints "no".

    The above discussion shows that if there is an algorithm CheckPrint which, for a fixed quantity a, an input algorithm X, and a data set D, can determine whether X prints a when run with data set D, then there is an algorithm CheckHalt that solves the halting problem. Since there is no algorithm that solves the halting problem, there is no algorithm with the property described.

Chapter 6 Counting

The primary aim of this chapter is to help students develop an intuitive understanding of the fundamental counting principles and an ability to apply them in a wide variety of situations. The chapter also introduces the notion of probability and gives students practice in computing probabilities of events in situations where all possible outcomes are equally likely.

Students seem most successful in solving counting problems when they have a clear mental image of the objects they are counting. It is helpful to encourage them to get into the habit of beginning a counting problem by listing (or at least imagining) some of the objects they are trying to count. If they see that all the objects to be counted can be matched up with the integers from m to n inclusive, then the total is $n - m + 1$ (Section 6.1). If they see that all the objects can be produced by a multi-step process, then the total can be found by counting the distinct paths from root to leaves in a possiblility tree that shows the outcomes of each successive step (Section 6.2). And in case each step of the process can be performed in a fixed number of ways (regardless of how the previous steps were performed), the total can be calculated by applying the multiplication rule (Section 6.2). If they see that the objects to be counted can be separated into disjoint categories, then the total is just the sum of the subtotals for each category (Section 6.3). And in case the categories are not disjoint, the total can be counted using the inclusion/exclusion rule (Section 6.3). If they see that the objects to be counted can be represented as all the subsets of size r of a set with n elements, then the total is $\binom{n}{r}$ for which there is a computational formula (Section 6.4). And if the objects can be represented as all the multisets of size r of a set with n elements, then the total is $\binom{n+r-1}{r}$.

The chapter ends with a discussion of Pascal's formula (Section 6.6) and the binomial theorem (Section 6.7), each of which is proved both algebraically and combinatorially. Exercise 21 of Section 6.4 is a warm-up for the combinatorial proof of the binomial theorem, and exercise 6 of Section 6.7 is intended to help students perceive how Pascal's formula is applied in the algebraic proof of the binomial theorem.

Section 6.1

3. *b.* {10♣, J♣, Q♣, K♣, A♣, 10♦, J♦, Q♦, K♦, A♦, 10♥, J♥, Q♥, K♥, A♥, 10♠, J♠, Q♠, K♠, A♠}
   probability $= 20/52 = 5/13 \cong 38.5\%$

4. *b.* {11, 22, 33, 44, 55}   probability $= 6/36 = 1/6 \cong 16.7\%$
   c. {11, 12, 13, 14, 21, 22, 23, 31, 32, 41}   probability $= 10/36 = 5/18 \cong 27.8\%$

5. *b.* (ii) {HHT, HTH, THH, HHH}   probability $= 4/8 = 1/2 = 50\%$
   (iii) {TTT}   probability $= 1/8 = 12.5\%$

6. *b.* (ii) {GGB, GBG, BGG, GGG}   probability $= 4/8 = 1/2 = 50\%$
   (iii) {BBB}   probability $= 1/8 = 12.5\%$

7. *b.* (ii) {CCW, CWC, WCC, CCC}   probability $= 4/8 = 1/2 = 50\%$
   (iii) {WWW}   probability $= 1/8 = 12.5\%$

8. *b.* $4/8 = 1/2 = 50\%$
   c. $1/8 = 12.5\%$

9. The methods used to compute the probabilities in exercises 6, 7, and 8 are exactly the same as those in exercise 5. The only difference in the solutions are the symbols used to denote the outcomes; the probabilities are identical. These exercises illustrate the fact that computing

various probabilities that arise in connection with tossing a coin is mathematically identical to computing probabilities in other, more realistic situations. So if the coin tossing model is completely understood, many other probabilities can be computed without difficulty.

11. *a.*

$$
\begin{array}{cccc}
100\ 101\ 102\ 103\ 104\ 105\ 106\ 107\ 108\ldots & 990\ 991\ 992\ 993\ 994\ 995\ 996\ 997\ 998\ 999 \\
\updownarrow \qquad\qquad \updownarrow \qquad\qquad \updownarrow & \updownarrow \\
6\cdot 17 \qquad\qquad 6\cdot 18 \qquad 6\cdot 165 & 6\cdot 166
\end{array}
$$

The above diagram shows that there are as many three-digit integers that are multiples of 6 as there are integers from 17 to 166 inclusive. But by Theorem 6.1.1, there are $166 - 17 + 1 = 150$ such integers.

b. The probability that a randomly chosen three-digit integer is a multiple of 6 is $150/(999-100+1) = 150/900 = 1/6 \cong 16.7\%$.

12. *a.* n

    b. $19 - 4 + 1 = 16$

13. *a.* When n is even, $\lfloor n/2 \rfloor = n/2$, and so the answer is $n/2$.

14. *a.* When n is odd, $\lfloor n/2 \rfloor = (n-1)/2$, and so the answer is $n - (n-1)/2 + 1 = (2n - n + 1 + 2)/2 = (n+3)/2$.

16. Let k be the 62nd integer in the list. By Theorem 6.1.1, $k - 29 + 1 = 62$, and so $k = 62 + 29 - 1 = 90$.

18. Let m be the smallest of the integers. By Theorem 6.1.1, $326 - m + 1 = 87$, and so $m = 326 - 87 + 1 = 240$.

19.

$$
\begin{array}{cccccccccc}
1 & 2 & 3 & 4 & 5 & 6 & \ldots & 998 & 999 & 1000 & 1001 \\
& \updownarrow & & \updownarrow & & \updownarrow & & \updownarrow & & \updownarrow \\
& 2\cdot 1 & & 2\cdot 2 & & 2\cdot 3 & & 2\cdot 499 & & 2\cdot 500
\end{array}
$$

The above diagram shows that there are as many even integers between 1 and 1001 as there are integers from 1 to 500 inclusive. There are 500 such integers.

20. *b.*

$$
\begin{array}{cccccccccccc}
\text{M} & \text{Tu} & \text{W} & \text{Th} & \text{F} & \text{Sa} & \text{Su} & \text{M} & \text{Tu} & \ldots & \text{F} & \text{Sa} & \text{Su} & \text{M} \\
1 & 2 & 3 & 4 & 5 & 6 & 7 & 8 & 9 & & 362 & 363 & 364 & 365 \\
\updownarrow & & & & & & & \updownarrow & & & & & & \updownarrow \\
7\cdot 0+1 & & & & & & & 7\cdot 1+1 & & & & & & 7\cdot 52+1
\end{array}
$$

In the above diagram Mondays occur on days numbered $7 \cdot k + 1$ where k is an integer from 0 to 52 inclusive. Thus there are as many Mondays in the year as there are such integers, namely $52 - 0 + 1 = 53$ of them.

22. Let S be any set consisting entirely of integers from 1 through 100, and suppose that no integer in S divides any other integer in S. Write each integer in S as $2^k \cdot m$, where m is an odd integer. Now consider any two such integers in S, say $2^r \cdot a$ and $2^s \cdot b$. Observe that $a \neq b$. The reason is that if $a = b$, then whichever integer contains the fewer number of factors of 2 divides the other integer. (For example, $2^2 \cdot 3 \mid 2^4 \cdot 3$.) Thus there can be no more integers in S than there are distinct odd integers from 1 through 100, namely 50. Furthermore, it is possible to find a set T of 50 integers from 1 through 100 no one of which divides any other. For instance, $T = \{51, 52, 53, \ldots, 99, 100\}$. Hence the largest number of elements that a set of integers from 1 through 100 can have so that no one element in the set is divisible by any other is 50.

23. *Proof:* Let m be any integer. We prove the theorem by induction on n.

The formula is true for n = m: There is just one integer, m, from m to m inclusive. So the number of such integers equals $1 = m - m + 1$ *[as was to be shown]*.

If the formula is true for n = k, then it is true for n = k+1: Suppose that the number of integers from m to k inclusive is $k - m + 1$. *[This is the inductive hypothesis.]* We must show that the number of integers from m to $k + 1$ inclusive is $m - (k + 1) + 1$.

Consider the sequence of integers from m to $k + 1$ inclusive:

$$\underbrace{m, \quad m+1, \quad m+2, \quad \ldots, \quad k,}_{k-m+1} \quad (k+1).$$

By inductive hypothesis there are $k - m + 1$ integers from m to k inclusive. So there are $(k - m + 1) + 1$ integers from m to $k + 1$ inclusive. But $(k - m + 1) + 1 = (k + 1) - m + 1$. So there are $(k + 1) - m + 1$ integers from m to $k + 1$ inclusive *[as was to be shown]*.

Section 6.2

2.

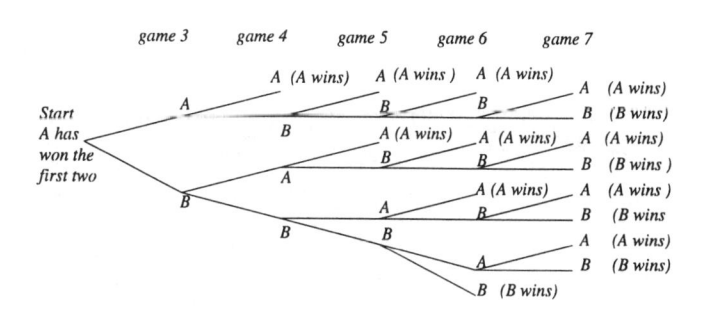

There are fifteen ways to complete the series.

5.

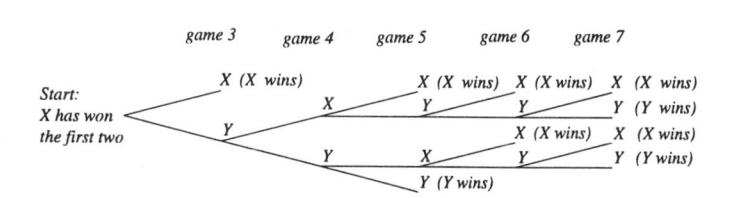

There are nine ways to play the competition.

7. a.

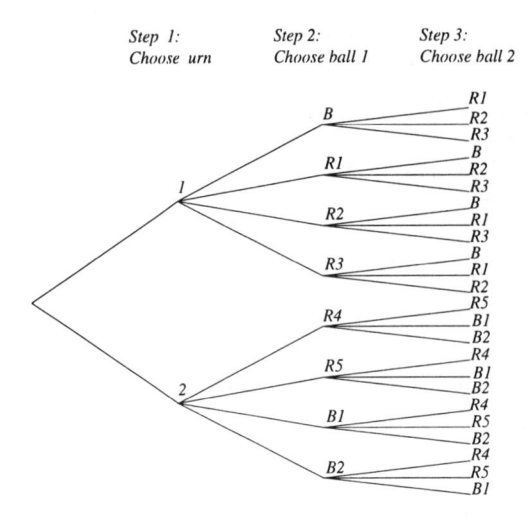

```
Step 1:          Step 2:          Step 3:
Choose urn       Choose ball 1    Choose ball 2
```

(tree diagram)

b. There are 24 outcomes of this experiment.

c. The probability that two red balls are chosen is $8/24 = 1/3$.

10. Sketch:

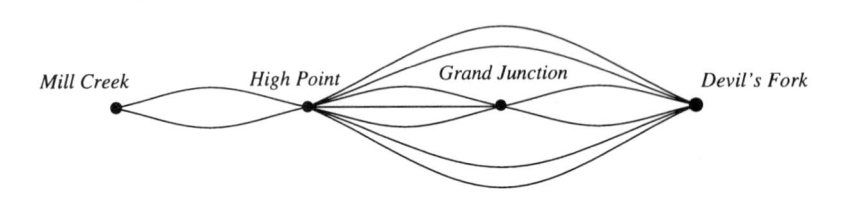

(graph with Mill Creek, High Point, Grand Junction, Devil's Fork)

a. $2 \cdot 3 \cdot 2 = 12$

b. $2 \cdot 4 = 8$

11. c. If a bit string of length 8 begins and ends with a 1, then the six middle positions can be filled with any bit string of length 6. Hence there are $2^6 = 64$ such strings.

d. $2^8 = 256$

12. c. $8 \cdot 16 = 128$

13. c. There are four outcomes in which exactly one head occurs: $HTTT, THTT, TTHT, TTTH$. So the probability of obtaining exactly one head is $4/16 = 1/4$.

14. c. $10 \cdot 10 \cdot 10 = 1000$

e. $24 \cdot 10 \cdot 9 \cdot 8 = 17,280$

15. a. $20^3 = 8,000$

b. $20 \cdot 19 \cdot 18 = 6,840$

16. *b.* Constructing a PIN that is obtainable by the same keystroke sequence as 6809 can be thought of as the following four-step process. Step 1 is to choose either the digit 6 or one of the three letters on the same key as the digit 6, step 2 is to choose either the digit 8 or one of the three letters on the same key as the digit 8, step 3 is to choose the digit 0, and step 4 is to choose either the digit 9 or one of the three letters on the same key as the digit 9. There are four ways to perform steps 1, 2, and 4 and one way to perform step 3. So by the multiplication rule there are $4 \cdot 4 \cdot 1 \cdot 4 = 64$ different PINs that are keyed the same as 6809.

c. Constructing a numeric PIN with no repeated digit can be thought of as the following four-step process. Step one is to choose the left-most digit, step two is to choose the next digit to the right, step three is to choose the next digit to the right of the one chosen in step two, and step four is to choose the right-most digit. Because no digit may be repeated, there are 10 ways to perform step one, 9 ways to perform step two, 8 ways to perform step three, and 7 ways to perform step four. Thus the number of numeric PINs with no repeated digit is $10 \cdot 9 \cdot 8 \cdot 7 = 5040$.

18. *b.* There are 10 ways to perform step one, 22 ways to perform step two (because we may choose any of the thirteen letters from N through Z or any of the nine digits not chosen in step 1), 34 ways to perform step three (because we may not use either of the two previously used symbols), and 33 ways to perform step four (because we may not use either of the three previously used symbols). So the total number of PINs is $10 \cdot 22 \cdot 34 \cdot 33 = 246,840$.

20. *a.* $9 \cdot 10 \cdot 10 = 900$, or $999 - 100 + 1 = 900$

    b. $9 \cdot 10 \cdot 5 = 450$

    c. $9 \cdot 9 \cdot 8 = 648$

    d. Step 1 is to pick the right-most digit, step 2 is to pick the left-most digit, and step 3 is to pick the middle digit. The answer is $5 \cdot 8 \cdot 8 = 320$.

    e. The probability that a randomly chosen three-digit integer has distinct digits is $648/900 = 72\%$. The probability that a randomly chosen three-digit integer has distinct digits and is odd is $320/900 \cong 35.6\%$.

22. $m \cdot n$

23. $p \cdot n \cdot m$

25. $(d - c + 1)(b - a + 1)$

26. Use five digits to represent each number from 1 through 99,999 by adding leading 0's as necessary. Constructing a five-digit number can then be thought of as placing five digits into five adjacent positions. Imagine constructing a number containing one each of the digits 2, 3, 4, and 5 as the following five-step process: Step one is to choose a position for the 2, step two is to choose a position for the 3, step three is to choose a position for the 4, step four is to choose a position for the 5, and step five is to choose an unused digit to fill in the remaining position. There are 5 ways to perform step one, 4 ways to perform step two, 3 ways to perform step three, 2 ways to perform step four, and 6 ways to perform step five (because there are six digits not equal to 2, 3, 4, or 5). So there are $5 \cdot 4 \cdot 3 \cdot 2 \cdot 6 = 720$ numbers containing one each of the digits 2, 3, 4, and 5.

27. *a.* Call one of the integers r and the other s. Since r and s have no common factors, if p_i is a factor of r, then p_i is not a factor of s. So for each $i = 1, 2, \ldots, m$, either $p_i^{k_i}$ is a factor of r or $p_i^{k_i}$ is a factor of s. Thus constructing r can be thought of as an m-step process in which step i is to decide whether $p_i^{k_i}$ is a factor of r or not. There are two ways to perform each step, and so the number of different possible r's is 2^m. Observe that once r is specified, s is completely determined because $s = n/r$. Hence the number of ways n can be written as a product of two positive integers $r \cdot s$ which have no common factors is 2^m. Note that this analysis assumes that

order matters because, for instance, $r = 1$ and $s = n$ will be counted separately from $r = n$ and $s = 1$.

b. Each time that we can write n as $r \cdot s$, where r and s have no common factors, we can also write $n = s \cdot r$. So if order matters, there are twice as many ways to write n as a product of two integers with no common factors as there are if order does not matter. Thus if order does not matter, there are $2^m/2 = 2^{m-1}$ ways to write n as a product of two integers with no common factors.

28. c. A divisor of $p^a q^b r^c$ is any one of the $(a+1)(b+1)$ divisors of $p^a q^b$ counted in part (b) times any one of the $c+1$ numbers $1, r, r^2, \ldots, r^c$. So by the multiplication rule, there are $(a+1)(b+1)(c+1)$ divisors in all.

d. By the multiplication rule, the answer is $(a_1+1)(a_2+1)(a_3+1)\ldots(a_m+1)$. (A full formal proof would use mathematical induction.)

e. Because $12 = 2 \cdot 2 \cdot 3 = 4 \cdot 3 = 6 \cdot 2$, by part (d) the possibilities are $k_1 = 1$, $k_2 = 1$, and $k_3 = 2$ (of which the smallest example is $2^2 \cdot 3^1 \cdot 5^1 = 60$), or $k_1 = 3$, $k_2 = 2$ (of which the smallest example is $2^3 \cdot 3^2 = 72$), or $k_1 = 5$, $k_2 = 1$ (of which the smallest example is $2^5 \cdot 3^1 = 96$). The smallest of the numbers obtained is 60, and so the answer is 60. (The twelve divisors of 60 are 1, 2, 3, 4, 5, 6, 10, 12, 15, 20, 30, and 60.)

29. c. $7! = 5040$

30. a. $6! = 720$ b. $5! = 120$ c. $3! = 6$

33. abc, abd, acb, acd, adb, adc, bac, bad, bca, bcd, bda, bdc, cab, cad, cba, cbd, cda, cdb, dab, dac, dba, dbc, dca, dcb

34. b. $P(6,6) = 6!/(6-6)! = 6!/0! = 6!/1 = 720$

c. $P(6,2) = 6!/(6-2)! = 6!/4! = 6 \cdot 5 = 30$

d. $P(6,1) = 6!/(6-1)! = 6!/5! = 6$

35. b. $P(7,2) = 7!/(7-2)! = 7!/5! = 7 \cdot 6 = 42$

36. b. $P(9,5) = 9!/(9-5)! = 9!/4! = 9 \cdot 8 \cdot 7 \cdot 6 \cdot 5 = 15,120$

d. $P(7,3) = 7!/(7-3)! = 7!/4! = 7 \cdot 6 \cdot 5 = 210$

37. *Proof:* Let n be any integer such that $n \geq 2$. By Theorem 6.2.3,

$$P(n+1, 3) = \frac{(n+1)!}{((n+1)-3)!} = \frac{(n+1)n(n-1)(n-2)!}{(n-2)!} = n^3 - n.$$

39. *Proof:* Let n be any integer such that $n \geq 3$. By Theorem 6.2.3,

$$\begin{aligned}
P(n+1,3) - P(n,3) &= \frac{(n+1)!}{((n+1)-3)!} - \frac{n!}{(n-3)!} \\
&= \frac{(n+1)!}{(n-2)!} - \frac{n!}{(n-3)!} \\
&= \frac{(n+1) \cdot n!}{(n-2)!} - \frac{(n-2) \cdot n!}{(n-2) \cdot (n-3)!} \\
&= \frac{n!((n+1)-(n-2))}{(n-2)!} \\
&= \frac{n!}{(n-2)!} \cdot 3 \\
&= 3P(n,2).
\end{aligned}$$

40. *Proof:* Let n be any integer such that $n \geq 2$. By Theorem 6.2.3,

$$P(n,n) = \frac{n!}{(n-n)!} = \frac{n!}{0!} = \frac{n!}{1} = n!.$$

On the other hand,

$$P(n, n-1) = \frac{n!}{(n-(n-1))!} = \frac{n!}{1!} = \frac{n!}{1} = n!$$

also. Hence $P(n,n) = P(n, n-1)$.

41. *Proof:* For each integer $k \geq 1$, let $P(k)$ be the property "If an operation consists of k steps and the first step can be performed in n_1 ways, the second step can be performed in n_2 ways, ..., and the kth step can be performed in n_k ways, then the entire operation can be performed in $n_1 \cdot n_2 \cdot \ldots \cdot n_k$ ways." We will prove by mathematical induction that the property is true for all integers $k \geq 1$.

The property is true for k = 1: If an operation consists of one step that can be performed in n_1 ways, then the entire operation can be performed in n_1 ways.

The property is true for k = 2 : Suppose an operation consists of two steps and the first step can be performed in n_1 ways and the second step can be performed in n_2 ways. Each of the n_1 ways of performing the first step can be paired with one of the n_2 ways of performing the second step. Thus the total number of ways to perform the entire operation is $\underbrace{n_2 + n_2 + \ldots + n_2}_{n_1 \text{ terms}}$, which equals $n_1 \cdot n_2$.

If the property is true for k = i then it is true for k = i + 1 : Let k be an integer such that $k \geq 2$ and suppose that $P(i)$ is true. *[This is the inductive hypothesis.]* Consider an operation that consists of $i + 1$ steps where the first step can be performed in n_1 ways, the second step can be performed in n_2 ways, ..., the ith step can be performed in n_i ways, and the $(i+1)st$ step can be performed in n_{i+1} ways. This operation can be regarded as a two-step operation in which the first step is an i-step operation that consists of the original first i steps and the second step is the original $(i+1)st$ step. By inductive hypothesis, the first step of the operation can be performed in $n_1 \cdot n_2 \cdot \ldots \cdot n_i$ ways and by assumption the second step can be performed in n_{i+1} ways. Therefore, by the same argument used to establish the case $k = 2$ above, the entire operation can be performed in $(n_1 \cdot n_2 \cdot \ldots \cdot n_i) \cdot n_{i+1} = n_1 \cdot n_2 \cdot \ldots \cdot n_i \cdot n_{i+1}$ ways. This is what was to be shown.

42. *Proof:* For each integer $n \geq 1$, let $P(n)$ be the property "The number of permutations of a set with n elements is $n!$."

The property is true for n = 1: If a set consists of one element there is just one way to order it, and $1! = 1$.

If the property is true for n = k then it is true for n = k + 1 : Let k be an integer with $k \geq 1$ and suppose that the number of permutations of a set with k elements is $k!$. *[This is the inductive hypothesis.]* Let X be a set with $k+1$ elements. The process of forming a permutation of the elements of X can be considered a two-step operation where step 1 is to choose the element to write first and step 2 is to write the remaining elements of X in some order. Since X has $k + 1$ elements, there are $k + 1$ ways to perform step 1, and by inductive hypothesis there are $k!$ ways to perform step 2. Hence by the product rule there are $(k+1) \cdot k! = (k+1)!$ ways to form a permutation of the elements of X. But this means that there are $(k+1)!$ permutations of X.

43. *Proof:* Let n be any integer with $n \geq 1$. We will prove the following statement by mathematical induction on r: For each integer r with $1 \leq r \leq n$, $P(n,r) = n!/(n-r)!$.

The formula holds for r = 1: For $r = 1$, $P(n, r)$ is the number of 1-permutations of a set with n elements which equals n because there are n ways to choose one element from a set with n elements. But when $r = 1$, $n!/(n - r)! = n!/(n - 1)! = n$ also. Hence the formula holds for $r = 1$.

If the formula holds for r = k then it holds for r = k + 1: Let k be an integer with $k \geq 1$ and suppose $P(n, k) = \dfrac{n!}{(n - k)!}$. *[We must show that $P(n, k + 1) = \dfrac{n!}{(n - (k + 1))!}$.]* Consider the process of forming a $(k + 1)$-permutation of a set X of n elements. This process can be thought of as a two-step operation as follows: Imagine $k+1$ positions spread out in a row. Step 1 is to choose k elements from X to place in the left-most k positions and step 2 is to choose the element to put in the right-most position. By inductive hypothesis, there are $n!/(n - k)!$ ways to perform step 1. Since step 1 involves choosing k elements from the n elements in X, after step 1 has been performed there are $n - k$ elements left to choose from. It follows that there are $n - k$ ways to perform step 2. Thus by Theorem 6.2.2, there are $(n!/(n - k)!) \cdot (n - k)$ ways to perform the entire operation, or, in other words, to form a $(k + 1)$-permutation of X. But

$$\frac{n!}{(n - k)!} \cdot (n - k) = \frac{n! \cdot (n - k)}{(n - k) \cdot (n - k - 1)!} = \frac{n!}{(n - k - 1)!} = \frac{n!}{(n - (k + 1))!}.$$

Thus the number of $(k+1)$-permutations of X equals $n!/(n-(k+1))!$, or, equivalently, $P(n, k+1) = \dfrac{n!}{(n - (k + 1))!}$ *[as was to be shown]*.

Section 6.3

2. a. $16 + 16^2 + 16^3 + 16^4 = 69,904$

   b. $16^3 + 16^4 + 16^5 = 1,118,208$

5. a. Such integers must end in a 0 or a 5. Therefore, the answer is $9 \cdot 10 \cdot 10 \cdot 2 = 1,800$.

   b. The total number of four-digit integers is $9999 - 1000 + 1 = 9000$. By part (a), 1800 of these are divisible by 5. So the probability that a randomly chosen four-digit integer is divisible by 5 is $1800/9000 = 1/5 = 20\%$.

7. Note that certain numbers are equal and should therefore not be counted twice. For instance, $001.90 = 1.9 = 1.90$ and so forth.

   Numbers that consist of one significant digit: Excluding zero, there are 9 such numbers that have the form x. and 9 that have the form $.x$. Thus, including zero, there are $2 \cdot 9 + 1 = 19$ such numbers.

   Numbers that consist of two significant digits: There are $9 \cdot 10$ such number of the form xx., $9 \cdot 9$ of the form $x.x$, and $10 \cdot 9$ of the form $.xx$. Thus there are $90 + 81 + 90 = 261$ such numbers in all.

   Numbers that consist of three significant digits: There are $9 \cdot 10^2$ such number of the form xxx., $9 \cdot 10 \cdot 9$ of the form $xx.x$, $9 \cdot 10 \cdot 9$ of the form $x.xx$ and $10^2 \cdot 9$ of the form $.xxx$. Thus there are $2 \cdot 900 + 2 \cdot 810 = 3420$ such numbers in all.

   Similar analysis shows that there are $2 \cdot 9 \cdot 10^3 + 3 \cdot 9^2 \cdot 10^2 = 42,300$ numbers with four significant digits, $2 \cdot 9 \cdot 10^4 + 4 \cdot 9^2 \cdot 10^3 = 504,000$ numbers with five significant digits, $2 \cdot 9 \cdot 10^5 + 5 \cdot 9^2 \cdot 10^4 = 5,850,000$ numbers with six significant digits, $2 \cdot 9 \cdot 10^6 + 6 \cdot 9^2 \cdot 10^5 = 66,600,000$ numbers with seven significant digits, and $2 \cdot 9 \cdot 10^7 + 7 \cdot 9^2 \cdot 10^6 = 747,000,000$ numbers with eight significant digits. Adding gives that there are 820,000,000 distinct numbers that can be displayed. Note that if the calculator has a $\pm$ indicator, the total is $2 \cdot 820000000 - 1 = 1,639,999,999$ (so as not to count zero twice).

8. *a.* $1 + 2 + 3 + 4 = 10$

   b. On the ith iteration of the outer loop, there are i iterations of the inner loop, and this is true for each $i = 1, 2, \ldots, n$. Therefore, the total number of iterations of the inner loop is $1 + 2 + 3 + \cdots + n = n(n+1)/2$.

10. *a.* $6! = 720$ *b.* $5! + 5! = 120 + 120 = 240$

13. $(26 + 26 \cdot 37 + 26 \cdot 37^2 + \cdots + 26 \cdot 37^{30}) - 82 = 26(1 + 37 + 37^2 + \cdots + 37^{30}) - 82 =$
$26 \cdot \sum_{k=0}^{30} 37^k - 82 = 26\left(\dfrac{37^{31} - 1}{37 - 1}\right) - 82 \cong 2.97 \times 10^{48}$.

15. *a.* $16 \cdot 15 \cdot 14 \cdot 13 = 43,680$ *b.* $16^4 - 16 \cdot 15 \cdot 14 \cdot 13 = 21,856$

    c. The total number of strings of four hexadecimal digits is $16^4 = 65,536$. By part (b), 21,856 of these have at least one repeated digit. So the probability that a randomly chosen string of four hexadecimal digits has at least one repeated digit is $21856/65536 \cong 33.3\%$.

16. *b. Proof:* Let A and B be events in a sample space S. By the inclusion/exclusion rule (Theorem 6.3.3), $n(A \cup B) = n(A) + n(B) - n(A \cap B)$. So by the equally likely probability formula,

$$P(A \cup B) = \dfrac{n(A \cup B)}{n(S)} = \dfrac{n(A) + n(B) - n(A \cap B)}{n(S)} = \dfrac{n(A)}{n(S)} + \dfrac{n(B)}{n(S)} - \dfrac{n(A \cap B)}{n(S)}$$
$$= P(A) + P(B) - P(A \cap B).$$

17. The number of triples of integers from 1 through 39 where the integers in the first two positions are the same and the integer in the third position is different is $39 \cdot 38$ (the number of ways to choose the integer to appear in the first and second positions times the number of ways to choose the integer in the third position). Similarly, the number of triples of integers from 1 through 39 where the integers in the first and third positions are the same and the integer in the second position is different is $39 \cdot 38$. Furthermore, there are 39 triples of integers from 1 through 39 where all three integers are the same. Therefore, by the addition rule the total number of triples in which two adjacent integers are the same is $2 \cdot 39 \cdot 38 + 39$, and so by the difference rule the number of triples in which no two adjacent integers are the same is $39^3 - (2 \cdot 39 \cdot 38 + 39) = 56,316$.

18. *b.* As in part (a), represent each integer from 1 through 99,999 as a string of five digits. The number of integers from 1 through 99,999 that do not contain the digit 6 is 9^5 because there are 9 choices of digit for each of the five positions (namely, all ten digits except 6). In addition, 100,000 does not contain the digit 6. So there are $9^5 + 1$ integers from 1 through 100,000 that do not contain the digit 6. Therefore, by the difference rule, there are $100,000 - (9^5 + 1) = 40,950$ integers from 1 through 100,000 that contain at least one occurrence of the digit 6.

    c. By parts (a) and (b) and the difference rule, the number of integers from 1 through 100,000 that contain two or more occurrences of the digit 6 is the difference between the number that contain at least one occurrence and the number that contain exactly one occurrence, namely, $40,950 - 32,805 = 8145$. Since there are 100,000 integers from 1 thorugh 100,000, the probability that a randomly chosen integer in this range contains two or more occurrences of the digit 6 is $8145/100000 = 8.145\%$.

19. Call the employees U, V, W, X, Y, and Z, and suppose that U and V are the married couple. Let A be the event that U and V have adjacent desks. Since the desks of U and V can be adjacent either in the order UV or in the order VU, the number of desk assignments with U and V adjacent is the same as the sum of the number of permutations of the symbols $\boxed{UV}$, W, X, Y, Z plus the number of permutions of the symbols $\boxed{VU}$, W, X, Y, Z. By the multiplication

rule each of these is $5!$, and so by the addition rule the sum is $2 \cdot 5!$. Since the total number of permutations of U, V, W, X, Y, Z is $6!$, $P(A) = 2 \cdot 5!/6! = 2/6 = 1/3$. Hence by the formula for the probability of the complement of an event, $P(A^c) = 1 - P(A) = 1 - 1/3 = 2/3$. So the probability that the married couple have nonadjacent desks is $2/3$.

20. *a.* By the multiplication rule, the number of strings of length n over $\{a, b, c, d\}$ with no two consecutive characters the same is $4 \cdot \underbrace{3 \cdot 3 \cdot \ldots \cdot 3}_{n-1} = 4 \cdot 3^{n-1}$ because any of the four letters may be chosen for the left-most character, and for each subsequent character any letter may be chosen except the one directly to its left. The total number of strings over $\{a, b, c, d\}$ of length n is 4^n, and so by the difference rule the number of such strings with at least two consecutive characters the same is $4^n - 4 \cdot 3^{n-1}$.

b. The probability is $\dfrac{4^{10} - 4 \cdot 3^9}{4^{10}} = 1 - \left(\dfrac{3}{4}\right)^9 \cong 92.5\%$.

21. *b.* By part (a) and the equally likely probability formula, the probability that an integer chosen at random from 1 thorugh 1000 is a multiple of 4 or a multiple of 7 is $\dfrac{n(A \cup B)}{1000} = \dfrac{357}{1000} = 35.7\%$.

c. By the difference rule the number of integers from 1 through 1000 that are neither multiples of 4 nor multiples of 7 is $1000 - 357 = 643$.

22. *a.* Let A and be B the sets of all integers from 1 through 1,000 that are multiples of 2 and 9 respectively. Then $n(A) = 500$ and $n(B) = 111$ (because $9 = 9 \cdot 1$ is the smallest integer in B and $999 = 9 \cdot 111$ is the largest). Also $A \cap B$ is the set of all integers from 1 through 1,000 that are multiples of 18, and $n(A \cap B) = 55$ (because $18 = 18 \cdot 1$ is the smallest integer in B and $990 = 18 \cdot 55$ is the largest). It follows from the inclusion/exclusion rule that the number of integers from 1 through 1,000 that are multiples of 2 or 9 equals $n(A \cup B) = n(A) + n(B) - n(A \cup B) = 500 + 111 - 55 = 556$.

b. The probability is $556/1000 = 55.6\%$.

c. $1000 - 556 = 444$

23. $n(T) = 28$, $n(N) = 26$, $n(U) = 14$, $n(T \cap N) = 8$, $n(T \cap U) = 4$, $n(N \cap U) = 3$, $n(T \cap N \cap U) = 2$.

a. $n(T \cup N \cup U) = n(T) + n(N) + n(U) - n(T \cap N) - n(T \cap U) - n(N \cap U) + n(T \cap N \cap U) = 28 + 26 + 14 - 8 - 4 - 3 + 2 = 55$

b. $n((T \cup N \cup U)^c) = 100 - n(T \cup N \cup U) = 100 - 55 = 45$

c.

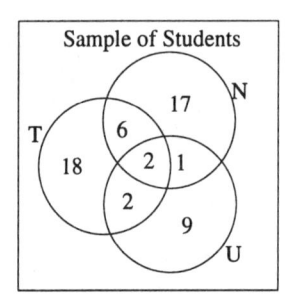

Sample of Students

T: 18, 6, 2, 2, 1, 17 (N), 9, U

e. 1   *f.* 17

24. Let A, B, and C be the sets of all people who reported relief from drugs A, B, and C respectively. Then $n(A) = 23$, $n(B) = 18$, $n(C) = 31$, $n(A \cap B) = 11$, $n(A \cap C) = 19$, $n(B \cap C) = 14$, and $n(A \cup B \cup C) = 37$.

a. $n((A \cup B \cup C)^c) = 40 - n(A \cup B \cup C) = 40 - 37 = 3$

b. $n(A \cap B \cap C) = n(A \cup B \cup C) - n(A) - n(B) - n(C) + n(A \cap B) + n(A \cap C) + n(B \cap C) = 37 - 23 - 18 - 31 + 11 + 19 + 14 = 9$

c.

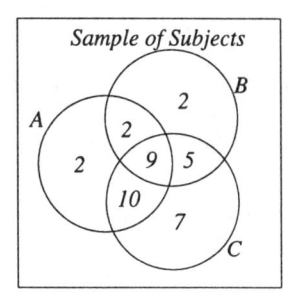

Sample of Subjects

A, B, C Venn diagram with values: A only = 2, B only = 2, C only = 7, A∩B only = 2, A∩C only = 10, B∩C only = 5, A∩B∩C = 9.

d. 2

26. a. by the difference rule b. by De Morgan's law c. by the inclusion/exclusion rule

28. Let U be the set of all permutations of $a, b, c, d,$ and e, let A be the set of all permutations of $a, b, c, d,$ and e in which the left-most character is $a, b,$ or c, and let B be the set of all permutations in which the right-most character is $c, d,$ or e. By the formula from exercise 26, $n(A \cap B) = n(U) - (n(A^c) + n(B^c) - n(A^c \cap B^c))$. Now by the multiplication rule, $n(U) = 5! = 120$. Also the number of permutations of $a, b, c, d,$ and e in which the left-most character is neither $a, b,$ or c is $2 \cdot 4 \cdot 3 \cdot 2 \cdot 1 = 48$ (because only d or e may be chosen as the left-most character). And the number of permutations of $a, b, c, d,$ and e in which the right-most character is neither $c, d,$ or e is $2 \cdot 4 \cdot 3 \cdot 2 \cdot 1 = 48$ (because only a or b may be chosen as the right-most character — imagine choosing the right-most character first and then the four on the left one after another). Thus $n(A^c) = 2 \cdot 4! = n(B^c)$. Furthermore, $A^c \cap B^c$ is the set of permutations of $a, b, c, d,$ and e in which the left-most character is neither $a, b,$ nor c and the right-most character is neither $c, d,$ nor e. In other words, the left-most character is d or e and the right-most character is a or b. Imagine constructing such a permutation as a five-step process in which the first step is to choose the left-most character, the second step is to choose the right-most character, and the third through fifth steps are to choose the middle characters one after another. By the multiplication rule there are $2 \cdot 2 \cdot 3 \cdot 2 \cdot 1 = 24$ such permutations, and so $n(A^c \cap B^c) = 24$. Thus

$$n(A \cap B) = n(U) - (n(A^c) + n(B^c) - n(A^c \cap B^c)) = 120 - (48 + 48 - 24) = 48.$$

29. Imagine each integer from 1 through 999,999 as a string of six digits with leading 0's allowed. For each $i = 1, 2, 3$, let A_i be the set of all integers from 1 through 999,999 that do not contain the digit i. We want to compute $n(A_1^c \cap A_2^c \cap A_3^c)$. By De Morgan's law,

$$A_1^c \cap A_2^c \cap A_3^c = (A_1 \cup A_2)^c \cap A_3^c = (A_1 \cup A_2 \cup A_3)^c = U - (A_1 \cup A_2 \cup A_3),$$

and so by the difference rule

$$n(A_1^c \cap A_2^c \cap A_3^c) = n(U) - n(A_1 \cup A_2 \cup A_3).$$

By the inclusion/exclusion rule,

$$n(A_1 \cup A_2 \cup A_3) = n(A_1) + n(A_2) + n(A_3) - n(A_1 \cap A_2) - n(A_1 \cap A_3) - n(A_2 \cap A_3) + n(A_1 \cap A_2 \cap A_3).$$

Now $n(A_1) = n(A_2) = n(A_3) = 9^6$ because in each case any of nine digits may be chosen for each character in the string (for A_i these are all the ten digits except i). Also each $n(A_i \cap A_j) = 8^6$ because in each case any of eight digits may be chosen for each character of the string (for

$A_i \cap A_j$ these are all the ten digits except i and j). Similarly, $n(A_1 \cap A_2 \cap A_3) = 7^6$ because any digit except 1, 2, and 3 may be chosen for each character in the string. Thus

$$n(A_1 \cup A_2 \cup A_3) = 3 \cdot 9^6 - 3 \cdot 8^6 + 7^6,$$

and so

$$n(A_1{}^c \cap A_2{}^c \cap A_3{}^c) = n(U) - n(A_1 \cup A_2 \cup A_3) = 10^6 - (3 \cdot 9^6 - 3 \cdot 8^6 + 7^6) = 74,460.$$

30. *Proof:* Let $P(k)$ be the property "If a finite set A equals the union of k distinct mutually disjoint subsets $A_1, A_2, \ldots, A_k$, then $n(A) = n(A_1) + n(A_2) + \ldots + n(A_k)$."

    The property is true for n = 1: Suppose a finite set A equals the union of one subset A_1, then $A = A_1$, and so $n(A) = n(A_1)$.

    If the property is true for n = i then it is true for n = i+1: Let i be an integer with $i \geq 1$ and suppose the property is true for $n = i$. *[This is the inductive hypothesis.]* Let A be a finite set that equals the union of $i+1$ distinct mutually disjoint subsets $A_1, A_2, \ldots, A_{i+1}$. Then $A = A_1 \cup A_2 \cup \ldots \cup A_{i+1}$ and $A_i \cap A_j = \emptyset$ for all integers i and j with $i \neq j$. Let B be the set $A_1 \cup A_2 \cup \ldots \cup A_i$. Then $A = B \cup A_{i+1}$ and $B \cap A_{i+1} = \emptyset$. *[For if $x \in B \cap A_{i+1}$, then $x \in A_1 \cup A_2 \cup \ldots \cup A_i$ and $x \in A_{i+1}$, which implies that $x \in A_j$, for some j with $1 \leq j \leq i$, and $x \in A_{i+1}$. But A_j and A_i are mutually disjoint. Thus no such x exists.]* Hence A is the union of the two mutually disjoint sets B and A_{i+1}. Since B and A_{i+1} have no elements in common, the total number of elements in $B \cup A_{i+1}$ can be obtained by first counting the elements in B, next counting the elements in A_{i+1}, and then adding the two numbers together. It follows that $n(B \cup A_{i+1}) = n(B) + n(A_{i+1})$ which equals $n(A_1) + n(A_2) + \ldots + n(A_i) + n(A_{i+1})$ by inductive hypothesis. Hence $P(i+1)$ is true *[as was to be shown]*.

31. *Proof:* Let A and B be sets. We first show that $A \cup B$ can be partitioned into $A - (A \cap B)$, $B - (A \cap B)$, and $A \cap B$.

    1. **A ∪ B ⊆ (A − (A ∩ B)) ∪ (B − (A ∩ B)) ∪ (A ∩ B)**: Let $x \in A \cup B$. Then $x \in A$ or $x \in B$.

    Case 1 ($x \in A$): Either $x \in B$ or $x \notin B$. If $x \in B$, then $x \in A \cap B$ by definition of intersection, and so by definition of union $x \in (A - (A \cap B)) \cup (B - (A \cap B)) \cup (A \cap B)$. If $x \notin B$, then $x \notin A \cap B$ either *[by definition of intersection]* and so $x \in A - (A \cap B)$. Hence by definition of union, $x \in (A - (A \cap B)) \cup (B - (A \cap B)) \cup (A \cap B)$.

    Case 2 ($x \in B$): Either $x \in A$ or $x \notin A$. If $x \in A$, then $x \in A \cap B$ by definition of intersection, and so by definition of union $x \in (A - (A \cap B)) \cup (B - (A \cap B)) \cup (A \cap B)$. If $x \notin A$, then $x \notin A \cap B$ either *[by definition of intersection]* and so $x \in B - (A \cap B)$. Hence by definition of union, $x \in (A - (A \cap B)) \cup (B - (A \cap B)) \cup (A \cap B)$.

    Thus in either case $x \in (A - (A \cap B)) \cup (B - (A \cap B)) \cup (A \cap B)$ *[and so $A \cup B$ subseteq $A - (A \cap B)) \cup (B - (A \cap B)) \cup (A \cap B)$ by definition of subset]*.

    (A − (A ∩ B)) ∪ (B − (A ∩ B)) ∪ (A ∩ B) subseteq A ∪ B: Let $x \in (A - (A \cap B)) \cup (B - (A \cap B)) \cup (A \cap B)$. By definition of union, $x \in (A - (A \cap B))$ or $x \in (B - (A \cap B))$ or $x \in (A \cap B)$. If $x \in (A - (A \cap B))$, then by definition of set difference $x \in A$ and $x \notin A \cap B$. In particular, $x \in A$, and so by definition of union, $x \in A \cup B$. If $x \in (B - (A \cap B))$, then by definition of set difference $x \in B$ and $x \notin A \cap B$. In particular, $x \in B$, and so by definition of union, $x \in A \cup B$. If $x \in (A \cap B)$, then by definition of intersection $x \in A$ and $x \in B$. In particular, $x \in A$, and so by definition of union, $x \in A \cup B$. Hence in all cases, $x \in A \cup B$ *[and so by definition of subset $(A - (A \cap B)) \cup (B - (A \cap B)) \cup (A \cap B)$ subseteq $A \cup B$.]*

    [Since both set containments $A \cup B$ subseteq $(A - (A \cap B)) \cup (B - (A \cap B)) \cup (A \cap B)$ and $(A - (A \cap B)) \cup (B - (A \cap B)) \cup (A \cap B)$ subseteq $A \cup B$ have been proved, $A \cup B = (A - (A \cap B)) \cup (B - (A \cap B)) \cup (A \cap B)$ by definition of set equality.]

    2. The sets $(A - (A \cap B))$, $(B - (A \cap B))$, and $(A \cap B)$ are mutually disjoint because if $x \in A \cap B$ then $x \notin A - (A \cap B)$ and $x \notin B - (A \cap B)$ by definition of set difference. Next if $x \in A - (A \cap B)$,

then $x \notin A \cap B$ and so $x \notin B$; consequently $x \notin B - (A \cap B)$. Finally, if $x \in B - (A \cap B)$, then $x \notin A \cap B$ and so $x \notin A$; consequently $x \notin A - (A \cap B)$.

Next we derive the inclusion/exclusion formula from this partition. Since $A \cup B$ can be partitioned into $A - (A \cap B)$, $B - (A \cap B)$, and $A \cap B$, then

$$\begin{aligned} n(A \cup B) &= n(A - (A \cap B)) + n(B - (A \cap B)) + n(A \cap B) &&\text{by the addition rule} \\ &= n(A) - n(A \cap B) + n(B) - n(A \cap B) + n(A \cap B) &&\text{by the difference rule and} \\ &&&\text{the fact that } A \cap B \text{ is a} \\ &&&\text{subset of both } A \text{ and } B \\ &= n(A) + n(B) - n(A \cap B) &&\text{by basic algebra.} \end{aligned}$$

32. Proof: Suppose A, B, and C are finite sets.

$$\begin{aligned} n(A \cup B \cup C) &= n(A \cup (B \cup C)) &&\text{by the associative law for } \cup \\ &= n(A) + n(B \cup C) - n(A \cap (B \cup C)) &&\text{by the inclusion/exclusion rule} \\ &&&\text{for two sets} \\ &= n(A) + n(B) + n(C) - n(B \cap C) &&\text{by the inclusion/exclusion rule} \\ &\quad - n(A \cap (B \cup C)) &&\text{for two sets} \\ &= n(A) + n(B) + n(C) - n(B \cap C) \\ &\quad - n((A \cap B) \cup (A \cap C)) &&\text{by the distributive law for sets} \\ &= n(A) + n(B) + n(C) - n(B \cap C) \\ &\quad - [n(A \cap B) + n(A \cap C) \\ &\quad - n((A \cap B) \cap (A \cap C))] &&\text{by the inclusion/exclusion rule} \\ &&&\text{for two sets} \\ &= n(A) + n(B) + n(C) - n(A \cap B) &&\text{by basic algebra and because} \\ &\quad - n(A \cap C) - n(B \cap C) &&(A \cap B) \cap (A \cap C) \\ &\quad + n(A \cap B \cap C) &&= (A \cap A) \cap B \cap C \\ &&&= A \cap B \cap C. \end{aligned}$$

33. *Proof (by mathematical induction)*:

The property is true for n = 2: This was proved in one way in the text preceding Theorem 6.3.3 and in another way in the solution to exercise 31.

If the property is true for n = r then it is true for n = r + 1: Let r be an integer with $r \geq 2$, and suppose that the formula holds for any collection of r finite sets. *[This is the inductive hypothesis.]* Let $A_1, A_2, \ldots, A_{r+1}$ be finite sets. Then $n(A_1 \cup A_2 \cup \ldots \cup A_{r+1})$

$= n(A_1 \cup (A_2 \cup A_3 \cup \ldots \cup A_{r+1}))$ by the associative law for $\cup$

$= n(A_1) + n(A_2 \cup A_3 \cup \ldots \cup A_{r+1}) - n(A_1 \cap (A_2 \cup A_3 \cup \ldots \cup A_{r+1}))$

by the inclusion/exclusion rule for two sets

$= n(A_1) + n(A_2 \cup A_3 \cup \ldots \cup A_{r+1}) - n((A_1 \cap A_2) \cup (A_1 \cap A_3) \cup \ldots \cup (A_1 \cap A_{r+1}))$

by the distributive law for sets

$= n(A_1) + \left(\sum_{2 \leq i \leq r+1} n(A_i) - \sum_{2 \leq i < j \leq r+1} n(A_i \cap A_j) \right.$

$\left. + \sum_{2 \leq i < j < k \leq r+1} n(A_i \cap A_j \cap A_k) - \ldots + (-1)^{r+1} n(A_2 \cap A_3 \cap \ldots \cap A_{r+1}) \right)$

$- \left(\sum_{2 \leq i \leq r+1} n(A_1 \cap A_i) - \sum_{2 \leq i < j \leq r+1} n((A_1 \cap A_i) \cap (A_1 \cap A_j)) + \ldots \right.$

$\left. + (-1)^{r+1} n((A_1 \cap A_2) \cap (A_1 \cap A_3) \cap \ldots \cap (A_1 \cap A_{r+1})) \right)$

by inductive hypothesis

$$= n(A_1) + \left(\sum_{2 \leq i \leq r+1} n(A_i) - \sum_{2 \leq i < j \leq r+1} n(A_i \cap A_j)\right.$$
$$\left. + \sum_{2 \leq i < j < k \leq r+1} n(A_i \cap A_j \cap A_k) - \ldots + (-1)^{r+1} n(A_2 \cap A_3 \cap \ldots \cap A_{r+1})\right)$$
$$- \left(\sum_{2 \leq i \leq r+1} n(A_1 \cap A_i) - \sum_{2 \leq i < j \leq r+1} n(A_1 \cap A_i \cap A_j) + \ldots \right.$$
$$\left. + (-1)^{r+1} n(A_1 \cap A_2 \cap A_3 \cap \ldots \cap A_{r+1})\right)$$
$$= \sum_{1 \leq i \leq r+1} n(A_i) - \sum_{1 \leq i < j \leq r+1} n(A_i \cap A_j) + \sum_{1 \leq i < j < k \leq r+1} n(A_i \cap A_j \cap A_k)$$
$$- \ldots + (-1)^{r+2} n(A_1 \cap A_3 \cap \ldots \cap A_{r+1}).$$

This is what was to be proved.

Section 6.4

2. *a.* The 3-combinations are $x_1x_2x_3$, $x_1x_2x_4$, $x_1x_2x_5$, $x_1x_3x_4$, $x_1x_3x_5$, $x_1x_4x_5$, $x_2x_3x_4$, $x_2x_3x_5$, $x_2x_4x_5$, $x_3x_4x_5$. Therefore, $\binom{5}{3} = 10$.

b. The unordered selections of three elements are $x_1x_2x_3$, $x_1x_2x_4$, $x_1x_2x_5$, $x_1x_2x_6$, $x_1x_2x_7$, $x_1x_3x_4$, $x_1x_3x_5$, $x_1x_3x_6$, $x_1x_3x_7$, $x_1x_4x_5$, $x_1x_4x_6$, $x_1x_4x_7$, $x_1x_5x_6$, $x_1x_5x_7$, $x_1x_6x_7$, $x_2x_3x_4$, $x_2x_3x_5$, $x_2x_3x_6$, $x_2x_3x_7$, $x_2x_4x_5$, $x_2x_4x_6$, $x_2x_4x_7$, $x_2x_5x_6$, $x_2x_5x_7$, $x_2x_6x_7$, $x_3x_4x_5$, $x_3x_4x_6$, $x_3x_4x_7$, $x_3x_5x_6$, $x_3x_5x_7$, $x_3x_6x_7$, $x_4x_5x_6$, $x_4x_5x_7$, $x_4x_6x_7$, $x_5x_6x_7$.
Therefore, $\binom{7}{3} = 35$.

4. $\binom{8}{5} = \dfrac{P(8,5)}{5!}$

5. *c.* $\binom{5}{2} = \dfrac{5!}{2!(5-2)!} = \dfrac{5!}{2! \cdot 3!} = \dfrac{5 \cdot 4 \cdot 3!}{2 \cdot 1 \cdot 3!} = 10$ *d.* $\binom{5}{3} = \dfrac{5!}{3!(5-3)!} = \dfrac{5 \cdot 4 \cdot 3!}{3! \cdot 2 \cdot 1} = 10$

   e. $\binom{5}{4} = \dfrac{5!}{4!(5-4)!} = \dfrac{5!}{4! \cdot 1!} = \dfrac{5 \cdot 4!}{4!} = 5$ *f.* $\binom{5}{5} = \dfrac{5!}{5!(5-5)!} = \dfrac{5!}{5! \cdot 0!} = 1$

7. *a.* $\binom{14}{7} = \dfrac{14 \cdot 13 \cdot 12 \cdot 11 \cdot 10 \cdot 9 \cdot 8 \cdot 7!}{7 \cdot 6 \cdot 5 \cdot 4 \cdot 3 \cdot 2 \cdot 1 \cdot 7!} = 3432$

   b. (i) $\binom{8}{4} \cdot \binom{6}{3} = \dfrac{8 \cdot 7 \cdot 6 \cdot 5 \cdot 4!}{4 \cdot 3 \cdot 2 \cdot 1 \cdot 4!} \cdot \dfrac{6 \cdot 5 \cdot 4}{3 \cdot 2} = 1400$ *[the number of subsets of four women chosen from eight times the number of subsets of three men chosen from six]*

   (ii) $\binom{14}{7} - \binom{8}{7} = 3432 - \dfrac{8 \cdot 7!}{1! \cdot 7!} = 3432 - 8 = 3424$ *[the total number of groups minus the number that contain no men]*

   (iii) $\binom{8}{1}\binom{6}{6} + \binom{8}{2}\binom{6}{5} + \binom{8}{3}\binom{6}{4} = 8 \cdot 1 + \dfrac{8 \cdot 7 \cdot 6!}{2 \cdot 1 \cdot 6!} \cdot \dfrac{6 \cdot 5!}{1 \cdot 5!} + \dfrac{8 \cdot 7 \cdot 6 \cdot 5!}{5! \cdot 3 \cdot 2 \cdot 1} \cdot \dfrac{6 \cdot 5 \cdot 4!}{4! \cdot 2 \cdot 1} = 8 + 168 + 840 = 1016$ *[the number of groups with one woman and six men plus the number with two women and five men plus the number with three women and four men — there are no groups with no women because there are only six men]*

   c. $\binom{12}{6} + \binom{12}{6} + \binom{12}{7} = 2\left(\dfrac{12 \cdot 11 \cdot 10 \cdot 9 \cdot 8 \cdot 7 \cdot 6!}{6 \cdot 5 \cdot 4 \cdot 3 \cdot 2 \cdot 1 \cdot 6!}\right) + \left(\dfrac{12 \cdot 11 \cdot 10 \cdot 9 \cdot 8 \cdot 7!}{7! \cdot 5 \cdot 4 \cdot 3 \cdot 2 \cdot 1}\right) = 1848 + 792 = 2640$ *[Let the people be A and B. The number of groups that do not contain both A and B equals the number of groups with A and six others (none B) plus the number of groups with B and six others (none A) plus the number of groups with neither A nor B.]*

d. $\binom{12}{5} + \binom{12}{7} = 792 + 792 = 1584$ [the number of groups with both A and B and five others plus the number of groups with neither A nor B]

8. a. $\binom{14}{10} = 1001$

b. (i) $\binom{6}{4} \cdot \binom{8}{6} = 15 \cdot 28 = 420$

(ii) Since only eight questions do not require proof, any group of ten questions contains at least two that require proof. Thus the answer is the same as for part (a): $\binom{14}{10} = 1001$.

(iii) By the reasoning of part (iii), the number of groups of ten questions with three or fewer that require proof is the sum of the number with two that require proof plus the number with three that require proof: $\binom{8}{8}\binom{6}{2} + \binom{8}{7}\binom{6}{3} = 1 \cdot 15 + 8 \cdot 20 = 175$.

c. Any set of questins containing at most one of questions 1 and 2 can be split into three subsets: those that contain question 1 and not 2, those that contain question 2 and not 1, and those that contain neither question 1 nor 2. There are $\binom{12}{9}$ choices of questions in the first set, $\binom{12}{9}$ choices of questions in the second set, and $\binom{12}{10}$ choices of questions in the third set. So there are $2 \cdot \binom{12}{9} + \binom{12}{10} = 2 \cdot 220 + 66 = 506$ choices of ten questions containing at most one of questions 1 and 2.

d. There are $\binom{12}{8}$ choices of questions that contain both questions 1 and 2, and there are $\binom{12}{10}$ choices of questions that contain neither question 1 nor 2. So by the addition rule, there are $\binom{12}{8} + \binom{12}{10} = 195 + 66 = 261$ choices of questions that contain either both questions 1 and 2 or neither question 1 nor 2.

9. a. $\binom{30}{6}$

10. $\binom{40}{15} \cdot \binom{25}{15} = \frac{40!}{15!25!} \cdot \frac{25!}{15!10!} = \frac{40!}{15! \cdot 15! \cdot 10!} \cong 1.315 \times 10^{17}$

[The assignment of treatments to mice can be considered a two-step operation. Step 1 is to choose 15 mice out of the 40 to receive treatment A, step 2 is to choose 15 mice out of the remaining 25 to receive treatment B. The remaining 10 mice are the controls.]

11. b. (i) The number of hands with a straight flush is $4 \cdot 9 = 36$ [the number of suits, namely 4, times the number of possible lowest denominated cards in the straight, which is 9 because aces can be low]

(ii) probability $= \frac{36}{\binom{52}{5}} = \frac{36}{2,598,960} \cong .0000139$

d. (i) The number of hands with a full house is $\binom{13}{1} \cdot \binom{4}{3} \cdot \binom{12}{1} \cdot \binom{4}{2} = 3744$ [because constructing a full house can be thought of as a four-step process where step one is to choose the denomination for the three of a kind, step two is to choose three cards out of the four of that denomination, step three is to choose the denomination for the pair, and step four is to choose two cards of that denomination].

(ii) probability $= \dfrac{3744}{\binom{52}{5}} = \dfrac{3744}{2{,}598{,}960} \cong .00144$

e. (i) The number of hands with a flush (including a royal or a straight flush) is $4 \cdot \binom{13}{5} = 5148$ [the number of suits times the number of ways to pick five cards from a suit]. Forty of these are royal or straight flushes, and so there are 5108 hands with a flush.

(ii) probability $= \dfrac{5148}{\binom{52}{5}} = \dfrac{5148}{2{,}598{,}960} \cong .00197$

g. (i) The number of hands with three of a kind is $\binom{13}{1} \cdot \binom{4}{3} \cdot \binom{12}{2} \cdot \binom{4}{1} \cdot \binom{4}{1} = 54{,}912$ [the number of ways to choose the denomination for the three of a kind times the number of ways three cars from that denomination can be chosen times the number of ways to choose two other denominations for the other two cards times the number of ways a card can be chosen from the lower ranked of the two denominations times the number of ways a card can be chosen from the higher ranked of the two denominations].

(ii) probability $= \dfrac{54912}{\binom{52}{5}} = \dfrac{54912}{2{,}598{,}960} \cong .021$

h. The number of hands with one pair is $\binom{13}{1} \cdot \binom{4}{2} \cdot \binom{12}{3} \cdot \binom{4}{1} \cdot \binom{4}{1} \cdot \binom{4}{1} = 1{,}098{,}240$ [the number of ways to choose the denomination for the pair times the number of ways two cars from that denomination can be chosen times the number of ways to choose three other denominations for the other three cards times the number of ways a card can be chosen from the lowest ranked of the three denominations times the number of ways a card can be chosen from the middle ranked of the three denominations times the number of ways a card can be chosen from the highest ranked of the three denominations].

(ii) probability $= \dfrac{1{,}098{,}240}{\binom{52}{5}} = \dfrac{1{,}098{,}240}{2{,}598{,}960} \cong .4226$

i. The sum of the answers from (a)–(h) plus the answer from Example 6.4.9 (the number of poker hands with two pairs) is 1,296,420, and so by the difference rule there are $\binom{52}{5} - 1{,}296{,}420 = 2{,}598{,}960 - 1{,}296{,}420 = 1{,}302{,}540$ hands whose cards neither contain a repeated denomination nor five adjacent denominations.

(ii) probability $= \dfrac{1{,}302{,}540}{\binom{52}{5}} = \dfrac{1{,}302{,}540}{2{,}598{,}960} \cong .5012$

The results of this exercise are summarized in the following table.

Type of Hand	Probability
royal flush	.0000015 = .00015%
straight flush	.000014 = .0014%
four of a kind	.00024 = .024%
full house	.00144 = .144%
flush	.00197 = .197%
straight	.00392 = .392%
three of a kind	.02113 = 2.113%
two pairs	.0475 = 4.75%
one pair	.4226 = 42.26%
none of the above	.5012 = 50.12%

12. Given a group of n people, imagine them lined up alphabetically by name. Assuming that each year has 365 days, the number of assignments of birthdays to individual people (with repetition allowed) is 365^n [because each position in the line can be assigned any one of 365

birthdays], and the number of assignments of birthdays to individual people without repetition is $365 \cdot 364 \cdot 363 \cdot \ldots \cdot (365 - n + 1)$. Assuming that all birthdays are equally likely, each of the 365^n assignments of birthdays to individual people is as likely as any other. So the probability that no two people have the same birthday is $\frac{365 \cdot 364 \cdot 363 \cdot \ldots \cdot (365 - n + 1)}{365^n}$, and thus by the formula for the probability of the complement of an event, the probability that at least two people have the same birthday is $p = 1 - \left(\frac{365 \cdot 364 \cdot 363 \cdot \ldots \cdot (365 - n + 1)}{365^n} \right)$. If $n = 22$, then $p \cong 1 - .5243 = .4757 < 1/2$, and if $n = 23$, then $p \cong 1 - .4927 = .5073 \geq 1/2$. Therefore n must be a minimum of 23 in order for the probability to be at least 50% that two or more people in the group have the same birthday.

13. b. $\binom{10}{5} = \frac{10 \cdot 9 \cdot 8 \cdot 7 \cdot 5 \cdot 5!}{5 \cdot 4 \cdot 3 \cdot 2 \cdot 1 \cdot 5!} = 252$

    c. $\binom{10}{9} + \binom{10}{10} = 10 + 1 = 11$

    e. $\binom{10}{0} + \binom{10}{1} = 1 + 10 = 11$

14. a. $\binom{16}{9} = \frac{16 \cdot 15 \cdot 14 \cdot 13 \cdot 12 \cdot 11 \cdot 10 \cdot 9!}{9! \cdot 7 \cdot 6 \cdot 5 \cdot 4 \cdot 3 \cdot 2 \cdot 1} = 11,440$

    b. $\binom{16}{14} + \binom{16}{15} + \binom{16}{16} = \frac{16 \cdot 15 \cdot 14!}{14! \cdot 2 \cdot 1} + \frac{16 \cdot 15!}{15! \cdot 1} + \frac{16!}{16! \cdot 1} = 120 + 16 + 1 = 137$

    c. $2^{16} - 1 = 65,535$ d. $\binom{16}{0} + \binom{16}{1} = 1 + 16 = 17$

16. a. $\binom{40}{4} = \frac{40 \cdot 39 \cdot 38 \cdot 37}{4 \cdot 3 \cdot 2 \cdot 1} = 91,390$

    b. $\binom{37}{3}\binom{3}{1} + \binom{37}{2}\binom{3}{2} + \binom{37}{1}\binom{3}{3} = 23310 + 1998 + 37 = 25,345$ *[The number of samples with at least one defective equals the number with one defective plus the number with two defectives plus the number with three defectives.]*

17. b. $\binom{8}{2} = \frac{8 \cdot 7}{2} = 28$ c. $\binom{9}{3} = \frac{9 \cdot 8 \cdot 7}{3 \cdot 2} = 84$ d. $\binom{8}{3} = \frac{8 \cdot 7 \cdot 6}{3 \cdot 2} = 56$

18. $\binom{11}{1}\binom{10}{4}\binom{6}{4}\binom{2}{2} = \frac{11!}{1! \cdot 10!} \cdot \frac{10!}{4! \cdot 6!} \cdot \frac{6!}{4! \cdot 2!} \cdot \frac{2!}{2! \cdot 0!} = \frac{11!}{1! \cdot 4! \cdot 4! \cdot 2!} = 34,650$ *[which agrees with the result in Example 6.4.11]*

20. a. $\binom{12}{2}\binom{10}{2}\binom{8}{1}\binom{7}{3}\binom{4}{2}\binom{2}{1}\binom{1}{1} = \frac{12!}{2! \cdot 2! \cdot 3! \cdot 2!} = 9,979,200$

    b. $\binom{10}{2}\binom{8}{2}\binom{6}{3}\binom{3}{2}\binom{1}{1} = \frac{10!}{2! \cdot 2! \cdot 3! \cdot 2!} = 75,600$ *[Once the T and the G have been fixed, there are ten positions left to fill in.]*

    c. $\binom{9}{1}\binom{8}{3}\binom{5}{2}\binom{3}{1}\binom{2}{1}\binom{1}{1} = \frac{9!}{3! \cdot 2!} = 30,240$ *[There are a total of nine symbol groups to arrange in order: one INT, three E's, two L's, one IG, one N, and one C.]*

21. c. $\binom{4}{2} = \frac{4 \cdot 3}{2 \cdot 1} = 6$ *[Or one could write the answer as $\binom{4}{2} \cdot \binom{2}{2} = 6.]*

22. $2 + 2^2 + 2^3 + 2^4 + 2^5 + 2^6 = 2(1 + 2 + 2^2 + 2^3 + 2^4 + 2^5) = 2\left(\frac{2^6 - 1}{2 - 1}\right) = 126$

24. a. $60 = 2 \cdot 2 \cdot 3 \cdot 5$, distinct factorizations: $1 \cdot 60$, $2 \cdot 30$, $3 \cdot 20$, $4 \cdot 15$, $5 \cdot 12$, $6 \cdot 10$, answer $= 6$

b. distinct factorizations: $1 \cdot (p_1 p_2 p_3 p_4 p_5)$, $p_1 \cdot (p_2 p_3 p_4 p_5)$, $p_2 \cdot (p_1 p_3 p_4 p_5)$, $p_3 \cdot (p_1 p_2 p_4 p_5)$, $p_4 \cdot (p_1 p_2 p_3 p_5)$, $p_5 \cdot (p_1 p_2 p_3 p_4)$, $(p_1 p_2) \cdot (p_3 p_4 p_5)$, $(p_1 p_3) \cdot (p_2 p_4 p_5)$, $(p_1 p_4) \cdot (p_2 p_3 p_5)$, $(p_1 p_5) \cdot (p_2 p_3 p_4)$, $(p_2 p_3) \cdot (p_1 p_4 p_5)$, $(p_2 p_4) \cdot (p_1 p_3 p_5)$, $(p_2 p_5) \cdot (p_1 p_3 p_4)$, $(p_3 p_4) \cdot (p_1 p_2 p_5)$, $(p_3 p_5) \cdot (p_1 p_2 p_4)$, $(p_4 p_5) \cdot (p_1 p_2 p_3)$

[Any additional listings would repeat those already noted. For instance, $(p_1 p_2 p_3) \cdot (p_4 p_5) = (p_4 p_5) \cdot (p_1 p_2 p_3)$.]

answer $= \binom{5}{0} + \binom{5}{1} + \binom{5}{2} = 16$

c. First note that for each way of writing n as a product of two positive integer factors, the product of i of the prime factors will be one factor (including $i = 0$ which corresponds to a factor of 1) and the product of the remaining prime factors will be the second factor. Next observe that for each integer i, there are $\binom{k}{i}$ ways to choose i prime factors out of the total of k. Finally, notice that if one of the factors consists of no more than $\lfloor k/2 \rfloor$ prime factors then the other consists of at least $\lfloor k/2 \rfloor + 1$. Thus the answer is $\binom{k}{0} + \binom{k}{1} + \binom{k}{2} + \ldots + \binom{k}{\lfloor k/2 \rfloor}$.

25. The answer is the total number of committees (which equals $\binom{15}{8}$) minus the number of committees that have no members from at least one class. Let A_1 be the set of committees with no freshmen, A_2 the set of committees with no sophomores, A_3 the set of committees with no juniors, and A_4 the set of committees with no seniors. By the inclusion/exclusion rule for four sets (see exercise 33 of Section 6.3),

$$n(A_1 \cup A_2 \cup A_3 \cup A_4) = \sum_{1 \le i \le 4} n(A_i) - \sum_{1 \le i < j \le 4} n(A_i \cap A_j)$$
$$+ \sum_{1 \le i < j < k \le 4} n(A_i \cap A_j \cap A_k)$$
$$- \sum_{1 \le i < j < k < l \le 4} n(A_i \cap A_j \cap A_k \cap A_l).$$

Now $n(A_1) = \binom{12}{8} = 495$ *[because a committee that contains no freshmen has its entire membership of eight taken from the $4 + 3 + 5 = 12$ sophomores, juniors, and seniors]*, $n(A_2) = \binom{11}{8} = 165$, $n(A_3) = \binom{12}{8} = 495$, $n(A_4) = \binom{10}{8} = 45$, $n(A_1 \cap A_2) = \binom{8}{8}$ *[because a committee that contains no freshmen or sophomores has its entire membership of eight taken from the $3 + 5 = 8$ juniors and seniors]*, $n(A_1 \cap A_3) = \binom{9}{8}$, $n(A_1 \cap A_4) = 0$ *[because if the freshmen and seniors are taken away, not enough students remain to form a committee of eight]*, $n(A_2 \cap A_3) = \binom{8}{8}$, $n(A_2 \cap A_4) = 0$, $n(A_3 \cap A_4) = 0$, $n(A_i \cap A_j \cap A_k) = 0$ for all possible i, j, and k *[because if students from three of the classes are taken away, not enough students remain to form a committee of eight]*, and $n(A_1 \cap A_2 \cap A_3 \cap A_4) = 0$ *[because every committee must contain students from some class]*. Consequently,

$$n(A_1 \cup A_2 \cup A_3 \cup A_4) = \binom{12}{8} + \binom{11}{8} + \binom{12}{8} + \binom{10}{8} - \binom{8}{8} - \binom{9}{8} - \binom{8}{8}$$
$$= 495 + 165 + 495 + 45 - 1 - 9 - 1$$
$$= 1189.$$

Thus the answer is $\binom{15}{8} - 1189 = 6435 - 1189 = 5246$.

26. Given nonnegative integers r and n with $r \le n$, $P(n, r)$ is the set of r-permutations that can be formed from a set of n elements. Partition this set of r-permutations into subsets so

that all the r-permutations in each subset are permutations of the same collection of elements. For instance, if $n = 5$ and $r = 3$ and $X = \{a, b, c, d, e\}$ is a set of $n = 5$ elements, then acd, adc, cda, cad, dac, and dca are all permutations of the same collection of elements, namely $\{a, c, d\}$. Thus each subset of the partition corresponds to a subset of X of size r. Furthermore, all subsets of the partition have the same size, namely $r!$ *[because there are $r!$ permutations of a set of r elements]*. Hence the number of subsets of X of size r equals the number of subsets of the partition. By the division rule, this equals the number of elements in the partition divided by the number of elements in each set of the partition, or $\dfrac{P(n,r)}{r!}$.

27. The false solution overcounts the number of poker hands with two pairs. For instance, consider the poker hand $\{4\clubsuit, 4\diamondsuit, J\heartsuit, J\spadesuit, 9\clubsuit\}$. This hand is counted twice if the steps outlined in the false solution in the exercise statement are followed. It is first counted when the denomination 4 is chosen in step one, the cards $4\clubsuit$ and $4\diamondsuit$ are chosen in step two, the denomination J is chosen in step three, the cards $J\heartsuit$ and $J\spadesuit$ are chosen in step four, and $9\clubsuit$ is chosen in step five. The hand is counted a second time when the denomination J is chosen in step one, the cards $J\heartsuit$ and $J\spadesuit$ are chosen in step two, the denomination 4 is chosen in step three, the cards $4\clubsuit$ and $4\diamondsuit$ are chosen in step four, and $9\clubsuit$ is chosen in step five.

Section 6.5

2. b. [a,a,a,a], [a,a,a,b], a,a,b,b], [a,a,b,c], [a,a,c,c], [a,b,b,b], [a,b,b,c], [a,b,c,c], [a,c,c,c], [b,b,b,b], [b,b,b,c], [b,b,c,c], [b,c,c,c], [c,c,c,c]

4. a. $\binom{30 + 10 - 1}{30} = \binom{39}{30} = 211,915,132$

   b. $\binom{26 + 10 - 1}{26} = \binom{35}{26} = 70,607,460$

   c. $\dfrac{\binom{35}{26}}{\binom{39}{30}} \cong .3332 = 33.32\%$

   d. $\dfrac{\binom{26 + 9 - 1}{26}}{\binom{39}{30}} = \dfrac{\binom{34}{26}}{\binom{39}{30}} = \dfrac{18,156,204}{211,915,132} \cong .0857 = 8.57\%$

6. $\binom{5 + n - 1}{5} = \binom{n + 4}{5} = \dfrac{(n+4)(n+3)(n+2)(n+1)n}{120}$

7. Consider any nonnegative integral solution $x_1, x_2, \ldots, x_n$ of the equation $x_1 + x_2 + \cdots + x_n = m$. For each $i = 1, 2, \ldots n$, let $y_i = x_1 + x_2 + \cdots + x_i$. Then $0 \leq y_1 \leq y_2 \leq \cdots \leq y_n = m$. Conversely, suppose $(y_1, y_2, \ldots, y_n)$ is any n-tuple of nonnegative integers such that $0 \leq y_1 \leq y_2 \leq \cdots \leq y_n = m$, and let $x_1 = y_1$, and $x_i = y_i - y_{i-1}$ for all integers $i = 1, 2, \ldots, n$. Then $x_1 + x_2 + \cdots + x_n = y_1 + (y_2 - y_1) + (y_3 - y_2) + \cdots + (y_n - y_{n-1}) = y_n = m$, and so $x_1 + x_2 + \cdots + x_n = m$. Consequently, the number of nonnegative integral solutions of the equation $x_1 + x_2 + \cdots + x_n = m$ is the same as the number of n-tuples of nonnegative integers $(y_1, y_2, \ldots, y_n)$ such that $0 \leq y_1 \leq y_2 \leq \cdots \leq y_n = m$. Since $y_n = m$, this is the same as the number of $(n-1)$-tuples of nonnegative integers $(y_1, y_2, \ldots, y_{n-1})$ where $0 \leq y_1 \leq y_2 \leq \cdots \leq y_{n-1} \leq m$. By reasoning similar to that of Example 6.5.3, this number is the same as the number of ways of placing $m + 1$ objects (the integers from 0 through m) into $n - 1$ categories (the elements of the $(n-1)$-tuple), which is $\binom{(n-1) + (m+1) - 1}{n-1} = \binom{(n-1) + m}{n-1}$. Thus

the number of nonnegative integral solutions of $x_1 + x_2 + \cdots + x_n = m$ is $\binom{(n-1)+m}{n-1}$. (In Section 6.6 we show that this number equals $\binom{(n-1)+m}{m}$, and so this result agrees with the one obtained in Example 6.5.5.)

9. The number of iterations of the inner loop is the same as the number of integer triples (i, j, k) where $1 \leq k \leq j \leq i \leq n$. By reasoning similar to that of Example 6.5.3, the number of such triples is $\binom{n+2}{3} = \dfrac{n(n+1)(n+2)}{6}$.

12. Think of the number 32 as divided into 32 individual units and the variables (y_1, y_2, y_3, y_4) as four categories into which these units are placed. The number of units in category x_i indicates the value of x_i in a solution of the equation. By Theorem 6.5.1, the number of ways to place 32 objects into four categories is $\binom{32+4-1}{32} = \binom{35}{32} = 6545$. So there are 6545 nonnegative integral solutions of the equation.

13. The analysis for this exercise is the same as for exercise 12 except that since each $y_i \geq 1$, we can imagine taking eight of the 32 units, placing two in each category (y_1, y_2, y_3, y_4), and then distributing the remaining 24 units among the four categories. The number of ways to do this is $\binom{24+4-1}{24} = \binom{27}{24} = 2925$. So there are 2925 integral solutions of the equation where each integer in the solution is at least two.

14. By reasoning similar to that of Example 6.5.6, after ten units have been placed in each category a, b, c, d, and e, 950 units remain to distribute among the five categories. The number of ways to do this is $\binom{950+5-1}{950} = \binom{954}{950} = 34{,}296{,}318{,}126$. So there are 34,296,318,126 solutions to $a + b + c + d + e = 1000$ for which each of a, b, c, d, and e is at least ten.

15. a. $\binom{24+30-1}{24} = \binom{53}{24} \cong 7.79255312 \times 10^{14}$

   b. probability $= \dfrac{\binom{(60-24)+24-1}{60-24}}{\binom{60+24-1}{60}} = \dfrac{\binom{59}{36}}{\binom{83}{60}} \cong .0126 = 1.26\%$

16. a. $\binom{20+4-1}{20} = \binom{23}{20} = 1771$

   b. probability $= \dfrac{\binom{(20-4 \cdot 4)+4-1}{20-4 \cdot 4}}{1771} = \dfrac{\binom{7}{4}}{1771} = \dfrac{25}{1771} \cong .0198 = 1.98\%$

17. Since the digits in 10,000 do not add up to 10, any number from 1 through 10,000 whose digits do add up to 10 must be a four-digit number. Imagine that the four digits are categories into which we place ten units. (For instance, $\times \times\,|\quad\,|\,\times \times \times \times \times \times\,|\,\times \times$ corresponds to the number 2062.) By Theorem 6.5.1, there are $\binom{10+4-1}{10} = \binom{13}{10} = 286$ ways to place the units into the categories. So there are 286 integers from 1 through 10,000 the sum of whose digits is 10.

19. a. For each selection of k A76 batteries (where $0 \leq k \leq 8$), $30 - k$ other batteries are obtained from the nine remaining types. The number of ways to select these other batteries is

$\binom{(30-k)+9-1}{30-k} = \binom{38-k}{30-k}$. So the total number of ways the inventory can be distributed is $\sum_{k=0}^{8} \binom{38-k}{30-k} = 197,607,982$.

b. For each selection of k A76 batteries and m D303 batteries, $30 - k - m$ other batteries are obtained from the eight remaining types. The number of ways to select these other batteries is $\binom{(30-k-m)+8-1}{30-k-m} = \binom{37-k-m}{30-k-m}$. So the total number of ways the inventory can be distributed is $\sum_{m=0}^{6}\sum_{k=0}^{8} \binom{37-k-m}{30-k-m} = 170,376,372$.

20. Consider those columns of a trace table corresponding to an arbitrary value of k. The values of j go from 1 to k, and for each value of j, the values of i go from 1 to j.

k	k					
j	1	2	3			k
i	1	1 2	1 2 3	. . .	1 2 3 . . . k

So for each value of k, there are $1 + 2 + 3 + \cdots + k$ columns of the table. Since k goes from 1 to n, the total number of columns in the table is

$1 + (1+2) + (1+2+3) + \cdots + (1+2+3+\cdots+n)$

$= \sum_{k=1}^{1} k + \sum_{k=1}^{2} k + \cdots + \sum_{k=1}^{n-1} k + \sum_{k=1}^{n} k$

$= \frac{1 \cdot 2}{2} + \frac{2 \cdot 3}{2} + \cdots + \frac{(n-1) \cdot n}{2} + \frac{n \cdot (n+1)}{2}$

$= \frac{1}{2}[1 \cdot 2 + 2 \cdot 3 + \cdots + (n-1) \cdot n + n \cdot (n+1)]$

$= \frac{1}{2}\left(\frac{n(n+1)(n+2)}{3}\right)$ by exercise 12 of Section 4.2

$= \frac{n(n+1)(n+2)}{6}$,

which agrees with the result of Example 6.5.4.

Section 6.6

2. $\binom{n}{1} = \frac{n!}{1!(n-1)!} = \frac{n \cdot (n-1)!}{(n-1)!} = n$

4. $\binom{n}{3} = \frac{n!}{3! \cdot (n-3)!} = \frac{n \cdot (n-1) \cdot (n-2) \cdot (n-3)!}{3 \cdot 2 \cdot 1 \cdot (n-3)!} = \frac{n(n-1)(n-2)}{6}$

7. By (6.6.5), $\binom{n}{n-2} = \frac{n(n-1)}{2}$ for $n \geq 2$. If $n \geq -1$, then $n+3 \geq 2$, and so substituting $n+3$ for n gives $\binom{n+3}{n+1} = \frac{(n+3)((n+3)-1)}{2} = \frac{(n+3)(n+2)}{2}$.

8. By (6.6.1), $\binom{n}{n} = 1$ for $n \geq 0$. If $k - r \geq 0$, then $k - r$ may be substituted for n giving $\binom{k-r}{k-r} = 1$.

10. $\binom{7}{0} = 1$, $\binom{7}{1} = 1 + 6 = 7$, $\binom{7}{2} = 6 + 15 = 21$, $\binom{7}{3} = 15 + 20 = 35$, $\binom{7}{4} = 20 + 15 = 35$, $\binom{7}{5} = 15 + 6 = 21$, $\binom{7}{6} = 6 + 1 = 7$, $\binom{7}{7} = 1$

11. $\binom{9}{0} = 1$, $\binom{9}{1} = 1 + 8 = 9$, $\binom{9}{2} = 8 + 28 = 36$, $\binom{9}{3} = 28 + 56 = 84$, $\binom{9}{4} = 56 + 70 = 126$, $\binom{9}{5} = 70 + 56 = 126$, $\binom{9}{6} = 56 + 28 = 84$, $\binom{9}{7} = 28 + 8 = 36$, $\binom{9}{8} = 8 + 1 = 9$, $\binom{9}{9} = 1$

12.
$$\begin{aligned}
\binom{n+3}{r} &= \binom{n+2}{r-1} + \binom{n+2}{r} \\
&= \left(\binom{n+1}{r-2} + \binom{n+1}{r-1}\right) + \left(\binom{n+1}{r-1} + \binom{n+1}{r}\right) \\
&= \binom{n+1}{r-2} + 2 \cdot \binom{n+1}{r-1} + \binom{n+1}{r} \\
&= \left(\binom{n}{r-3} + \binom{n}{r-2}\right) + 2 \cdot \left(\binom{n}{r-2} + \binom{n}{r-1}\right) + \left(\binom{n}{r-1} + \binom{n}{r}\right) \\
&= \binom{n}{r-3} + 3 \cdot \binom{n}{r-2} + 3 \cdot \binom{n}{r-1} + \binom{n}{r}
\end{aligned}$$

13. *Proof:* Suppose n and r are nonnegative integers with $r + 1 \leq n$. Then
$$\begin{aligned}
\frac{n-r}{r+1} \cdot \binom{n}{r} &= \frac{n-r}{r+1} \cdot \frac{n!}{r!(n-r)!} \\
&= \frac{n-r}{r+1} \cdot \frac{n!}{r!(n-r) \cdot (n-r-1)!} \\
&= \frac{n!}{(r+1)! \cdot (n-r-1)!} \\
&= \frac{n!}{(r+1)! \cdot (n-(r+1))!} \\
&= \binom{n}{r+1}
\end{aligned}$$
This is what was to be shown.

15. *Proof:* Let n be an integer with $n \geq 1$. By exercise 14, $\sum_{i=1}^{n+1} \binom{i}{2} = \binom{n+2}{3}$. But for each $i = 2, 3, \ldots, n+1$,
$$\binom{i}{2} = \frac{i!}{i! \cdot (i-2)!} = \frac{i \cdot (i-1)}{2} = \frac{(i-1) \cdot i}{2}.$$
So
$$\begin{aligned}
\binom{n+2}{3} &= \sum_{i=2}^{n+1} \binom{i}{2} \\
&= \frac{1 \cdot 2}{2} + \frac{2 \cdot 3}{2} + \frac{3 \cdot 4}{2} + \cdots + \frac{n \cdot (n+1)}{2} \\
&= \frac{1 \cdot 2 + 2 \cdot 3 + 3 \cdot 4 + \cdots + n \cdot (n+1)}{2}.
\end{aligned}$$
Multiplying both sides by 2 gives
$$1 \cdot 2 + 2 \cdot 3 + 3 \cdot 4 + \cdots + n \cdot (n+1) = 2 \cdot \binom{n+2}{3}.$$

17. b. *Proof*: Let n be an integer with $n \geq 1$. Then

$$\begin{aligned}
\frac{1}{4n+2}\binom{2n+2}{n+1} &= \left(\frac{1}{2(2n+1)}\right) \cdot \left(\frac{(2n+2)!}{(n+1)!((2n+2)-(n+1))!}\right) \\
&= \left(\frac{1}{2(2n+1)}\right) \cdot \left(\frac{(2n+2)!}{(n+1)!(n+1)!}\right) \\
&= \left(\frac{1}{2(2n+1)}\right) \cdot \left(\frac{(2n+2)(2n+1)(2n)!}{(n+1)\cdot n! \cdot (n+1)\cdot n!}\right) \\
&= \frac{1}{2}\cdot\left(\frac{2(n+1)}{(n+1)\cdot(n+1)}\right) \cdot \left(\frac{(2n)!}{n!\cdot n!}\right) \\
&= \frac{1}{n+1}\binom{2n}{n} \\
&= C_n.
\end{aligned}$$

18. Suppose m and n are positive integers and r is a nonnegative integer that is less than or equal to the minimum value of m and n. Let S be a set of $m+n$ elements, and write $S = A \cup B$, where A is a subset of S with m elements, B is a subset of S with n elements, and $A \cap B = \emptyset$. The collection of subsets of r elements chosen from S can be partitioned as follows: those consisting of 0 elements chosen from A and r elements chosen from B, those consisting of 1 element chosen from A and $r-1$ elements chosen from B, those consisting of 2 elements chosen from A and $r-2$ elements chosen from B, and so forth, up to those consisting of r elements chosen from A and 0 elements chosen from B. By the multiplication rule, there are $\binom{m}{i}\binom{n}{r-i}$ ways to choose i objects from m and $r-i$ objects from n, and so the number of subsets of size r in which i elements are from A and $r-i$ objects are from B is $\binom{m}{i}\binom{n}{r-i}$. Hence by the addition rule, the total number of subsets of S of size r is

$$\binom{m}{0}\binom{n}{r} + \binom{m}{1}\binom{n}{r-1} + \binom{m}{2}\binom{n}{r-2} + \cdots + \binom{m}{r}\binom{n}{0}.$$

But also since S has $m+n$ elements, the number of subsets of size r of S is $\binom{m+n}{r}$. So

$$\binom{m+n}{r} = \binom{m}{0}\binom{n}{r} + \binom{m}{1}\binom{n}{r-1} + \binom{m}{2}\binom{n}{r-2} + \cdots + \binom{m}{r}\binom{n}{0}.$$

19. *Proof*: Let n be any integer with $n \geq 0$. By exercise 18 with $m = r = n$,

$$\binom{n+n}{n} = \binom{n}{0}\binom{n}{n} + \binom{n}{1}\binom{n}{n-1} + \binom{n}{2}\binom{n}{n-2} + \cdots + \binom{n}{n}\binom{n}{0}.$$

But by Example 6.6.2, $\binom{n}{n-k} = \binom{n}{k}$ for all $k = 0, 1, 2, \ldots, n$. Hence

$$\binom{2n}{n} = \binom{n}{0}^2 + \binom{n}{1}^2 + \binom{n}{2}^2 + \cdots + \binom{n}{n}^2,$$

as was to be shown.

20. Let m be any nonnegative integer. We will show by mathematical induction that the given formula holds for all integers $n \geq 0$.

The formula holds for n = 0: For $n = 0$, the formula asserts that $\binom{m}{0} = \binom{n+0+1}{0} = \binom{n+1}{0}$. But this is true because by exercise 1 both sides of this equation equal 1.

If the formula holds for n = k then it holds for n = k+1: Let k be an integer with $k \geq 0$, and suppose

$$\binom{m}{0} + \binom{m+1}{1} + \binom{m+2}{2} + \cdots + \binom{m+k}{k} = \binom{m+k+1}{k}.$$

[This is the inductive hypothesis.]

We must show that

$$\binom{m}{0} + \binom{m+1}{1} + \binom{m+2}{2} + \cdots + \binom{m+(k+1)}{k+1} = \binom{m+(k+1)+1}{k+1},$$

or, equivalently,

$$\binom{m}{0} + \binom{m+1}{1} + \binom{m+2}{2} + \cdots + \binom{m+k+1}{k+1} = \binom{m+k+2}{k+1}.$$

But

$$\binom{m}{0} + \binom{m+1}{1} + \binom{m+2}{2} + \cdots + \binom{m+k+1}{k+1} = \binom{m+k+1}{k} + \binom{m+k+1}{k+1}$$

by inductive hypothesis

$$= \binom{m+k+2}{k+1}$$

by Pascal's formula (with $m+k$ in place of n and $k+1$ in place of r).

[This is what was to be shown.]

21. *Proof:* Let p be a prime number and r an integer with $0 < r < p$. Then $\binom{p}{r} = \frac{p!}{r! \cdot (p-r)!} = \frac{p \cdot (p-1)!}{r! \cdot (p-r)!}$, or, equivalently, $p \cdot (p-1)! = \binom{p}{r} \cdot (r! \cdot (p-r)!)$. Now $\binom{p}{r}$ is an integer because it equals the number of subsets of size r that can be formed from a set with p elements. Thus we can apply the unique factorization theorem to express each side of this equation as a product of prime numbers. Clearly, p is a factor of the left-hand side, and so p must also be a factor of the right-hand side. But $0 < r < p$, and so p does not appear as one of the prime factors in either $r!$ or $(p-r)!$. Therefore, p must occur as one of the prime factors of $\binom{p}{r}$, and hence $\binom{p}{r}$ is divisible by p.

Section 6.7

3. $(p + 3q)^4 = p^4 + \binom{4}{1}p^3(3q)^1 + \binom{4}{2}p^2(3q)^2 + \binom{4}{3}p^1(3q)^3 + (3q)^4 = p^4 + 12p^3q + 54p^2q^2 + 108pq^3 + 81q^4$

4. $(u^2 - 2v)^4 = (u^2)^4 + \binom{4}{1}(u^2)^3(-2v)^1 + \binom{4}{2}(u^2)^2(-2v)^2 + \binom{4}{3}(u^2)^1(-2v)^3 + (-2v)^4$
$= u^8 - 8u^6v + 24u^4v^2 - 32u^2v^3 + 16v^4$

5. $\left(\dfrac{2}{a} - \dfrac{a}{2}\right)^5$

$= \left(\dfrac{2}{a}\right)^5 + \binom{5}{1}\left(\dfrac{2}{a}\right)^4\left(-\dfrac{a}{2}\right)^1 + \binom{5}{2}\left(\dfrac{2}{a}\right)^3\left(-\dfrac{a}{2}\right)^2 + \binom{5}{3}\left(\dfrac{2}{a}\right)^2\left(-\dfrac{a}{2}\right)^3 + \binom{5}{4}\left(\dfrac{2}{a}\right)^1\left(-\dfrac{a}{2}\right)^4 + \left(-\dfrac{a}{2}\right)^5$

$= \dfrac{32}{a^5} - \dfrac{40}{a^3} + \dfrac{20}{a} - 5a + \dfrac{5}{8}a^3 - \dfrac{a^5}{32}$

6. $(a+b)^6$

$\begin{aligned}
&= (a+b)\cdot(a+b)^5 \\
&= (a+b)\cdot(a^5 + 5a^4b + 10a^3b^2 + 10a^2b^3 + 5ab^4 + b^5) \\
&= a^6 + 5a^5b + 10a^4b^2 + 10a^3b^3 + 5a^2b^4 + ab^5 + ba^5 + 5a^4b^2 + 10a^3b^3 + 10a^2b^4 + 5ab^5 + b^6 \\
&= a^6 + (5+1)a^5b + (10+5)a^4b^2 + (10+10)a^3b^3 + (5+10)a^2b^4 + (1+5)ab^5 + b^6 \\
&= a^6 + 6a^5b + 15a^4b^2 + 20a^3b^3 + 15a^2b^4 + 6ab^5 + b^6
\end{aligned}$

8. Term is $\binom{10}{3}(2x)^7 3^3$. Coefficient is $\dfrac{10!}{3!\cdot 7!}\cdot 2^7 \cdot 3^3 = 414{,}720$.

10. Term is $\binom{10}{2}(u^2)^8(-v^2)^2$. Coefficient is $\dfrac{10!}{2!\cdot 8!}\cdot (-1)^2 = 45$.

12. $(1.2)^{4000} = 1^{4000} + \binom{4000}{1}\cdot 1^{3999}\cdot (0.2)^1 +$ other positive terms $= 801+$ other positive terms

Therefore, $(1.2)^{4000}$ is greater than 800.

14. *Proof:* Let n be an integer with $n \geq 0$. Apply the binomial theorem with $a = 5$ and $b = -1$ to obtain

$\begin{aligned}
3^n &= (1+2)^n \\
&= \binom{n}{0}\cdot 1^n \cdot 2^0 + \binom{n}{1}\cdot 1^{n-1}\cdot 2^1 + \cdots + \binom{n}{k}\cdot 1^{n-k}\cdot 2^k + \cdots + \binom{n}{n}\cdot 1^{n-n}\cdot 2^n \\
&= \binom{n}{0} + 2\binom{n}{1} + 2^2\binom{n}{2} + \cdots + 2^n\binom{n}{n}
\end{aligned}$

because $2^0 = 1$ and $1^{n-k} = 1$ for all integers k.

15. Let n be an integer with $n \geq 0$. Apply the binomial theorem with $a = 5$ and $b = -1$ to obtain

$\begin{aligned}
4^n &= (5+(-1))^n = \sum_{k=0}^{n}\binom{n}{k}5^{n-k}(-1)^k \\
&= \binom{n}{0}5^n(-1)^0 + \binom{n}{1}5^{n-1}(-1)^1 + \binom{n}{2}5^{n-2}(-1)^2 + \cdots + \binom{n}{n}5^0(-1)^n \\
&= 5^n - \binom{n}{1}5^{n-1} + \binom{n}{2}5^{n-2} - \binom{n}{3}5^{n-3} + \cdots + \binom{n}{n-1}5(-1)^{n-1} + \binom{n}{n}(-1)^n.
\end{aligned}$

16. Let n be an integer with $n \geq 0$. Apply the binomial theorem with $a = -1$ and $b = 4$ to obtain

$\begin{aligned}
3^n &= ((-1)+4)^n = \sum_{k=0}^{n}\binom{n}{k}(-1)^{n-k}4^k \\
&= \binom{n}{0}(-1)^n 4^0 + \binom{n}{1}(-1)^{n-1}4^1 + \cdots + \binom{n}{n-1}(-1)^1 4^{n-1} + \binom{n}{n}(-1)^0 4^n \\
&= (-1)^n\binom{n}{0} + (-1)^{n-1}\binom{n}{1}4 + \cdots - \binom{n}{n-1}4^{n-1} + \binom{n}{n}4^n \\
&= \binom{n}{n}4^n - \binom{n}{n-1}4^{n-1} + \cdots + (-1)^{n-1}\binom{n}{n-1}4^{n-1} + (-1)^n\binom{n}{n}4^n.
\end{aligned}$

17. *a.* Let n be an integer with $n \geq 0$. Apply the binomial theorem with $a = 1$ and $b = x$ to obtain

$$(1+x)^n = \sum_{k=0}^{n} \binom{n}{k} 1^{n-k} x^k = \sum_{k=0}^{n} \binom{n}{k} x^k$$

because any power of 1 is 1.

c. (ii) Let n be an integer with $n \geq 1$. Apply the formula from part (b) with $x = -1$ to obtain

$$0 = n(1+(-1))^{n-1} = \sum_{k=1}^{n} \binom{n}{k} k(-1)^{k-1}.$$

d. Apply the formula of part (b) with $x = 3$ to obtain

$$n \cdot 4^{n-1} = n(1+3)^{n-1} = \sum_{k=1}^{n} \binom{n}{k} k 3^{k-1} = \sum_{k=1}^{n} \binom{n}{k} k 3^k 3^{-1} = \frac{1}{3} \sum_{k=1}^{n} \binom{n}{k} k 3^k.$$

So $\sum_{k=1}^{n} \binom{n}{k} k 3^k = 3n 4^{n-1}$.

Chapter 7 Functions

Students often come out of high school mathematics courses identifying functions with formulas. The aim of Section 7.1 is to promote a broad view of the function concept and to give students experience with the wide variety of functions that arise in discrete mathematics. Representation of functions by arrow diagrams is emphasized to prepare the way for the discussion of one-to-one and onto functions in Section 7.3. Students do not usually find Section 7.1 difficult, but many seem to need the practice working with functions defined on, say, power sets or formal languages to be able to reason effectively with such functions in later sections of the chapter. For instance, if you are planning to assign exercise 20 in Section 7.3, it is desirable to have previously assigned exercise 13 in Section 7.1.

The discussion of finite-state automata in Section 7.2 is intended to be a natural sequel to the discussion of Boolean functions in Section 7.1. The next-state function of an automaton governs the operation of sequential circuit in much the same way that a Boolean function governs the operation of a combinatorial circuit. Students seem genuinely to enjoy working with automata. When you present this section in class, you might add the following kind of example: for a particular finite-state automaton under discussion, give a list of strings and ask students to determine whether or not these strings are accepted by the automaton. Such an example makes a nice lead in to the concept of the language accepted by an automaton.

As they are learning about one-to-one and onto functions in Section 7.3, students appreciate some explicit review of logical principles such as the negation of ∀, ∃, and if-then statements and the equivalence of a conditional statement and its contrapositive. These logical principles are needed, of course, to understand the equivalence of the two forms of the definition of one-to-one and what it means for a function not to be one-to-one or onto. This is a good opportunity to solicit student participation since at this point in the course students in theory know the logic, and so you can ask them to recall it and apply it to the study of function properties themselves. Because the techniques used to test for injectivity and surjectivity and to find inverse functions are quite different for functions with finite and infinite domains, examples using both kinds of functions are discussed in this section.

Section 7.4 on the pigeonhole principle is another section that students seem especially to enjoy. The range of difficulty of the problems is deliberately broad to enable you to tailor your choice of exercises to the abilities of your students.

Sections 7.5 and 7.6 go together in the sense that the relations between one-to-one and onto functions and composition of functions developed in Section 7.5 are used to prove the fundamental theorem about cardinality in Section 7.6. The proofs that a composition of one-to-one functions is one-to-one or that a composition of onto functions is onto (and the related exercises) test the degree to which students have learned to instantiate mathematical definitions in abstract contexts, apply the method of generalizing from the generic particular in a sophisticated setting, develop mental models of mathematical concepts that are both vivid and generic enough to reason with, and create moderately complex chains of deductions.

There are always students who respond with enthusiasm to the idea of different sizes of infinity discussed in Section 7.6. When covering this section, it is of interest to point out the connections among the Cantor diagonalization argument used in the proof of the uncountablity of the reals and the arguments used in the discussion of Russell's paradox and to solve the halting problem in Section 5.4.

Comments on Exercises:

Exercises #5 in Section **7.1** and #6 and #7 in Section **7.3** explore the question of how many functions of certain types there are from a finite set of one size to a finite set of a (possibly different) size. Exercise #10 in Section **7.1** is somewhat helpful for #12 of **Section 7.5**. Exercise #12 in Section **7.1** is preparation for #19 and #40 in Section **7.3**, and Exercise #13 in Section **7.1** is preparation for #18, #20, #21, #39, and #41 in Section 7.3. Exercise #24 in Section 7.1 integrates topics from Chapters 6 and 7 by relating permutations and functions. Exercises 1b, 1c, and 11 in Section 7.3 are designed to counteract a common misunderstanding about the definition of one-to-one (that a function $f: X \to Y$ is one-to-one if each element of X is sent to exactly one element of Y). Exercise #46 in Section 7.3 integrates topics from Chapters 5, 6, and 7. The fact that a set with n elements has 2^n subsets is first proved by induction in Section 5.3. In Section 6.7 a second proof is given using the binomial theorem. This exercise leads students through a third proof that works by setting up a one-to-one correspondence between subsets and strings of 0's and 1's of length n.

Section 7.1

2. *a.* domain of $g = \{1, 3, 5\}$, co-domain of $g = \{s, t, u, v\}$ *b.* $g(1) = g(3) = g(5) = t$
   c. range of $g = \{t\}$ *d.* inverse image of $t = \{1, 3, 5\}$, inverse image of $u = \emptyset$
   e. $\{(1, t), (3, t), (5, t)\}$

3. The arrow diagram in *d* determines a function; those in *a*, *b*, *c*, and *e* do not.

4. *b.* There is just one function from X to Y. It is represented by the arrow diagram shown below.

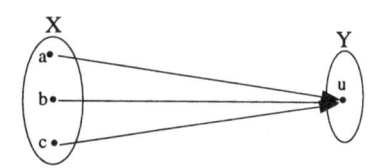

   c. There are eight functions from X to Y. They are represented by the arrow diagrams shown below.

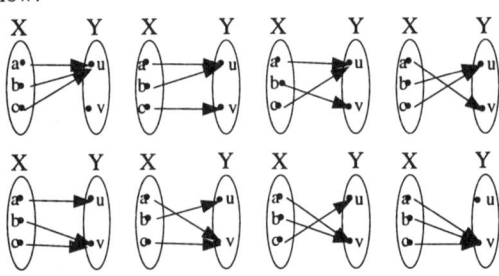

5. *b.* $2 \cdot 2 \cdot 2 \cdot 2 \cdot 2 = 2^5$

   c. The answer is n^m because each of the m elements of the domain can be sent to any one of n possible elements in the co-domain. In other words, the m elements of the domain can be placed in order and the process of constructing a function can be thought of as an m-step operation where, for each i from 1 to m, the ith step is to choose one of the n elements of the co-domain to be the image of the ith element of the domain. Since there are n ways to perform each of the m steps of the operation, the entire operation can be performed in $n \cdot n \cdot \ldots \cdot n$ (m factors) ways.

7. No. For instance, $H(3) = \lfloor 3 \rfloor + 1 = 3 + 1 = 4$, whereas $K(3) = \lceil 3 \rceil = 3$.

9. No. For instance, let F and G be defined by the rules $F(x) = x$ and $G(x) = 0$ for all real numbers x. Then $(F - G)(2) = F(2) - G(2) = 2 - 0 = 2$, whereas $(G - F)(2) = G(2) - F(2) = 0 - 2 = -2$, and $2 \neq -2$.

10. $i_{\mathbf{Z}}(e) = e$, $i_{\mathbf{Z}}(b_i^{jk}) = b_i^{jk}$, $i_{\mathbf{Z}}(K(t)) = K(t)$

11. b. Define $F: \mathbf{Z}^{nonneg} \to \mathbf{R}$ as follows: for each nonnegative integer n, $F(n) = (-1)^n(2n)$.

12. b. $F(\emptyset) = 0$ (because 0 is an even number)   d. $F(\{2,3,4,5\}) = 0$

13. b. $g(aba) = aba$, $g(bbab) = babb$, $g(b) = b$ The range of g is Σ^*.

14. b. $3^{-2} = 1/9$

   d. the exponent to which 5 must be raised to obtain 5^n is n

   e. $4^0 = 1$

15. b. $\log_2 1024 = 10$ because $2^{10} = 1024$

   d. $\log_2 1 = 0$ because $2^0 = 1$

   e. $\log_4 \dfrac{1}{4} = -1$ because $4^{-1} = \dfrac{1}{4}$

   f. $\log_2 2 = 1$ because $2^1 = 2$

   g. $\log_2 2^k = k$ because the exponent to which 2 must be raised to obtain 2^k is k

17. *Proof*: Let b be any positive real number with $b \neq 1$. Then $b^0 = 1$, and so $\log_b 1 = 0$. (Note that we do not allow b to equal 1 here because $\log_b$ is not defined for $b = 1$.)

19. Since $\log_b y = 2$, then $b^2 = y$. Hence $(b^2)^1 = y$, and so $\log_{b^2}(y) = 1$.

20. b. $p_2(2, y) = y$, $p_2(5, x) = x$ The range of p_2 is $\{x, y\}$.

21. b. $mod(57, 8) = 1$, $div(57, 8) = 7$ c. $mod(32, 4) = 0$, $div(32, 4) = 8$

22. b. $E(1010) = 111000111000$, $D(000000111111) = 0011$

23. b. $H(00110, 10111) = 2$

24. b.

```
1 2 3 4         1 2 3 4
↓ ↓ ↓ ↓         ↓ ↓ ↓ ↓
1 2 3 4         3 2 1 4
```

   d.

```
1 2 3 4         1 2 3 4         1 2 3 4
↓ ↓ ↓ ↓         ↓ ↓ ↓ ↓         ↓ ↓ ↓ ↓
2 3 4 1         3 4 1 2         4 1 2 3

1 2 3 4         1 2 3 4         1 2 3 4
↓ ↓ ↓ ↓         ↓ ↓ ↓ ↓         ↓ ↓ ↓ ↓
2 1 4 3         3 1 4 2         4 3 1 2

1 2 3 4         1 2 3 4         1 2 3 4
↓ ↓ ↓ ↓         ↓ ↓ ↓ ↓         ↓ ↓ ↓ ↓
2 4 1 3         3 4 2 1         4 3 2 1
```

25. *b.*

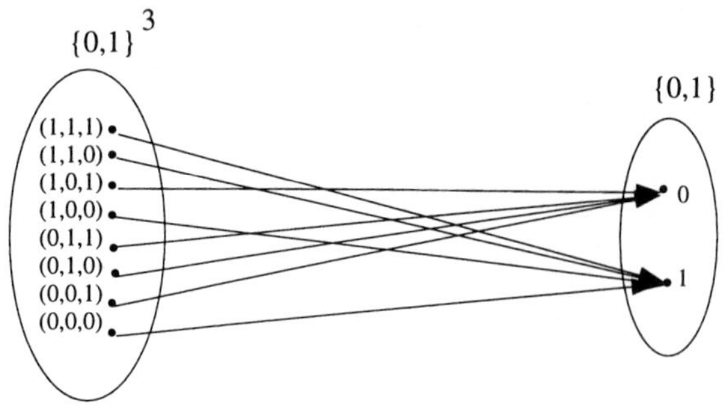

26.

input		f_1	f_2	f_3	f_4	f_5	f_6	f_7	f_8	f_9	f_{10}	f_{11}	f_{12}	f_{13}	f_{14}	f_{15}	f_{16}
1	1	0	0	0	0	0	0	0	0	1	1	1	1	1	1	1	1
1	0	0	0	0	0	1	1	1	1	0	0	0	0	1	1	1	1
0	1	0	0	1	1	0	0	1	1	0	0	1	1	0	0	1	1
1	1	0	1	0	1	0	1	0	1	0	1	0	1	0	1	0	1

27. *b.*

input			**output**
x_1	x_2	x_3	f
1	1	1	0
1	1	0	0
1	0	1	1
1	0	0	1
0	1	1	1
0	1	0	1
0	0	1	0
0	0	0	0

29. No. Suppose h were well-defined. Then $h(\frac{1}{2}) = h(\frac{2}{4})$ because $\frac{1}{2} = \frac{2}{4}$. But $h(\frac{1}{2}) = \frac{1}{2}$ and $h(\frac{2}{4}) = \frac{2^2}{4} = 1$, and $\frac{1}{2} \neq 1$. *[This contradiction shows that the supposition that h is well-defined is false, and so h is not well-defined.]*

30. f is not well defined because $f(n) \notin S$ for many values of n in S. For instance, $f(100\,000) = (100\,000)^2 = 10\,000\,000\,000 \notin S$.

31. *a.*

$$\begin{aligned}\chi_A(u)\cdot\chi_B(u) &= \begin{cases} 1\cdot 1, & \text{if } u \in A \text{ and } u \in B \\ 1\cdot 0, & \text{if } u \in A \text{ and } u \notin B \\ 0\cdot 1, & \text{if } u \notin A \text{ and } u \in B \\ 0\cdot 0, & \text{if } u \notin A \text{ and } u \notin B \end{cases} \\ &= \begin{cases} 1, & \text{if } u \in A\cap B \\ 0, & \text{if } u \notin A\cap B \end{cases} \\ &= \chi_{A\cap B}(u)\end{aligned}$$

b.

$$\chi_A(u) + \chi_B(u) - \chi_A(u) \cdot \chi_B(u) = \begin{cases} 1+1-1\cdot 1, & \text{if } u \in A \text{ and } u \in B \\ 1+0-1\cdot 0, & \text{if } u \in A \text{ and } u \notin B \\ 0+1-0\cdot 1, & \text{if } u \notin A \text{ and } u \in B \\ 0+0-0\cdot 0, & \text{if } u \notin A \text{ and } u \notin B \end{cases}$$

$$= \begin{cases} 1, & \text{if } u \in A \text{ and } u \in B \\ 1, & \text{if } u \in A \text{ and } u \notin B \\ 1, & \text{if } u \notin A \text{ and } u \in B \\ 0, & \text{if } u \notin A \text{ and } u \notin B \end{cases}$$

$$= \begin{cases} 1, & \text{if } u \in A \cup B \\ 0, & \text{if } u \notin A \cup B \end{cases}$$

$$= \chi_{A \cup B}(u)$$

32. $\phi(12) = 4$ *[because 1, 5, 7, and 11 have no common factors with 12 other than ±1]*

$\phi(11) = 10$ *[because 1, 2, 3, 4, 5, 6, 7, 8, 9, 10, and 11 have no common factors with 11 other than ±1]*

$\phi(1) = 1$ *[because 1 is the only positive integer which has no common factors with 1 other than ±1]*

34. *Proof*: By exercise 33 with $p = 2$, for any integer n with $n \geq 1$, $\phi(2^n) = 2^n - 2^{n-1} = 2^{n-1}(2-1) = 2^{n-1}$. Given any odd integer $n \geq 3$, $n = 2k+1$ for some integer $k \geq 1$. Hence $\phi(2^n) = 2^{n-1} = 2^{(2k+1)-1} = 2^{2k} = (2^k)^2$, which is a perfect square. Thus $\phi(2^n)$ is a perfect square for each of the infinitely many odd integers $n \geq 3$.

35. *Proof*: Given any integer n with $n = pq$, where p and q are prime numbers, let A be the set of all divisors of n that are divisible by p and let B be the set of all divisors of n that are divisible by q. Note that $A = \{p, 2p, 3p, \ldots, qp\}$, $B = \{q, 2q, 3q, \ldots, pq\}$, and $A \cap B = \{pq\}$. By definition of ϕ,

$$\begin{aligned} \phi(n) &= n - [n(A \cup B)] \\ &= n - [n(A) + n(B) - n(A \cap B)] \quad \text{by the inclusion/exclusion formula} \\ &= pq - [q + p - 1] \\ &= (p-1)(q-1) \end{aligned}$$

36. *Proof*: Given any integer n with $n = pqr$, where p, q, and r are prime numbers, let A be the set of all divisors of n that are divisible by p, B the set of all divisors of n that are divisible by q, and C the set of all divisors of n that are divisible by r. Note that $A = \{p, 2p, 3p, \ldots, qr \cdot p\}$, $B = \{q, 2q, 3q, \ldots, pr \cdot q\}$, $C = \{r, 2r, 3r, \ldots, pq \cdot r\}$, $A \cap B = \{pq, 2pq, 3pq, \ldots, r \cdot pq\}$, $A \cap C = \{pr, 2pr, 3pr, \ldots, q \cdot pr\}$, $B \cap C = \{qr, 2qr, 3qr, \ldots, p \cdot qr\}$, and $A \cap B \cap C = \{pqr\}$. By definition of ϕ,

$$\begin{aligned} \phi(n) &= n - [n(A \cup B \cup C)] \\ &= n - [n(A) + n(B) + n(C) - n(A \cap B) - n(A \cap C) - n(B \cap C) + n(A \cap B \cap C)] \\ & \quad \text{by the inclusion/exclusion formula} \\ &= pqr - [qr + pr + pq - r - q - p + 1] \\ &= pqr - qr - pr - pq + r + q + p - 1 \\ &= (p-1)(q-1)(r-1) \end{aligned}$$

38. This property is true. *Proof*: Let $f: X \to Y$ be any function, and suppose $A \subseteq X$ and $B \subseteq X$.

$f(A \cup B) \subseteq f(A) \cup f(B)$: Let $y \in f(A \cup B)$. Then $y = f(x)$ for some $x \in A \cup B$. By definition of union, $x \in A$ or $x \in B$. So $y = f(x)$ for some $x \in A$ (in which case $y \in f(A)$) or $y = f(x)$ for some $x \in B$ (in which case $y \in f(B)$). Hence by definition of union, $y \in f(A) \cup f(B)$.

$f(A) \cup f(B) \subseteq f(A \cup B)$: Let $y \in f(A) \cup f(B)$. By definition of union, $y \in f(A)$ or $y \in f(B)$. If $y \in f(A)$, then $y = f(x)$ for some $x \in A$. In this case, since $A \subseteq B$, $x \in A \cup B$, and so $y \in f(A \cup B)$. If $y \in f(B)$, then $y = f(x)$ for some $x \in B$. In this case, since $B \subseteq B$, $x \in A \cup B$, and so $y \in f(A \cup B)$. Hence, in either case, $y \in f(A \cup B)$.

40. This property is false. *Counterexample*: Let $X = \{1, 2, 3\}$, $Y = \{a, b\}$, $A = \{1, 2\}$, $B = \{3\}$, and let $f(1) = f(3) = a$ and $f(2) = b$. Then $f(A) = \{a, b\}$, $f(B) = \{a\}$, $f(A - B) = f(\{1, 2\}) = \{a, b\}$, and $f(A) - f(B) = \{a, b\} - \{a\} = \{b\}$. So $f(A - B) \neq f(A) - f(B)$.

42. This property is true. *Proof*: Let $f: X \to Y$ be any function, and suppose that $C \subseteq Y$ and $D \subseteq Y$. For any element x in X, by definition of inverse image and union,

$$x \in f^{-1}(C \cup D) \Leftrightarrow f(x) \in C \cup D \Leftrightarrow f(x) \in C \text{ or } f(x) \in D$$
$$\Leftrightarrow x \in f^{-1}(C) \text{ or } \in f^{-1}(D) \Leftrightarrow x \in f^{-1}(C) \cup f^{-1}(D).$$

Hence $f^{-1}(C \cup D) = f^{-1}(C) \cup f^{-1}(D)$.

43. This property is true. *Proof*: Let $f: X \to Y$ be any function, and suppose that $C \subseteq Y$ and $D \subseteq Y$. For any element x in X, by definition of inverse image and intersection,

$$x \in f^{-1}(C \cap D) \Leftrightarrow f(x) \in C \cap D \Leftrightarrow f(x) \in C \text{ and } f(x) \in D$$
$$\Leftrightarrow x \in f^{-1}(C) \text{ and } x \in f^{-1}(D) \Leftrightarrow x \in f^{-1}(C) \cap f^{-1}(D).$$

Hence $f^{-1}(C \cap D) = f^{-1}(C) \cap f^{-1}(D)$.

44. This property is true. *Proof*: Let $f: X \to Y$ be any function, and suppose that $C \subseteq Y$ and $D \subseteq Y$. For any element x in X, by definition of inverse image and set difference,

$$x \in f^{-1}(C - D) \Leftrightarrow f(x) \in C - D \Leftrightarrow f(x) \in C \text{ and } f(x) \notin D$$
$$\Leftrightarrow x \in f^{-1}(C) \text{ and } x \notin f^{-1}(D) \Leftrightarrow x \in f^{-1}(C) - f^{-1}(D).$$

Hence $f^{-1}(C - D) = f^{-1}(C) - f^{-1}(D)$.

Section 7.2

1. b. 20¢ or more deposited   c. 15¢ deposited

3. a. U_0, U_1, U_2, U_3 b. a, b c. U_0 d. U_3

   e.

	input a	input b
→ U_0	U_2	U_1
U_1	U_2	U_3
U_2	U_2	U_2
◎ U_3	U_3	U_3

4. a. s_0, s_1, s_2 b. 0, 1   c. s_0 d. s_2

   e.

	input 0	input 1
→ s_0	s_1	s_0
s_1	s_2	s_0
◎ s_2	s_2	s_0

6. a. s_0, s_1, s_2, s_3 b. $0, 1$ c. s_0 d. s_0

e.

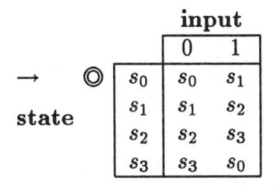

	input	
state	0	1
→ ◎ s_0	s_0	s_1
s_1	s_1	s_2
s_2	s_2	s_3
s_3	s_3	s_0

9. a. s_0, s_1, s_2, s_3 b. $0, 1$ c. s_0 d. s_1

e.

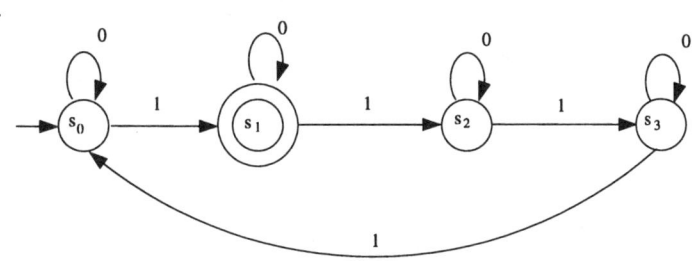

10. b. $N(s_2, 0) = s_3$, $N(s_1, 0) = s_3$ d. $N^*(s_2, 11010) = s_3$, $N^*(s_0, 01000) = s_3$

11. b. $N(s_0, 0) = s_1$, $N(s_4, 1) = s_3$ d. $N^*(s_0, 1111) = s_3$, $N^*(s_2, 00111) = s_2$

13. a. (i) U_3 (ii) U_2 (iii) U_2 (iv) U_3

    b. *bb* and *bbaaaabaa*

    c. The language accepted by this automaton is the set of all strings of a's and b's that begin *bb*.

14. The language accepted by this automaton is the set of all strings of 0's and 1's that end 00.

16. The language accepted by this automaton is the set of all strings in which the number of 1's is divisible by 4.

19. The language accepted by this automaton is the set of all strings in which the number of 1's has the form $4k + 1$ for some integer k.

21.

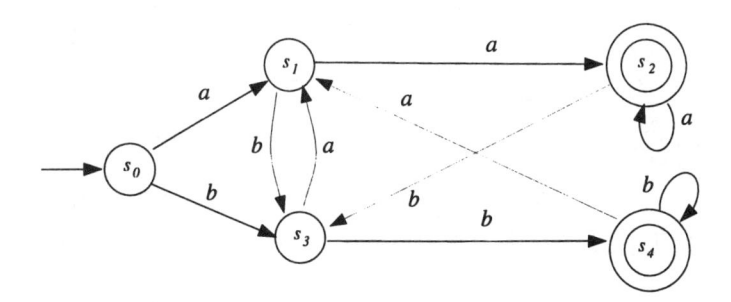

141

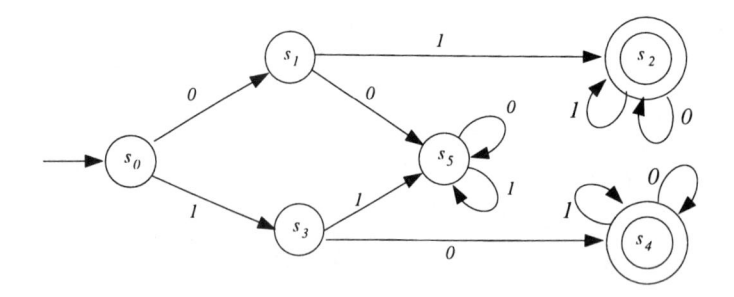

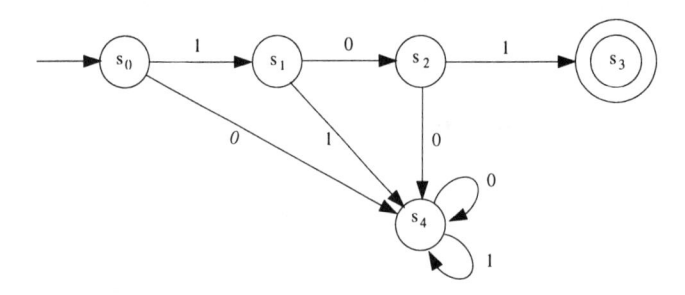

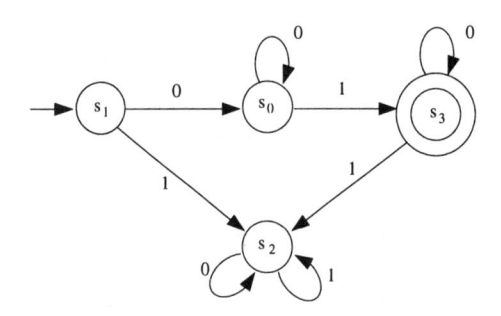

22.

25.

28.

30.

[Transition diagram with states s_0 through s_{12}. From s_0: on 1 to s_8, on 0 to s_{12}. From s_8: on 2,3,4,5,6,7,8,9 to s_9. From s_{12}: on 2,3,4,5,6,7,8,9 to s_9. From s_9: on 0,1 to s_{10}. From s_{10}: on 0,1,2,3,4,5,6,7,8,9 to s_{11}. From s_{11}: on 2,3,4,5,6,7,8,9 to s_1. From s_0: on 2,3,4,5,6,7,8,9 to s_1. From s_1: on 2,3,4,5,6,7,8,9 to s_2. From s_2: on 0,1,2,3,4,5,6,7,8,9 to s_4. From s_4: on 0,1,2,3,4,5,6,7,8,9 to s_3. From s_3: on 0,1,2,3,4,5,6,7,8,9 to s_5. From s_5: on 0,1,2,3,4,5,6,7,8,9 to s_6. From s_6: on 0,1,2,3,4,5,6,7,8,9 to s_7. States s_7 and s_{12} are accepting.]

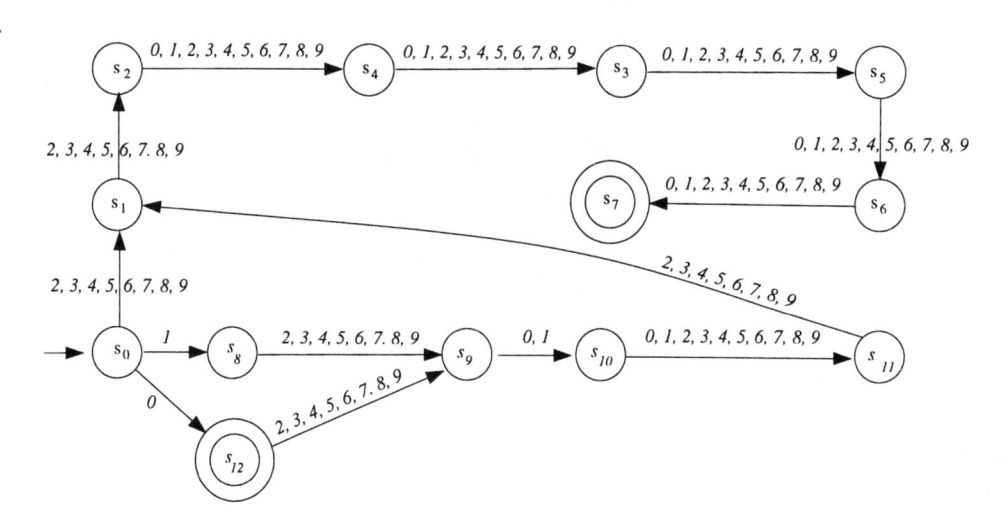

31.

Algorithm Finite-State Automaton of Exercise 2

[This algorithm simulates the action of the finite-state automaton of exercise 2 by mimicking the functioning of the transition diagram. The states are denoted 0, 1, and 2.]

Input: *string [a string of 0's and 1's plus an end marker e]*

Algorithm Body:

$state := 0$

$symbol :=$ first symbol in the input string

while $(symbol \neq e)$

if $state = 0$ **then if** $symbol = 0$

                            then $state := 1$

                            else $state := 0$

else if $state = 1$ **then if** $symbol = 0$

                            then $state := 1$

                            else $state := 2$

else if $state = 2$ **then if** $symbol = 0$

                            then $state := 2$

                            else $state := 2$

$symbol :=$ next symbol in the input string

end while

Output: *state*

end Algorithm

32.

Algorithm Finite-State Automaton of Exercise 8

[This algorithm simulates the action of the finite-state automaton of exercise 8 by repeated application of the next-state function. The states are denoted 0, 1, and 2.]

Input: *string [a string of 0's and 1's plus an end marker e]*

Algorithm Body:

$N(0,0) := 1$, $N(0,1) := 2$, $N(1,0) := 1$, $N(1,1) := 2$, $N(2,0) := 1$, $N(2,1) := 2$,

$state := 0$

$symbol :=$ first symbol in the input string

while $(symbol \neq e)$

    $state := N(state, symbol)$

    $symbol :=$ next symbol in the input string

end while

Output: *state*

end Algorithm

Section 7.3

3. *b*. G is not one-to-one because, for example, $G(a) = y = G(b)$ and $a \neq b$. G is not onto because z is in Y but $z \neq G(x)$ for any x in X.

4. *a*. H is not one-to-one because $H(b) = y = H(c)$ and $b \neq c$. G is not onto because, for example, $x \in Y$ and $x \neq H(x)$ for any $x \in X$.

   b. K is one-to-one because no two elements of X are sent by K to the same element of Y. K is not onto because z is in Y and $z \neq K(x)$ for any x in X.

5. In each case below there are a number of correct answers.

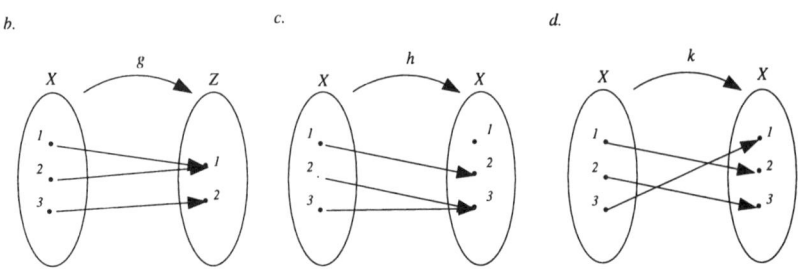

   b.   *c.*   *d.*

6. *c*. $3 \cdot 2 \cdot 1 = 6$ *[There are three choices for where to send the first element of the domain, two choices for where to send the second element (since the function is one-to-one, the second element cannot go the same place as the first), and one choice for where to send the third, which cannot go the same place as the first two.]*

   d. $5 \cdot 4 \cdot 3 = 60$ *[There are five choices for where to send the first element of the domain, four choices for where to send the second element, and three choices for where to send the third, which cannot go the same place as the first two.]*

   e. The answer is $n \cdot (n-1) \cdot (n-2) \cdot \ldots \cdot (n-m+1)$ because there are n choices for where to send the first element of the domain, $n-1$ for where to send the second (since it cannot go the same place as the first), $n-2$ for where to send the third (since it cannot go the same place as either of the first two), and so forth. At the time an image is chosen for the mth element of the domain, the other $m-1$ elements of the domain have all been sent to $m-1$ distinct elements of the co-domain, and so there are $n-(m-1) = n-m+1$ choices for where to send the mth element.

7. b. None.

c. Let the elements of the co-domain be called u and v. Either three elements of the domain are sent to u and one to v (there are $\binom{4}{3}$ ways to do this), or two elements are sent to u and two to v (there are $\binom{4}{2}$ ways to do this), or one element is sent to u and three to v (there are $\binom{4}{1}$ ways to do this). Thus the number of onto functions from a set with four elements to a set with two elements is $\binom{4}{3} + \binom{4}{2} + \binom{4}{1} = 4 + 6 + 4 = 14$.

d. Let the elements of the co-domain be called u, v and w. Either three elements of the domain are sent to u, one is sent to v, and one to w (there are $\binom{5}{3}\binom{2}{1}\binom{1}{1}$ ways to do this), or two elements are sent to u, two to v, and one to w (there are $\binom{5}{2}\binom{3}{2}\binom{1}{1}$ ways to do this), or two elements are sent to u, one to v, and two to w (there are $\binom{5}{2}\binom{3}{1}\binom{2}{2}$ ways to do this), or one element is sent to u, three to v, and one to w (there are $\binom{5}{1}\binom{4}{3}\binom{1}{1}$ ways to do this), or one element is sent to u, two to v, and two to w (there are $\binom{5}{1}\binom{4}{2}\binom{2}{2}$ ways to do this), or one element is sent to u, one to v, and three to w (there are $\binom{5}{1}\binom{4}{1}\binom{3}{3}$ ways to do this). Thus the number of onto functions from a set with four elements to a set with two elements is

$$\binom{5}{3}\binom{2}{1}\binom{1}{1} + \binom{5}{2}\binom{3}{2}\binom{1}{1} + \binom{5}{2}\binom{3}{1}\binom{2}{2} + \binom{5}{1}\binom{4}{3}\binom{1}{1} + \binom{5}{1}\binom{4}{2}\binom{2}{2} + \binom{5}{1}\binom{4}{1}\binom{3}{3}$$
$$= 10 \cdot 2 + 10 \cdot 3 + 10 \cdot 3 + 5 \cdot 4 + 5 \cdot 6 + 5 \cdot 4 = 150.$$

e. Let X be a set with m elements and Y be a set with n elements, where $m \geq n \geq 1$, and let x be any particular element of X. If f is a function from X to Y, then either $f^{-1}(f(x))$ has more than one element or $f^{-1}(f(x))$ has only the single element x. Thus the number of onto functions from X to Y equals the number of onto functions for which $f^{-1}(f(x))$ has more than one element plus the number of onto functions for which $f^{-1}(f(x))$ has only the single element x. Now constructing an onto function from X to Y for which $f^{-1}(f(x))$ has more than one element can be regarded as a two-step operation: step 1 is to choose where to send x and step two is to choose an onto function from $X - \{x\}$ to Y. Thus the number of such functions equals the number of choices for where to send x times the number of onto functions from $X - \{x\}$ to Y, which equals $n \cdot c_{m-1,n}$. Similarly, constructing an onto function from X to Y for which $f^{-1}(f(x))$ has only the single element x can be regarded as a two-step operation: step 1 is to choose where to send x and step two is to choose an onto function from $X - \{x\}$ to $Y - \{f(x)\}$. Thus the number of such functions equals the number of choices for where to send x times the number of onto functions from $X - \{x\}$ to $Y - \{f(x)\}$, which equals $n \cdot c_{m-1,n-1}$. It follows that the total number of onto functions from X to $Y = c_{m,n} = n \cdot c_{m-1,n} + n \cdot c_{m-1,n-1}$.

9. a. (i) g *is one-to-one*: Suppose n_1 and n_2 are in $\mathbf{Z}$ and $g(n_1) = g(n_2)$. By definition of g, $3n_1 - 2 = 3n_2 - 2$. Adding 2 to both sides and dividing by 3 gives $n_1 = n_2$.

(ii). g *is not onto*: Let $m = 0$. Then m is in $\mathbf{Z}$ but $m \neq g(n)$ for any integer n. [For if $m = g(n)$ then $0 = 3n - 2$, and so $n = 2/3$. But $2/3$ is not in $\mathbf{Z}$.]

10. b. K *is onto*: Suppose y is any element of $\mathbf{R}^{nonneg}$. Let $x = \sqrt{y}$. Then x is a real number because $y \geq 0$, and by definition of K, $K(x) = K(\sqrt{y}) = (\sqrt{y})^2 = y$.

11. To say that for each integer n there is only one possible value for $f(x)$ is just another way of saying that f satisfies one of the conditions necessary for it to be a function. To show that f is

one-to-one, one must show that any integer n_1 has a *different* function value from that of the integer n_2 if $n_2 \neq n_1$.

14. f is one-to-one. *Proof*: Let x_1 and x_2 be any nonzero real numbers such that $f(x_1) = f(x_2)$. By definition of f, $\dfrac{2x_1 + 1}{x_1} = \dfrac{2x_2 + 1}{x_2}$. Cross-multiplying gives $(2x_1 + 1)x_2 = (2x_2 + 1)x_1$, or, equivalently, $2x_1x_2 + x_2 = 2x_1x_2 + x_1$. Subtracting $2x_1x_2$ from both sides gives $x_1 = x_2$.

15. f is one-to-one. *Proof*: Let x_1 and x_2 be any real numbers other than -1, and suppose that $f(x_1) = f(x_2)$. By definition of f, $\dfrac{x_1 - 1}{x_1 + 1} = \dfrac{x_2 - 1}{x_2 + 1}$. Cross-multiplying gives $(x_1 - 1)(x_2 + 1) = (x_2 - 1)(x_1 + 1)$, or, equivalently, $x_1x_2 + x_1 - x_2 - 1 = x_1x_2 + x_2 - x_1 - 1$. Adding $1 - x_1x_2$ to both sides gives $x_1 - x_2 = x_2 - x_1$, or, equivalently, $2x_1 = 2x_2$. Dividing both sides by 2 gives $x_1 = x_2$.

16. b. $h(364\text{-}98\text{-}1703) = 2$. Since position 2 is occupied, the next position is examined. That also is occupied, but the following position is free. So 364-98-1703 is placed in position 4.

    c. $h(283\text{-}09\text{-}0787) = 0$. Since position 0 is occupied but the next position is free, 283-09-0787 is placed in position 1.

19. a. *F is not one-to-one*: Let $A = \{a\}$ and $B = \{b\}$. Then $F(A) = F(B) = 1$ but $A \neq B$.

    b. *F is not onto*: The number 4 is in $\mathbf{Z}$ but $F(A) \neq 4$ for any set A in $\mathcal{P}(\{a, b, c\})$ because no subset of $\{a, b, c\}$ has four elements.

20. a. *N is not one-to-one*: Let $s_1 = a$ and $s_2 = ab$. Then $N(s_1) = N(s_2) = 1$ but $s_1 \neq s_2$.

21. a. *C is one-to-one*: Suppose s_1 and s_2 are strings in Σ^* and $C(s_1) = C(s_2)$. By definition of C, this means that $as_1 = as_2$. But strings are just n-tuples written without parentheses or commas. By definition of equality of ordered n-tuples, therefore, for each integer $n \geq 0$, the nth character from the left in as_1 equals the nth character from the left in as_2. It follows that for each integer $n \geq 0$, the nth character from the left in s_1 equals the nth character from the left in s_2, and so $s_1 = s_2$.

    b. *C is not onto*: The string b is in Σ^* but $b \neq as$ for any string a in Σ^*. Hence $b \neq C(s)$ for any s in Σ^*.

22. a. *F is one-to-one*: Suppose $F(a, b) = F(c, d)$ for some ordered pairs (a, b) and (c, d) in $\mathbf{Z}^+ \times \mathbf{Z}^+$. By definition of F, $3^a 6^b = 3^c 6^d$, and so $3^a 3^b 2^b = 3^c 2^d 3^d$, or, equivalently, $3^{a+b} 2^b = 3^{c+d} 3^d$. Thus, by the unique factorization theorem (Theorem 3.3.3), $a + b = c + d$ and $b = d$. Solving these equations gives $a = b$ and $c = d$, or $(a, b) = (c, d)$.

    b. *G is one-to-one*: Suppose $G(a, b) = G(c, d)$ for some ordered pairs (a, b) and (c, d) in $\mathbf{Z}^+ \times \mathbf{Z}^+$. By definition of G, $3^a 5^b = 3^c 5^d$. Thus, by the unique factorization theorem (Theorem 3.3.3), $a = b$ and $c = d$, or $(a, b) = (c, d)$.

23. b. Let $x = \log_9 25$ and $y = \log_3 5$. By definition of logarithm, $9^x = 25$ and $3^y = 5$. Since $9 = 3^2$ and $25 = 5^2$, substitution gives $(3^2)^x = 25 = 5^2 = (3^y)^2$. So by one of the laws of exponents (property (7.3.2)), $3^{2x} = 3^{2y}$. Hence by property (7.3.4), $2x = 2y$, and thus $x = y$. Therefore the answer is yes.

25. Suppose b, x, and y are any positive real numbers with $b \neq 1$. Let $u = \log_b(x)$ and $v = \log_b(y)$. By definition of logarithm, $x = b^u$ and $y = b^v$. Then $x \cdot y = b^u \cdot b^v = b^{u+v}$ by property (7.3.1). Applying the definition of logarithm to the extreme parts of this last equation gives $\log_b(x \cdot y) = u + v = \log_b(x) + \log_b(y)$.

28. When $f: \mathbf{R} \to \mathbf{R}$ and $g: \mathbf{R} \to \mathbf{R}$ are both onto, it need not be the case that $f + g$ is onto. *Counterexample*: Let $f: \mathbf{R} \to \mathbf{R}$ and $g: \mathbf{R} \to \mathbf{R}$ be defined by $f(x) = x$ and $g(x) = -x$ for all $x \in \mathbf{R}$. Then both f and g are onto, but $(f + g)(x) = f(x) + g(x) = x + (-x) = 0$ for all x, and so $f + g$ is not onto.

30. If $f\colon \mathbf{R} \to \mathbf{R}$ is onto and c is any nonzero real number, then $c \cdot f$ is also onto.

  Proof: Suppose $f\colon \mathbf{R} \to \mathbf{R}$ is onto and c is any nonzero real number. Let $y \in \mathbf{R}$. Since $c \neq 0$, y/c is a real number, and since f is onto, there is an $x \in \mathbf{R}$ with $f(x) = y/c$. Then $y = c \cdot f(x) = (c \cdot f)(x)$. So $c \cdot f$ is onto.

32.

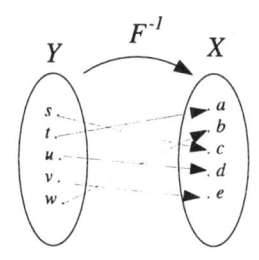

36. By the result of exercise 9(b), G is one-to-one. G is also onto. For if y is any real number, let $x = (y+2)/3$. Then x is a real number, and by definition of G, $G(x) = G((y+2)/3) = 3 \cdot ((y+2)/3) - 2 = (y+2) - 2 = y$. Hence G is onto. Since G is both one-to-one and onto, it is a one-to-one correspondence, and the above calculation shows that for all $y \in \mathbf{R}$, $G^{-1}(y) = (y+2)/3$.

37. K is one-to-one and onto. Hence it is a one-to-one correspondence. For all $y \in \mathbf{R}^{nonneg}$, $K^{-1}(y) = \sqrt{y}$.

40. F is neither one-to-one nor onto. Hence it is not a one-to-one correspondence.

41. N is neither one-to-one nor onto. Hence it is not a one-to-one correspondence.

43. This function is not a one-to-one correspondence because it is not one-to-one.

44. By the result of exercise 14, f is one-to-one. f is also onto. For if y is any real number other than 2, let $x = \dfrac{1}{y-2}$. Then x is a real number (because $y \neq 2$) and

$$f(x) = f\left(\frac{1}{y-2}\right) = \frac{2\left(\frac{1}{y-2}\right)+1}{\frac{1}{y-2}} = \frac{2\left(\frac{1}{y-2}\right)+1}{\frac{1}{y-2}} \cdot \frac{(y-2)}{(y-2)} = \frac{2+(y-2)}{1} = y.$$

So f is onto. This calculation also shows that $f^{-1}(y) = \dfrac{1}{y-2}$ for all real numbers $y \neq 2$.

45. By the result of exercise 15, f is one-to-one. f is also onto. For if y is any real number other than -1, let $x = \dfrac{1+y}{1-y}$. Then x is a real number (because $y \neq -1$) and

$$f(x) = f\left(\frac{1+y}{1-y}\right) = \frac{\left(\frac{1+y}{1-y}\right)-1}{\left(\frac{1+y}{1-y}\right)+1} = \frac{\left(\frac{1+y}{1-y}\right)-1}{\left(\frac{1+y}{1-y}\right)+1} \cdot \frac{(1-y)}{(1-y)} = \frac{1+y-(1-y)}{1+y+(1-y)} = \frac{2y}{2} = y.$$

So f is onto. This calculation also shows that $f^{-1}(y) = \dfrac{1+y}{1-y}$ for all real numbers $y \neq 1$.

46. *a.* Let $X = \{x_1, x_2, \ldots, x_n\}$ and let Σ^n be the set of all strings of 0's and 1's that have length n. Define a function $F\colon \mathcal{P}(X) \to \Sigma^n$ as follows: for each A in $\mathcal{P}(X)$, $F(A)$ = the string of 0's and 1's for which the character in the ith position is a 1 if $x_i \in A$ and the character in the ith

position is a 0 if $x_i \notin A$. For instance, if $n = 10$ and $A = \{x_2, x_4, x_9\}$, then $F(A) = 0101000010$; the 1's in positions 2, 4, and 9 indicate that x_2, x_4, and x_9 are the elements in A.

F is one-to-one: Suppose $F(A_1) = F(A_2)$ for some sets A_1 and A_2 in $\mathcal{P}(X)$. By definition of equality of strings, for each integer $i = 1, 2, \ldots, n$, the ith character of $F(A_1)$ is a 1 if, and only if, the ith character of $F(A_2)$ is a 1. By definition of F, this implies that $x_i \in A_1$ if, and only if, $x_i \in A_2$. It follows that every element of A_1 is in A_2 and every element of A_2 is in A_1. Consequently, $A_1 = A_2$ by definition of set equality.

F is onto: Suppose y is a string of 0's and 1's of length n. Define a subset A of X as follows: Let A consist of the set of all x_i in X for which the character in position i of y is a 1; otherwise $x \notin A$. (For instance, if $n = 10$ and $y = 0001110100$, then $A = \{x_4, x_5, x_6, x_8\}$ because there are 1's in positions 4, 5, 6, and 8 and 0's in all other positions.) Then $F(A) = y$ by definition of F.

Since F is one-to-one and onto, F is a one-to-one correspondence.

47. **Algorithm** **Checking Whether a Function is One-to-One**

    [For a given function F with domain $X = \{a[1], a[2], \ldots, a[n]\}$, this algorithm discovers whether or not F is one-to-one. Initially, answer is set equal to "one-to-one". Then the values of $F(a[i])$ and $F(a[j])$ are systematically compared for indices i and j with $1 \leq i < j \leq n$. If at any point it is found that $F(a[i]) = F(a[j])$ and $a[i] \neq a[j]$, then F is not one-to-one, and so answer is set equal to "not one-to-one" and execution ceases. If after all possible values of i and j have been examined, the value of answer is still "one-to-one", then F is one-to-one.]

    Input: n *[a positive integer]*, $a[1], a[2], \ldots, a[n]$ *[a one-dimensional array representing the set X]*, F *[a function with domain X]*

    Algorithm Body:

    answer := "one-to-one"
    $i := 1$
    while ($i \leq n - 1$ and answer = "one-to-one")
        $j := i + 1$
        while ($j \leq n$ and answer = "one-to-one")
            if ($F(a[i]) = F(a[j])$ and $a[i] \neq a[j]$) **then** answer := "not one-to-one"
            $j := j + 1$
        end while
        $i := i + 1$
    end while

    Output: answer *[a string]*

    end Algorithm

48. **Algorithm** **Checking Whether a Function is Onto**

    [For a given function F with domain $X = \{a[1], a[2], \ldots, a[n]\}$ and co-domain $Y = \{b[1], b[2], \ldots, b[m]\}$, this algorithm discovers whether or not F is onto. Initially, answer is set equal to "onto", and then successive elements of Y are considered. For each such element, $b[i]$, a search is made through elements of the domain to determine if any is sent to $b[i]$. If not, the value of answer is changed to "not onto" and execution of the algorithm ceases. If so, the next successive element of Y is considered. If all elements of Y have been considered and the value of answer has not been changed from its initial value, then F is onto.]

    Input: n *[a positive integer]*, $a[1], a[2], \ldots, a[n]$ *[a one-dimensional array representing the set X]*, m *[a positive integer]*, $b[1], b[2], \ldots, b[m]$ *[a one-dimensional array representing the set X]*, F *[a function with domain X]*

Algorithm Body:

$answer := $ "onto"

$i := 1$

while ($i \leq m$ and $answer = $ "onto")

    $j := 1$

    $found := $ "no"

    while ($j \leq n$ and $answer = $ "onto")

        if $F(a[j]) = b[i]$ **then** $found := $ "yes"

        $j := j + 1$

    end while

    if $found = $ "no" **then** $answer := $ "not onto"

    $i := i + 1$

end while

Output: $answer$ [a string]

end Algorithm

Section 7.4

2. *a.* No. For example, thirteen hearts could be selected: $2, 3, 4, 5, 6, 7, 8, 9, 10, J, Q, K, A$. No two of these are of the same denomination.

   b. Yes. Let X be the set consisting of the 20 selected cards, and let Y be the 13 possible denominations of cards. Define a function D from X (the pigeons) to Y (the pigeonholes) by the rule $D(x) = $ the denomination of x for all x in X. Now X has 20 elements and Y has 13 and $20 > 13$. So by the pigeonhole principle, D is not one-to-one. Hence $D(x_1) = D(x_2)$ for some cards x_1 and x_2 with $x_1 \neq x_2$. Then x_1 and x_2 are two distinct cards out of the 20 selected cards that have the same denomination.

4. Yes. Let X be the set of the 700 people and Y the set of all possible ordered pairs of first and last initials, and consider the function I from X (the pigeons) to Y (the pigeonholes) defined by the rule: $I(x) = $ the ordered pair of initials of person x. Now $n(Y)$, the number of possible ordered pairs of initials, is $26 \cdot 26 = 676$. By the pigeonhole principle, since $700 > 676$, I is not one-to-one, and so at least two people must have the same first and last initials.

6. *a.* Yes. Let X be the set of seven integers and Y the set of all possible remainders obtained through division by 6, and consider the function R from X (the pigeons) to Y (the pigeonholes) defined by the rule: $R(n) = n \bmod 6$ ($=$ the remainder obtained by the integer division of n by 6). Now X has 7 elements and Y has 6 elements (0, 1, 2, 3, 4, and 5). Hence by the pigeonhole principle, R is not one-to-one: $R(n_1) = R(n_2)$ for some integers n_1 and n_2 with $n_1 \neq n_2$. But this means that n_1 and n_2 have the same remainder when divided by 6.

   b. No. Consider the set $\{1, 2, 3, 4, 5, 6, 7\}$. This set has seven elements no two of which have the same remainder when divided by 8.

7. Yes. Let Y be the set of all pairs of integers from S that add up to 15. There are 5 elements in $Y - \{3, 12\}, \{4, 11\}, \{5, 10\}, \{6, 9\}, \{7, 8\}-$ and each integer in S occurs in exactly one such pair. Let X be the set of six integers chosen from S, and consider the function from X to Y defined by the rule: $P(x) = $ the pair to which x belongs. Since X has 6 elements and Y has 5 elements and $6 > 5$, then by the pigeonhole principle, P is not one-to-one. Thus $P(x_1) = P(x_2)$ for some integers x_1 and x_2 in X with $x_1 \neq x_2$. This means that x_1 and x_2 are distinct integers in the same pair, which implies that $x_1 + x_2 = 15$.

8. No. For instance, the five integers 1, 2, 3, 4, 5 could be chosen. The sum of any two of these is less than 10 and so no two have a sum of 10.

11. Yes. There are n odd integers in the set $\{1, 2, \ldots, 2n\}$, namely, $1 \ (= 2 \cdot 1 - 1)$, $3 \ (= 2 \cdot 2 - 1)$, $\ldots$, $2n - 1 \ (= 2 \cdot n - 1)$. So the maximum number of odd integers that can be chosen is n. Thus if $n + 1$ integers are chosen, at least one of them must be even.

13. Seven. Since there are only six pairs of boots in the pile, if at most one boot is chosen from each pair, the maximum number of boots chosen would be six. It follows that if seven boots are chosen, at least two must be from the same pair.

15. There are $n+1$ even integers from 0 to $2n$ inclusive: $0 \ (= 2 \cdot 0), 2 \ (= 2 \cdot 1), 4 \ (= 2 \cdot 2), \ldots, 2n \ (= 2 \cdot n)$. So a maximum of $n + 1$ even integers can be chosen, and thus if at least $n + 2$ integers are chosen, one is sure to be odd. Similarly, since there are n odd integers from 0 to $2n$ inclusive $-1 \ (= 2 \cdot 1 - 1), 3 \ (= 2 \cdot 2 - 1), \ldots, 2n - 1 \ (= 2 \cdot n - 1)$ – if at least $n + 1$ integers are chosen, one is sure to be even.

16. There are 20 integers from 1 to 100 inclusive that are divisible by 5: $5 \ (= 5 \cdot 1), 10 \ (= 5 \cdot 2), 15 \ (= 5 \cdot 3), \ldots, 100 \ (= 5 \cdot 20)$. Hence there are 80 that are not divisible by 5, and so it is necessary to pick at least 81 in order to be sure to get one that is divisible by 5.

18. There are 15 distinct remainders that can be obtained through integer division by 15 (0, 1, 2, $\ldots$, 14). Hence at least 16 integers must be chosen in order to be sure that at least two have the same remainder when divided by 15.

19. Each number from 100 through 999 contains at least one of the nine digits 1, 2, 3, 4, 5, 6, 7, 8, or 9. Therefore, if ten are selected, at least two of them must have a digit in common.

21. Irrational. The decimal expansion of a rational number must either terminate or repeat, and the decimal expansion of this number does neither (because the number of 0's between successive 1's continually increases).

22. Let A be the set of the thirteen chosen numbers, and let B be the set of all prime numbers between 1 and 40. Note that $B = \{2, 3, 5, 7, 11, 13, 17, 19, 23, 29, 31, 37\}$. For each x in A, let $F(x)$ be the smallest prime number that divides x. Since A has 13 elements and B has 12 elements, by the pigeonhole principle F is not one-to-one. So $F(x_1) = F(x_2)$ for some $x_1 \neq x_2$ in A. By definition of F, this means that the smallest prime number that divides x_1 equals the smallest prime number that divides x_2. So two numbers in A, namely x_1 and x_2, have a common divisor greater than 1.

23. *Proof (by contradiction):* Suppose there is a finite-state automaton A that accepts L. Consider all strings of the form a^i for some integer $i \geq 0$. Since the set of all such strings is infinite and the number of states of A is finite, by the pigeonhole principle at least two of these strings, say a^p and a^q with $p < q$, must send A to the same state, say s, when input to A starting in its initial state. (The strings of the given form are the pigeons, the states are the pigeonholes, and each string is associated with the state to which A goes when the string is input to A starting in its initial state.) Because A accepts L, A accepts $a^q b^q$ but not $a^p b^q$. But since $a^q b^q$ is accepted by A, inputting b^q to A when it is in state s (after input of a^q) sends A to an accepting state. Because A also goes to state s after input of a^p, inputting b^q to A after inputting a^p also sends A to an accepting state. Thus $a^p b^q$ is accepted by A and yet it is not accepted by A, which is a contradiction. Hence the supposition is false: there is no finite-state automaton that accepts L.

24. *Proof (by contradiction):* Suppose there is a finite-state automaton A that accepts L. Consider all strings of the form a^i for some integer $i \geq 0$. Since the set of all such strings is infinite and the number of states of A is finite, by the pigeonhole principle at least two of these strings, say a^p and a^q with $p < q$, must send A to the same state, say s, when input to A starting in its initial state. Because A accepts L, A accepts $a^p b^p$ but not $a^q b^p$. But since $a^p b^p$ is accepted

by A, inputting b^p to A when it is in state s (after input of a^p) sends A to an accepting state. Because A also goes to state s after input of a^q, inputting b^p to A after inputting a^q also sends A to an accepting state. Thus $a^q b^p$ is accepted by A and yet it is not accepted by A, which is a contradiction. Hence the supposition is false: there is no finite-state automaton that accepts L.

25. *Proof 1 (by contradiction)*: Suppose there is a finite-state automaton A that accepts L. Let A have N states. Choose an integer m with $(m+1)^2 - m^2 > N$. *[Such an integer exists because $(m+1)^2 - m^2 = 2m+1$, and $2m+1 > N$ if, and only if, $m > (N-1)/2$. So any integer m with $m > (N-1)/2$ will work.]* Consider the set of all strings a^i with $m^2 \le i \le (m+1)^2$. Since $(m+1)^2 - m^2 > N$ and A has N states, there exist integers p and q so that $m^2 \le p < q \le (m+1)^2$ and both a^p and a^q send A to the same state s_i. *[This follows from the pigeonhole principle: the strings a^i with $m^2 \le i \le (m+1)^2$ are the pigeons and the N states are the pigeonholes.]* Now $a^{(m+1)^2}$ is in L and hence sends A to an accepting state. But $(m+1)^2 = q + ((m+1)^2 - q)$, and so $a^{(m+1)^2} = a^q a^{(m+1)^2 - q}$. This implies that when A is in state s_i, input of $a^{(m+1)^2 - q}$ sends A to an accepting state. Since a^p also sends A to state s_i, it follows that $a^p a^{(m+1)^2 - q} = a^{p + (m+1)^2 - q} = a^{(m+1)^2 - (q-p)}$ sends A to an accepting state also. Let $k = (m+1)^2 - (q-p)$. Then a^k is accepted by A. However, a^k is not in L because $(m+1)^2 - (q-p)$ is not a perfect square. (The reason is that since $m^2 \le p < q \le (m+1)^2$, then $q - p < (m+1)^2 - m^2$, and so $m^2 < (m+1)^2 - (q-p)$. Furthermore $(m+1)^2 - (q-p) < (m+1)^2$ because $q - p > 0$. Hence $k = (m+1)^2 - (q-p)$ is in between two successive perfect squares and so is not a perfect square.) This result contradicts the supposition that A is an automaton that accepts L. Hence the supposition that A exists is false.

Proof 2 (by contradiction): Suppose that there is a finite-state automaton A that accepts all strings of the form a^n where $n = m^2$ for some positive integer m. Since there are infinitely many strings of the form a^i for some integer $i \ge 1$ and A has only finitely many states, at least two strings a^p and a^q with $0 < p < q$ must go to the same state s. Consider the strings $a^p a^{(q-1)q}$ and $a^q a^{(q-1)q} = a^{q^2}$. Since A accepts strings of the form a^{m^2}, it will accept $a^q a^{(q-1)q}$. But since A will be in the same state after processing a^p as it is after processing a^q, it will also accept $a^p a^{(q-1)q} = a^{q^2 - (q-p)}$. Now $q^2 - (q-p) \ne m^2$ for any integer m. The reason is that since p and q are integers with $0 < p < q$, then $q + p > 1$. It follows that $q^2 - 2q + 1 < q^2 - (q-p)$, and so $(q-1)^2 < q^2 - (q-p)$. Furthermore, since $q - p > 0$, $q^2 - (q-p) < q^2$. Hence $(q-1)^2 < q^2 - (q-p) < q^2$. Thus on the one hand A accepts $a^{q^2 - (q-p)}$ but on the other hand A does not accept $a^{q^2 - (q-p)}$ because $q^2 - (q-p)$ is not a perfect square. This is a contradiction. It follows that the supposition that A exists is false.

28. Yes. Let X be the set of 2,000 people (the pigeons) and Y the set of all 366 possible birthdays (the pigeonholes). Define a function $B: X \to Y$ by the rule $B(x) = x$'s birthday. Now $2000 > 4 \cdot 366 = 1464$, and so by the generalized pigeonhole principle, there must be some birthday y such that $B^{-1}(y)$ has at least $4 + 1 = 5$ elements. Hence at least 5 people must share the same birthday.

29. Yes. This follows from the generalized pigeonhole principle with 500 pigeons (the lines of code), 17 pigeonholes (the days), and $k = 29$, using the fact that $500 > 29 \cdot 17 = 493$.

31. Consider the maximum number of pennies that can be chosen without getting at least five from the same year. This maximum, which is 12, is obtained when four pennies are chosen from each of the three years. Hence at least thirteen pennies must be chosen to be sure of getting at least five from the same year.

32. *Proof (by contradiction)*: Suppose that two or fewer secretaries are each assigned to three or more executives. Then the remaining secretaries are each assigned to two or fewer executives. Since the maximum number of executives to which any secretary can be assigned is four, the maximum number of executives that can be served by the secretaries is obtained when 2 secretaries (the maximum possible) are each assigned to 4 executives (the maximum possible) and

the remaining 3 secretaries are each assigned to 2 executives (the maximum possible for that group). The maximum number of executives that can be served by the secretaries is, therefore, $2 \cdot 4 + 3 \cdot 2 = 14$. Thus the five secretaries are assigned to at most 14 executives, and so at least one executive does not have a secretary, which contradicts the fact that each executive is assigned a secretary. Consequently, the supposition that two or fewer secretaries are assigned to three or more executives is false *[and so at least three secretaries are assigned to three or more executives]*.

Proof (direct): Let k be the number of secretaries assigned to three or more executives. Because no secretary is assigned to more than four executives, these secretaries are assigned to at most $4k$ executives. Each of the remaining $5-k$ secretaries is assigned to at most two executives, and so together they are assigned to at most $2(5-k) = 10 - 2k$ executives. Therefore the number of executives assigned secretaries is at most $4k + (10-2k) = 10 + 2k$. Since 15 executives are assigned secretaries, $15 \leq 10 + 2k$, or $k \geq 5/2$. So, since k is an integer, $k \geq 3$. Hence at least three secretaries are assigned to three or more executives.

33. *Proof*: Let S be the set of all possible sums of elements of subsets of A and define a function F from $\mathcal{P}(A)$ to S as follows: for each subset X of A, let $F(X)$ be the sum of the elements of X. By Theorem 5.3.6, $\mathcal{P}(A)$ has $2^6 = 64$ elements. Also because A has six elements each of which is less than thirteen, the maximum possible sum of elements of any subset of A is 57 ($= 12 + 11 + 10 + 9 + 8 + 7$). Hence S has 58 elements (the numbers from 0 to 57 inclusive). Since $64 > 58$, the pigeonhole principle guarantees that F is not one-to-one. Thus there exist distinct subsets S_1 and S_2 of S such that $F(S_1) = F(S_2)$, which implies that the elements of S_1 add up to the same sum as the elements of S_2.

34. *Proof*: Let X be the set consisting of the given 52 positive integers and let Y be the set containing the following elements: $\{00\}, \{50\}, \{01, 99\}, \{02, 98\}, \{03, 97\}, \ldots, \{49, 51\}$. Define a function F from X to Y by the rule $F(x) =$ the set containing the last two digits of x. Now X has 52 elements and Y has 51 elements. So by the pigeonhole principle, F is not one-to-one: there exist elements x_1 and x_2 in X such that $F(x_1) = F(x_2)$ and $x_1 \neq x_2$.

Case 1 (x_1 and x_2 have the same last two digits): In this case $x_1 - x_2$ is divisible by 100.

Case 2 (x_1 and x_2 do not have the same last two digits): In this case since $F(x_1) = F(x_2)$ and $x_1 \neq x_2$, $F(x_1) = F(x_2)$ must be one of the two-element sets in Y, and since x_1 and x_2 do not have the same last two digits, the last two digits of x_1 must be one of the numbers in this set and the last two digits of x_2 must be the other number in this set. But the numbers in each of the two-element sets of Y add up to 100. Consequently the sum of the last two digits of x_1 and x_2 add up to 100, which implies that $x_1 + x_2$ is divisible by 100.

35. *Proof 1*: Suppose that 101 integers are chosen from 1 to 200 inclusive. Call them $x_1, x_2, \ldots, x_{101}$. Represent each of these integers in the form $x_i = 2^{k_i} \cdot a_i$ where a_i is the uniquely determined odd integer obtained by dividing x_i by the highest possible power of 2. Because each x_i satisfies the condition $1 \leq x_i \leq 200$, each a_i satisfies the condition $1 \leq a_i \leq 199$. Define a function F from $X = \{x_1, x_2, \ldots, x_{101}\}$ to the set Y of all odd integers from 1 to 199 inclusive by the rule $F(x_i) =$ that odd integer a_i such that x_i equals $2^{k_i} \cdot a_i$. Now X has 101 elements and Y has 100 elements ($1 = 2 \cdot 1 - 1, 3 = 2 \cdot 2 - 1, 5 = 2 \cdot 3 - 1, \ldots, 199 = 2 \cdot 100 - 1$). Hence by the pigeonhole principle, F is not one-to-one: $F(x_i) = F(x_j)$ and $x_i \neq x_j$. But $x_i = 2^{k_i} \cdot a_i$ and $x_j = 2^{k_j} \cdot a_j$ and $F(x_i) = a_i$ and $F(x_j) = a_j$. Thus $x_i = 2^{k_i} \cdot a_i$ and $x_j = 2^{k_j} \cdot a_i$. If $k_j > k_i$, then $x_j = 2^{k_j} \cdot a_i = 2^{k_j - k_i} \cdot 2^{k_i} \cdot a_i = 2^{k_j - k_i} \cdot x_i$, and so x_j is divisible by x_i. Similarly, if $k_j < k_i$, x_i is divisible by x_j. Thus in either case, one of the numbers is divisible by another.

Proof 2: Suppose that 101 integers are selected from 1 to 200 inclusive. Call them $x_1, x_2, \ldots, x_{101}$. Let k be the number of integers selected that are less than or equal to 100. Then there are $101 - k$ integers selected that are greater than 100. Consider partitioning all the 100 numbers from 101 to 200 inclusive into two groups: those that are multiples of the k selected integers and those that are not. Call these two groups A and B respectively. The size of A must be at least k because every integer less than or equal to 100 has at least one multiple in the 101–200 range. Thus the number of elements in B is at most $100 - k$. Since there are $101 - k$ selected numbers above 100, at least one must come from the set A (because $101 - k > 100 - k$), and so is divisible by one of the selected numbers. Therefore there must be two elements with the property that one is divisible by the other.

36. *a. Proof:* Suppose $a_1, a_2, \ldots, a_n$ is a sequence of n integers none of which is divisible by n. Define a function F from $X = \{a_1, a_2, \ldots, a_n\}$ to $Y = \{1, 2, \ldots, n-1\}$ by the rule $F(x) = x \bmod n$ (the remainder obtained through integer division of x by n). Since no element of X is divisible by n, F is well-defined. Now X has n elements and Y has $n-1$ elements, and so by the pigeonhole principle $F(a_i) = F(a_j)$ for some elements a_i and a_j in X with $a_i \neq a_j$. By definition of F, both a_i and a_j have the same remainder when divided by n, and so $a_i - a_j$ is divisible by n. *[More formally, by the quotient-remainder theorem we can write $a_i = n \cdot q_i + r_i$ and $a_j = n \cdot q_j + r_j$ where $0 \leq r_i < n$ and $0 \leq r_j < n$, and since $F(a_i) = F(a_j)$, $r_i = r_j$. Thus $a_i - a_j = (n \cdot q_i + r_i) - (n \cdot q_j + r_j) = n \cdot (q_i - q_j) + (r_i - r_j) = n \cdot (q_i - q_j)$ because $r_i = r_j$. So by definition of divisibility, $a_i - a_j$ is divisible by n.]*

   b. Proof: Suppose $x_1, x_2, \ldots, x_n$ is a sequence of n integers. For each $k = 1, 2, \ldots, n$, let $a_k = x_1 + x_2 + \ldots + x_k$. If some a_k is divisible by n, the problem is solved: the sum of the numbers in the consecutive subsequence $x_1, x_2, \ldots, x_k$ is divisible by n. If no a_k is divisible by n, then $a_1, a_2, \ldots, a_n$ satisfies the hypothesis of part *a*, and so $a_j - a_i$ is divisible by n for some integers i and j with $j > i$. But $a_j - a_i = x_{i+1} + x_{i+2} + \ldots + x_j$. Thus the sum of the numbers in the consecutive subsequence $x_{i+1}, x_{i+2}, \ldots, x_j$ is divisible by n.

37. Let $a_1, a_2, \ldots, a_{n^2+1}$ be any sequence of $n^2 + 1$ distinct real numbers, and suppose that $a_1, a_2, \ldots, a_{n^2+1}$ does not contain a strictly increasing or a strictly decreasing subsequence of length at least $n + 1$. That is, suppose that every such subsequence has length at most n. Let S be the set of ordered pairs of integers (i, d) where $1 \leq i \leq n$ and $1 \leq d \leq n$. Define $F: \{a_1, a_2, \ldots, a_{n^2+1}\} \to S$ as follows:

$$F(a_k) = (i_k, d_k)$$

where
   i_k is the length of the longest increasing subsequence starting at a_k,

and
   d_k is the length of the longest decreasing subsequence starting at a_k.

Since there are $n^2 + 1$ elements in $\{a_1, a_2, \ldots, a_{n^2+1}\}$ and n^2 elements in S, by the pigeonhole principle F is not one-to-one. So $F(a_k) = F(a_m)$ for some integers k and m with $k \neq m$. By definition of F, then, $i_k = i_m$ and $d_k = d_m$. Now if $a_k < a_m$, then the longest strictly increasing sequence starting at a_k is at least one more than the longest strictly increasing sequence starting at a_m (because a_k can be added onto the front of any increasing sequence that starts at a_m). So, in this case, $i_k > i_m$, which is a contradiction. Similarly, if $a_k > a_m$, then the longest strictly decreasing sequence starting at a_k is at least one more than the longest strictly decreasing sequence starting at a_m (because a_k san be added onto the front of any decreasing sequence that starts at a_m). So, in this case, $d_k > d_m$, which is a contradiction. Hence $a_k \not< a_m$ and $a_k \not> a_m$, and so $a_k = a_m$. But this also is impossible because all the numbers in $\{a_1, a_2, \ldots, a_{n^2+1}\}$ are distinct. Thus the supposition is false, and so $\{a_1, a_2, \ldots, a_{n^2+1}\}$ contains a strictly increasing or a strictly decreasing sequence of length at least $n + 1$.

38. **Algorithm**   **Finding Pigeons in the Same Pigeonhole**

   [For a given function F with domain $X = \{x[1], x[2], \ldots, x[n]\}$ and co-domain $Y = \{y[1], y[2], \ldots, y[m]\}$ with $n > m$, this algorithm finds elements a and b so that $F(a) = F(b)$ and $a \neq b$. The existence of such elements is guaranteed by the pigeonhole principle because $n > m$. Initially, the variable done is set equal to "no". Then the values of $F(x[i])$ and $F(x[j])$ are systematically compared for indices i and j with $1 \leq i < j \leq n$. When it is found that $F(x[i]) = F(x[j])$ and

$x[i] \neq x[j]$, a is set equal to $x[i]$, b is set equal to $x[j]$, done is set equal to "yes", and execution ceases.]

Input: n [a positive integer], m [a positive integer with $m < n$], $x[1], x[2], \ldots, x[n]$ [a one-dimensional array representing the set X], $y[1], y[2], \ldots, y[m]$ [a one-dimensional array representing the set Y], F [a function from X to Y]

Algorithm Body:

    done := "no"

    $i := 1$

    while ($i \leq n-1$ and done = "no")

        $j := i+1$

        while ($j \leq n$ and done = "no")

            if ($F(x[i]) = F(x[j])$ and $x[i] \neq x[j]$)

                then do $a := x[i]$, $b := y[j]$, done := "yes" **end do**

            $j := j+1$

        end while

        $i := i+1$

    end while

Output: a, b [positive integers]

end Algorithm

Section 7.5

2.

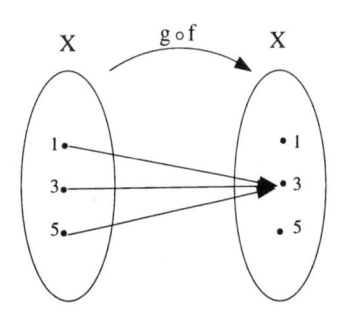

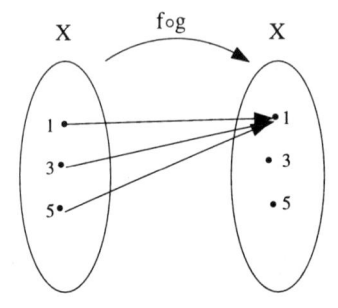

Then $f \circ g \neq g \circ f$ because, for instance, $(g \circ f)(1) = 3$ whereas $(f \circ g)(1) = 1$.

4. $G \circ F$ is defined by $(G \circ F)(x) = x^{20}$, for all $x \in \mathbf{R}$, because for any real number x, $(G \circ F)(x) = G(F(x)) = G(x^5) = (X^5)^4 = x^{20}$. $F \circ G$ is defined by $(F \circ G)(x) = x^{20}$, for all $x \in \mathbf{R}$, because for any real number x, $(F \circ G)(x) = F(G(x)) = F(x^4) = (x^4)^5 = x^{20}$. Thus $G \circ F = F \circ G$.

5. $G \circ F$ is defined by $(G \circ F)(n) = n$, for all integers n, because for any integer n, $(G \circ F)(n) = G(F(n)) = G(2n) = \left\lfloor \dfrac{2n}{2} \right\rfloor = \lfloor n \rfloor = n$. $F \circ G$ is defined by $(F \circ G)(n) = 2 \cdot \left\lfloor \dfrac{n}{2} \right\rfloor$, for all integers n, because for any integer n, $(F \circ G)(n) = F(G(n)) = F(\left\lfloor \dfrac{n}{2} \right\rfloor) = 2 \cdot \left\lfloor \dfrac{n}{2} \right\rfloor$. Then $G \circ F \neq F \circ G$ because, for instance, $(G \circ F)(3) = 3$, whereas $(F \circ G)(3) = 2 \cdot \left\lfloor \dfrac{3}{2} \right\rfloor = 2 \cdot 1 = 2$.

7. $(G \circ F)(2) = G(2^2/3) = G(4/3) = \lfloor 4/3 \rfloor = 1$
   $(G \circ F)(-3) = G((-3)^2/3) = G(3) = \lfloor 3 \rfloor = 3$
   $(G \circ F)(5) = G(5^2/3) = G(25/3) = \lfloor 25/3 \rfloor = 8$

9. For each x in $\mathbf{R}^+$, $(G \circ G^{-1})(x) = G(G^{-1}(x)) = G(\sqrt{x}) = (\sqrt{x})^2 = x$. Hence $G \circ G^{-1} = i_{\mathbf{R}^+}$ by definition of equality of functions.

   For each x in $\mathbf{R}^+$, $(G^{-1} \circ G)(x) = G^{-1}(G(x)) = G(x^2) = \sqrt{x^2} = x$ (because $x > 0$). Hence $G^{-1} \circ G = i_{\mathbf{R}^+}$ by definition of equality of functions.

10. Since $H = H^{-1}$, for all real numbers $x \neq 1$,

$$(H \circ H^{-1})(x) = (H^{-1} \circ H)(x) = H\left(\frac{x+1}{x-1}\right)$$

$$= \frac{\left(\frac{x+1}{x-1}\right) + 1}{\left(\frac{x+1}{x-1}\right) - 1} = \frac{\left(\frac{x+1}{x-1}\right) + 1}{\left(\frac{x+1}{x-1}\right) - 1} \cdot \frac{(x-1)}{(x-1)}$$

$$= \frac{(x+1) + (x-1)}{(x+1) - (x-1)} = \frac{2x}{2} = x.$$

11. b. For all positive real numbers b and x, $\log_b x$ is the exponent to which b must be raised to obtain x. So if b is raised to this exponent, x is obtained. In other words, $b^{\log_b x} = x$.

12. *Proof:* Suppose f is a function from a set X to a set Y. For each x in X, $(i_Y \circ f)(x) = i_Y(f(x)) = f(x)$ by definition of i_Y. Hence $i_Y \circ f = f$.

13. *Proof:* Suppose $f: X \to Y$ is a one-to-one and onto function with inverse function $f^{-1}: Y \to X$. Then for all $y \in Y$, $(f \circ f^{-1})(y) = f(f^{-1}(y)) = f$ (that element x in X for which $f(x)$ equals y) $= y = i_Y(y)$. Hence $f \circ f^{-1} = i_Y$.

14. b. $z/2 = t/2$ c. $f(x_1) = f(x_2)$

16. No. *Counterexample*: For the functions f and g defined by the arrow diagrams below, $g \circ f$ is onto but f is not onto.

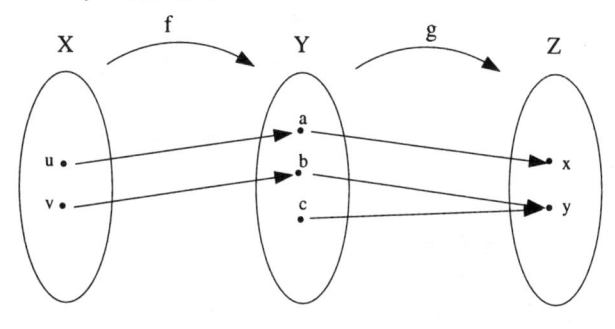

17. Yes. *Proof:* Suppose $f: X \to Y$ and $g: Y \to Z$ are functions and $g \circ f: X \to Z$ is one-to-one. To show that f is one-to-one, suppose x_1 and x_2 are in X and $f(x_1) = f(x_2)$. [We must show that $x_1 = x_2$.] Then $g(f(x_1)) = g(f(x_2))$, and so $g \circ f(x_1) = g \circ f(x_2)$. But $g \circ f$ is one-to-one. Hence $x_1 = x_2$ [as was to be shown].

18. Yes. *Proof:* Suppose $f: X \to Y$ and $g: Y \to Z$ are functions and $g \circ f: X \to Z$ is onto. To show that g is onto, let $z \in Z$. [We must show that $z = g(y)$ for some $y \in Y$.] Since $g \circ f$ is onto, there is an x in X such that $g \circ f(x) = z$. But $g \circ f(x) = g(f(x))$; so if $y = f(x)$ then $g(y) = z$. Hence there is an element y in Y such that $g(y) = z$ [as was to be shown].

19. Yes. *Proof:* Suppose $f: W \to X$, $g: X \to Y$, and $h: Y \to Z$ are functions. For each $w \in W$, $[h \circ (g \circ f)](w) = h((g \circ f)(w)) = h(g(f(w))) = h \circ g(f(w)) = [(h \circ g) \circ f](w)$. Hence $h \circ (g \circ f) = (h \circ g) \circ f$ by definition of equality of functions.

21. $g \circ f: \mathbf{R} \to \mathbf{R}$ is defined by $(g \circ f)(x) = g(f(x)) = g(x+2) = -(x+2)$ for all $x \in \mathbf{R}$.

    Since $z = -(x+2)$ if, and only if, $x = -z-2$, $(g \circ f)^{-1}: \mathbf{R} \to \mathbf{R}$ is defined by $(g \circ f)^{-1}(z) = -z-2$ for all $z \in \mathbf{R}$.

    Since $z = -y$ if, and only if, $y = -z$, $g^{-1}: \mathbf{R} \to \mathbf{R}$ is defined by $g^{-1}(z) = -z$ for all $z \in \mathbf{R}$.

    Since $y = x + 2$ if, and only if, $x = y - 2$, $f^{-1}: \mathbf{R} \to \mathbf{R}$ is defined by $f^{-1}(y) = y - 2$.

    $f^{-1} \circ g^{-1}: \mathbf{R} \to \mathbf{R}$ is defined by $(f^{-1} \circ g^{-1})(z) = f^{-1}(g^{-1}(z)) = f^{-1}(-z) = (-z) - 2 = -z - 2$ for all $x \in \mathbf{R}$.

    By the above and the definition of equality of functions, $(g \circ f)^{-1} = f^{-1} \circ g^{-1}$.

22. *Proof:* Suppose $f: X \to Y$ and $g: Y \to Z$ are functions that are both one-to-one and onto. By Theorems 7.5.3 and 7.5.4, $g \circ f$ is one-to-one and onto, and so by Theorem 7.3.1 and the definition of inverse function, $g \circ f$ has an inverse function $(g \circ f)^{-1}: Z \to X$. To show that $(g \circ f)^{-1} = f^{-1} \circ g^{-1}$, let z be any element of Z and let $x = (g \circ f)^{-1}(z)$. By definition of inverse function, $(g \circ f)^{-1}(z) = x$ if, and only if, $g \circ f(x) = z$. Hence $z = g \circ f(x) = g(f(x))$. Let $f(x) = y$. Then $z = g(y)$. Now since f and g are one-to-one and onto, by Theorem 7.3.1 and the definition of inverse function, f and g have inverse functions $f^{-1}: Y \to X$ and $g^{-1}: Z \to Y$. Then $g^{-1}(z) = y$ because $g(y) = z$ and $f^{-1}(y) = x$ because $f(x) = y$. Consequently, $f^{-1} \circ g^{-1}(z) = f^{-1}(g^{-1}(z)) = f^{-1}(y) = x$. But then $(g \circ f)^{-1}(z) = x = (f^{-1} \circ g^{-1})(z)$. Since the choice of z was arbitrary, $(g \circ f)^{-1}(z) = x = (f^{-1} \circ g^{-1})(z)$ for all $z \in Z$, and so $(g \circ f)^{-1} = f^{-1} \circ g^{-1}$ by definition of equality of functions.

23. *Proof 1:* Suppose $f: X \to Y$ and $g: Y \to Z$ are functions such that $g \circ f = i_X$ and $f \circ g = i_Y$. Since both i_X and i_Y are one-to-one and onto, by the results of exercises 15 and 16, both f and g are one-to-one and onto, and so by Theorem 7.3.1 and the definition of inverse function, both have inverse functions. Now for each $y \in Y$, $f^{-1}(y) = $ that element of X such that $f(x)$ equals y. But $f(g(y)) = f(g(y) = i_Y(y) = y$. Hence $g(y)$ is that element of X such that $f(x)$ equals y, and so $g(y) = f^{-1}(y)$. It follows by definition of equality of functions that $g = f^{-1}$.

    Proof 2: Suppose $f: X \to Y$ and $g: Y \to Z$ are functions such that $g \circ f = i_X$ and $f \circ g = i_Y$. Since both i_X and i_Y are one-to-one and onto, by the results of exercises 15 and 16, both f and g are one-to-one and onto, and so by Theorem 7.3.1 and the definition of inverse function, both have inverse functions. By Theorem 7.5.2(b), $f \circ f^{-1} = i_Y$. Since $f \circ g = i_Y$ also, $f \circ f^{-1} = f \circ g$, and so for all $y \in Y$, $(f \circ f^{-1})(y) = (f \circ g)(y)$. This implies that for all $y \in Y$, $f(f^{-1}(y)) = f(g(y))$. Since f is one-to-one, it follows that $f^{-1}(y) = g(y)$ for all $y \in Y$, and so by definition of equality of functions $f^{-1} = g$.

Section 7.6

2. *a.* Define $f: \mathbf{Z} \to 3\mathbf{Z}$ by the rule $f(n) = 3n$ for all integers n. The function f is one-to-one because for any integers n_1 and n_2, if $f(n_1) = f(n_2)$ then $3n_1 = 3n_2$ and so $n_1 = n_2$. Also f is onto because if m is any element in $3\mathbf{Z}$, then $m = 3k$ for some integer k. But then $f(k) = 3k = m$ by definition of f. So since there is a function $f: \mathbf{Z} \to 3\mathbf{Z}$ that is one-to-one and onto, $\mathbf{Z}$ has the same cardinality as $3\mathbf{Z}$.

    b. It was shown in Example 7.6.1 that $\mathbf{Z}$ is countably infinite, which means that $\mathbf{Z}^+$ has the same cardinality as $\mathbf{Z}$. By part (a), $\mathbf{Z}$ has the same cardinality as $3\mathbf{Z}$. It follows by the transitive property of cardinality (Theorem 7.6.1(c)) that $\mathbf{Z}^+$ has the same cardinality as $3\mathbf{Z}$. So $3\mathbf{Z}$ is countably infinite *[by definition of countably infinite]*, and hence $3\mathbf{Z}$ is countable *[by definition of countable]*.

4. a. $f(1) = -(1-1)/2 = 0$, $f(2) = 2/2 = 1$, $f(3) = -(3-1)/2 = -1$, $f(4) = 4/2 = 2$. All these values agree with those indicated in Figure 7.6.4.

5. Start in the middle of the integers with 0, letting $h(1) = 0$ and $h(2) = 0$, and work outwards using a back-and-forth motion as in Example 7.6.1 but "counting" each integer twice. Thus $h(3) = h(4) = 1$, $h(5) = h(6) = -1$, $h(7) = h(8) = 2$, $h(9) = h(10) = -2$, and so forth. In general,

$$h(n) = \begin{cases} -\dfrac{n-1}{4} & \text{if } n = 4k+1 \text{ for some integer } k \\ -\dfrac{n-2}{4} & \text{if } n = 4k+2 \text{ for some integer } k \\ \dfrac{n+1}{4} & \text{if } n = 4k+3 \text{ for some integer } k \\ \dfrac{n}{4} & \text{if } n = 4k \text{ for some integer } k \end{cases} = (-1)^{\lfloor \frac{n+1}{2} \rfloor} \left\lfloor \dfrac{n+1}{4} \right\rfloor.$$

6. Let B be the set of all bit strings (strings of 0's and 1's). Define a function $F: \mathbf{Z}^+ \to B$ as follows: $F(1) = \varepsilon$, $F(2) = 0$, $F(3) = 1$, $F(4) = 00$, $F(5) = 01$, $F(6) = 10$, $F(7) = 11$, $F(8) = 000$, $F(9) = 001$, $F(10) = 010$, and so forth. At each stage, all the strings of length k are counted before the strings of length $k+1$, and the strings of length k are counted in order of increasing magnitude when interpreted as binary representations of integers. We can define F more formally as follows:

$$F(n) = \begin{cases} \varepsilon & \text{if } n = 1 \\ \text{the } k\text{-bit binary representation of } n - 2^k & \text{if } \lfloor \log_2 n \rfloor = k. \end{cases}$$

For instance, $F(7) = 11$ because $\lfloor \log_2 7 \rfloor = 2$ and the two-bit binary representation of $7 - 2^2$ ($= 3$) is 11.

8. Suppose r_1 and r_2 are any two rational numbers with $r_1 < r_2$. Let $x = (r_1 + r_2)/2$. Now $r_1 + r_2$ is rational by Theorem 3.2.2, and so $x = (r_1 + r_2)/2$ is rational by exercise 17 of Section 3.2. Furthermore, since $r_1 < r_2$, $r_1 + r_1 < r_2 + r_1$, which implies that $2r_1 < r_1 + r_2$, or equivalently $r_1 < (r_1 + r_2)/2$. Similarly, since $r_1 < r_2$, $r_1 + r_2 < r_2 + r_2$, which implies that $r_1 + r_2 < 2r_2$, or equivalently $(r_1 + r_2)/2 < r_2$. Putting the inequalities together gives $r_1 < x < r_2$.

9. No. For instance, both $\sqrt{2}$ and $-\sqrt{2}$ are irrational, and yet their average is $(\sqrt{2} + (-\sqrt{2}))/2$ which equals 0 and is rational. (More generally, if r is any rational number and x is any irrational number, then both $r + x$ and $r - x$ are irrational and yet their average, $((r+x) + (r-x))/2 = r$ is rational.)

10. *Proof:* Suppose r and s are real numbers with $s > r > 0$. Let n be an integer such that $n > \dfrac{\sqrt{2}}{s-r}$. Then $s - r > \dfrac{\sqrt{2}}{n}$. Let $m = \dfrac{nr}{\sqrt{2}} + 1$. Then m is an integer and $m > \dfrac{nr}{\sqrt{2}} \geq m - 1$. Multiply the inequality by $\sqrt{2}$ and divide by n to obtain $\dfrac{m \cdot \sqrt{2}}{n} > r \geq \dfrac{\sqrt{2} \cdot (m-1)}{n}$. Now since $s = r + (s-r)$ and $s - r > \dfrac{\sqrt{2}}{n}$, then $s > r + \dfrac{\sqrt{2}}{n} \geq \dfrac{\sqrt{2} \cdot (m-1)}{n} + \dfrac{\sqrt{2}}{n} = \dfrac{m \cdot \sqrt{2}}{n}$. Hence $r < \dfrac{m \cdot \sqrt{2}}{n} < s$. Note that $\dfrac{m \cdot \sqrt{2}}{n}$ is irrational because m and n are integers, $\dfrac{m \cdot \sqrt{2}}{n} = \dfrac{m}{n} \cdot \sqrt{2}$, and the product of a nonzero rational number and an irrational number is irrational (exercise 17 of Section 3.6).

11. *Two examples:* Define f and g from $\mathbf{Z}$ to $\mathbf{Z}$ as follows: $f(n) = 2n$ and $g(n) = 3n - 2$ for all integers n. By exercises 8 and 9 of Section 7.3, these functions are one-to-one but not onto.

12. *Two examples:* Define $F: \mathbf{Z} \to \mathbf{Z}$ by the rule $F(n) = \begin{cases} n/2 & \text{if } n \text{ is even} \\ 0 & \text{if } n \text{ is odd} \end{cases}$. Then F is onto because given any integer m, $m = F(2m)$. But F is not one-to-one because, for instance, $F(1) = F(3) = 0$.

Define $G: \mathbf{Z} \to \mathbf{Z}$ by the rule $G(n) = \lfloor n/2 \rfloor$ for all integers n. Then G is onto because given any integer m, $m = \lfloor m \rfloor = \lfloor (2m)/2 \rfloor$. But G is not one-to-one because, for instance, $G(2) = \lfloor 2/2 \rfloor = 1$ and $G(3) = \lfloor 3/2 \rfloor = 1$ also.

13. Observe that g is one-to-one. For suppose $g(a,b) = g(c,d)$ for some ordered pairs (a,b) and (c,d) in $\mathbf{Z}^+ \times \mathbf{Z}^+$. By definition of g, $2^a 3^b = 2^c 3^d$, and so by the unique factorization theorem (Theorem 3.3.3), $a = b$ and $c = d$, or, equivalently, $(a,b) = (c,d)$. Hence there is a one-to-one correspondence between $\mathbf{Z}^+ \times \mathbf{Z}^+$ and a subset S (the range of g) of $\mathbf{Z}^+$. But by Theorem 7.6.3, any subset of a countable set is countable, and thus S is countable. It follows from the transitive property of cardinality that $\mathbf{Z}^+ \times \mathbf{Z}^+$ is also countable.

14. *b.* The fundamental observation is that if one adds up the numbers of ordered pairs along successive diagonals starting from the upper left corner, one obtains a sum of successive integers. The reason is that the number of pairs in the $(m+1)$st diagonal is $m+1$ more than the number in the mth diagonal. We show below that the value of H for a given pair (m,n) in the diagram is the sum of the numbers of pairs in the diagonals preceding the one containing (m,n) plus the number of the position of (m,n) in its diagonal counting down from the top starting from 0.

Starting in the upper left corner, number the diagonals of the diagram so that the diagonal containing only $(0,0)$ is 0, the diagonal containing $(1,0)$ and $(0,1)$ is 1, the diagonal containing $(2,0)$, $(1,1)$, and $(0,2)$ is 2, and so forth. Within each diagonal, number each ordered pair starting at the top. Thus within diagonal 2, for example, the pair $(2,0)$ is 0, the pair $(1,1)$ is 1 and the pair $(0,2)$ is 2. Each ordered pair of nonnegative integers can be uniquely specified by giving the number of the diagonal that contains it and stating its numerical position within that diagonal. For instance, the pair $(1,1)$ is in position 1 of diagonal 2, and the pair $(0,1)$ is in position 0 of diagonal 1. In general, each pair of the form (m,n) lies in diagonal $m+n$, and its position within diagonal $m+n$ is n. Observe that if the arrows in the diagram of exercise 14(a) are followed, the number of ordered pairs that precede $(m+n, 0)$, the top pair of the $(m+n)$th diagonal, is the sum of the numbers of pairs in each of the diagonals from the zeroth through the $(m+n-1)$st. Since there are k pairs in the diagonal numbered $k-1$, the number of pairs that precede $(m+n, 0)$ is

$$1 + 2 + 3 + \cdots + (m+n) = \frac{(m+n)(m+n+1)}{2}$$

by Theorem 4.2.2. Then $\frac{(m+n)(m+n+1)}{2} + n$ is the sum of the number of pairs that precede the top pair of the $(m+n)$th diagonal plus the numerical position of the pair (m,n) within the $(m+n)$th diagonal. Hence $H(n) = n + \frac{(m+n)(m+n+1)}{2}$ is the numerical position of the pair (m,n) in the total ordering of all the pairs if the ordering is begun with 0 at $(0,0)$ and is continued by following the arrows in the diagram of exercise 14(a).

15. The proof given below is adapted from one in *Foundations of Modern Analysis* by Jean Dieudonné, New York: Academic Press, 1969, page 14.

Proof: Suppose (a,b) and (c,d) are in $\mathbf{Z}^+ \times \mathbf{Z}^+$ and $(a,b) \neq (c,d)$.

Case 1, $a+b \neq c+d$: By interchanging (a,b) and (c,d) if necessary, we may assume that $a+b < c+d$. Then

$$H(a,b) = b + \frac{(a+b)(a+b+1)}{2} \quad \text{by definition of } H$$

$$\Rightarrow H(a,b) \leq a + b + \frac{(a+b)(a+b+1)}{2} \quad \text{because } a \geq 0$$

$$\Rightarrow H(a,b) < (a+b+1) + \frac{(a+b)(a+b+1)}{2} \quad \text{because } a+b < a+b+1$$

$$\Rightarrow H(a,b) < \frac{2(a+b+1)}{2} + \frac{(a+b)(a+b+1)}{2}$$

$$\Rightarrow H(a,b) < \frac{(a+b+1)(a+b+2)}{2} \quad \text{by factoring out } (a+b+1)$$

$$\Rightarrow H(a,b) < \frac{(c+d)(c+d+1)}{2} \quad \text{because since } a+b < c+d \text{ and } a, b, c,$$
$$\text{and } d \text{ are integers, } a+b \leq (c+d)+1$$

$$\Rightarrow H(a,b) < d + \frac{(c+d)(c+d+1)}{2} \quad \text{because } d \geq 0$$

$$\Rightarrow H(a,b) < H(c,d) \quad \text{by definition of } H.$$

Therefore, $H(a,b) \neq H(c,d)$.

Case 2, $a+b = c+d$: First observe that in this case $b \neq d$. For if $b = d$, then subtracting b from both sides of $a+b = c+d$ gives $a = c$, and so $(a,b) = (c,d)$, which contradicts our assumption that $(a,b) \neq (c,d)$. Hence,

$$H(a,b) = b + \frac{(a+b)(a+b+1)}{2} = b + \frac{(c+d)(c+d+1)}{2} \neq d + \frac{(c+d)(c+d+1)}{2} = H(c,d),$$

and so $H(a,b) \neq H(c,d)$.

Thus both in case 1 and in case 2, $H(a,b) \neq H(c,d)$, and hence H is one-to-one.

16. *Proof 1*: Let $x = .199999\ldots$. Then $10x = 1.99999\ldots$, and so $10x - x = 1.8$. But also $10x - x = 9x$. Hence $9x = 1.8$, or equivalently $x = 0.2$.

    Proof 2 (by contradiction): Let $x = 0.1999999\ldots$ and suppose $x \neq 0.2$. Then $0.2 - x = \epsilon$ for some positive number ϵ. Let n be a positive integer such that $10^n > \frac{1}{\epsilon}$, or equivalently such that $\epsilon > \frac{1}{10^n}$, and let $a = 0.19999\ldots 9$ ($n-1$ *nines*). Then $a < x < 0.2$, and so $0.2 - x < 0.2 - a$. This implies that $0.2 - a > \epsilon$. But $0.2 - a = 0.2 - 0.199999\ldots 9$ ($n-1$ *nines*) $= 0.000000\ldots 01$ (n *decimal places*) $= \frac{1}{10^n} < \epsilon$. Thus $0.2 - a > \epsilon$ and $0.2 - a < \epsilon$, which is a contradiction. Hence the supposition is false, and so $x = 0.2$.

18. *Proof*: Define $h: S \to V$ by the rule $h(x) = 3x + 2$ for all real numbers x in S. Then h is one-to-one because if x_1 and x_2 are in S and $h(x_1) = h(x_2)$, then $3x_1 + 2 = 3x_2 + 2$ and so *[by subtracting 2 and dividing by 3]* $x_1 = x_2$. Furthermore, h is onto because if y is any element in V, then $2 < y < 5$ and so $0 < (y-2)/3 < 1$. Consequently, $(y-2)/3 \in S$ and $h((y-2)/3) = 3((y-2)/3) + 2 = y$. Hence h is a one-to-one correspondence, and so S and V have the same cardinality.

19. *Proof*: Define $F: S \to W$ by the rule $F(x) = (b-a)x + a$ for all real numbers x in S. Then F is well-defined because if $0 < x < 1$, then $a < (b-a)x + a < b$. In addition, F is one-to-one because if x_1 and x_2 are in S and $F(x_1) = F(x_2)$, then $(b-a)x_1 + a = (b-a)x_2 + a$ and so *[by subtracting a and dividing by $b-a$]* $x_1 = x_2$. Furthermore, F is onto because if y is any element in W, then $a < y < b$ and so $0 < (y-a)/(b-a) < 1$. Consequently, $(y-a)/(b-a) \in S$ and $h((y-a)/(b-a)) = (b-a)[(y-a)/(b-a)] + a = y$. Hence F is a one-to-one correspondence, and so S and W have the same cardinality.

21. *Proof (without calculus):* The function g is defined by the rule $g(x) = \frac{1}{2} \cdot \left(\frac{x}{1+|x|}\right) + \frac{1}{2}$ for all real numbers x. To show that g is one-to-one, suppose x_1 and x_2 are in $\mathbf{R}$ and $g(x_1) = g(x_2)$. It follows that x_1 and x_2 have the same sign because if $x_1 < 0$ and $x_2 > 0$, then $g(x_1) < \frac{1}{2}$ and $g(x_2) > \frac{1}{2}$ and if $x_1 > 0$ and $x_2 < 0$, then $g(x_1) > \frac{1}{2}$ and $g(x_2) < \frac{1}{2}$. Now if $x_1 > 0$ and $x_2 > 0$, then $|x_1| = x_1$ and $|x_2| = x_2$, and so the condition $g(x_1) = g(x_2)$ implies that $\frac{x_1}{1+x_1} = \frac{x_2}{1+x_2}$. Cross multiplying gives $x_1 + x_1 x_2 = x_2 + x_1 x_2$, which implies that $x_1 = x_2$. Furthermore, if $x_1 < 0$ and $x_2 < 0$, then $|x_1| = -x_1$ and $|x_2| = -x_2$, and so the condition $g(x_1) = g(x_2)$ implies that $\frac{x_1}{1-x_1} = \frac{x_2}{1-x_2}$. Cross multiplying gives $x_1 - x_1 x_2 = x_2 - x_1 x_2$, which implies that $x_1 = x_2$. Therefore, in all cases if $g(x_1) = g(x_2)$ then $x_1 = x_2$, and so g is one-to-one.

To show that g is onto, let $y \in S = \{x \in \mathbf{R} \mid 0 < x < 1\}$. In case $0 < y < \frac{1}{2}$, let $x = \frac{2y-1}{2y}$ and note that $x < 0$. Then

$$g(x) = g\left(\frac{2y-1}{2y}\right)$$
$$= \frac{1}{2} \cdot \left(\frac{\frac{2y-1}{2y}}{1+\left|\frac{2y-1}{2y}\right|}\right) + \frac{1}{2}$$
$$= \frac{1}{2} \cdot \left(\frac{\frac{2y-1}{2y}}{1-\frac{2y-1}{2y}}\right) + \frac{1}{2}$$
$$= \frac{1}{2} \cdot \left(\frac{2y-1}{2y-(2y-1)}\right) + \frac{1}{2} = y - \frac{1}{2} + \frac{1}{2} = y.$$

In case $\frac{1}{2} < y < 1$, let $x = \frac{2y-1}{2(1-y)}$ and note that $x > 0$. Then

$$g(x) = g\left(\frac{2y-1}{2(1-y)}\right)$$
$$= \frac{1}{2} \cdot \left(\frac{\frac{2y-1}{2(1-y)}}{1+\left|\frac{2y-1}{2(1-y)}\right|}\right) + \frac{1}{2}$$
$$= \frac{1}{2} \cdot \left(\frac{\frac{2y-1}{2(1-y)}}{1+\frac{2y-1}{2(1-y)}}\right) + \frac{1}{2}$$
$$= \frac{1}{2} \cdot \left(\frac{2y-1}{2-2y+2y-1}\right) + \frac{1}{2}$$
$$= y - \frac{1}{2} + \frac{1}{2}$$
$$= y.$$

Therefore, in either case there exists a real number x such that $g(x) = y$, and so g is onto. Since g is both one-to-one and onto, g is a one-to-one correspondence. It follows that the set of all real numbers and S have the same cardinality.

Proof (with calculus): To show that g is one-to-one, note that for all $x > 0$,

$$g(x) = \frac{1}{2} \cdot \frac{x}{1+x} + \frac{1}{2}.$$

Then
$$g'(x) = \frac{1}{2}\left(\frac{(1+x)-x}{(1+x)^2}\right) = \frac{1}{2(1+x)^2} > 0.$$

Hence g is increasing on the interval $(0, \infty)$. In addition, for all $x < 0$,
$$g(x) = \frac{1}{2} \cdot \frac{x}{1-x} + \frac{1}{2}.$$

Then
$$g'(x) = \frac{1}{2}\left(\frac{(1-x)-x}{(1-x)^2}\right) = \frac{1-2x}{2(1-x)^2} > 0.$$

Hence g is increasing on the interval $(-\infty, 0)$. Now if $x < 0$, then $g(x) < 1/2$, and if $x \geq 0$, then $g(x) \geq 1/2$. Putting this information together with the fact that g is increasing on the intervals $(-\infty, 0)$ and $(0, \infty)$ gives that g is increasing on the entire set of real numbers, and so g is one-to-one. (See the solution to exercise 18 of Section 9.1 for a proof of this last fact.) To show g is onto, note that $\lim_{x \to \infty} g(x) = 1$ and $\lim_{x \to -\infty} g(x) = 0$. Also g is continuous on its entire domain (because g is obviously differentiable for all $x \neq 0$ and $\lim_{x \to 0^-} g(x) = \frac{1}{2} = \lim_{x \to 0^+} g(x)$). Hence by definition of limit and the intermediate value theorem, g takes every value strictly between 0 and 1.

23. *Proof*: Let A and B be countable sets.

    Case 1 (both A and B are finite): In this case $A \cup B$ is also finite and hence countable. *[In fact, $n(A \cup B) = n(A) + n(B) - n(A \cap B)$ by the inclusion/exclusion rule.]*

    Case 2 (A is countably infinite and B is finite): In this case there is a one-to-one correspondence $f: \mathbf{Z}^+ \to A$. Let $C = B - A$. Then $A \cup B = A \cup C$ and $A \cap C = \emptyset$. Since $C \subseteq B$ and B is finite, so is C. Then $C = \emptyset$ or there is a one-to-one correspondence from $\{1, 2, \ldots, m\}$ to C, where m is a positive integer. If $C = \emptyset$, then $A \cup C = A$ which is countably infinite. If $C \neq \emptyset$, then there is a one-to-one correspondence h from $\{1, 2, \ldots, m\}$ to C, where m is a positive integer. Define a function $F: \mathbf{Z}^+ \to A \cup C$ as follows: for each integer n, $h(n) = \begin{cases} h(n) & \text{if } 1 \leq n \leq m \\ f(n-m) & \text{if } n \geq m+1 \end{cases}$.
    Then F is one-to-one because $A \cap C = \emptyset$ and f and h are one-to-one. And F is onto because both f and h are onto and every positive integer can be written in the form $n - m$ for some integer $n \geq m+1$. Since h is one-to-one and onto, h is a one-to-one correspondence. Therefore, $A \cup C = A \cup B$ is countably infinite and hence countable.

    Case 3 (B is countably infinite and A is finite): The proof that $A \cup B$ is countable is virtually the same as that given in Case 2.

    Case 4 (A and B are countably infinite and $A \cap B = \emptyset$): In this case there are one-to-one correspondences $f: \mathbf{Z}^+ \to A$ and $g: \mathbf{Z}^+ \to B$. Define $h: \mathbf{Z}^+ \to A \cup B$ as follows: for all integers $n \geq 1$, $h(n) = \begin{cases} f(n/2) & \text{if } n \text{ is even} \\ g((n+1)/2) & \text{if } n \text{ is odd} \end{cases}$. Then h is one-to-one because if $h(n_1) = h(n_2)$ then (since $A \cap B = \emptyset$) either both n_1 and n_2 are even or both n_1 and n_2 are odd. If both n_1 and n_2 are even, it follows by definition of h that $f(n_1/2) = f(n_2/2)$, and so since f is one-to-one, $n_1/2 = n_2/2$, which implies that $n_1 = n_2$. If both n_1 and n_2 are odd, it follows by definition of h that $g((n_1+1)/2) = g((n_2+1)/2)$, and so since g is one-to-one, $(n_1+1)/2 = (n_2+1)/2$, which implies that $n_1 = n_2$. Hence in either case $n_1 = n_2$, and so h is one-to-one.

    To show that h is onto, let $y \in A \cup B$ be given. By definition of union, $y \in A$ or $y \in B$. If $y \in A$, then since f is onto, there is a positive integer m such that $f(m) = y$, and so by definition of h, $y = h(2m)$. If $y \in B$, then since g is onto, there is a positive integer m such that $g(m) = y$, and so by definition of h, $y = h(2m - 1)$. Now when m is a positive integer, then both $2m$ and $2m-1$ are positive integers. Thus in either case, there is a positive integer whose image under h is y. So h is onto.

The above arguments show that there is a one-to-one correspondence from $\mathbf{Z}^+$ to $A \cup B$, and so $A \cup B$ is countably infinite and hence countable.

Case 5 (A and B are countably infinite and $A \cap B \neq \emptyset$): Let $C = B - A$. Then $A \cup B = A \cup C$ and $A \cap C = \emptyset$. Since $C \subseteq B$, C is either finite or countably infinite by Theorem 7.6.3. If C is countably infinite, then $A \cup B = A \cup C$ is countable by case 4 above. If C is finite, $A \cup B = A \cup C$ is countable by case 2 above. Hence in either case $A \cup B$ is countable.

Note: An alternative proof uses a similar technique somewhat like the one shown in Example 7.6.3. Imagine arranging the distinct elements of A and B in two sequences: $a_1, a_2, a_3, \ldots$ and $b_1, b_2, b_3, \ldots$ and counting the elements in the union by counting alternate elements of $a_1, a_2, a_3, \ldots$ and $b_1, b_2, b_3, \ldots$ skipping over those that have already been counted (because A and B may have some elements in common). More formally, define a one-to-one correspondence F from $\mathbf{Z}^+$ to $A \cup B$ as follows: Let $F(1) = a_1$, and for each integer $n > 1$, assume $F(1), F(2), \ldots, F(n-1)$ have been defined. Let i be the largest subscript such that $a_i = F(m)$ for some $m \leq n-1$ and j be the largest subscript such that $b_j = F(l)$ for some $l \leq n-1$. If n is odd, $F(n) = a_k$ where k is the smallest subscript such that $k > i$ and $a_k \notin \{b_1, b_2, \ldots, b_j\}$, and if n is even, $F(n) = b_k$ where k is the smallest subscript such that $k > j$ and $b_k \notin \{a_1, a_2, \ldots, a_i\}$.

24. *Proof*: Suppose not. That is, suppose the set of all irrational numbers were countable. Then the set of all real numbers could be written as a union of two countable sets: the set of all rational numbers and the set of all irrational numbers. By exercise 23 this union is countable, and so the set of all real numbers would be countable. But this contradicts the fact that the set of all real numbers is uncountable (which follows immediately from Theorems 7.6.2 and 7.6.3). Hence the set of all irrational number is uncountable.

25. Two pieces of the proof of this statement can be written as separate lemmas.

    Lemma 1: $\mathbf{Z} \times \mathbf{Z}$ is countable.

    Proof: Use the one-to-one correspondence $f: \mathbf{Z}^+ \to \mathbf{Z}$ of Example 7.6.1 to define a function $G: \mathbf{Z}^+ \times \mathbf{Z}^+ \to \mathbf{Z} \times \mathbf{Z}$ by the equation $G(m, n) = (f(m), f(n))$.

    G *is one-to-one*: Suppose $G(a, b) = G(c, d)$. Then $(f(a), f(b)) = (f(c), f(d))$ by definition of G. So $f(a) = f(c)$ and $f(b) = f(d)$ by definition of equality of ordered pairs. Since f is one-to-one, then, $a = c$ and $b = d$, and so $(a, b) = (c, d)$.

    G *is onto*: Suppose $(r, s) \in \mathbf{Z} \times \mathbf{Z}$. Then $r \in \mathbf{Z}$ and $s \in \mathbf{Z}$. Since f is onto, there exist an $m \in \mathbf{Z}^+$ and an $n \in \mathbf{Z}^+$ with $f(m) = r$ and $f(n) = s$. But, then, by definition of G, $G(m, n) = (r, s)$.

    So G is a one-to-one correspondence from $\mathbf{Z}^+ \times \mathbf{Z}^+$ to $\mathbf{Z} \times \mathbf{Z}$, and thus the two sets have the same cardinality. But by exercise 13, $\mathbf{Z}^+ \times \mathbf{Z}^+$ has the same cardinality as $\mathbf{Z}^+$. So by the transitive property of cardinality, $\mathbf{Z} \times \mathbf{Z}$ has the same cardinality as $\mathbf{Z}^+$, and hence $\mathbf{Z} \times \mathbf{Z}$ is countable.

    Lemma 2: If A is any countable set, B is any set, and $g: A \to B$ is onto, then B is countable.

    Proof: Suppose A is any countable set, B is any set, and $g: A \to B$ is onto. Since A is countable, there is a one-to-one correspondence $f: \mathbf{Z}^+ \to A$. Then, in particular, f is onto, and so by Theorem 7.5.4, $g \circ f$ is an onto function from $\mathbf{Z}^+$ to B. Define a function $h: B \to \mathbf{Z}^+$ as follows: Suppose x is any element of B. Since $g \circ f$ is onto, $\{m \in \mathbf{Z}^+ \mid (g \circ f)(m) = x\} \neq \emptyset$. So by the well-ordering principle for the integers, this set has a least element. In other words, there is a least positive integer n with $(g \circ f)(n) = x$. Let $h(x)$ be this integer.

    We claim that h is one-to-one. For suppose $h(x_1) = h(x_2) = n$. By definition of h, n is the least positive integer with $(g \circ f)(n) = x_1$. But also by definition of h, n is the least positive integer with $(g \circ f)(n) = x_2$. So $x_1 = (g \circ f)(n) = x_2$.

    Thus h is a one-to-one correspondence between B and a subset S of positive integers (the range of h). Since any subset of a countable set is countable (Theorem 7.6.3), S is countable, and so there is a one-to-one correspondence between B and a countable set. Hence by the transitive property of cardinality B is countable.

Proof of exercise statement: First note that there are as many equations of the form $x^2 + ax + b = 0$ as there are pairs (a, b) where a and b are in $\mathbf{Z}$. By Lemma 1, the set of all such pairs is countable, and so the set of equations of the form $x^2 + ax + b = 0$ is countable.

Next observe that by the quadratic formula, each equation $x^2 + ax + b = 0$ has at most two solutions:
$$x = \frac{-a + \sqrt{a^2 - 4b}}{2} \quad \text{and} \quad x = \frac{-a - \sqrt{a^2 - 4b}}{2}.$$

Let
$$S_1 = \left\{ x \;\Big|\; x = \frac{-a + \sqrt{a^2 - 4b}}{2} \text{ for some integers } a \text{ and } b \right\},$$

$$S_2 = \left\{ x \;\Big|\; x = \frac{-a - \sqrt{a^2 - 4b}}{2} \text{ for some integers } a \text{ and } b \right\},$$

and $S = S_1 \cup S_2$. Then S is the set of all solutions of equations of the form $x^2 + ax + b = 0$ where a and b are integers.

Define functions F_1 and F_2 from the set of equations of the form $x^2 + ax + b = 0$ to the sets S_1 and S_2 as follows:
$$F_1(x^2 + ax + b = 0) = \frac{-a + \sqrt{a^2 - 4b}}{2} \quad \text{and} \quad F_2(x^2 + ax + b = 0) = \frac{-a - \sqrt{a^2 - 4b}}{2}.$$

Then F_1 and F_2 are onto functions defined on countable sets, and so by Lemma 2, S_1 and S_2 are countable. Since any union of two countable sets is countable (exercise 23), $S = S_1 \cup S_2$ is countable.

26. *Proof 1*: Define a function $f: \mathcal{P}(S) \to T$ as follows: For each subset A of S, let $f(A) = \chi_A(x)$, the characteristic function of A, where $\chi_A(x) = \begin{cases} 1 & \text{if } x \in A \\ 0 & \text{if } x \notin A \end{cases}$ for all $x \in S$. Then f is one-to-one because if $f(A_1) = f(A_2)$ then $\chi_{A_1}(x) = \chi_{A_2}(x)$ for all $x \in S$, which implies that $x \in A_1$ if, and only if, $x \in A_2$ *[for instance, if $x \in A_1$, then $\chi_{A_1}(x) = 1 = \chi_{A_2}(x)$ and so $x \in A_2$]*, or equivalently $A_1 = A_2$. Furthermore, f is onto because given any function $g: S \to \{0, 1\}$, let A be the set of all x in S such that $g(x) = 1$. Then $g = \chi_A = f(A)$. Since f is one-to-one and onto, $\mathcal{P}(S)$ and T have the same cardinality.

Proof 2: Define $H: T \to \mathcal{P}(S)$ by letting $H(f) = \{x \in S \mid f(x) = 1\}$.

H is one-to-one: Suppose $H(f_1) = H(f_2)$. By definition of H, $\{x \in S \mid f_1(x) = 1\} = \{x \in S \mid f_2(x) = 1\}$. So for all $x \in S$, $f_1(x) = 1 \Leftrightarrow f_2(x) = 1$. This implies that for all $x \in S$, $f_1(x) = f_2(x)$ (because f_1 and f_2 only take the values 1 and 0, and so if they do not have the value 1 they must have the value 0). Thus $f_1 = f_2$.

H is onto: Suppose $A \subseteq S$. Define $g: A \to 0, 1$ as follows: for all $x \in S$,
$$g(x) = \begin{cases} 1 & \text{if } x \in A \\ 0 & \text{if } x \notin A \end{cases}.$$

Then $x \in A$ if, and only if, $g(x) = 1$, and so $A = H(g)$.

Since we have found a function $H: T \to \mathcal{P}(S)$ that is one-to-one and onto, we conclude that T and $\mathcal{P}(S)$ have the same cardinality.

27. *Proof (by contradiction)*: Suppose not. Suppose S and $\mathcal{P}(S)$ have the same cardinality. This means that there is a one-to-one, onto function $f: S \to \mathcal{P}(S)$. Let $A = \{x \in S \mid x \notin f(x)\}$. Then $A \in \mathcal{P}(S)$, and since f is onto, there is a $z \in S$ such that $A = f(z)$. Now either $z \in A$ or $z \notin A$. In case $z \in A$, then by definition of A, $z \notin f(z) = A$. Hence in this case $z \in A$ and $z \notin A$ which is impossible. In case $z \notin A$, then since $A = f(z)$, $z \notin f(z)$ and so z satisfies the

condition of membership for the set A which implies that $z \in A$. Hence in this case $z \notin A$ and $z \in A$ which is impossible. Thus in both cases a contradiction is obtained. It follows that the supposition is false, and so S and $\mathcal{P}(S)$ do not have the same cardinality.

28. *Proof:* Let B be the set of all functions from $\mathbf{Z}^+$ to $\{0,1\}$ and let D be the set of all functions from $\mathbf{Z}^+$ to $\{0,1,2,3,4,5,6,7,8,9\}$. Elements of B can be represented as infinite sequences of 0's and 1's (for instance, 01101010110...) and elements of D can be represented as infinite sequences of digits from 0 to 9 inclusive (for instance, 20775931124...).

We define a function $H\colon B \to D$ as follows: For each function f in B, consider the representation of f as an infinite sequence of 0's and 1's. Such a sequence is also an infinite sequence of digits chosen from 0 to 9 inclusive (one formed without using 2,3,...,9), which represents a function in D. We define this function to be $H(f)$. More formally, for each $f \in B$, let $H(f)$ be the function in D defined by the rule $H(f)(n) = f(n)$ for all $n \in \mathbf{Z}^+$. It is clear from the definition that H is one-to-one.

We define a function $K\colon D \to B$ as follows: For each function g in D, consider the representation of g as a sequence of digits from 0 to 9 inclusive. Replace each of these digits by its 4-bit binary representation adding leading 0's if necessary to make a full four bits. (For instance, 2 would be replaced by 0010.) The result is an infinite sequence of 0's and 1's, which represents a function in B. This function is defined to be $K(g)$. Note that K is one-to-one because if $g_1 \neq g_2$ then the sequences representing g_1 and g_2 must have different digits in some position m, and so the corresponding sequences of 0's and 1's will differ in at least one of the positions $4m-3, 4m-2, 4m-1$, or $4m$, which are the locations of the 4-bit binary representations of the digits in position m.

It can be shown that whenever there are one-to-one functions from one set to a second and from the second set back to the first, then the two sets have the same cardinality. This fact is known as the Schröder-Bernstein theorem after its two discoverers. For a proof see, for example, *Set Theory and Metric Spaces* by Irving Kaplansky, *A Survey of Modern Algebra*, Third Edition, by Garrett Birkhoff and Saunders MacLane, *Naive Set Theory* by Paul Halmos, or *Topology* by James R. Munkres. The above discussion shows that there are one-to-one functions from B to D and from D to B, and hence by the Schröder-Bernstein theorem the two sets have the same cardinality.

29. *Proof:* Let A and B be countable sets. If both A and B are finite, then $A \times B$ is also finite (it has $n(A) \cdot n(B)$ elements) and is therefore countable. So assume that at least one of A or B is infinite and represent the elements of A and B as infinite sequences $A\colon a_1, a_2, a_3, \ldots$ and $B\colon b_1, b_2, b_3, \ldots$ where if one of A or B is finite, the sequence is filled out by repeating one of the elements forever. Then all the elements of $A \times B$ are listed in the rectangular array below (possibly more than once).

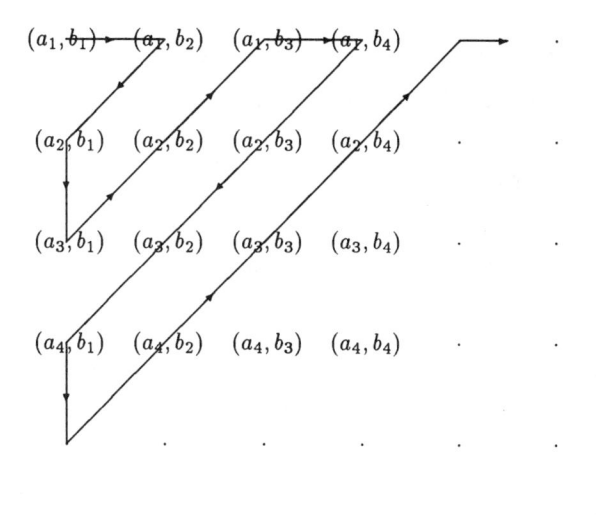

$(a_1, b_1) \rightarrow (a_1, b_2) \quad (a_1, b_3) \rightarrow (a_1, b_4)$

$(a_2, b_1) \quad (a_2, b_2) \quad (a_2, b_3) \quad (a_2, b_4)$

$(a_3, b_1) \quad (a_3, b_2) \quad (a_3, b_3) \quad (a_3, b_4)$

$(a_4, b_1) \quad (a_4, b_2) \quad (a_4, b_3) \quad (a_4, b_4)$

Define a function F from $\mathbf{Z}^+$ to $A \times B$ as follows: Let $F(1) = (a_1, b_1)$ and let each successive value of $F(n)$ be the next successive ordered pair obtained by following the arrows unless that pair has already been made the image of some integer in which case it is skipped. Thus (if there is no repetition) $F(2) = (a_1, b_2)$, $F(3) = (a_2, b_1)$, $F(4) = (a_3, b_1)$, $F(5) = (a_2, b_2)$, and so forth. It is clear that F is one-to-one because ordered pairs are skipped if they have already been used as function values, and F is onto because every ordered pair in $A \times B$ appears in the array. Hence F is a one-to-one correspondence from $\mathbf{Z}^+$ to $A \times B$, and so $A \times B$ is countably infinite and hence countable.

30. *Proof:* Let sets $A_1, A_2, A_3, \ldots$ be countable sets. If all A_i are finite and there are only finitely many A_i, then the union of all the A_i is a finite set and hence countable. So suppose that at least one A_i is infinite or there are infinitely many A_i. Represent the elements of each A_i as an infinite sequence $A_i : a_{i1}, a_{i2}, a_{i3}, \ldots$ where if any A_i is finite, the sequence is filled out by repeating one of the elements forever. If there are only finitely many A_i, say $A_1, A_2, \ldots, A_n$, let $A_j = A_n$ for all integers $j > n$. Consider the rectangular array of elements whose ith row is the sequence representing A_i for each $i = 1, 2, 3, \ldots$

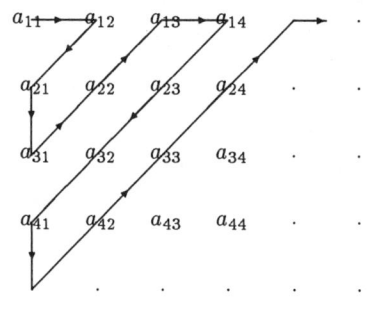

$a_{11} \rightarrow a_{12} \quad a_{13} \rightarrow a_{14}$

$a_{21} \quad a_{22} \quad a_{23} \quad a_{24}$

$a_{31} \quad a_{32} \quad a_{33} \quad a_{34}$

$a_{41} \quad a_{42} \quad a_{43} \quad a_{44}$

Note that each element of the array is in $\bigcup_{i=1}^{\infty} A_i$. Define a function G from $\mathbf{Z}^+$ to $\bigcup_{i=1}^{\infty} A_i$ as follows: Let $G(1) = a_{11}$ and let each successive value of $G(n)$ be the next successive element of the array obtained by following the arrows unless that element has already been made the image of some integer in which case it is skipped. Thus (if there is no repetition) $G(2) = a_{12}$, $G(3) = a_{21}$, $G(4) = a_{31}$, $G(5) = a_{22}$, and so forth. It is clear that G is one-to-one because elements are skipped if they have already been used as function values, and G is onto because every element in $\bigcup_{i=1}^{\infty} A_i$ appears in the array. Hence G is a one-to-one correspondence from $\mathbf{Z}^+$ to $\bigcup_{i=1}^{\infty} A_i$, and so $\bigcup_{i=1}^{\infty} A_i$ is countably infinite and hence countable.

Chapter 8 Recursion

This chapter can be covered at any time after Chapter 4. The first three sections discuss sequences that are defined recursively and the fourth section explores recursively defined sets and functions and recursive definitions for sum, product, union, and intersection.

Section 8.1 has two aims. One is to familiarize students with the concept and notation of recursively defined sequence. To help students distinguish recursive definitions from explicit formulas, the letter k is used to denote the variable in a recursive definition and the letter n is used to denote the variable in an explicit formula. The other aim of this section is to introduce students to the idea of "recursive thinking," namely assuming the answer to a problem is known for certain smaller cases and expressing the answer for a given case in terms of the answers to these smaller cases.

Sections 8.2 and 8.3 treat the questions of how to find an explicit formula for a sequence that has been defined recursively and how to use mathematical induction to verify that this explicit formula correctly describes the given sequence. In Section 8.2, the method used is "iteration," which consists of writing down successive terms of the sequence and looking for a pattern. In Section 8.3, explicit formulas for second order linear homogeneous recurrence relations are derived. The main difficulty students have with these sections is related to a lack of understanding of the extent to which definitions are universal. For instance, given a recurrence relation that expresses a_k for general k in terms of a_{k-1} and a_{k-2}, quite a few students have difficulty writing, say, a_{k-1} in terms of a_{k-2} and a_{k-3}. In addition, more students than you might expect get tripped up by the algebra of successive substitution, neglecting to substitute carefully and/or making errors in multiplying out or in regrouping terms.

Section 8.4 is not difficult for students, and at this point in the course many are able to appreciate the elegance and effectiveness of recursive definitions for describing sets, such as the set of Boolean expressions, that they are familiar with from other contexts.

Comments on Exercises

Section 8.1: #1–17, #22–26, and #46–47 develop students' skill in handling the notation and verifying properties of recursively defined sequences. Exercises #18–21, #27–38, #45, and #48–50 explore recursive thinking. Exercises #39–44 are intended to help students develop facility with Stirling numbers of the second kind.
Section 8.2: #1 and #2 are warm-up exercises which review formulas used to simplify expressions that arise in solving recurrence relations by iteration.
Section 8.3: A number of exercises in this section are designed to give students practice in the kind of thinking used to derive the main theorems of the section. They are meant to bridge the gap between mechanical application of the theorems and formal derivation of the theorems themselves.

Section 8.1

2. $b_1 = 2$, $b_2 = b_1 + 2 \cdot 2 = 6$, $b_3 = b_2 + 2 \cdot 3 = 12$, $b_4 = b_3 + 2 \cdot 4 = 20$

4. $d_0 = 2$, $d_1 = 1 \cdot d_0^2 = 4$, $d_2 = 2 \cdot d_1^2 = 32$, $d_3 = 3 \cdot d_2^2 = 3072$

6. $t_0 = -1$, $t_1 = 1$, $t_2 = t_1 + 2 \cdot t_0 = 1 + 2 \cdot (-1) = -1$, $t_3 = t_2 + 2 \cdot t_1 = -1 + 2 \cdot 1 = 1$

8. $v_1 = 1$, $v_2 = 2$, $v_3 = v_2 + v_1 + 1 = 2 + 1 + 1 = 4$, $v_4 = v_3 + v_2 + 1 = 4 + 2 + 1 = 7$

10. By definition of $b_0, b_1, b_2, \ldots$, for all integers $k \geq 1$, $b_k = 5^k$ and $b_{k-1} = 5^{k-1}$. So for all integers $k \geq 1$, $5 \cdot b_{k-1} = 5 \cdot 5^{k-1} = 5^k = b_k$.

167

12. Call the nth term of the sequence s_n. Then $s_n = \dfrac{(-1)^n}{n!}$ for all integers $n \geq 0$. So for all integers $k \geq 1$, $s_k = \dfrac{(-1)^k}{k!}$ and $s_{k-1} = \dfrac{(-1)^{k-1}}{(k-1)!}$. It follows that for all integers $k \geq 1$,

$$\dfrac{-s_{k-1}}{k} = \dfrac{-\frac{(-1)^{k-1}}{(k-1)!}}{k} = \dfrac{-(-1)^{k-1}}{k \cdot (k-1)!} = \dfrac{(-1)^k}{k!} = s_k.$$

14. Call the nth term of the sequence d_n. Then $d_n = 2^n - 1$ for all integers $n \geq 0$. So for all integers $k \geq 2$, $d_k = 2^k - 1$, $d_{k-1} = 2^{k-1} - 1$, and $d_{k-2} = 2^{k-2} - 1$. It follows that for all integers $k \geq 2$, $3d_{k-1} - 2d_{k-2} = 3(2^{k-1} - 1) - 2(2^{k-2} - 1) = 3 \cdot 2^{k-1} - 3 - 2 \cdot 2^{k-2} + 2 = 3 \cdot 2^{k-1} - 3 - 2^{k-1} + 2 = 2 \cdot 2^{k-1} - 1 = 2^k - 1 = d_k.$

16. According to exercise 17 of Section 6.6, for each integer $n \geq 1$, $C_n = \dfrac{1}{4n+2}\binom{2n+2}{n+1}$. Substituting $k - 1$ in place of n gives

$$C_{k-1} = \dfrac{1}{4(k-1)+2}\binom{2(k-1)+2}{(k-1)+1} = \dfrac{1}{4k-2}\binom{2k}{k}.$$

Then for each integer $k \geq 2$,

$$C_k = \dfrac{1}{k+1}\binom{2k}{k} = \dfrac{1}{k+1} \cdot \dfrac{4k-2}{4k-2}\binom{2k}{k} = \dfrac{4k-2}{k+1}C_{k-1}.$$

17. $m_7 = 2m_6 + 1 = 2 \cdot 63 + 1 = 127$, $m_8 = 2m_7 + 1 = 2 \cdot 127 + 1 = 255$

18. b. $a_4 = 26 + 1 + 26 + 1 + 26 = 80$

19. a. $b_1 = 1$, $b_2 = 1 + 1 + 1 + 1 = 4$, $b_3 = 4 + 4 + 1 + 4 = 13$

c. Note that it takes just as many moves to move a stack of disks from the middle pole to an outer pole as from an outer pole to the middle pole: the moves are the same except that their order and direction are reversed. For all integers $k \geq 2$,

$$\begin{aligned} b_k &= a_{k-1} \quad \text{(moves to transfer the top } k-1 \text{ disks from pole } A \text{ to pole } C\text{)} \\ &\quad +1 \quad \text{(move to transfer the bottom disk from pole } A \text{ to pole } B\text{)} \\ &\quad +b_{k-1} \quad \text{(moves to transfer the top } k-1 \text{ disks from pole } C \text{ to pole } B\text{)}. \\ &= a_{k-1} + 1 + b_{k-1}. \end{aligned}$$

d. One way to transfer a tower of k disks from pole A to pole B is to first transfer the top $k-1$ disks from pole A to pole B (this requires b_{k-1} moves), then transfer the top $k-1$ disks from pole B to pole C (this requires b_{k-1} moves), then transfer the bottom disk from pole A to pole B (this requires one move), and finally transfer the top $k-1$ disks from pole C to pole B (this requires b_{k-1} moves). This sequence of steps need not, however, result in a minimal number of moves. Therefore, for all integers $k \geq 2$,

$$b_k \leq b_{k-1} + b_{k-1} + 1 + b_{k-1} = 3b_{k-1} + 1.$$

20. c. Name the poles A, B, C, and D going from left to right. Because disks can be moved from one pole to any other pole, the number of moves needed to move a tower of disks from any one pole to any other pole is the same for any two poles. One way to transfer a tower of k disks from pole A to pole D is to first transfer the top $k-2$ disks from pole A to pole B, then transfer the

second largest disk from pole A to pole C, then transfer the largest disk from pole A to pole D, then transfer the second largest disk from pole C to pole D, and finally transfer the top $k-2$ disks from pole B to pole D. This need not result in a minimal number of moves, however. So for all integers $k \geq 3$,

$$
\begin{aligned}
s_k \quad &\leq \quad s_{k-2} \quad &&\text{(moves to transfer the top } k-2 \text{ disks from pole } A \text{ to pole } B) \\
& \quad +1 \quad &&\text{(move to transfer the second largest disk from pole } A \text{ to pole } C) \\
& \quad +1 \quad &&\text{(move to transfer the largest disk from pole } A \text{ to pole } D) \\
& \quad +1 \quad &&\text{(move to transfer the second largest disk from pole } C \text{ to pole } D) \\
& \quad +s_{k-2} \quad &&\text{(moves to transfer the top } k-2 \text{ disks from pole } B \text{ to pole } D) \\
&\leq \quad 2s_{k-2}+3.
\end{aligned}
$$

21. *a.* $t_1 = 2$, $\quad t_2 = 2+2+2 = 6$

 c. For all integers $k \geq 2$,

$$
\begin{aligned}
t_k \quad &= \quad t_{k-1} \quad &&\text{(moves to transfer the top } 2k-2 \text{disks from pole } A \text{ to pole } B) \\
& \quad +2 \quad &&\text{(moves to transfer the bottom two disks from pole } A \text{ to pole } C) \\
& \quad +t_{k-1} \quad &&\text{(moves to transfer the top } 2k-2 \text{ disks from pole } B \text{ to pole } C) \\
&= \quad 2t_{k-1}+2.
\end{aligned}
$$

22. $F_{13} = F_{12} + F_{11} = 233 + 144 = 377$, $\quad F_{14} = F_{13} + F_{12} = 377 + 233 = 610$

23. *b.* $F_{k+2} = F_{k+1} + F_k$ $\quad$ *c.* $F_{k+3} = F_{k+2} + F_{k+1}$

24. *b.* According to the definition of the Fibonacci sequence, for any integer $k \geq 1$, $F_{k+1}^2 - F_k^2 - F_{k-1}^2 = (F_k + F_{k-1})^2 - F_k^2 - F_{k-1}^2 = F_k^2 + 2F_k F_{k-1} + F_{k-1}^2 - F_k^2 - F_{k-1}^2 = 2F_k F_{k-1}$.

 c. According to the definition of the Fibonacci sequence, for any integer $k \geq 1$, $F_{k+1}^2 - F_k^2 = (F_{k+1} - F_k) \cdot (F_{k+1} + F_k) = (F_{k+1} - F_k) \cdot F_{k+2}$. But since $F_{k+1} = F_k + F_{k-1}$, then $F_{k+1} - F_k = F_{k-1}$. By substitution, $F_{k+1}^2 - F_k^2 = F_{k-1} \cdot F_{k+2}$.

 d. Proof:

 The formula holds for n = 0: For $n = 0$, the left-hand side of the equation is $F_{0+2}F_0 - F_1^2 = 2 \cdot 1 - 1^2 = 1$, and the right-hand side is $(-1)^0 = 1$ also.

 If the formula holds for n = k then it holds for n = k+1: Suppose that $F_{k+2}F_k - F_{k+1}^2 = (-1)^k$ for some integer $k \geq 0$. *[Inductive Hypothesis]* We must show that $F_{k+3}F_{k+1} - F_{k+2}^2 = (-1)^{k+1}$. But by inductive hypothesis,

$$F_{k+1}^2 = F_{k+2}F_k - (-1)^k = F_{k+2}F_k + (-1)^{k+1}. (*)$$

Hence,

$$\begin{aligned}
F_{k+3}F_{k+1} - F_{k+2}^2 &= (F_{k+1} + F_{k+2})F_{k+1} - F_{k+2}^2 & \text{by definition of the Fibonacci} \\
& & \text{sequence} \\
&= F_{k+1}^2 + F_{k+2}F_{k+1} - F_{k+2}^2 \\
&= F_{k+2}F_k + (-1)^{k+1} + F_{k+2}F_{k+1} - F_{k+2}^2 & \text{by (*)} \\
&= F_{k+2}(F_k + F_{k+1} - F_{k+2}) + (-1)^{k+1} & \text{by factoring out } F_{k+2} \\
&= F_{k+2}(F_{k+2} - F_{k+2}) + (-1)^{k+1} & \text{by definition of the Fibonacci} \\
& & \text{sequence} \\
&= F_{k+2} \cdot 0 + (-1)^{k+1} \\
&= (-1)^{k+1}.
\end{aligned}$$

26. Let $L = \lim_{n \to \infty} x_n$. By definition of $x_0, x_1, x_2, \ldots$ and by the continuity of the square root function,
$$L = \lim_{n \to \infty} x_n = \lim_{n \to \infty} \sqrt{2 + x_{n-1}} = \sqrt{2 + \lim_{n \to \infty} x_{n-1}} = \sqrt{2 + L}.$$
Hence $L^2 = 2 + L$, and so $L^2 - L - 2 = 0$. Factoring gives $(L-2)(L+1) = 0$, and so $L = 2$ or $L = -1$. But $L \geq 0$ because each $x_i \geq 0$. Thus $L = 2$.

27. a. $r_k = r_{k-1} + 4r_{k-3}$ for all integers $k \geq 3$

    c. $r_7 = r_6 + 4r_4 = 181 + 4 \cdot 29 = 297$, $r_8 = r_7 + 4r_5 = 297 + 4 \cdot 65 = 557$, $r_9 = r_8 + 4r_6 = 557 + 4 \cdot 181 = 1281$, $r_{10} = r_9 + 4r_7 = 1281 + 4 \cdot 297 = 2469$, $r_{11} = r_{10} + 4r_8 = 2469 + 4 \cdot 557 = 4697$, $r_{12} = r_{11} + 4r_9 = 4697 + 4 \cdot 1281 = 9821$

    At the end of the year there will be $r_{12} = 9821$ rabbit pairs or 19,642 rabbits.

28. a. $s_k = s_{k-1} + 3s_{k-3}$ for all integers $k \geq 3$

    b. $s_0 = 1$, $s_1 = 1$, $s_2 = 1$, $s_3 = 1 + 3 \cdot 1 = 4$, $s_4 = 4 + 3 \cdot 1 = 7$, $s_5 = 7 + 3 \cdot 1 = 10$

    c. $s_6 = s_5 + 3s_3 = 22$, $s_7 = s_6 + 3s_4 = 43$, $s_8 = s_7 + 3s_5 = 73$, $s_9 = s_8 + 3s_6 = 139$, $s_{10} = s_9 + 3s_7 = 268$, $s_{11} = s_{10} + 3s_9 = 487$, $s_{12} = s_{11} + 3s_{10} = 904$. So at the end of the year there are $2(904) = 1808$ rabbits.

30. a. When 6% interest is compounded monthly, the interest per month is $0.06/12 = 0.005$. If S_k is the amount on deposit at the end of month k, then $S_k = S_{k-1} + 0.005 S_{k-1} = (1 + 0.005) S_{k-1} = (1.005) S_{k-1}$ for each integer $k \geq 1$.

    b. $S_{12} = (1.005)S_{11} = (1.005)[(1.005)S_{10}] = (1.005)^2 S_{10} = (1.005)^2[(1.005)S_9] = (1.005)^3 S_9 = (1.005)^3[(1.005)S_8] = (1.005)^4 S_8 = (1.005)^4[(1.005)S_7] = (1.005)^5 S_7 = (1.005)^5[(1.005)S_6] = (1.005)^6 S_6 = (1.005)^6[(1.005)S_5] = (1.005)^7 S_5 = (1.005)^7[(1.005)S_4] = (1.005)^8 S_4 = (1.005)^8[(1.005)S_3] = (1.005)^9 S_3 = (1.005)^9[(1.005)S_2] = (1.005)^{10} S_2 = (1.005)^{10}[(1.005)S_1] = (1.005)^{11} S_1 = (1.005)^{11}[(1.005)S_0] = (1.005)^{12} S_0 = (1.005)^{12} \cdot 1000 \cong 1061.68$ dollars.

32. a. length 0: ε
    length 1: a, b, c
    length 2: $ab, ac, ba, bb, bc, ca, cb, cc$
    length 3: $aba, abb, abc, aca, acb, acc, bab, bac, bba, bbb, bbc, bca, bcb, bcc, cab, cac, cba, cbb, cbc, cca, ccb, ccc$

    b. By part (a), $s_0 = 1$, $s_1 = 3$, $s_2 = 8$, and $s_3 = 22$

    c. Let k be an integer with $k \geq 2$. Any string of length k that does not contain the pattern aa starts with a, with b, or with c. If it starts with b or c, this can be followed by any string of length $k-1$ that does not contain the pattern aa. There are s_{k-1} such strings, and so there are $2s_{k-1}$ strings that start either with b or with c. If the string starts with a, then the first two characters must be ab or ac. In either case, the remaining $k-2$ characters can be any string of length $k-2$ that does not contain the pattern aa. There are s_{k-2} such strings, and so there are $2s_{k-2}$ strings that start either ab or ac. It follows that for all integers $k \geq 2$, $s_k = 2s_{k-1} + 2s_{k-2}$.

    c. By part (b) $s_2 = 8$ and $s_3 = 22$, and so $s_4 = 2s_3 + 2s_2 = 44 + 16 = 60$.

33. *a.* Let k be an integer with $k \geq 3$. The set of bit strings of length k that do not contain the pattern 101 can be partitioned into $k+1$ subsets: the subset of strings that start with 0 and continue with any bit string of length $k-1$ not containing 101 (there are a_{k-1} of these), the subset of strings that start with 100 and continue with any bit string of length $k-3$ not containing 101 (there are a_{k-3} of these), the subset of strings that start with 1100 and continue with any bit string of length $k-4$ not containing 101 (there are a_{k-4} of these), the subset of strings that start with 11100 and continue with any bit string of length $k-5$ not containing 101 (there are a_{k-5} of these), and so forth up to $\{\underbrace{11\ldots1}_{k-3\ 1\text{'s}}001, \underbrace{11\ldots1}_{k-3\ 1\text{'s}}000\}$ (there are $2 = a_1$ of these). In addition, the three single-element sets $\{\underbrace{11\ldots1}_{k-2\ 1\text{'s}}00\}$, $\{\underbrace{11\ldots1}_{k-1\ 1\text{'s}}0\}$, and $\{\underbrace{11\ldots1}_{k-1\ 1\text{'s}}1\}$ are in the partition. Thus by the addition rule,

$$a_k = a_{k-1} + a_{k-2} + a_{k-3} + \cdots + a_1 + 3.$$

b. By part (a), if $k \geq 4$,

$$\begin{aligned} a_k &= a_{k-1} + a_{k-2} + a_{k-3} + \cdots + a_1 + 3 \\ a_{k-1} &= a_{k-2} + a_{k-3} + a_{k-4} + \cdots + a_1 + 3. \end{aligned}$$

Subtracting the second equation from the first gives

$$\begin{aligned} a_k - a_{k-1} &= a_{k-1} + a_{k-3} - a_{k-2} \\ \Rightarrow \qquad a_k &= 2a_{k-1} + a_{k-3} - a_{k-2}. \end{aligned}$$

35. Imagine a tower of height k inches. If the bottom block has height one inch, then the remaining blocks make up a tower of height $k-1$ inches. There are t_{k-1} such towers. If the bottom block has height two inches, then the remaining blocks make up a tower of height $k-2$ inches. There are t_{k-2} such towers. If the bottom block has height four inches, then the remaining blocks make up a tower of height $k-4$ inches. There are t_{k-4} such towers. Therefore, $t_k = t_{k-1} + t_{k-2} + t_{k-4}$ for all integers $k \geq 5$.

36. *b.* Let $k \geq 3$ and consider a permutation of $\{1, 2, \ldots, k\}$ that does not move any number more than one place from its "natural" position. Such a permutation either leaves 1 fixed or it interchanges 1 and 2. If it leaves 1 fixed, then the remaining $k-1$ numbers can be permuted in any way except that they must not be moved more than one place from their natural positions. There are a_{k-1} ways to do this. If it interchanges 1 and 2, then the remaining $k-2$ numbers can be permuted in any way except that they must not be moved more than one place from their natural positions. There are a_{k-2} ways to do this. Therefore, $a_k = a_{k-1} + a_{k-2}$ for all integers $k \geq 2$.

37. To get a sense of the problem, we compute s_4 directly. If there are four seats in the row, there can be a single student in any one of the four seats or there can be a pair of students in seats 1&3, 1&4, or 2&4. No other arrangements are possible because with more than two students, two would have to sit next to each other. Thus $s_4 = 4 + 3 = 7$. In general, if there are k chairs in a row, then

$$\begin{aligned} s_k &= s_{k-1} \quad \text{(the number of ways a nonempty set of students can sit} \\ &\qquad \text{in the row with no two students adjacent and chair } k \text{ empty)} \\ &+s_{k-2} \quad \text{(the number of ways students can sit in the row with chair } k \\ &\qquad \text{occupied, chair } k-1 \text{ empty, and chairs 1 through} \\ &\qquad k-2 \text{ occupied by a nonempty set of students in such a} \\ &\qquad \text{way that no two students are adjacent)} \\ &+1 \quad \text{(for the seating in which chair } k \text{ is occupied} \\ &\qquad \text{and all the other chairs are empty)} \\ &= s_{k-1} + s_{k-2} + 1 \text{ for all integers } k \geq 3. \end{aligned}$$

39. The partitions are

$\{x_1\}\{x_2\}\{x_3\}\{x_4,x_5\}$ $\{x_1\}\{x_2\}\{x_4\}\{x_3,x_5\}$ $\{x_1\}\{x_3\}\{x_4\}\{x_2,x_5\}$ $\{x_2\}\{x_3\}\{x_4\}\{x_1,x_5\}$
$\{x_1\}\{x_2\}\{x_5\}\{x_3,x_4\}$ $\{x_1\}\{x_3\}\{x_5\}\{x_2,x_4\}$ $\{x_2\}\{x_3\}\{x_5\}\{x_1,x_4\}$ $\{x_1\}\{x_4\}\{x_5\}\{x_2,x_3\}$
$\{x_2\}\{x_4\}\{x_5\}\{x_1,x_3\}$ $\{x_3\}\{x_4\}\{x_5\}\{x_1,x_2\}$

So $S_{5,4} = 10$.

41. By the recurrence relation from Example 8.1.11 and the values computed in Example 8.1.10, $S_{5,3} = S_{4,2} + 3 \cdot S_{4,3} = 7 + 3 \cdot 6 = 25$.

42. By the definition and initial conditions for Stirling numbers of the second kind and the results of exercises 39–41, the total number of partitions of a set with five elements is $S_{5,1} + S_{5,2} + S_{5,3} + S_{5,4} + S_{5,5} = 1 + 15 + 25 + 10 + 1 = 52$.

44. *Proof*:

    The formula holds for n = 2: For $n = 2$ the left-hand side of the equation is $\sum_{k=2}^{2} 3^{2-k} S_{k,2} = 3^{2-2} S_{2,2} = 1$, and the right-hand side is $S_{2+1,3} = S_{3,3} = 1$ also.

    If the formula holds for n = m then it holds for n = m+1: Suppose

    $$\sum_{k=2}^{m} 3^{m-k} S_{k,2} = S_{m+1,3}.$$ [*This is the inductive hypothesis.*]

    We must show that
    $$\sum_{k=2}^{m+1} 3^{(m+1)-k} S_{k,2} = S_{m+2,3}.$$

    But

    $$\begin{aligned}
    \sum_{k=2}^{m+1} 3^{(m+1)-k} S_{k,2} &= \sum_{k=2}^{m} 3 \cdot 3^{m-k} S_{k,2} + 3^0 S_{m+1,2} \\
    &= 3 \sum_{k=2}^{m} 3^{m-k} S_{k,2} + S_{m+1,2} \\
    &= S_{m+1,2} + 3 S_{m+1,3} \qquad \text{by inductive hypothesis} \\
    &= S_{m+2,3} \qquad \text{by the recurrence relation for Stirling numbers of the second kind.}
    \end{aligned}$$

45. If X is a set with n elements and Y is a set with m elements, then the number of onto functions from X to Y is $m! S_{n,m}$, where $S_{n,m}$ is a Stirling number of the second kind. The reason is that we can construct all possible onto functions from X to Y as follows: For each partition of X into m subsets, order the subsets of the partition; call them, say, $S_1, S_2, \ldots, S_m$. Define an onto function from X to Y by first choosing an element of Y to be the image of all the elements in S_1 (there are m ways to do this), then choosing another element of Y to be the image of all the elements in S_2 (there are $m - 1$ ways to do this), then choosing another element of Y to be the image of all the elements in S_3 (there are $m - 2$ ways to do this), and so forth. Each of the $m!$ functions constructed in this way is onto because since Y has m elements and there are m subsets in the partition, eventually every element in Y will be the image of at least one element in X. Thus for each partition of X into m subsets, there are $m!$ onto functions, and so the total number of onto functions is the number of partitions, $S_{n,m}$, times $m!$, or $m! S_{n,m}$.

46. *Proof (by strong mathematical induction)*: Consider the inequality $F_n < 2^n$ where F_n is the nth Fibonacci number.

    The inequality is true for n = 1 and n = 2: $F_1 = 1 < 2 = 2^1$ and $F_2 = 3 < 4 = 2^2$.

    If the inequality is true for all i with $0 \leq i < k$ then it is true for k: Suppose that $F_i < 2^i$ for all integers i with $0 \leq i < k$. [*This is the inductive hypothesis.*] We must show that $F_k < 2^k$. Now by definition of the Fibonacci numbers, $F_k = F_{k-1} + F_{k-2}$. But by inductive hypothesis [since $k > 2$], $F_{k-1} < 2^{k-1}$ and $F_{k-2} < 2^{k-2}$. Hence $F_k = F_{k-1} + F_{k-2} < 2^{k-1} + 2^{k-2} = 2^{k-2} \cdot (2+1) = 3 \cdot 2^{k-2} < 4 \cdot 2^{k-2} = 2^k$. Thus $F_k < 2^k$ [as was to be shown].

[Since the basis and inductive steps have both been proved, we conclude that $F_n < 2^n$ for all integers $n > 1$.]

47. Proof (by mathematical induction):

   The property is true for n = 0: To prove the property for $n = 0$, we must show that $\gcd(F_1, F_0) = 1$. But $F_1 = 1$ and $F_0 = 1$ and $\gcd(1,1) = 1$.

   If the property is true for n = k then it is true for n = k+1: Let k be an integer with $k \geq 0$. Suppose $\gcd(F_{k+1}, F_k) = 1$. *[This is the inductive hypothesis.]* We must show that $\gcd(F_{k+2}, F_{k+1}) = 1$. But by definition of the Fibonacci sequence $F_{k+2} = F_{k+1} + F_k$. It follows from Lemma 3.8.2 that $\gcd(F_{k+2}, F_{k+1}) = \gcd(F_{k+1}, F_k)$. But by inductive hypothesis, $\gcd(F_{k+1}, F_k) = 1$. Hence $\gcd(F_{k+2}, F_{k+1}) = 1$ [as was to be shown].

   [Since both the basis and the inductive steps have been proved, we conclude that $\gcd(F_{n+1}, F_n) = 1$ for all integers $n \geq 0$.]

48. a. $g_3 = 1$, $g_4 = 1$, $g_5 = 2$ ($LWLLL$ and $WWLLL$)

   b. $g_6 = 4$ ($WWWLLL$, $WLWLLL$, $LWWLLL$, $LLWLLL$)

   c. If $k \geq 6$, then any sequence of k games must begin with exactly one of the possibilities: W, LW, or LLW. The number of sequences of k games that begin with W is g_{k-1} because the succeeding $k-1$ games can consist of any sequence of wins and losses except that the first sequence of three consecutive losses occurs at the end. Similarly, the number of sequences of k games that begin with LW is g_{k-2} and the number of sequences of k games that begin with LLW is g_{k-3}. Therefore, $g_k = g_{k-1} + g_{k-2} + g_{k-3}$ for all integers $k \geq 6$.

49. a. $d_1 = 0$, $d_2 = 1$, $d_3 = 2$(231 and 312)

   b. $d_4 = 9$ (2143, 3412, 4321, 3142, 4123, 2413, 4312, 2341, 3421)

   c. Divide the set of all derangements into two subsets: one subset, S, consists of all derangements in which the number 1 changes places with another number, and the other subset, T, consists of all derangements in which the number 1 goes to position $i \neq 1$ but i does not go to position 1. Forming a derangement in S can be regarded as a two-step process: step 1 is to choose a position i and to interchange 1 and i and step 2 is to derange the remaining $k-2$ numbers. Now there are $k-1$ numbers with which 1 can trade places in step 1, and so by the product rule there are $(k-1) \cdot d_{k-2}$ derangements in S. Forming a derangement in T can also be regarded as a two-step process: step 1 is to derange the $k-1$ numbers $2, 3, \ldots, k$ in positions $2, 3, \ldots, k$, and step 2 is to interchange the number 1 in position 1 with any of the numbers in the derangement. Now there are $k-1$ choices of numbers to interchange 1 with in step 2, and so by the product rule there are $d_{k-1} \cdot (k-1)$ derangements in T. It follows that the total number of derangements of the given k numbers is $d_k = (k-1)d_{k-1} + (k-1)d_{k-2}$ for all integers $k \geq 3$.

50. For each integer $k = 1, 2, \ldots, n-1$, consider the product $(x_1 x_2 \ldots x_k)(x_{k+1} x_{k+2} \ldots x_n)$. The factor $x_1 x_2 \ldots x_k$ can be parenthesized in P_k ways, and the factor $x_{k+1} x_{k+2} \ldots x_n$ can be parenthesized in P_{n-k} ways. Therefore, the product $x_1 x_2 x_3 \ldots x_{n-1} x_n$ can be parenthesized in $P_k P_{n-k}$ ways if the final multiplication is $(x_1 x_2 \ldots x_k) \cdot (x_{k+1} x_{k+2} \ldots x_n)$. Now when $x_1 x_2 x_3 \ldots x_{n-1} x_n$ is fully parenthesized, the final multiplication can be any one of the following: $(x_1) \cdot (x_2 x_3 \ldots x_n)$, $(x_1 x_2) \cdot (x_3 x_4 \ldots x_n)$, $(x_1 x_2 x_3) \cdot (x_4 x_5 \ldots x_n), \ldots, (x_1 x_2 \ldots x_{n-2}) \cdot (x_{n-1} x_n)$, $(x_1 x_2 \ldots x_{n-1}) \cdot (x_n)$. So the total number of ways to parenthesize the product is the sum of all the numbers $P_k P_{n-k}$ for $k = 1, 2, \ldots, n-1$. In symbols: $P_n = \sum_{k=1}^{n-1} P_k P_{n-k}$.

Section 8.2

1. c. $3+3\cdot 2+3\cdot 3+\cdots+3\cdot n+n = 3(1+2+3+\ldots+n)+n = 3\left(\dfrac{n(n+1)}{2}\right)+n = \dfrac{3n(n+1)}{2}+\dfrac{2n}{2} = \dfrac{3n^2+5n}{2}.$

2. b. $3^{n-1}+3^{n-2}+\cdots+3^2+3+1 = 1+3+3^2+\cdots+3^{n-2}+3^{n-1} = \dfrac{3^{(n-1)+1}-1}{3-1} = \dfrac{3^n-1}{2}.$

   d. Note that $\dfrac{1}{(-1)^n} = (-1)^n$ and $(-1)^n \cdot (-1)^{n+1} = (-1)^{2n+1} = -1$. Thus

   $2^n - 2^{n-1} + 2^{n-2} - 2^{n-3} + \cdots + (-1)^{n-1}\cdot 2 + (-1)^n$

   $\quad = \dfrac{1}{(-1)^n}(1 - 2 + 2^2 + \cdots + (-1)^{n-1}2^{n-1} + (-1)^n 2^n)$

   $\quad = (-1)^n(1 + (-2) + (-2)^2 + \cdots + (-2)^{n-1} + (-2)^n)$

   $\quad = (-1)^n\left(\dfrac{(-2)^{n+1}-1}{(-2)-1}\right)$

   $\quad = (-1)^n\left(\dfrac{1-(-2)^{n+1}}{3}\right)$

   $\quad = \dfrac{(-1)^n + 2^{n+1}}{3}.$

4.
$b_0 = 1$
$b_1 = \dfrac{b_0}{1+b_0} = \dfrac{1}{1+1} = \dfrac{1}{2}$
$b_2 = \dfrac{b_1}{1+b_1} = \dfrac{\frac{1}{2}}{1+\frac{1}{2}} = \dfrac{1}{2+1} = \dfrac{1}{3}$
$b_3 = \dfrac{b_2}{1+b_2} = \dfrac{\frac{1}{3}}{1+\frac{1}{3}} = \dfrac{1}{3+1} = \dfrac{1}{4}$
$b_4 = \dfrac{b_3}{1+b_3} = \dfrac{\frac{1}{4}}{1+\frac{1}{4}} = \dfrac{1}{4+1} = \dfrac{1}{5}$
$\vdots$

Guess: $b_n = \dfrac{1}{n+1}$ for all integers $n \geq 0$.

6.
$d_1 = 2$
$d_2 = 2d_1 + 3 = 2\cdot 2 + 3 = 2^2 + 3$
$d_3 = 2d_2 + 3 = 2(2\cdot 2 + 3) + 3 = 2^3 + 2\cdot 3 + 3$
$d_4 = 2d_3 + 3 = 2(2^3 + 2\cdot 3 + 3) + 3 = 2^4 + 2^2\cdot 3 + 2\cdot 3 + 3$
$d_5 = 2d_4 + 3 = 2(2^4 + 2^2\cdot 3 + 2\cdot 3 + 3) + 3 = 2^5 + 2^3\cdot 3 + 2^2\cdot 3 + 2\cdot 3 + 3$
$\vdots$

Guess: $d_n = 2^n + 2^{n-2} \cdot 3 + 2^{n-3} \cdot 3 + \ldots + 2^2 \cdot 3 + 2 \cdot 3 + 3$
$= 2^n + 3(2^{n-2} + 2^{n-3} + \ldots + 2^2 + 2 + 1)$
$= 2^n + 3\left(\dfrac{2^{(n-2)+1} - 1}{2 - 1}\right) = 2^n + 3(2^{n-1} - 1)$
$= 2^{n-1}(2 + 3) - 3 = 5 \cdot 2^{n-1} - 3$ for all integers $n \geq 1$

8.

$f_0 = 0$
$f_1 = f_0 + 3 \cdot 1 + 1 = 3 + 1$
$f_2 = f_1 + 3 \cdot 2 + 1 = (3 + 1) + 3 \cdot 2 + 1 = 3 + 3 \cdot 2 + 2$
$f_3 = f_2 + 3 \cdot 3 + 1 = ((3 + 3 \cdot 2) + 2) + 3 \cdot 3 + 1 = 3 + 3 \cdot 2 + 3 \cdot 3 + 3$
$f_4 = f_3 + 3 \cdot 4 + 1 = (3 + 3 \cdot 2 + 3 \cdot 3 + 3) + 3 \cdot 4 + 1 = 3 + 3 \cdot 2 + 3 \cdot 3 + 3 \cdot 4 + 4$
.
.
.

Guess: $f_n = 3 + 3 \cdot 2 + 3 \cdot 3 + 3 \cdot 4 + \ldots + 3 \cdot n + n$
$= 3(1 + 2 + 3 + 4 + \ldots + n) + n = 3\left(\dfrac{n(n+1)}{2}\right) + n$
$= \dfrac{3n^2 + 3n + 2n}{2} = \dfrac{3n^2 + 5n}{2}$ for all integers $n \geq 0$

10.

$v_1 = 1$
$v_2 = v_1 + 2^2 = 1 + 2^2$
$v_3 = v_2 + 2^3 = 1 + 2^2 + 2^3$
$v_4 = v_3 + 2^4 = 1 + 2^2 + 2^3 + 2^4$
$v_5 = v_4 + 2^5 = 1 + 2^2 + 2^3 + 2^4 + 2^5$
.
.
.

Guess: $v_n = 1 + 2^2 + 2^3 + \ldots + 2^n = \left(\dfrac{2^{n+1} - 1}{2 - 1}\right) = 2^{n+1} - 1$ for all integers $n \geq 1$

12.

$u_1 = 1$
$u_2 = u_1 + 2^2 = 1 + 2^2$
$u_3 = u_2 + 3^2 = (1 + 2^2) + 3^2 = 1 + 2^2 + 3^2$
$u_4 = u_3 + 4^2 = (1 + 2^2 + 3^2) + 4^2$
$== 1 + 2^2 + 3^2 + 4^2$
.
.
.

Guess:
$u_n = 1 + 2^2 + 3^2 + \ldots + n^2 = \dfrac{n(n+1)(2n+1)}{6}$ by exercise 9 of Section 4.2

14. The recurrence relation in exercise 21 of Section 8.1 is $t_k = 2t_{k-1} + 2$. The initial condition is $t_1 = 2$.

$$\begin{aligned} t_1 &= 2 \\ t_2 &= 2t_1 + 2 = 2 \cdot 2 + 2 = 2^2 + 2 \\ t_3 &= 2t_2 + 2 = 2(2^2 + 2) + 2 = 2^3 + 2^2 + 2 \\ t_4 &= 2t_3 + 2 = 2(2^3 + 2^2 + 2) + 2 = 2^4 + 2^3 + 2^2 + 2 \\ t_5 &= 2t_4 + 2 = 2(2^4 + 2^3 + 2^2 + 2) + 2 = 2^5 + 2^4 + 2^3 + 2^2 + 2 \end{aligned}$$

$$\begin{aligned} \text{Guess: } t_n &= 2^n + 2^{n-1} + 2^{n-2} + \cdots + 2^2 + 2 \\ &= 2(2^{n-1} + 2^{n-2} + \cdots + 2^2 + 2 + 1) \\ &= 2\left(\frac{2^{(n-1)+1} - 1}{2 - 1}\right) = 2(2^n - 1) = 2^{n+1} - 2 \quad \text{for all integers } n \geq 1 \end{aligned}$$

17. Let t_n be the runner's target time on day n. Then $t_k = t_{k-1} - 3$ seconds for all integers $k \geq 1$. Hence $t_0, t_1, t_2, \ldots$ is an arithmetic sequence with constant adder -3. It follows that $t_n = t_0 + n \cdot (-3)$ for all integers $n \geq 0$. Now $t_0 = 3$ minutes, and 3 minutes equals 180 seconds. Hence the runner's target time on day 14 is $t_{14} = 180 + (-3)14 = 180 - 42 = 138$ seconds $= 2$ minutes 18 seconds.

18. *Proof:* Let r be a fixed constant and $a_0, a_1, a_2, \ldots$ a sequence that satisfies the recurrence relation $a_k = r \cdot a_{k-1}$ for all integers $k \geq 1$. We will show by mathematical induction that $a_n = a_0 \cdot r^n$ for all integers $n \geq 0$.

    The formula holds for n = 0: For $n = 0$ the formula gives $a_0 \cdot r^0 = a_0 \cdot 1 = a_0$ which is the first term of $a_0, a_1, a_2, \ldots$.

    If the formula holds for n = k then it holds for n = k + 1: Suppose that $a_k = a_0 \cdot r^k$ for some integer $k \geq 0$. *[This is the inductive hypothesis.]* We must show that $a_{k+1} = a_0 \cdot r^{k+1}$. But

$$\begin{aligned} a_{k+1} &= r \cdot a_k && \text{by definition of } a_0, a_1, a_2, \ldots \\ &= r \cdot (a_0 \cdot r^k) && \text{by substitution from the inductive hypothesis} \\ &= a_0 \cdot r^{k+1} && \text{by the laws of algebra.} \end{aligned}$$

    [This is what was to be shown.]

19. The recurrence relation $P_k = (1 + i/m)P_{k-1}$ defines a geometric sequence with constant multiplier $1 + i/m$. Therefore, $P_n = P_0 \cdot (1 + i/m)^n$ for all integers $n \geq 0$.

20. For each integer $n \geq 0$, let P_n be the population at the end of year n. Then for all integers $k \geq 1$, $P_k = P_{k-1} + (0.03)P_{k-1} = (1.03)P_{k-1}$. Hence $P_0, P_1, P_2, \ldots$ is a geometric sequence with constant multiplier 1.03, and so $P_n = P_0 \cdot (1.03)^n$ for all integers $n \geq 0$. Since $P_0 = 50$ million, it follows that the population at the end of 25 years is $P_{25} = 50 \cdot (1.03)^{25} \cong 104.7$ million.

22. For each integer $n \geq 1$, let s_{n-1} be the number of operations the algorithm executes when it is run with an input of size n. Then $s_0 = 7$ and $s_k = 2 \cdot s_{k-1}$ for each integer $k \geq 1$. Therefore, $s_0, s_1, s_2, \ldots$ is a geometric sequence with constant multiplier 2, and so $s_n = s_0 \cdot 2^n = 7 \cdot 2^n$ for all integers $n \geq 0$. For an input of size 25, the number of operations executed by the algorithm is $s_{25-1} = s_{24} = 7 \cdot 2^{24} = 117,440,512$.

23. *a.* For each integer $k \geq 1$, the amount in the account at the end of k months equals the amount in the account at the end of the $(k-1)st$ month plus the interest earned on that amount during the month plus the $100 monthly addition to the account. Therefore, $A_k = A_{k-1} + (0.06/12)A_{k-1} + 100 = (1.005)A_{k-1} + 100$.

b.

$$A_0 = 1000$$
$$A_1 = (1.005)A_0 + 100 = 1000(1.005) + 100$$
$$A_2 = (1.005)A_1 + 100 = (1.005)[1000(1.005) + 100] + 100$$
$$= 1000(1.005)^2 + 100(1.005) + 100$$
$$A_3 = (1.005)A_2 + 100 = (1.005)[1000(1.005)^2 + 100(1.005) + 100] + 100$$
$$= 1000(1.005)^3 + 100(1.005)^2 + 100(1.005) + 100$$
$$A_4 = (1.005)A_3 + 100$$
$$= (1.005)[1000(1.005)^3 + 100(1.005)^2 + 100(1.005) + 100] + 100$$
$$= 1000(1.005)^4 + 100(1.005)^3 + 100(1.005)^2 + 100(1.005) + 100$$

⋮

Guess: $A_n = 1000(1.005)^n + [100(1.005)^{n-1} + 100(1.005)^{n-2} + \ldots$
$\qquad + 100(1.005)^2 + 100(1.005) + 100]$
$\qquad = 1000(1.005)^n + 100[(1.005)^{n-1} + (1.005)^{n-2} + \ldots + (1.005)^2 + (1.005) + 1]$
$\qquad = 1000(1.005)^n + 100\left(\dfrac{(1.005)^n - 1}{1.005 - 1}\right)$
$\qquad = (1.005)^n(1000) + \dfrac{100}{0.005}((1.005)^n - 1)$
$\qquad = (1000 + 20000)(1.005)^n - 20000 = 21000(1.005)^n - 20000$

c. Proof: Let $A_0, A_1, A_2, \ldots$ be a sequence that satisfies the recurrence relation $A_k = (1.005)A_{k-1} + 100$ for all integers $k \geq 1$ with initial condition $A_0 = 1000$. We will show by mathematical induction that $A_n = 21000(1.005)^n - 20000$ for all integers $n \geq 0$.

The formula holds for n = 0 : For $n = 0$ the formula gives $21000(1.005)^0 - 20000 = 1000$, which equals A_0.

If the formula holds for n = k then it holds for n = k + 1: Suppose that $A_k = 21000(1.005)^k - 20000$ for some integer $k \geq 0$. *[This is the inductive hypothesis.]* We must show that $A_{k+1} = 21000(1.005)^{k+1} - 20000$. But

$$\begin{aligned} A_{k+1} &= (1.005)A_k + 100 && \text{by definition of } A_0, A_1, A_2, \ldots \\ &= (1.005)[21000(1.005)^k - 20000] + 100 && \text{by substitution from the inductive hypothesis} \\ &= 21000(1.005)^{k+1} - 20100 + 100 \\ &= 21000(1.005)^{k+1} - 20000 && \text{by the laws of algebra.} \end{aligned}$$

[This is what was to be shown.]

e. By parts (b) and (c), $A_n = 21000(1.005)^n - 20000$, and so we need to find the value of n for which

$$21000(1.005)^n - 20000 = 10000.$$

But this equation holds

$$\Leftrightarrow \quad 21000(1.005)^n = 30000$$

$$\Leftrightarrow \quad (1.005)^n = \frac{30000}{21000} = \frac{10}{7}$$

$$\Leftrightarrow \quad \log_{10}(1.005)^n = \log_{10}\left(\frac{10}{7}\right) \quad \text{by property (7.3.5)}$$

$$\Leftrightarrow \quad n \cdot \log_{10}(1.005) = \log_{10}\left(\frac{10}{7}\right) \quad \text{by exercise 26 of Section 7.3}$$

$$\Leftrightarrow \quad n = \frac{\log_{10}(10/7)}{\log_{10}(1.005)} \cong 71.5.$$

So $n \cong 71.5$ months. If interest is only paid at the end of each month, then 72 months, or six years, will be required for the account to grow to just over \$10,000.

25. *Proof:* Let $b_0, b_1, b_2, \ldots$ be a sequence that satisfies the recurrence relation $b_k = \frac{b_{k-1}}{1 + b_{k-1}}$ for all integers $k \geq 1$ and the initial condition $b_0 = 1$. We will show by mathematical induction that $b_n = \frac{1}{n+1}$ for all integers $n \geq 0$.

The formula holds for n = 0: For $n = 0$ the formula gives $\frac{1}{0+1} = 1$, which equals b_0.

If the formula holds for n = k then it holds for n = k + 1: Suppose that $b_k = \frac{1}{k+1}$ for some integer $k \geq 0$. *[This is the inductive hypothesis.]* We must show that $b_{k+1} = \frac{1}{(k+1)+1}$, or, equivalently, that $b_{k+1} = \frac{1}{k+2}$. But

$$\begin{aligned}
b_{k+1} &= \frac{\frac{1}{k+1}}{1 + \frac{1}{k+1}} \quad \text{by definition of } b_0, b_1, b_2, \ldots \\
&= \frac{1}{(k+1)+1} \quad \text{by substitution from the inductive hypothesis} \\
&= \frac{1}{k+2}.
\end{aligned}$$

[This is what was to be shown.]

27. *Proof:* Let $d_1, d_2, d_3, \ldots$ be a sequence that satisfies the recurrence relation $d_k = 2d_{k-1} + 3$ for all integers $k \geq 2$ and the initial condition $d_1 = 2$. We will show by mathematical induction that $d_n = 5 \cdot 2^{n-1} - 3$ for all integers $n \geq 1$.

The formula holds for n = 1: For $n = 1$ the formula gives $5 \cdot 2^{1-1} - 3 = 5 - 3 = 2$, which equals d_1.

If the formula holds for n = k then it holds for n = k + 1: Suppose that $d_k = 5 \cdot 2^{k-1} - 3$ for some integer $k \geq 1$. *[This is the inductive hypothesis.]* We must show that $d_{k+1} = 5 \cdot 2^{(k+1)-1} - 3$. But

$$\begin{aligned}
d_{k+1} &= 2d_k + 3 \quad &\text{by definition of } d_1, d_2, d_3, \ldots \\
&= 2(5 \cdot 2^{k-1} - 3) + 3 \quad &\text{by substitution from the inductive hypothesis} \\
&= 5 \cdot 2^k - 6 + 3 \\
&= 5 \cdot 2^{(k+1)-1} - 3 \quad &\text{by the laws of algebra.}
\end{aligned}$$

[This is what was to be shown.]

28. *Proof:* Let $e_0, e_1, e_2, \ldots$ be a sequence that satisfies the recurrence relation $e_k = e_{k-1} + 2k$ for all integers $k \geq 1$ and the initial condition $e_0 = 3$. We will show by mathematical induction that $e_n = 3 + n(n+1)$ for all integers $n \geq 0$.

The formula holds for n = 0: For $n = 0$ the formula gives $3 + 0(0+1) = 3$, which equals e_0.

If the formula holds for n = k then it holds for n = k+1: Suppose that $e_k = 3 + k(k+1)$ for some integer $k \geq 0$. *[This is the inductive hypothesis.]* We must show that $e_{k+1} = 3 + (k+1)((k+1)+1)$, or, equivalently, that $e_{k+1} = 3 + (k+1)(k+2)$. But

$$\begin{aligned} e_{k+1} &= e_k + 2(k+1) & \text{by definition of } e_0, e_1, e_2, \ldots \\ &= (3 + k(k+1)) + 2(k+1) & \text{by substitution from the inductive hypothesis} \\ &= 3 + k^2 + 3k + 2 \\ &= 3 + (k+1)(k+2) & \text{by the laws of algebra.} \end{aligned}$$

[This is what was to be shown.]

29. *Proof:* Let $f_0, f_1, f_2, \ldots$ be a sequence that satisfies the recurrence relation $f_k = f_{k-1} + 3k + 1$ for all integers $k \geq 1$ and the initial condition $f_0 = 0$. We will show by mathematical induction that $f_n = \dfrac{3n^2 + 5n}{2}$ for all integers $n \geq 0$.

The formula holds for n = 0: For $n = 0$ the formula gives $\dfrac{3 \cdot 0^2 + 5 \cdot 0}{2} = 0$, which equals f_0.

If the formula holds for n = k then it holds for n = k+1: Suppose that $f_k = \dfrac{3k^2 + 5k}{2}$ for some integer $k \geq 0$. *[This is the inductive hypothesis.]* We must show that $f_{k+1} = \dfrac{3(k+1)^2 + 5(k+1)}{2}$, or, equivalently, that $f_{k+1} = \dfrac{3k^2 + 11k + 8}{2}$ *[because*

$$\dfrac{3(k+1)^2 + 5(k+1)}{2} = \dfrac{3(k^2 + 2k + 1) + 5k + 5}{2} = \dfrac{3k^2 + 11k + 8}{2}].$$

But

$$\begin{aligned} f_{k+1} &= f_k + 3(k+1) + 1 & \text{by definition of } f_0, f_1, f_2, \ldots \\ &= \dfrac{3k^2 + 5k}{2} + 3(k+1) + 1 & \text{by substitution from the inductive hypothesis} \\ &= \dfrac{3k^2 + 5k + 6k + 6 + 2}{2} \\ &= \dfrac{3k^2 + 11k + 8}{2} & \text{by the laws of algebra.} \end{aligned}$$

[This is what was to be shown.]

30. *Proof:* Let $v_1, v_2, v_3, \ldots$ be a sequence that satisfies the recurrence relation $v_k = v_{k-1} + 2^k$ for all integers $k \geq 2$ and the initial condition $v_1 = 1$. We will show by mathematical induction that $v_n = 2^{n+1} - 3$ for all integers $n \geq 1$.

The formula holds for n = 1: For $n = 1$ the formula gives $2^{1+1} - 3 = 4 - 3 = 1$, which equals v_1.

If the formula holds for n = k then it holds for n = k+1: Suppose that $v_k = 2^{k+1} - 3$ for some integer $k \geq 1$. *[This is the inductive hypothesis.]* We must show that $v_{k+1} = 2^{(k+1)+1} - 3$, or, equivalently, that $v_{k+1} = 2^{k+2} - 3$. But

$$\begin{aligned} v_{k+1} &= v_k + 2^{k+1} & \text{by definition of } v_1, v_2, v_3, \ldots \\ &= 2^{k+1} - 3 + 2^{k+1} & \text{by substitution from the inductive hypothesis} \\ &= 2 \cdot 2^{k+1} - 3 \\ &= 2^{k+2} - 3 & \text{by the laws of algebra.} \end{aligned}$$

[This is what was to be shown.]

31. *Proof:* Let $w_0, w_1, w_2, \ldots$ be a sequence that satisfies the recurrence relation $w_k = 2^k - w_{k-1}$ for all integers $k \geq 1$ and the initial condition $w_0 = 1$. We will show by mathematical induction that $w_n = \dfrac{2^{n+1} - (-1)^{n+1}}{3}$ for all integers $n \geq 0$.

The formula holds for n = 0: For $n = 0$ the formula gives $\dfrac{2^{0+1} - (-1)^{0+1}}{3} = \dfrac{2 - (-1)}{3} = 1$, which equals w_0.

If the formula holds for n = k then it holds for n = k + 1: Suppose that
$$w_k = \frac{2^{k+1} - (-1)^{k+1}}{3} \quad \text{for some integer } k \geq 0.$$

[This is the inductive hypothesis.]

We must show that
$$w_{k+1} = \frac{2^{(k+1)+1} - (-1)^{(k+1)+1}}{3},$$
or, equivalently, that
$$w_{k+1} = \frac{2^{k+2} - (-1)^{k+2}}{3}.$$

But
$$\begin{aligned}
w_{k+1} &= 2^{k+1} - w_k && \text{by definition of } w_0, w_1, w_2, \ldots \\
&= 2^{k+1} - \frac{2^{k+1} - (-1)^{k+1}}{3} && \text{by substitution from the inductive hypothesis} \\
&= \frac{3 \cdot 2^{k+1} - 2^{k+1} + (-1)^{k+1}}{3} \\
&= \frac{2 \cdot 2^{k+1} - (-1)^{k+2}}{3} \\
&= \frac{2^{k+2} - (-1)^{k+2}}{3} && \text{by the laws of algebra.}
\end{aligned}$$

[This is what was to be shown.]

33. *Proof:* Let $u_1, u_2, u_3, \ldots$ be a sequence that satisfies the recurrence relation $u_k = u_{k-1} + k^2$ for all integers $k \geq 2$ and the initial condition $u_1 = 1$. We will show by mathematical induction that $u_n = \dfrac{n(n+1)(2n+1)}{6}$ for all integers $n \geq 1$.

The formula holds for n = 1: For $n = 1$ the formula gives $\dfrac{1 \cdot (1+1) \cdot (2 \cdot 1 + 1)}{6} = 1$, which equals u_1.

If the formula holds for n = k then it holds for n = k + 1: Suppose that
$$u_k = \frac{k(k+1)(2k+1)}{6} \quad \text{for some integer } k \geq 1.$$

[This is the inductive hypothesis.]

We must show that
$$u_{k+1} = \frac{(k+1)((k+1)+1)(2(k+1)+1)}{6},$$
or, equivalently, that $u_{k+1} = \dfrac{(k+1)(k+2)(2k+3)}{6}$. But

180

$$\begin{aligned}
u_{k+1} &= u_k + (k+1)^2 & \text{by definition of } u_1, u_2, u_3, \ldots \\
&= \frac{k(k+1)(2k+1)}{6} + (k+1)^2 & \text{by substitution from the inductive hypothesis} \\
&= \frac{k(k+1)(2k+1) + 6(k+1)^2}{6} \\
&= \frac{(k+1)[k(2k+1) + 6(k+1)]}{6} \\
&= \frac{(k+1)(2k^2 + 7k + 6)}{6} \\
&= \frac{(k+1)(k+2)(2k+3)}{6} & \text{by the laws of algebra.}
\end{aligned}$$

[This is what was to be shown.]

34. *Proof:* Let $a_1, a_2, a_3, \ldots$ be a sequence that satisfies the recurrence relation $a_k = 3a_{k-1} + 2$ for all integers $k \geq 2$ and the initial condition $a_1 = 2$. We will show by mathematical induction that $a_n = 3^n - 1$ for all integers $n \geq 1$.

The formula holds for n = 1: For $n = 1$ the formula gives $3^1 - 1 = 2$, which equals a_1.

If the formula holds for n = k then it holds for n = k + 1: Suppose that $a_k = 3^k - 1$ for some integer $k \geq 1$. *[This is the inductive hypothesis.]* We must show that $a_{k+1} = 3^{k+1} - 1$. But

$$\begin{aligned}
a_{k+1} &= 3a_k + 2 & \text{by definition of } a_1, a_2, a_3, \ldots \\
&= 3(3^k - 1) + 2 & \text{by substitution from the inductive hypothesis} \\
&= 3^{k+1} - 3 + 2 \\
&= 3^{k+1} - 1 & \text{by the laws of algebra.}
\end{aligned}$$

[This is what was to be shown.]

35. *Proof*:

The formula holds for n = 1: For $n = 1$ the formula gives $t_1 = 2^2 - 2 = 2$, which is true.

If the formula holds for n = k then it holds for n = k+1: Suppose that $t_k = 2^{k+1} - 2$, for some integer $k \geq 1$. *[This is the inductive hypothesis.]* We must show that $t_{k+1} = 2^{(k+1)+1} - 2$, or, equivalently, $t_{k+1} = 2^{k+2} - 2$. But

$$\begin{aligned}
t_{k+1} &= 2t_k + 2 & \text{by definition of } t_1, t_2, t_3, \ldots \\
&= 2(2^{k+1} - 2) + 2 & \text{by inductive hypothesis} \\
&= 2^{k+2} - 4 + 2 \\
&= 2^{k+2} - 2
\end{aligned}$$

37. a.

$$\begin{aligned}
s_0 &= 1 \\
s_1 &= 2 \\
s_2 &= 2s_0 = 2 \\
s_3 &= 2s_1 = 2 \cdot 2 = 2^2 \\
s_4 &= 2s_2 = 2 \cdot 2 = 2^2 \\
s_5 &= 2s_3 = 2 \cdot 2^2 = 2^3 \\
s_6 &= 2s_4 = 2 \cdot 2^2 = 2^3 \\
s_7 &= 2s_5 = 2 \cdot 2^3 = 2^4 \\
s_8 &= 2s_6 = 2 \cdot 2^3 = 2^4
\end{aligned}$$

.
.
.

Guess: $s_n = \begin{cases} 2^{(n+1)/2} & \text{if } n \text{ is odd} \\ 2^{n/2} & \text{if } n \text{ is even} \end{cases}$

b. *Proof:* Let $s_0, s_1, s_2, \ldots$ be a sequence that satisfies the recurrence relation $s_k = 2s_{k-2}$ for all integers $k \geq 2$ and the initial conditions $s_0 = 1$ and $s_1 = 2$. We will show by strong mathematical induction that

$$s_n = \begin{cases} 2^{(n+1)/2} & \text{if } n \text{ is odd} \\ 2^{n/2} & \text{if } n \text{ is even} \end{cases} \quad \text{for all integers } n \geq 0.$$

The formula holds for n = 0 and n = 1: For $n = 0$ and $n = 1$ the formula gives $2^{0/2} = 2^0 = 1$ and $2^{(1+1)/2} = 2^1 = 2$, which equal s_0 and s_1 respectively.

If the formula holds for all i with $0 \leq i < k$ then it holds for k: Let k be an integer with $k \geq 2$ and suppose

$$s_i = \begin{cases} 2^{(i+1)/2} & \text{if } i \text{ is odd} \\ 2^{i/2} & \text{if } i \text{ is even} \end{cases} \quad \text{for all integers } i \text{ with } 1 \leq i < k.$$

[This is the inductive hypothesis.]

We must show that

$$s_k = \begin{cases} 2^{(k+1)/2} & \text{if } k \text{ is odd} \\ 2^{k/2} & \text{if } k \text{ is even.} \end{cases}$$

But

$$\begin{aligned} s_k &= 2s_{k-2} && \text{by definition of } s_0, s_1, s_2, \ldots \\ &= \begin{cases} 2 \cdot 2^{((k-2)+1)/2} & \text{if } k-2 \text{ is odd} \\ 2 \cdot 2^{(k-2)/2} & \text{if } k-2 \text{ is even} \end{cases} && \text{by substitution from the inductive hypothesis} \\ &= \begin{cases} 2^{(k-1)/2+1} & \text{if } k \text{ is odd} \\ 2^{(k-2)/2+1} & \text{if } k \text{ is even} \end{cases} && \text{because } k-2 \text{ and } k \text{ have the same parity} \\ &= \begin{cases} 2^{(k+1)/2} & \text{if } k \text{ is odd} \\ 2^{k/2} & \text{if } k \text{ is even.} \end{cases} \end{aligned}$$

[This is what was to be shown.]

38. a.

$$\begin{aligned} t_0 &= 0 \\ t_1 &= 1 - t_0 = 1 - 0 = 1 \\ t_2 &= 2 - t_1 = 2 - 1 = 1 \\ t_3 &= 3 - t_2 = 3 - 1 = 2 \\ t_4 &= 4 - t_3 = 4 - 2 = 2 \\ t_5 &= 5 - t_4 = 5 - 2 = 3 \\ t_6 &= 6 - t_5 = 6 - 3 = 3 \end{aligned}$$

.
.
.

Guess: $t_n = \lceil n/2 \rceil$ for all integers $n \geq 0$

b. *Proof:* Let $t_0, t_1, t_2, \ldots$ be a sequence that satisfies the recurrence relation $t_k = k - t_{k-1}$ for all integers $k \geq 1$ and the initial condition $t_0 = 0$. We will show by strong mathematical induction that $t_n = \lceil n/2 \rceil$ for all integers $n \geq 0$.

The formula holds for n = 0: For $n = 0$ the formula gives $\lceil 0/2 \rceil = 0$, which equals t_0.

If the formula holds for all i with $1 \le i < k$ then it holds for k: Let k be an integer with $k > 0$ and suppose $t_i = \lceil i/2 \rceil$ for all integers i with $1 \le i < k$. *[This is the inductive hypothesis.]* We must show that $t_k = \lceil k/2 \rceil$. But

$$
\begin{aligned}
t_k &= k - t_{k-1} && \text{by definition of } t_0, t_1, t_2, \ldots \\
&= k - \lceil (k-1)/2 \rceil && \text{by substitution from the inductive hypothesis} \\
&= \begin{cases} k - k/2 & \text{if } k \text{ is even} \\ k - (k-1)/2 & \text{if } k \text{ is odd} \end{cases} \\
&= \begin{cases} k/2 & \text{if } k \text{ is even} \\ (k+1)/2 & \text{if } k \text{ is odd} \end{cases} && \text{by the laws of algebra} \\
&= \lceil k/2 \rceil && \text{by definition of ceiling}
\end{aligned}
$$

[This is what was to be shown.]

39. a.

$$
\begin{aligned}
w_1 &= 1 \\
w_2 &= 2 \\
w_3 &= w_1 + 3 = 1 + 3 \\
w_4 &= w_2 + 4 = 2 + 4 \\
w_5 &= w_3 + 5 = 1 + 3 + 5 \\
w_6 &= w_4 + 6 = 2 + 4 + 6 \\
w_7 &= w_5 + 7 = 1 + 3 + 5 + 7
\end{aligned}
$$

$\vdots$

$$
\begin{aligned}
\text{Guess: } w_n &= \begin{cases} 1 + 3 + 5 + \ldots + n & \text{if } n \text{ is odd} \\ 2 + 4 + 6 + \ldots + n & \text{if } n \text{ is even} \end{cases} \\
&= \begin{cases} \left(\dfrac{n+1}{2}\right)^2 & \text{if } n \text{ is odd} \\ 2\left(1 + 2 + 3 + \ldots + \dfrac{n}{2}\right) & \text{if } n \text{ is even} \end{cases} && \text{by exercise 15 of Section 4.2} \\
&= \begin{cases} \left(\dfrac{n+1}{2}\right)^2 & \text{if } n \text{ is odd} \\ 2\left(\dfrac{\frac{n}{2}\left(\frac{n}{2}+1\right)}{2}\right) & \text{if } n \text{ is even} \end{cases} && \text{by Theorem 4.2.1} \\
&= \begin{cases} \dfrac{(n+1)^2}{4} & \text{if } n \text{ is odd} \\ \dfrac{n(n+1)}{4} & \text{if } n \text{ is even} \end{cases} && \text{by the laws of algebra}
\end{aligned}
$$

b. *Proof:* Let $w_1, w_2, w_3, \ldots$ be a sequence that satisfies the recurrence relation $w_k = w_{k-2} + k$ for all integers $k \ge 2$ and the initial conditions $w_1 = 1$ and $w_2 = 2$. We will show by strong mathematical induction that

$$
w_n = \begin{cases} \dfrac{(n+1)^2}{4} & \text{if } n \text{ is odd} \\ \dfrac{n(n+2)}{4} & \text{if } n \text{ is even} \end{cases} \quad \text{for all integers } n \ge 1.
$$

The formula holds for n = 1 and n = 2: For $n = 1$ and $n = 2$ the formula gives $(1+1)^2/4 = 1$ and $2(2+2)/4 = 2$, which equal w_1 and w_2 respectively.

If the formula holds for all i with $1 \leq i < k$ then it holds for k: Let k be an integer with $k > 2$ and suppose

$$w_i = \begin{cases} \dfrac{(i+1)^2}{4} & \text{if } i \text{ is odd} \\ \dfrac{i(i+2)}{4} & \text{if } i \text{ is even} \end{cases} \quad \text{for all integers } i \text{ with } 1 \leq i < k.$$

[This is the inductive hypothesis.]

We must show that

$$w_k = \begin{cases} \dfrac{(k+1)^2}{4} & \text{if } k \text{ is odd} \\ \dfrac{k(k+2)}{4} & \text{if } k \text{ is even.} \end{cases}$$

But

$$\begin{aligned}
w_k &= w_{k-2} + k & &\text{by definition of } w_1, w_2, w_3, \ldots \\
&= \begin{cases} \dfrac{((k-2)+1)^2}{4} + k & \text{if } k-2 \text{ is odd} \\ \dfrac{(k-2)((k-2)+2)}{4} + k & \text{if } k-2 \text{ is even} \end{cases} & &\text{by substitution from the inductive hypothesis} \\
&= \begin{cases} \dfrac{(k-1)^2}{4} + \dfrac{4k}{4} & \text{if } k \text{ is odd} \\ \dfrac{(k-2)\cdot k}{4} + \dfrac{4k}{4} & \text{if } k \text{ is even} \end{cases} & &\text{because } k-2 \text{ and } k \text{ have the same parity} \\
&= \begin{cases} \dfrac{k^2 - 2k + 1 + 4k}{4} & \text{if } k \text{ is odd} \\ \dfrac{k^2 - 2k + 4k}{4} & \text{if } k \text{ is even} \end{cases} \\
&= \begin{cases} \dfrac{k^2 + 2k + 1}{4} & \text{if } k \text{ is odd} \\ \dfrac{k^2 + 2k}{4} & \text{if } k \text{ is even} \end{cases} \\
&= \begin{cases} \dfrac{(k+1)^2}{4} & \text{if } k \text{ is odd} \\ \dfrac{k(k+2)}{4} & \text{if } k \text{ is even} \end{cases} & &\text{by the laws of algebra}
\end{aligned}$$

[This is what was to be shown.]

40.

$$\begin{aligned}
u_0 &= 2 \\
u_1 &= 2 \\
u_2 &= u_0 \cdot u_1 = 2 \cdot 2 = 2^{1+1} = 2^2 \\
u_3 &= u_1 \cdot u_2 = 2 \cdot 2^2 = 2^{1+2} = 2^3 \\
u_4 &= u_2 \cdot u_3 = 2^2 \cdot 2^3 = 2^{2+3} = 2^5 \\
u_5 &= u_3 \cdot u_4 = 2^3 \cdot 2^5 = 2^{3+5} = 2^8 \\
u_6 &= u_4 \cdot u_5 = 2^5 \cdot 2^8 = 2^{5+8} = 2^{13}
\end{aligned}$$

$$\vdots$$

Guess: $u_n = 2^{F_n}$, where F_n is the nth Fibonacci number, for all integers $n \geq 0$.

b. *Proof:* Let $u_0, u_1, u_2, \ldots$ be a sequence that satisfies the recurrence relation $u_k = u_{k-2} \cdot u_{k-1}$ for all integers $k \geq 2$ and the initial conditions $u_0 = 2$ and $u_1 = 2$. We will show by strong mathematical induction that $u_n = 2^{F_n}$, where F_n is the nth Fibonacci number, for all integers $n \geq 0$.

The formula holds for n = 0 and n = 1: For $n = 0$ and $n = 1$, $F_0 = 1$ and $F_1 = 1$, and so the formula gives $2^1 = 2$ and $2^1 = 2$, which equal u_0 and u_1 respectively.

If the formula holds for all i with $1 \leq i < k$ then it holds for k: Let k be an integer with $k > 1$ and suppose $u_i = 2^{F_i}$, where F_i is the ith Fibonacci number, for all integers i with $1 \leq i < k$. *[This is the inductive hypothesis.]* We must show that $u_k = 2^{F_k}$, where F_k is the kth Fibonacci number. But

$$\begin{aligned}
u_k &= u_{k-2} \cdot u_{k-1} & \text{by definition of } u_0, u_1, u_2, \ldots \\
&= 2^{F_{k-2}} \cdot 2^{F_{k-1}} & \text{by substitution from the inductive hypothesis} \\
&= 2^{F_{k-2}+F_{k-1}} & \text{by the laws of exponents} \\
&= 2^{F_k} & \text{by definition of the Fibonacci sequence.}
\end{aligned}$$

[This is what was to be shown.]

42. The sequence does not satisfy the formula. By definition of $a_1, a_2, a_3, \ldots$, $a_1 = 0$, $a_2 = (a_1 + 1)^2 = 1^2 = 1$, $a_3 = (a_2 + 1)^2 = (1 + 1)^2 = 4$, $a_4 = (a_3 + 1)^2 = (4 + 1)^2 = 25$. But according to the formula $a_4 = (4 - 1)^2 = 9 \neq 25$.

43. Suppose there are $k - 1$ lines already drawn in the plane in such a way that they divide the plane into a maximum number P_{k-1} of regions. If addition of a new line is to create a maximum number of regions, it must cross all the $k - 1$ lines that are already drawn. But if all $k - 1$ lines are crossed by the new line, then one can imagine traveling along the new line from a point before it reaches the first line it crosses to a point after it reaches the last line it crosses. One sees that for each integer $i = 1, 2, \ldots, k - 1$, the region just before the ith line is reached is divided in half. This creates $k - 1$ new regions. But the final region after the last line is passed is also divided in half. This creates one additional new region, which brings the total number of new regions to k. Therefore, $P_k = P_{k-1} + k$ for all integers $k \geq 1$.

44.
$$\begin{bmatrix} 1 & 1 \\ 1 & 0 \end{bmatrix}^2 = \begin{bmatrix} 2 & 1 \\ 1 & 1 \end{bmatrix} = \begin{bmatrix} F_2 & F_1 \\ F_1 & F_0 \end{bmatrix}$$

$$\begin{bmatrix} 1 & 1 \\ 1 & 0 \end{bmatrix}^3 = \begin{bmatrix} 3 & 2 \\ 2 & 1 \end{bmatrix} = \begin{bmatrix} F_3 & F_2 \\ F_2 & F_1 \end{bmatrix}$$

$$\begin{bmatrix} 1 & 1 \\ 1 & 0 \end{bmatrix}^4 = \begin{bmatrix} 5 & 3 \\ 3 & 2 \end{bmatrix} = \begin{bmatrix} F_4 & F_3 \\ F_3 & F_2 \end{bmatrix}$$

$$\begin{bmatrix} 1 & 1 \\ 1 & 0 \end{bmatrix}^5 = \begin{bmatrix} 8 & 5 \\ 5 & 3 \end{bmatrix} = \begin{bmatrix} F_5 & F_4 \\ F_4 & F_3 \end{bmatrix}$$

Guess: For all integers $n \geq 1$,

$$\begin{bmatrix} 1 & 1 \\ 1 & 0 \end{bmatrix}^n = \begin{bmatrix} F_n & F_{n-1} \\ F_{n-1} & F_{n-2} \end{bmatrix}$$

Proof:

The formula holds for n = 2: For $n = 2$ the formula simply states that

$$\begin{bmatrix} 1 & 1 \\ 1 & 0 \end{bmatrix}^2 = \begin{bmatrix} F_2 & F_1 \\ F_1 & F_0 \end{bmatrix}$$

which the calculations above show to be true.

If the formula holds for n = k then it holds for n = k+1: Suppose that for some integer $k \geq 2$,

$$\begin{bmatrix} 1 & 1 \\ 1 & 0 \end{bmatrix}^k = \begin{bmatrix} F_k & F_{k-1} \\ F_{k-1} & F_{k-2} \end{bmatrix}.$$

Then

$$\begin{bmatrix} 1 & 1 \\ 1 & 0 \end{bmatrix}^{k+1} = \begin{bmatrix} 1 & 1 \\ 1 & 0 \end{bmatrix} \begin{bmatrix} F_k & F_{k-1} \\ F_{k-1} & F_{k-2} \end{bmatrix} = \begin{bmatrix} F_k + F_{k-1} & F_{k-1} + F_{k-2} \\ F_k & F_{k-1} \end{bmatrix} = \begin{bmatrix} F_{k+1} & F_k \\ F_k & F_{k-1} \end{bmatrix}.$$

45. a.

$Y_1 = E + c + mY_0$
$Y_2 = E + c + mY_1 = E + c + m(E + c + mY_0) = (E + c) + m(E + c) + m^2 Y_0$
$Y_3 = E + c + mY_2 = E + c + m((E+c) + m(E+c) + m^2 Y_0) = (E+c) + m(E+c) + m^2(E+c) + m^3 Y_0$
$Y_4 = E + c + mY_3 = E + c + m((E + c) + m(E + c) + m^2(E + c) + m^3 Y_0)$
$\quad = (E + c) + m(E + c) + m^2(E + c) + m^3(E + c) + m^4 Y_0$

.
.
.

Guess: $Y_n = (E + c) + m(E + c) + m^2(E + c) + \cdots + m^{n-1}(E + c) + m^n Y_0$
$\quad = (E + c)[1 + m + m^2 + \cdots + m^{n-1}] + m^n Y_0$
$\quad = (E + c)\left(\dfrac{m^n - 1}{m - 1}\right) + m^n Y_0$, for all integers $n \geq 1$.

b. Suppose $0 < m < 1$. Then

$$\lim_{n \to \infty} Y_n = \lim_{n \to \infty}\left((E + c)\left(\dfrac{m^n - 1}{m - 1}\right) + m^n Y_0\right)$$
$$= (E + c)\left(\dfrac{\lim_{n \to \infty} m^n - 1}{m - 1}\right) + \lim_{n \to \infty} m^n Y_0$$
$$= (E + c)\left(\dfrac{0 - 1}{m - 1}\right) + 0 \cdot Y_0 \qquad \text{because when } 0 < m < 1$$
$$\qquad \qquad \qquad \qquad \qquad \qquad \qquad \text{. then } \lim_{n \to \infty} m^n = 0$$
$$= \dfrac{E + c}{1 - m}.$$

Section 8.3

2. *b* and *f*

3. *b*.
$$\begin{cases} a_0 = C \cdot 2^0 + D = C + D = 0 \\ a_1 = C \cdot 2^1 + D = 2C + D = 2 \end{cases} \Leftrightarrow \begin{cases} D = -C \\ 2C + (-C) = 2 \end{cases} \Leftrightarrow \begin{cases} D = -C \\ C = 2 \end{cases} \Leftrightarrow \begin{cases} C = 2 \\ D = -2 \end{cases}$$
$$a_2 = C \cdot 2^2 + D = 2 \cdot 2^2 + (-2) = 6$$

4. b.
$$\begin{cases} b_0 = C \cdot 3^0 + D \cdot (-2)^0 = C + D = 3 \\ b_1 = C \cdot 3^1 + D \cdot (-2)^1 = 3C - 2D = 4 \end{cases} \Leftrightarrow \begin{cases} 2C + 2D = 6 \\ 3C - 2D = 4 \end{cases} \Leftrightarrow \begin{cases} C = 2 \\ D = 3 - 2 = 1 \end{cases}$$

$$b_2 = C \cdot 3^2 + D \cdot (-2)^2 = 2 \cdot 3^2 + 1 \cdot (-2)^2 = 18 + 4 = 22$$

6. *Proof:* Given that $b_n = C \cdot 3^n + D \cdot (-2)^n$, then for any choice of C and D and integer $k \geq 2$, $b_k = C \cdot 3^k + D \cdot (-2)^k$, $b_{k-1} = C \cdot 3^{k-1} + D \cdot (-2)^{k-1}$, and $b_{k-2} = C \cdot 3^{k-2} + D \cdot (-2)^{k-2}$. Hence, $b_{k-1} + 6b_{k-2} = (C \cdot 3^{k-1} + D \cdot (-2)^{k-1}) + 6(C \cdot 3^{k-2} + D \cdot (-2)^{k-2}) = C \cdot (3^{k-1} + 6 \cdot 3^{k-2}) + D \cdot ((-2)^{k-1} + 6 \cdot (-2)^{k-2}) = C \cdot 3^{k-2}(3 + 6) + D \cdot (-2)^{k-2}(-2 + 6) = C \cdot 3^{k-2} \cdot 3^2 + D \cdot (-2)^{k-2} 2^2 = C \cdot 3^k + D \cdot (-2)^k = b_k$.

7.
$$\begin{cases} C + D = 1 \\ C \cdot \left(\dfrac{1+\sqrt{5}}{2}\right) + D \cdot \left(\dfrac{1-\sqrt{5}}{2}\right) = 1 \end{cases}$$

$$\Leftrightarrow \begin{cases} C \cdot \left(\dfrac{1+\sqrt{5}}{2}\right) + D \cdot \left(\dfrac{1+\sqrt{5}}{2}\right) = \dfrac{1+\sqrt{5}}{2} \\ C \cdot \left(\dfrac{1+\sqrt{5}}{2}\right) + D \cdot \left(\dfrac{1-\sqrt{5}}{2}\right) = 1 \end{cases}$$

$$\Leftrightarrow \begin{cases} D \cdot \left(\dfrac{1+\sqrt{5}}{2} - \dfrac{1-\sqrt{5}}{2}\right) = \dfrac{1+\sqrt{5}}{2} - 1 \\ C + D = 1 \end{cases} \Leftrightarrow \begin{cases} D \cdot \sqrt{5} = \dfrac{1+\sqrt{5}-2}{2} \\ C + D = 1 \end{cases}$$

$$\Leftrightarrow \begin{cases} D = \dfrac{-(1-\sqrt{5})}{2\sqrt{5}} \\ C + \dfrac{-(1-\sqrt{5})}{2\sqrt{5}} = 1 \end{cases} \Leftrightarrow \begin{cases} D = \dfrac{-(1-\sqrt{5})}{2\sqrt{5}} \\ C = \dfrac{2\sqrt{5}+(1-\sqrt{5})}{2\sqrt{5}} = \dfrac{1+\sqrt{5}}{2\sqrt{5}} \end{cases}$$

9. a. If for all integers $k \geq 2$, $t^k = 7t^{k-1} - 10t^{k-2}$ and $t \neq 0$, then $t^2 = 7t - 10$ and so $t^2 - 7t + 10 = 0$. But $t^2 - 7t + 10 = (t-2)(t-5)$. Thus $t = 2$ or $t = 5$.

b. It follows from part (a) and the distinct roots theorem that for some constants C and D, $b_0, b_1, b_2, \ldots$ satisfies the equation $b_n = C \cdot 2^n + D \cdot 5^n$ for all integers $n \geq 0$. Since $b_0 = 2$ and $b_1 = 2$, then

$$\begin{cases} b_0 = C \cdot 2^0 + D \cdot 5^0 = C + D = 2 \\ b_1 = C \cdot 2^1 + D \cdot 5^1 = 2C + 5D = 2 \end{cases} \Leftrightarrow \begin{cases} D = 2 - C \\ 2C + 5(2 - C) = 2 \end{cases} \Leftrightarrow$$

$$\begin{cases} D = 2 - C \\ C = 8/3 \end{cases} \Leftrightarrow \begin{cases} D = 2 - (8/3) = -(2/3) \\ C = 8/3 \end{cases}$$

Thus $b_n = \dfrac{8}{3} \cdot 2^n - \dfrac{2}{3} \cdot 5^n$ for all integers $n \geq 0$.

10. a. If for all integers $k \geq 2$, $t^k = t^{k-1} + 6t^{k-2}$ and $t \neq 0$, then $t^2 = t + 6$ and so $t^2 - t - 6 = 0$. But $t^2 - t - 6 = (t-3)(t+2)$. Thus $t = 3$ or $t = -2$.

b. It follows from part (a) and the distinct roots theorem that for some constants C and D, $c_0, c_1, c_2, \ldots$ satisfies the equation $c_n = C \cdot 3^n + D \cdot (-2)^n$ for all integers $n \geq 0$. Since $c_0 = 0$ and $c_1 = 3$, then

$$\begin{cases} c_0 = C \cdot 3^0 + D \cdot (-2)^0 = C + D = 0 \\ c_1 = C \cdot 3^1 + D \cdot (-2)^1 = 3C - 2D = 3 \end{cases} \Leftrightarrow \begin{cases} D = -C \\ 3C - 2(-C) = 3 \end{cases} \Leftrightarrow \begin{cases} D = -3/5 \\ C = 3/5 \end{cases}$$

Thus $c_n = \dfrac{3}{5} \cdot 3^n - \dfrac{3}{5} \cdot (-2)^n$ for all integers $n \geq 0$.

12. The characteristic equation is $t^2 - 9 = 0$. Since $t^2 - 9 = (t-3)(t+3)$, the roots are $t = 3$ and $t = -3$. By the distinct roots theorem, for some constants C and D, $e_n = C \cdot 3^n + D \cdot (-3)^n$ for all integers $n \geq 0$. Since $e_0 = 0$ and $e_1 = 2$, then

$$\begin{cases} e_0 = C \cdot 3^0 + D \cdot (-3)^0 = C + D = 0 \\ e_1 = C \cdot 3^1 + D \cdot (-3)^1 = 3C - 3D = 2 \end{cases} \Leftrightarrow \begin{cases} D = -C \\ 3C - 3(-C) = 2 \end{cases} \Leftrightarrow \begin{cases} D = -1/3 \\ C = 1/3 \end{cases}$$

Thus $e_n = \dfrac{1}{3} \cdot 3^n - \dfrac{1}{3} \cdot (-3)^n = 3^{n-1} + (-3)^{n-1} = 3^{n-1}(1 + (-1)^{n-1}) = \begin{cases} 2 \cdot 3^{n-1} & \text{if } n \text{ is odd} \\ 0 & \text{if } n \text{ is even} \end{cases}$

for all integers $n \geq 0$.

14. The characteristic equation is $t^2 + 4t + 4 = 0$. Since $t^2 + 4t + 4 = (t+2)^2$, there is only one root, $t = -2$. By the single root theorem, for some constants C and D,

$$s_n = C \cdot (-2)^n + D \cdot n \cdot (-2)^n \quad \text{for all integers } n \geq 0.$$

Since $s_0 = 0$ and $s_1 = -1$, then

$$\begin{cases} s_0 = C \cdot (-2)^0 + D \cdot 0 \cdot (-2)^0 = C = 0 \\ s_1 = C \cdot (-2)^1 + D \cdot 1 \cdot (-2)^1 = -2C - 2D = -1 \end{cases} \Leftrightarrow \begin{cases} C = 0 \\ -2D = -1 \end{cases} \Leftrightarrow \begin{cases} C = 0 \\ D = 1/2 \end{cases}$$

Thus $s_n = 0 \cdot (-2)^n + \dfrac{1}{2} \cdot n \cdot (-2)^n = -n \cdot 2^{n-1}$ for all integers $n \geq 0$.

15. The characteristic equation is $t^2 - 6t + 9 = 0$. Since $t^2 - 6t + 9 = (t-3)^2$, there is only one root, $t = 3$. By the single root theorem, for some constants C and D, $t_n = C \cdot 3^n + D \cdot n \cdot 3^n$ for all integers $n \geq 0$. Since $t_0 = 1$ and $t_1 = 3$, then

$$\begin{cases} t_0 = C \cdot 3^0 + D \cdot 0 \cdot 3^0 = C = 1 \\ t_1 = C \cdot 3^1 + D \cdot 1 \cdot 3^1 = 3C + 3D = 3 \end{cases} \Leftrightarrow \begin{cases} C = 1 \\ C + D = 1 \end{cases} \Leftrightarrow \begin{cases} C = 1 \\ D = 0 \end{cases}$$

Thus $t_n = 1 \cdot 3^n + 0 \cdot n \cdot 3^n = 3^n$ for all integers $n \geq 0$.

16. The characteristic equation is $t^2 - 2t - 2 = 0$. By the quadratic formula, the roots are $t = \dfrac{2 \pm \sqrt{4+8}}{2} = \dfrac{2 \pm 2\sqrt{3}}{2} = 1 \pm \sqrt{3}$. By the distinct roots theorem, for some constants C and D, $s_n = C \cdot (1+\sqrt{3})^n + D \cdot (1-\sqrt{3})^n$ for all integers $n \geq 0$. Since $s_0 = 1$ and $s_1 = 3$, then

$$\begin{cases} s_0 = C \cdot (1+\sqrt{3})^0 + D \cdot (1-\sqrt{3})^0 = C + D = 1 \\ s_1 = C \cdot (1+\sqrt{3})^1 + D \cdot (1-\sqrt{3})^1 = 3 \end{cases} \Leftrightarrow$$

$$\begin{cases} D = 1 - C \\ C \cdot (1+\sqrt{3}) + (1-C) \cdot (1-\sqrt{3}) = 3 \end{cases} \Leftrightarrow$$

$$\begin{cases} D = 1 - C \\ C(1 + \sqrt{3} - 1 + \sqrt{3}) = 3 - 1 + \sqrt{3} \end{cases} \Leftrightarrow$$

$$\begin{cases} D = 1 - \dfrac{2+\sqrt{3}}{2\sqrt{3}} = \dfrac{\sqrt{3}-2}{2\sqrt{3}} \\ C = \dfrac{\sqrt{3}+2}{2\sqrt{3}} \end{cases}$$

Thus $s_n = C \cdot (1+\sqrt{3})^n + D \cdot (1-\sqrt{3})^n = \dfrac{\sqrt{3}+2}{2\sqrt{3}} \cdot (1+\sqrt{3})^n + \dfrac{\sqrt{3}-2}{2\sqrt{3}} \cdot (1-\sqrt{3})^n$ for all integers $n \geq 0$.

17. Given the set-up of exercise 34 from Section 8.1, $c_1 = 1$ and $c_2 = 2$ and $c_k = c_{k-1} + c_{k-2}$ for all integers $k \geq 3$. Define $c_0 = 1$. Then $c_2 = c_0 + c_1$ and so the recurrence relation holds for all integers $k \geq 2$. The characteristic equation is $t^2 - t - 1 = 0$, which is the same as the characteristic equation for the Fibonacci sequence. In addition, the first two terms of this

sequence are the same as the Fibonacci sequence. Hence $c_0, c_1, c_2, \ldots$ satisfies the same explicit formula as the Fibonacci sequence, namely, $c_n = \dfrac{1}{\sqrt{5}} \cdot \left(\dfrac{1+\sqrt{5}}{2}\right)^{n+1} - \dfrac{1}{\sqrt{5}} \cdot \left(\dfrac{1-\sqrt{5}}{2}\right)^{n+1}$.

An alternative solution is to substitute the roots of the characteristic equation into the formula $c_n = C \cdot \left(\dfrac{1+\sqrt{5}}{2}\right)^n + D \cdot \left(\dfrac{1-\sqrt{5}}{2}\right)^n$ for $n = 1$ and $n = 2$ and solve for C and D from the resulting set of simultaneous equations.

18. *Proof:* Suppose that $s_0, s_1, s_2, \ldots$ and $t_0, t_1, t_2, \ldots$ are sequences such that $s_k = 5s_{k-1} - 4s_{k-2}$ and $t_k = 5t_{k-1} - 4t_{k-2}$ for all integers $k \geq 2$. Then for all integers $k \geq 2$, $5 \cdot (2s_{k-1} + 3t_{k-1}) - 4(2s_{k-2} + 3t_{k-2}) = (5 \cdot 2s_{k-1} - 4 \cdot 2s_{k-2}) + (5 \cdot 3t_{k-1} - 4 \cdot 3t_{k-2}) = 2(5s_{k-1} - 4s_{k-2}) + 3(5t_{k-1} - 4t_{k-2}) = 2s_k + 3t_k$. This is what was to be shown.

20. *Proof:* Suppose that r is a nonzero real number, k and l are distinct integers, and a_k and a_l are any real numbers. Consider the system of equations
$Cr^k + kDr^k = a_k$
$Cr^l + lDr^l = a_l$.

Without loss of generality, we may assume that $k > l$. Multiply the bottom equation by r^{k-l} to obtain the equivalent system
$Cr^k + kDr^k = a_k$
$Cr^l \cdot r^{k-l} + lDr^l \cdot r^{k-l} = Cr^k + lDr^k = a_l \cdot r^{k-l}$.

Subtracting the bottom equation from the top one gives $(k-l)Dr^k = a_k - a_l \cdot r^{k-l}$, or $D = \dfrac{a_k - a_l \cdot r^{k-l}}{(k-l)r^k}$ since $k - l \neq 0$ and $r \neq 0$. Substituting into the top equation gives $Cr^k + k\left(\dfrac{a_k - a_l \cdot r^{k-l}}{(k-l)r^k}\right) = a_k$, and solving for C gives $C = \dfrac{1}{r^k}\left(a_k - k\left(\dfrac{a_k - a_l \cdot r^{k-l}}{(k-l)r^k}\right)\right)$.
These calculations show that the given system of equations has the unique solutions C and D that are shown.

Alternatively, the determinant of the given system of two linear equations in the two unknowns C and D is $r^k \cdot l \cdot r^l - r^l \cdot k \cdot r^k = r^{k+l}(l - k)$. This is nonzero because $l \neq k$ and $r \neq 0$, and therefore the given system has a unique solution.

21. *Proof:* Suppose a sequence $a_0, a_1, a_2, \ldots$ satisfies the recurrence relation $a_k = A \cdot a_{k-1} + B \cdot a_{k-2}$ for some real numbers A and B and for all integers $k \geq 2$, and suppose the characteristic equation $t^2 - At - B = 0$ has a single root r. We will show that for all integers $n \geq 0$, $a_n = C \cdot r^n + n \cdot D \cdot r^n$ where C and D are numbers such that $a_0 = C \cdot r^0 + 0 \cdot D \cdot r^0$ and $a_1 = C \cdot r^1 + 1 \cdot D \cdot r^1$. Consider the formula $a_n = C \cdot r^n + n \cdot D \cdot r^n$. We use strong mathematical induction to prove that this formula holds for all integers $n \geq 0$.

The formula holds for n = 0 and n = 1: The truth of the formula for $n = 0$ and $n = 1$ is automatic because C and D are exactly those numbers that make the following equations true: $a_0 = C \cdot r^0 + 0 \cdot D \cdot r^0$ and $a_1 = C \cdot r^1 + 1 \cdot D \cdot r^1$.

If $k \geq 2$ and the formula holds for all integers i with $0 \leq i < k$ then it holds for k: Suppose that $k \geq 2$ and for all integers k with $0 \leq i < k$, $a_i = C \cdot r^i + i \cdot D \cdot r^i$. *[This is the inductive hypothesis.]* We must show that $a_k = C \cdot r^k + k \cdot D \cdot r^k$. Now by the inductive hypothesis, $a_{k-1} = C \cdot r^{k-1} + (k-1) \cdot D \cdot r^{k-1}$ and $a_{k-2} = C \cdot r^{k-2} + (k-2) \cdot D \cdot r^{k-2}$. So

$\begin{aligned}
a_k &= A \cdot a_{k-1} + B \cdot a_{k-2} & \text{by definition of } a_0, a_1, a_2, \ldots \\
&= A(C \cdot r^{k-1} + (k-1) \cdot D \cdot r^{k-1}) & \text{by substitution from the} \\
&\quad + B(C \cdot r^{k-2} + (k-2) \cdot D \cdot r^{k-2}) & \text{inductive hypothesis} \\
&= C \cdot (Ar^{k-1} + Br^{k-2}) \\
&\quad + D \cdot (A(k-1)r^{k-1} + B(k-2)r^{k-2}) & \text{by algebra} \\
&= C \cdot r^k + D \cdot kr^k & \text{by Lemma 8.3.4.}
\end{aligned}$

189

[This is what was to be shown.]

23. The characteristic equation is $t^2 - 2t + 5 = 0$. By the quadratic formula the roots of this equation are $t = \dfrac{2 \pm \sqrt{4-20}}{2} = \begin{cases} 1+2i \\ 1-2i \end{cases}$. By the distinct roots theorem, for some constants C and D, $b_n = C \cdot (1+2i)^n + D \cdot (1-2i)^n$ for all integers $n \geq 0$. Since $b_0 = 1$ and $b_1 = 1$, then

$$\begin{cases} b_0 = C \cdot (1+2i)^0 + D \cdot (1-2i)^0 = C + D = 1 \\ b_1 = C \cdot (1+2i)^1 + D \cdot (1-2i)^1 = 1 \end{cases} \Leftrightarrow \begin{cases} (1-2i)C + (1-2i)D = 1-2i \\ (1+2i)C + (1-2i)D = 1 \end{cases}$$

$$\Leftrightarrow \begin{cases} C + D = 1 \\ [(1+2i)-(1-2i)]C = 1-(1-2i) = 2i \end{cases} \Leftrightarrow \begin{cases} C + D = 1 \\ C = 2i/4i = 1/2 \end{cases}$$

$$\Leftrightarrow \begin{cases} D = 1 - 1/2 = 1/2 \\ C = 1/2 \end{cases}$$

Thus $b_n = (1/2) \cdot (1+2i)^n + (1/2) \cdot (1-2i)^n$ for all integers $n \geq 0$.

24. a. If $\dfrac{\phi}{1} = \dfrac{1}{\phi - 1}$, then $\phi(\phi - 1) = 1$, or, equivalently, $\phi^2 - \phi - 1 = 0$ and so ϕ satisfies the equation $t^2 - t - 1 = 0$.

    b. By the quadratic formula, the solution to $t^2 - t - 1 = 0$ are $t = \dfrac{1 \pm \sqrt{1+4}}{2} = \begin{cases} (1+\sqrt{5})/2 \\ (1-\sqrt{5})/2 \end{cases}$.

    Let $\phi_1 = (1+\sqrt{5})/2$ and $\phi_2 = (1-\sqrt{5})/2$.

    c. $F_n = \dfrac{1}{\sqrt{5}} \cdot \phi_1^{n+1} - \dfrac{1}{\sqrt{5}} \cdot \phi_2^{n+1} = \dfrac{1}{\sqrt{5}} \cdot (\phi_1^{n+1} - \phi_2^{n+1})$

25. The given recurrence relation can be rewritten in the form $\dfrac{1}{6} P_k = P_{k-1} - \dfrac{5}{6} P_{k-2}$, or $P_k = 6P_{k-1} - 5P_{k-2}$. Thus the characteristic equation is $t^2 - 6t + 5 = 0$. Since $t^2 - 6t + 5 = (t-1)(t-5)$, this equation has roots $t = 1$ and $t = 5$. By the distinct roots theorem, for some constants C and D, $P_n = C \cdot 1^n + D \cdot 5^n = C + D \cdot 5^n$ for all integers $n \geq 0$. Since $P_0 = 1$ and $P_{300} = 0$, then

$$\begin{cases} P_0 = C + D \cdot 5^0 = C + D = 1 \\ P_{300} = C + D \cdot 5^{300} = C + 5^{300} \cdot D = 0 \end{cases} \Leftrightarrow \begin{cases} C + D = 1 \\ (1 - 5^{300})D = 1 \end{cases} \Leftrightarrow$$

$$\begin{cases} C = 1 - D = 1 - \dfrac{1}{1 - 5^{300}} \\ D = \dfrac{1}{1 - 5^{300}} \end{cases} \Leftrightarrow \begin{cases} C = \dfrac{-5^{300}}{1 - 5^{300}} \\ D = \dfrac{1}{1 - 5^{300}} \end{cases}$$

Thus $P_n = \left(\dfrac{-5^{300}}{1 - 5^{300}}\right) + \left(\dfrac{1}{1 - 5^{300}}\right) \cdot 5^n = \left(\dfrac{5^{300} - 5^n}{5^{300} - 1}\right)$ for all integers $n \geq 0$.

$P_{20} = \left(\dfrac{5^{300} - 5^{20}}{5^{300} - 1}\right) \cong 1$

26. a. Let k be an integer with $k \geq 3$, call the k sectors of the disk $1, 2, 3, \ldots, k$, and suppose that the values of $S_1, S_2, \ldots, S_{k-1}$ are known. Note that $S_1 = 4$ and $S_2 = 4 \cdot 3 = 12$.

    Case 1, Sectors 1 and 3 are painted the same color: In this case $k > 3$ because otherwise sectors 1 and 3 would be both adjacent and painted the same color which is not allowed. Imagine sectors 1 through 3 made into a single unit painted the color of sectors 1 and 3. Then the disk would contain a total of $k - 2$ sectors: $1 - 3, 4, 5, \ldots, k$, and the number of ways to paint the disk would be S_{k-2}. Each of these ways corresponds to exactly three ways to paint the disk when sectors 1-3 are not united, for that is the number of choices of colors to paint sector 2 to contrast with sectors 1 and 3. Hence there are $3S_{k-2}$ ways to paint the disk in this case.

    Case 2, Sectors 1 and 3 are painted different colors: In this case, imagine shrinking sector 2 to nothing. Then there would be S_{k-1} ways to paint the resulting disk. Now imagine expanding sector 2 back to its original size and giving it its original color. Since there are two ways that sector 2 could be colored that would contrast with the colors of both sectors 1 and 3, each way to paint the disk leaving sector 2 out corresponds to exactly two ways to paint the disk when sector 2 is present. Hence there are $2S_{k-1}$ ways to paint the disk in this case.

If $k = 3$, then case 1 does not occur, and so $S_3 = 2S_2 = 24$. If $k \geq 4$, then the total number of ways to paint the disk is the sum of the ways counted in cases 1 and 2. Therefore, $S_k = 2S_{k-1} + 3S_{k-2}$ for all integers $k \geq 4$.

b. Let $T_0, T_1, T_2, \ldots$ be the sequence defined by $T_n = S_{n+2}$ for all integers $n \geq 0$. Then for all integers $k \geq 2$, $T_k = 2T_{k-1} + 3T_{k-2}$ and $T_0 = 12$ and $T_1 = 24$. The characteristic equation of the relation is $t^2 - 2t - 3 = 0$. Since $t^2 - 2t - 3 = (t-3)(t+1)$, the roots of this equation are $t = 3$ and $t = -1$. By the distinct roots theorem, for some constants C and D, $T_n = C \cdot 3^n + D \cdot (-1)^n$ for all integers $n \geq 0$. Since $T_0 = 12$ and $T_1 = 24$, then

$$\left\{ \begin{array}{l} T_0 = C \cdot 3^0 + D \cdot (-1)^0 = C + D = 12 \\ T_1 = C \cdot 3^1 + D \cdot (-1)^1 = 3C - D = 24 \end{array} \right\} \Leftrightarrow \left\{ \begin{array}{l} 4C = 36 \\ C + D = 12 \end{array} \right\} \Leftrightarrow \left\{ \begin{array}{l} C = 9 \\ D = 3 \end{array} \right\}$$

Thus $T_n = 9 \cdot 3^n + 3 \cdot (-1)^n$ for all integers $n \geq 0$, and hence $S_n = T_{n-2} = 9 \cdot 3^{n-2} + 3 \cdot (-1)^{n-2}$ for all integers $n \geq 2$.

Section 8.4

1. b. (1) $p, q, r,$ and s are Boolean expressions by I.

   (2) $(p \vee q)$ and $\sim s$ are Boolean expressions by (1), $II(b)$, and $II(c)$

   (3) $(p \wedge \sim s)$ is a Boolean expression by (1), (2), and II(a)

   (4) $((p \wedge \sim s) \wedge r)$ is a Boolean expression by (1), (3), and II(a)

   (5) $\sim ((p \wedge \sim s) \wedge r)$ is a Boolean expression by (4) and $II(c)$

   (6) $(p \vee q) \vee \sim ((p \wedge \sim s) \wedge r)$ is a Boolean expression by (2), (5) and $II(b)$

2. b. (1) $\epsilon \in \Sigma^*$ by I.

   (2) $b = \epsilon b \in \Sigma^*$ by (1) and $II(b)$.

   (3) $bb \in \Sigma^*$ by (2) and $II(b)$

3. b. (1) MI is in the MIU-system by I.

   (2) MII is in the MIU-system by (1) and $II(b)$.

   (3) MIIII is in the MIU-system by (2) and $II(b)$.

   (4) MIIIIIIII is in the MIU-system by (3) and $II(b)$.

   (5) MUIIIII is in the MIU-system by (4) and $II(c)$.

   (6) MUIIU is in the MIU-system by (5) and $II(c)$.

4. The string MU is not in the system. For if it were, it would be obtainable from the string MI through a certain number of applications of the rules of the recursion. But none of the recursion rules takes a string in which the number of I's is not divisible by 3 and produces a string in which the number of I's is divisible by 3. To see this, let s be a string, let n be the number of I's in s, and suppose $3 \not| n$. Consider the effect of acting upon s by each recursion rule in turn.

   a. Any string produced from s by applying rule (a) has the same number of I's as s.

   b. Let s' be the string obtained from s by applying rule (b). The number of I's in s' is twice the number of I's in s, which is n. Since $3 \not| n$, we have $n = 3k + 1$ or $n = 3k + 2$ for some integer k. In case $n = 3k + 1$, the number of I's in s' is $2(3k+1) = 3(2k) + 2$, which is not divisible by 3. In case $n = 3k + 2$, the number of I's in s' is $2(3k+2) = 6k + 4 = 3(2k+1) + 1$, which is not divisible by 3 either.

c. Let s' be the string obtained from s by applying rule (c). The number of I's in s' is three fewer than the number of I's in s, which is n. Because $3 \nmid n$, we have that $3 \nmid (n-3)$ either [for if $3 \mid (n-3)$ then $n - 3 = 3k$, for some integer k. Hence $n = 3k + 3 = 3(k+1)$, and so n would be divisible by 3, which it is not]. Thus the number of I's in s' is not divisible by 3.

d. Any string produced from s by applying rule d has the same number of I's as s.

Therefore, repeated application of the recursion rules cannot produce a string in which the number of I's is divisible by 3 unless the number of I's in the string being acted upon is also divisible by 3. Now the only string in the MIU-system that is not obtained through application of the recursion rules is MI, and the number of I's in MI is not divisible by 3. It follows that there is no string in the MIU-system in which the number of I's is divisible by 3. But the number of I's in MU is 0, which is divisible by 3. Hence MU is not in the MIU-system.

5. b. (1) ( ) is in P by I.

    (2) (( )) is in P by (1) and II(a).

    (3) (( ))( ) is in P by (2) and $II(b)$.

6. b. Even though the number of its left parentheses equals the number of its right parentheses, this structure is not in P either. Roughly speaking, the reason is that given any parenthesis structure derived from the base structure by repeated application of the rules of the recursion, as you move from left to right along the structure, the total of right parentheses you encounter will never be larger than the number of left parentheses you have already passed by. But if you move along (( )( )))(( ) from left to right, you encounter an extra right parenthesis in the fifth position.

    More formally, let A be the set of all finite sequences of integers and define a function $g: P \to A$ as follows: for each parenthesis structure S in P, let $g[S] = (a_1, a_2, \ldots, a_n)$ where a_i is the number of left parentheses in S minus the number of right parentheses in S counting from left to right through position i. For instance, if $S = (( )( ))$, then $g[S] = (1, 2, 1, 2, 1, 0)$. By the same argument as in part (a), the final component in $g[S]$ will always be 0. We claim that for all parenthesis structures S in P, each component of $g[S]$ is nonnegative.

    (1) Note that $g[( )] = (1, 0)$, and so each component of ( ) is nonnegative.

    (2) Suppose that S is any structure in P such that all components of $g[S]$ are nonnegative. Observe that that the first component in $g[(S)]$ is 1 (because (S) starts with a left parenthesis), every subsequent component of $g[(S)]$ except the last is one more than a corresponding (nonnegative) component of $g[S]$, and the final component of $g[(S)]$ is 0. So each component of $g[(S)]$ is nonnegative.

    (3) Suppose that S and T are any structures in P such that all components of $g[S]$ and of $g[T]$ are nonnegative. Consider $g[(S)(T)]$. For concreteness, suppose $g[S]$ has m components and $g[T]$ has n components. Then the first component of $g[(S)(T)]$ is 1 (because $(S)(T)$ begins with a left parenthesis), the second through $(m+1)$st components are each one more than the first m components of $g[S]$ (each of which is nonnegative), and the $(m+2)$nd component of $g[(S)(T)]$ is 0. Furthermore, the $(m+3)$d component of $g[(S)(T)]$ is 1 (because of the left parenthesis to the left of the T in $(S)(T)$), the next n components are each one more than the corresponding n components of $g[T]$, and the final component of $g[(S)(T)]$ is 0. So each component of $g[(S)(T)]$ is nonnegative.

    Items (1), (2), and (3) show that all parenthesis structures obtainable from the base structure ( ) by repeated application of II(a) and II(b) are sent by g to finite sequences all of whose components are nonnegative. But by III (the restriction condition), there are no other elements of P besides those obtainable from the base element by applying II(a) and II(b). Hence all components of $g[S]$ are nonnegative for all S in P.

    Now if (( )( )))(( ) were in P, then all components of $g[(( )( )))(( )]$ would be nonnegative. But $g[(( )( )))(( )] = (1, 2, 1, 3, 2, 1, 0, -1, 0, 1, 0)$, and one of these components is negative.

7. b (1) 9, 6.1, 2, 4, 7, and 6 are arithmetic expressions by I.

   (2) $(6.1 + 2)$ and $(4-7)$ are arithmetic expressions by (1), $II(c)$, and $II(d)$

   (3) $(9 \cdot (6.1 + 2))$ and $((4 - 7) \cdot 6)$ are arithmetic expressions by (1), (2), and $II(e)$

   (4) $\left(\dfrac{(9 \cdot (6.1 + 2))}{((4 - 7) \cdot 6)} \right)$ is an arithmetic expression by (3) and $II(f)$

9. Let S be the set of all strings of 0's and 1's in which all the 0's precede all the 1's. The following is a recursive definition of S.

   I. BASE: $\varepsilon \in S$, where ε is the null string

   II. RECURSION: If $s \in S$, then

   a. $0s \in S$ b. $s1 \in S$

   III. RESTRICTION: There are no elements of S other than those obtained from I and II.

11. Let S be the set of all strings of a's and b's that contain exactly one a. The following is a recursive definition of S.

    I. BASE: $a \in S$

    II. RECURSION: If $s \in S$, then

    a. $bs \in S$ b. $sb \in S$

    III. RESTRICTION: There are no elements of S other than those obtained from I and II.

13. *Proof (by mathematical induction)*:

    The formula holds for n = 1: Let a_1 and b_1 be any real numbers. By the recursive definition of product, $\prod_{i=1}^{1}(a_i \cdot b_i) = a_1 \cdot b_1$, $\prod_{i=1}^{1} a_i = a_1$, and $\prod_{i=1}^{1} b_i = b_1$. Therefore,

    $$\prod_{i=1}^{1}(a_i \cdot b_i) = \left(\prod_{i=1}^{1} a_i \right) \cdot \left(\prod_{i=1}^{1} b_i \right)$$

    and so the formula holds for $n = 1$.

    If the formula holds for n = k then it holds for n = k + 1: Let k be an integer such that $k \geq 1$. Suppose that if $a_1, a_2, \ldots, a_k$ and $b_1, b_2, \ldots, b_k$ are any real numbers, then $\prod_{i=1}^{k}(a_i \cdot b_i) = \left(\prod_{i=1}^{k} a_i \right) \cdot \left(\prod_{i=1}^{k} b_i \right)$. [This is the inductive hypothesis.] [We must show that if $a_1, a_2, \ldots, a_{k+1}$ and $b_1, b_2, \ldots, b_{k+1}$ are any real numbers, then

    $$\prod_{i=1}^{k+1}(a_i \cdot b_i) = \left(\prod_{i=1}^{k+1} a_i \right) \cdot \left(\prod_{i=1}^{k+1} b_i \right) .]$$

    Let $a_1, a_2, \ldots, a_{k+1}$ and $b_1, b_2, \ldots, b_{k+1}$ be any real numbers. Then $\prod_{i=1}^{k+1}(a_i \cdot b_i)$

    $= \left(\prod_{i=1}^{k}(a_i \cdot b_i) \right) \cdot (a_{k+1} \cdot b_{k+1})$ by the recursive definition of product

    $= \left(\left(\prod_{i=1}^{k} a_i \right) \cdot \left(\prod_{i=1}^{k} b_i \right) \right) \cdot (a_{k+1} \cdot b_{k+1})$ by substitution from the inductive hypothesis

    $= \left(\left(\prod_{i=1}^{k} a_i \right) \cdot a_{k+1} \right) \cdot \left(\prod_{i=1}^{k} b_i \right) \cdot b_{k+1})$ by the associative and commutative laws of algebra

    $= \left(\prod_{i=1}^{k+1} a_i \right) \cdot \left(\prod_{i=1}^{k+1} b_i \right)$ by the recursive definition of product.

    [This is what was to be shown.]

14. **Proof (by mathematical induction):**

The formula holds for n = 1: Let c and a_1 be any real numbers. By the recursive definition of product, both $\prod_{i=1}^{1}(c \cdot a_i)$ and $c^1 \cdot \prod_{i=1}^{1} a_i$ equal $c \cdot a_1$, and so the formula holds for $n = 1$.

If the formula holds for n = k then it holds for n = k + 1: Let k be an integer such that $k \geq 1$. Suppose that if c and $a_1, a_2, \ldots, a_k$ are any real numbers, then $\prod_{i=1}^{k}(c \cdot a_i) = c^k \cdot \left(\prod_{i=1}^{k} a_i\right)$. *[This is the inductive hypothesis.]* *[We must show that if c and $a_1, a_2, \ldots, a_{k+1}$ are any real numbers, then $\prod_{i=1}^{k+1}(c \cdot a_i) = c^{k+1} \cdot \left(\prod_{i=1}^{k+1} a_i\right)$.]* Let c and $a_1, a_2, \ldots, a_{k+1}$ be any real numbers. Then

$$\prod_{i=1}^{k+1}(c \cdot a_i)$$

$$= \left(\prod_{i=1}^{k}(c \cdot a_i)\right) \cdot (c \cdot a_{k+1}) \quad \text{by the recursive definition of product}$$

$$= \left(c^k \cdot \left(\prod_{i=1}^{k} a_i\right)\right) \cdot (c \cdot a_{k+1}) \quad \text{by substitution from the inductive hypothesis}$$

$$= (c^k \cdot c) \cdot \left(\left(\prod_{i=1}^{k} a_i\right) \cdot a_{k+1}\right) \quad \text{by the associative and commutative laws of algebra}$$

$$= c^{k+1} \cdot \left(\prod_{i=1}^{k+1} a_i\right) \quad \text{by the laws of exponents and the recursive definition of product.}$$

[This is what was to be shown.]

15. **Proof (by mathematical induction):**

The formula holds for n = 1: Let a_1 be any real number. By definition of absolute value

$$|a_1| = \begin{cases} a_1 & \text{if } a_1 \geq 0 \\ -a_1 & \text{if } a_1 < 0 \end{cases}.$$

Thus in case $a_1 \geq 0$ then $a_1 = |a_1|$, and so $a_1 \leq |a_1|$. In case $a_1 < 0$ then $a_1 < 0 < -a_1 = |a_1|$, and so $a_1 \leq |a_1|$ also. Hence the formula holds for $n = 1$.

If the formula holds for n = k then it holds for n = k + 1: Let k be an integer such that $k \geq 1$. Suppose that if $a_1, a_2, \ldots, a_k$ are any real numbers, then $\left|\sum_{i=1}^{k} a_i\right| \leq \sum_{i=1}^{k} |a_i|$. *[This is the inductive hypothesis.]* *[We must show that if $a_1, a_2, \ldots, a_{k+1}$ are any real numbers, then*

$$\left|\sum_{i=1}^{k+1} a_i\right| \leq \sum_{i=1}^{k+1} |a_i|.]$$

Let $a_1, a_2, \ldots, a_{k+1}$ be any real numbers. Then

$$\left|\sum_{i=1}^{k+1} a_i\right| = \left|\left(\sum_{i=1}^{k} a_i\right) + a_{k+1}\right| \quad \text{by the recursive definition of summation}$$

$$\Rightarrow \left|\sum_{i=1}^{k+1} a_i\right| \leq \left|\left(\sum_{i=1}^{k} a_i\right)\right| + |a_{k+1}| \quad \text{by the triangle inequality for absolute value}$$

$$\Rightarrow \left|\sum_{i=1}^{k+1} a_i\right| \leq \sum_{i=1}^{k} |a_i| + |a_{k+1}| \quad \text{by substitution from the inductive hypothesis}$$

$$\Rightarrow \left|\sum_{i=1}^{k+1} a_i\right| \leq \sum_{i=1}^{k+1} |a_i| \quad \text{by the recursive definition of summation.}$$

[This is what was to be shown.]

17. *Proof (by mathematical induction)*:

The formula holds for n = 1: Let A and B_1 be any sets. By the recursive definition of intersection, both $A \cup (\bigcap_{i=1}^{1} B_i)$ and $\bigcap_{i=1}^{1}(A \cup B_i)$ equal $A \cup B_1$. Hence the formula holds for $n = 1$.

If the formula holds for n = k then it holds for n = k + 1: Let k be an integer such that $k \geq 1$. Suppose that if A and $B_1, B_2, \ldots, B_k$ are any sets then $A \cup (\bigcap_{i=1}^{k} B_i) = \bigcap_{i=1}^{k}(A \cup B_i)$. *[This is the inductive hypothesis.]* *[We must show that if A and $B_1, B_2, \ldots, B_{k+1}$ are any sets then $A \cup (\bigcap_{i=1}^{k+1} B_i) = \bigcap_{i=1}^{k+1}(A \cup B_i)$.]* Let A and $B_1, B_2, \ldots, B_{k+1}$ be any sets. Then

$$A \cup \left(\bigcap_{i=1}^{k+1} B_i\right) = A \cup \left(\left(\bigcap_{i=1}^{k} B_i\right) \cap B_{k+1}\right)$$

by the recursive definition of intersection

$$= \left(A \cup \left(\bigcap_{i=1}^{k} B_i\right)\right) \cap (A \cup B_{k+1})$$

by the distributive laws for sets (Theorem 5.2.2(3))

$$= \bigcap_{i=1}^{k}(A \cup B_i) \cap (A \cup B_{k+1})$$

by inductive hypothesis

$$= \bigcap_{i=1}^{k+1}(A \cup B_i)$$

by the recursive definition of intersection.

[This is what was to be shown.]

18. *Proof (by mathematical induction)*:

The formula holds for n = 1: Let A_1 be any set. By the recursive definitions of intersection and union, both $(\bigcap_{i=1}^{1} A_i)^c$ and $\bigcup_{i=1}^{1} A_i{}^c$ equal $A_1{}^c$. Hence the formula holds for $n = 1$.

If the formula holds for n = k then it holds for n = k + 1: Let k be an integer such that $k \geq 1$. Suppose that if $A_1, A_2, \ldots, A_k$ are any sets then $(\bigcap_{i=1}^{k} A_i)^c = \bigcup_{i=1}^{k} A_i{}^c$. *[This is the inductive hypothesis.]* *[We must show that if $A_1, A_2, \ldots, A_{k+1}$ are any sets then $(\bigcap_{i=1}^{k+1} A_i)^c = \bigcup_{i=1}^{k+1} A_i{}^c$.]* Let $A_1, A_2, \ldots, A_{k+1}$ be any sets. Then

$$\left(\bigcap_{i=1}^{k+1} A_i\right)^c = \left(\left(\bigcap_{i=1}^{k} A_i\right) \cap A_{k+1}\right)^c$$

by the recursive definition of intersection

$$= \left(\bigcap_{i=1}^{k} A_i\right)^c \cup (A_{k+1})^c$$

by De Morgan's law for sets (Theorem 5.2.2(7))

$$= \left(\bigcup_{i=1}^{k} A_i{}^c\right) \cup (A_{k+1})^c$$

by inductive hypothesis

$$= \bigcup_{i=1}^{k+1} A_i{}^c$$

by the recursive definition of union.

[This is what was to be shown.]

19. $M(91) = M(M(102)) = M(92) = M(M(103)) = M(93) = M(M(104)) = M(94)$
$= M(M(105)) = M(95) = M(M(106)) = M(96) = M(M(107)) = M(97) = M(M(108))$
$= M(98) = M(M(109)) = M(99) = M(M(110)) = M(100) = M(M(111)) = M(101) = 91$

20. *Proof (by a variation of strong mathematical induction)*: Consider the property "$M(n) = 91$."

The property holds for all integers n with $91 \leq n \leq 101$: This statement is proved above in the solution to exercise 19.

If $1 \leq k < 91$ and the property holds for all i with $k < i \leq 101$ then it holds for k: Let k be an integer such that $1 \leq k < 91$ and suppose $M(i) = 91$ for all i with $k < i \leq 101$. *[This is the inductive hypothesis.]* Then $M(k) = M(M(k+11))$ by definition of M and $12 \leq k+11 < 102$. Thus $k < k + 11 \leq 101$, and so by inductive hypothesis $M(k + 11) = 91$. It follows that $M(k) = M(91)$, which equals 91 by the basis step above. Hence $M(91) = 91$ *[as was to be shown]*.

[Since the basis and inductive steps have been proved, it follows that $M(n) = 91$ for all integers $1 \leq n \leq 101$.]

21. b. $A(2,1) = A(1, A(2,0)) = A(1, A(1,1)) = A(1,3)$ *[by part (a)]* $= A(0, (A(1,2)) = A(0,4)$ *[by Example 8.4.8]* $= 4 + 1 = 5$

22. b. *Proof (by mathematical induction)*:

The formula holds for n = 0: When $n = 0$, $A(2,n) = A(2,0) = A(1,1)$ *[by 8.4.2]* $= 3$ *[by exercise 21]*. But also $3 = 3 + 2 \cdot 0 = 3 + 2n$. So the formula holds for $n = 0$.

If the formula holds for n = k then it holds for n = k + 1: Let k be an integer with $k \geq 0$ and suppose $A(2,k) = 3 + 2k$. *[This is the inductive hypothesis.]* We must show that $A(2, k+1) = 3 + 2(k+1)$. But

$$\begin{aligned} A(2, k+1) &= A(1, A(2,k)) \quad \text{by (8.4.3)} \\ &= A(1, 3 + 2k) \quad \text{by inductive hypothesis} \\ &= (3 + 2k) + 2 \quad \text{by part (a)} \\ &= 3 + 2(k+1) \quad \text{by the laws of algebra.} \end{aligned}$$

[This is what was to be shown.]

c. *Proof (by mathematical induction)*:

The formula holds for n = 0: When $n = 0$, $A(3,n) = A(3,0) = A(2,1)$ *[by 8.4.2]* $= 5$ *[by exercise 21]*. But also $5 = 8 \cdot 2^0 - 3 = 8 \cdot 2^n - 3$. So the formula holds for $n = 0$.

If the formula holds for n = k then it holds for n = k + 1: Let k be an integer with $k \geq 0$ and suppose $A(3,k) = 8 \cdot 2^k - 3$. *[This is the inductive hypothesis.]* We must show that $A(3, k+1) = 8 \cdot 2^{k+1} - 3$. But

$$\begin{aligned} A(3, k+1) &= A(2, A(3,k)) &&\text{by (8.4.3)} \\ &= A(2, 8 \cdot 2^k - 3) &&\text{by inductive hypothesis} \\ &= 3 + 2 \cdot (8 \cdot 2^k - 3) &&\text{by part b} \\ &= 3 + 8 \cdot 2^{k+1} - 6 \\ &= 8 \cdot 2^{k+1} - 3 &&\text{by the laws of algebra.} \end{aligned}$$

[This is what was to be shown.]

23. $H(2) = 1 + H(1) = 1 + 1 = 2$

$H(3) = H(10) = 1 + H(5) = 1 + H(16) = 1 + (1 + H(8)) = 2 + (1 + H(4)) = 3 + (1 + H(2)) = 3 + (1 + 2) = 6$

$H(4) = 1 + H(2) = 1 + (1 + H(1)) = 1 + 1 + 1 = 3$

$H(5) = H(16) = 1 + H(8)) = 1 + (1 + H(4)) = 1 + (1 + 3) = 5$

$H(6) = 1 + H(3) = 1 + 6 = 7$

$H(7) = H(22) = 1 + H(11) = 1 + H(34) = 1 + (1 + H(17)) = 2 + H(52) = 2 + (1 + H(26)) = 3 + (1 + H(13)) = 4 + H(40) = 4 + (1 + H(20)) = 5 + (1 + H(10)) = 6 + (1 + H(5)) = 7 + 5 = 12$

25. G is not well-defined. For each odd integer $n > 1$, $3n - 2$ is odd and $3n - 2 > n$. Thus the values of G for odd integers greater than 1 can never be found because each is defined in terms of values of G for even larger odd integers.

Chapter 9
Functions and the Efficiency of Algorithms

The focus of Chapter 9 is the analysis of algorithm efficiency.. The chapter opens with a brief review of those properties of function graphs that are especially important for understanding O-notation, which is introduced in Section 9.2. For simplicity, the examples in Section 9.2 are restricted to polynomial and rational functions. In Section 9.3 the analysis of algorithm efficiency is introduced with examples which include sequential search, insertion sort, selection sort (in the exercises), and polynomial evaluation (in the exercises). Section 9.4 treats those properties of logarithms that are particularly important in the analysis of algorithms and other areas of computer science are discussed, and Section 9.5 discusses algorithms whose orders involve logarithmic functions. Examples in Section 9.5 include binary search and merge sort.

The exercises in this chapter are designed to give you considerable latitude as to how thoroughly to cover both O-notation and algorithm analysis. The exercises for Sections 9.2–9.4 encompass an especially wide range of difficulty levels of problems. If you want to move rapidly through the chapter, just avoid the most demanding exercises. Section 9.5 is not particularly difficult and illustrates how various topics studied previously (recursive thinking, solving recurrence relations, strong mathematical induction, logarithms, and O-notation) combine to give useful information about interesting and practical algorithms.

Comments on Exercises:

Section 9.1: #20: This exercise is needed for various of the more theoretical exercises in Sections 9.2 and 9.4. #26 and #27 are warm-up exercises for the definition of O-notation.

Section 9.2: Core exercises for this section are #1-6 (for the concept of O-notation), #9-14 (which require proofs using the definition of O-notation), and #19-26 (which apply the theorem on polynomial orders).

Section 9.3: #45-46, and #48-51 are quite difficult.

Section 9.4: The result of #14 is needed for #17.

Section 9.5: #26 introduces the fast multiplication algorithm.

Section 9.1

3.

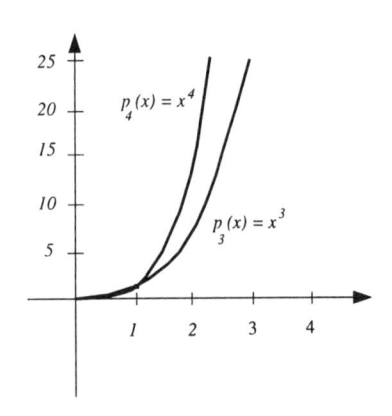

When $0 < x < 1$, $x^3 > x^4$. (For instance, $(1/2)^3 = 1/8 > 1/16 = (1/2)^4$.) When $x > 1$, $x^4 > x^3$.

6.

x	I(x)
0	0
0<x<1	1
1	0
-1	0
-1<x<0	1

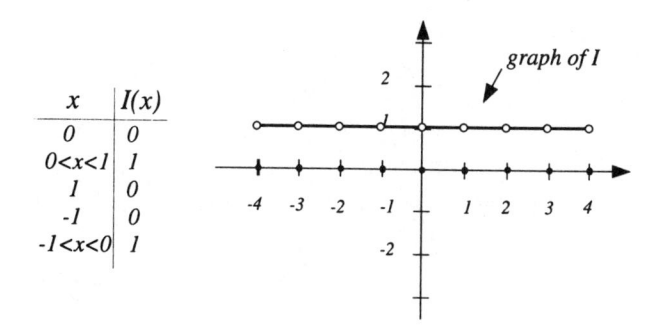

graph of I

8.

x	H(x)
0	0
1/2	1/2
3/4	3/4
1	0
1 1/2	1/2
1 3/4	3/4
2	0

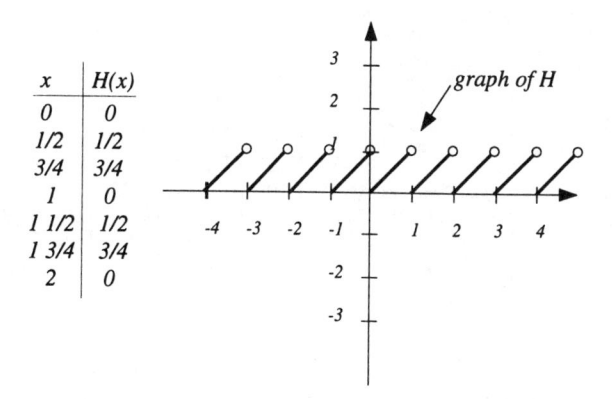

graph of H

10.

n	g(n) = (n/2) + 1
0	1
1	3/2
2	2
3	5/2
4	3
-1	1/2
-2	0
-3	-1/2

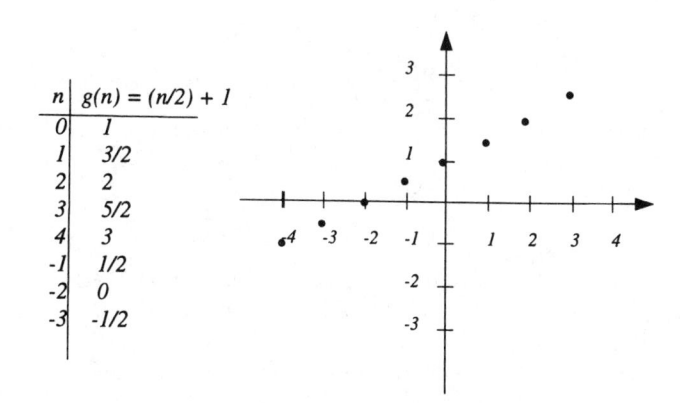

12.

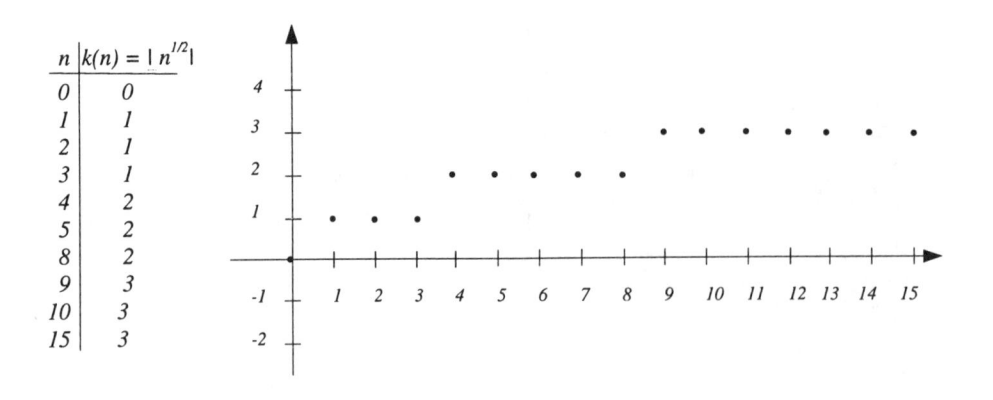

n	$k(n) = \lfloor n^{1/2} \rfloor$
0	0
1	1
2	1
3	1
4	2
5	2
8	2
9	3
10	3
15	3

15. *Proof*: Suppose that x_1 and x_2 are particular but arbitrarily chosen real numbers such that $x_1 < x_2$. *[We must show that $g(x_1) > g(x_2)$.]* Multiplying the inequality $x_1 < x_2$ by $-1/3$ gives $(-1/3)x_1 > (-1/3)x_2$. Adding 1 to both sides gives $(-1/3)x_1 + 1 > (-1/3)x_2 + 1$. So by definition of g, $g(x_1) > g(x_2)$ *[as was to be shown]*.

16. b. Let $h: \mathbf{R} \to \mathbf{R}$ be the function defined by the formula $h(x) = x^2$. We will show that h is increasing on the set of all real numbers greater than zero. Suppose x_1 and x_2 are real numbers greater than zero and such that $x_1 < x_2$. Multiply both sides of $x_1 < x_2$ by x_1 to obtain $x_1^2 < x_1 x_2$, and multiply both sides of $x_1 < x_2$ by x_2 to obtain $x_1 x_2 < x_2^2$. By transitivity of order *[Appendix A, T17]*, $x_1^2 < x_2^2$, and so by definition of h, $h(x_1) < h(x_2)$.

17. b. When $x < 0$, k is increasing.

    Proof: Suppose $x_1 < x_2 < 0$. Multiplying both sides of this inequality by -1 gives $-x_1 > -x_2$, and adding $x_1 x_2$ to both sides gives $x_1 x_2 - x_1 > x_1 x_2 - x_2$. Thus $\dfrac{x_1 x_2 - x_1}{x_1 x_2} > \dfrac{x_1 x_2 - x_2}{x_1 x_2}$ (because since x_1 and x_2 are both negative $x_1 x_2$ is positive). Simplifying the two fractions gives $\dfrac{x_2 - 1}{x_2} > \dfrac{x_1 - 1}{x_1}$.

19. *Proof*: Suppose x_1 and x_2 are any real numbers in D and $x_1 < x_2$. We must show that $(f+g)(x_1) < (f+g)(x_2)$. Since f and g are both increasing, $f(x_1) < f(x_2)$ and $g(x_1) < g(x_2)$. Adding the two inequalities gives $f(x_1) + g(x_1) < f(x_2) + g(x_2)$, and so by definition of $f+g$, $(f+g)(x_1) < (f+g)(x_2)$. Consequently, $f+g$ is increasing.

20. b. *Proof*: Let x_1 and x_2 be real numbers with $0 < x_1 < x_2$. By part (a), $x_1^m < x_2^m$. Now suppose that $g(x_1) \geq g(x_2)$, or, equivalently, $x_1^{\frac{m}{n}} \geq x_2^{\frac{m}{n}}$. Then, again by part (a), $\left(x_1^{\frac{m}{n}}\right)^n \geq \left(x_2^{\frac{m}{n}}\right)^n$. By the laws of exponents, this implies that $x_1^{\frac{m}{n} \cdot n} = x_1^m \geq x_2^m = x_2^{\frac{m}{n} \cdot n}$. But this contradicts the fact that $x_1^m < x_2^m$. Hence the supposition is false, and thus $g(x_1) < g(x_2)$. So g is increasing.

22.

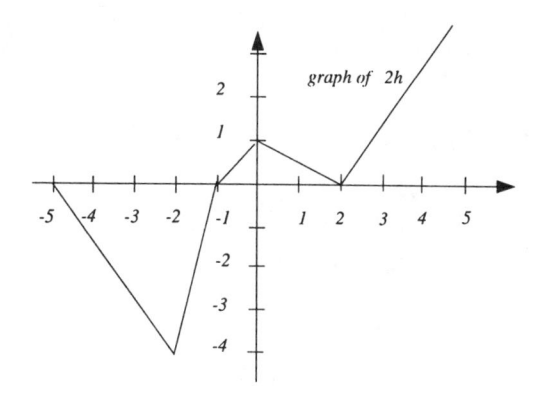

graph of 2h

24. *Proof:* Suppose that f is a real-valued function of a real variable, f is increasing on a set S, and M is any negative real number. *[We must show that $M \cdot f$ is decreasing on S. In other words, we must show that for all x_1 and x_2 in the set S, if $x_1 < x_2$ then $(M \cdot f)(x_1) > (M \cdot f)(x_2)$.]* Suppose x_1 and x_2 are in S and $x_1 < x_2$. Since f is increasing on S, $f(x_1) < f(x_2)$. Since M is negative, $M \cdot f(x_1) > M \cdot f(x_2)$ *[because when both sides of an inequality are multiplied by a negative number, the direction of the inequality is reversed]*. It follows by definition of $M \cdot f$ that $(M \cdot f)(x_1) > (M \cdot f)(x_2)$ *[as was to be shown]*.

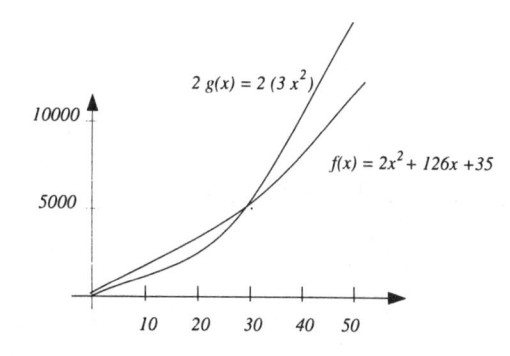

25. *Proof:* Suppose that f is a real-valued function of a real variable, f is decreasing on a set S, and M is any negative real number. *[We must show that $M \cdot f$ is increasing on S. In other words, we must show that for all x_1 and x_2 in the set S, if $x_1 < x_2$ then $(M \cdot f)(x_1) < (M \cdot f)(x_2)$.]* Suppose x_1 and x_2 are in S and $x_1 < x_2$. Since f is decreasing on S, $f(x_1) > f(x_2)$. Since M is negative, $M \cdot f(x_1) < M \cdot f(x_2)$ *[because when both sides of an inequality are multiplied by a negative number, the direction of the inequality is reversed]*. It follows by definition of $M \cdot f$ that $(M \cdot f)(x_1) < (M \cdot f)(x_2)$ *[as was to be shown]*.

26.

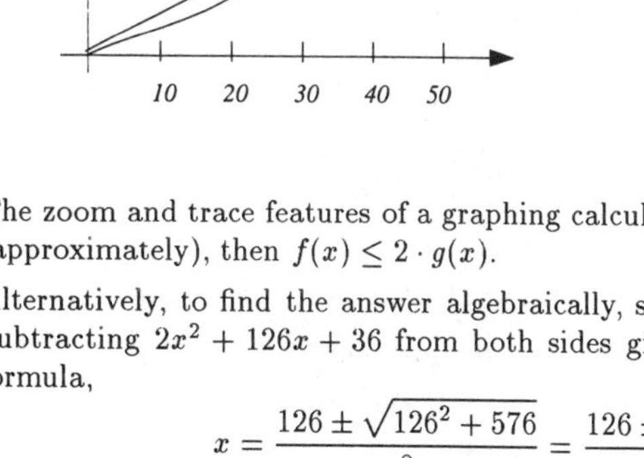

$2g(x) = 2(3x^2)$

$f(x) = 2x^2 + 126x + 35$

27.

The zoom and trace features of a graphing calculator or computer indicate that when $x \geq 31.8$ (approximately), then $f(x) \leq 2 \cdot g(x)$.

Alternatively, to find the answer algebraically, solve the equation $2(3x^2) = 2x^2 + 126x + 36$. Subtracting $2x^2 + 126x + 36$ from both sides gives $4x^2 - 126x - 36 = 0$. By the quadratic formula,

$$x = \frac{126 \pm \sqrt{126^2 + 576}}{8} = \frac{126 \pm \sqrt{16452}}{8} \cong \frac{126 \pm 128.2635}{8},$$

and so $x \cong 31.78317$ or $x \cong -0.28317$. Since f and g are only defined for positive values of x, the only place where the two graphs cross is at approximately $x \cong 31.78317$. Thus if $x_0 = 31.7832$, then for all $x > x_0$, $f(x) \leq 2g(x)$.

Section 9.2

4. Let $M = 4$ and $x_0 = 3$. Then $\left|\dfrac{(x^2-1)(12x+25)}{3x^2+4}\right| \leq M \cdot |x|$ for all real numbers $x > x_0$, and so by definition of O-notation, $\dfrac{(x^2-1)(12x+25)}{3x^2+4}$ is $O(x)$.

6. a. For all real numbers $x > 1$,

$$|23x^4 + 8x^3 + 4x| = 23x^4 + 8x^3 + 4x \quad \text{because } 23x^4,\ 8x^3,\ \text{and } 4x \text{ are all positive since } x > 1$$

$$\Rightarrow |23x^4 + 8x^3 + 4x| \leq 23x^4 + 8x^4 + 4x^4 \quad \text{because } x^3 < x^4 \text{ and } x < x^4 \text{ for } x > 1$$

$$\Rightarrow |23x^4 + 8x^3 + 4x| \leq 35x^4 \quad \text{by combining like terms}$$

$$\Rightarrow |23x^4 + 8x^3 + 4x| \leq 35|x^4| \quad \text{because } x^4 \text{ is positive.}$$

b. Let $M = 35$ and $x_0 = 1$. Then by substitution, $|23x^4 + 8x^3 + 4x| \leq M|x^4|$ for all $x > x_0$. Hence by definition of O-notation, $23x^4 + 8x^3 + 4x$ is $O(x^4)$.

8. a. For all real numbers $x > 1$,

$$\dfrac{12x^2 + 60}{5x^2 + 4} \leq 3$$

$$\Leftrightarrow \quad 12x^2 + 60 \leq 3(5x^2 + 4) \quad \text{by multiplying or dividing by } 5x^2 + 4, \text{ which is positive}$$

$$\Leftrightarrow \quad 4x^2 + 20 \leq 5x^2 + 4 \quad \text{by multiplying or dividing by 3}$$

$$\Leftrightarrow \quad 16 \leq x^2 \quad \text{by adding or subtracting } 4x^2 + 4.$$

But if $x > 4$, then $x^2 > 16$, and so, by the above equivalences, $\dfrac{12x^2 + 60}{5x^2 + 4} < 3$.

b. For all real numbers $x > 4$,

$$\dfrac{12x^2 + 60}{5x^2 + 4} \leq 3 \quad \text{by part (a)}$$

$$\Rightarrow \quad \left|\dfrac{12x^2 + 60}{5x^2 + 4}\right| \leq 3 \quad \text{because } \dfrac{12x^2 + 60}{5x^2 + 4} > 0 \text{ for all real numbers } x$$

$$\Rightarrow \quad \left|\dfrac{12x^2 + 60}{5x^2 + 4}\right| |x^{3/2}| \leq 3|x^{3/2}| \quad \text{by multiplying both sides by } |x^{3/2}|$$

$$\Rightarrow \quad \left|\dfrac{12x^{7/2} + 60x^{3/2}}{5x^2 + 4}\right| \leq 3|x^{3/2}| \quad \text{because } x^2 x^{3/2} = x^{4/2} x^{3/2} = x^{7/2}.$$

c. Let $M = 3$ and $x_0 = 4$. Then $\left|\dfrac{12x^{7/2} + 60x^{3/2}}{5x^2 + 4}\right| \leq M \cdot |x^{3/2}|$ for all real numbers $x > x_0$, and so by definition of O-notation, $\dfrac{12x^{7/2} + 60x^{3/2}}{5x^2 + 4}$ is $O(x^{3/2})$.

10. For all real numbers $x > 1$,

$$|6x^4 + x^2 + 13| \leq 6x^4 + x^2 + 13 \quad \text{because } 6x^4,\ x^2,\ \text{and } 13 \text{ are positive}$$

$$\Rightarrow |6x^4 + x^2 + 13| \leq 6x^4 + x^4 + 13x^4 \quad \text{because } x^2 < x^4 \text{ and } 13 < 13x^4 \text{ for } x > 1$$

$$\Rightarrow |6x^4 + x^2 + 13| \leq 20x^4 \quad \text{by combining like terms}$$

$$\Rightarrow |6x^4 + x^2 + 13| \leq 20|x^4| \quad \text{because } x^4 \text{ is positive.}$$

Let $M = 20$ and $x_0 = 1$. Then $|6x^4 + x^2 + 13| \leq M|x^4|$ for all real numbers $x > x_0$, and so by definition of O-notation, $6x^4 + x^2 + 13$ is $O(x^4)$.

12. For all real numbers $x > 1$,

$$\begin{aligned}
|10x^3 + x^2 - 5x + 6| &\leq |10x^3| + |x^2| + |5x| + |6| &&\text{by the triangle inequality} \\
\Rightarrow |10x^3 + x^2 - 5x + 6| &\leq 10x^3 + x^2 + 5x + 6 &&\text{because } 10x^3, x^2, 5x, \\
& && \text{and 6 are positive} \\
\Rightarrow |10x^3 + x^2 - 5x + 6| &\leq 10x^3 + x^3 + 5x^3 + 6x^3 &&\text{because } x^2 < x^3, x < x^3, \\
& && \text{and } 1 < x^3 \text{ for } x > 1 \\
\Rightarrow |10x^3 + x^2 - 5x + 6| &\leq 22x^3 &&\text{by combining like terms} \\
\Rightarrow |10x^3 + x^2 - 5x + 6| &\leq 22|x^3| &&\text{because } x^3 \text{ is positive.}
\end{aligned}$$

Let $M = 22$ and $x_0 = 1$. Then $|10x^3 + x^2 - 5x + 6| \leq M|x^3|$ for all real numbers $x > x_0$, and so by definition of O-notation, $10x^3 + x^2 - 5x + 6$ is $O(x^3)$.

14. For any integer $n > 1$,

$$\begin{aligned}
|\lfloor\sqrt{n}\rfloor| &= \lfloor\sqrt{n}\rfloor &&\text{because since } n > 1 > 0, \text{ then } \lfloor\sqrt{n}\rfloor > 0 \\
\Rightarrow |\lfloor\sqrt{n}\rfloor| &\leq \sqrt{n} &&\text{because } \lfloor x \rfloor \leq x \text{ for all real numbers } x \\
\Rightarrow |\lfloor\sqrt{n}\rfloor| &\leq |\sqrt{n}| &&\text{because } \sqrt{n} \geq 0.
\end{aligned}$$

Let $M = 1$ and $x_0 = 1$. Then $|\lfloor\sqrt{n}\rfloor| \leq M|\sqrt{n}|$ for all integers $n > x_0$, and so by definition of O-notation, $\lfloor\sqrt{n}\rfloor$ is $O(\sqrt{n})$.

15. *Lemma*: Let x be a real number with $x > 1$. Then for all integers $n \geq 1$, $x^n > 1$.

Proof (by mathematical induction):

The inequality holds for n = 1: For $n = 1$ the inequality is $x^1 > 1$ which is true since $x > 1$.

If the inequality holds for n = k then it holds for n = k + 1: Let k be an integer with $k \geq 1$ and suppose $x^k > 1$. *[This is the inductive hypothesis.]* We must show that $x^{k+1} > 1$. But we know that $x > 1$, and also by inductive hypothesis $x^k > 1$. So we may multiply the inequalities to obtain $x \cdot x^k > 1 \cdot 1$, or equivalently, $x^{k+1} > 1$ *[as was to be shown]*.

Proof of the exercise statement: Suppose $x > 1$ and m and n are nonnegative integers with $m < n$. Then $n - m$ is an integer with $n - m \geq 1$, and so by the lemma $x^{n-m} > 1$. Multiplying both sides by x^m gives $x^m \cdot x^{n-m} = x^n > x^m$ *[as was to be shown]*.

16. a. *Lemma*: Let x be any real number with $0 \leq x \leq 1$. For any positive integer $n \geq 1$, $x^n \leq 1$.

Proof 1 (by mathematical induction):

The inequality holds for n = 1: We must show that $x^1 \leq 1$. But $x \leq 1$ by assumption and $x^1 = x$. So the inequality holds for $n = 1$.

If the inequality holds for n = k then it holds for n = k+1: Let k be an integer with $k \geq 1$, and suppose $x^k \leq 1$. *[This is the inductive hypothesis.]* Multiplying both sides of this inequality by x (which is positive by assumption) gives $x \cdot x^k \leq x \cdot 1$, and so by the laws of exponents, $x^{k+1} \leq x$. Then $x^{k+1} \leq x$ and $x^1 \leq 1$, and hence by the transitive property of $\leq$, $x^{k+1} \leq 1$.

Proof 2: By exercise 20(a) of Section 9.1, for any positive integer n, the function f defined by $f(x) = x^n$ is increasing on the set of all nonnegative real numbers. Hence if $0 \leq x \leq 1$, then $x^n \leq 1^n = 1$.

Proof of exercise statement (by contraposition): Let n be a positive integer, and suppose x is a positive real number with $x^{1/n} \leq 1$. By Lemma 1 $\left(x^{1/n}\right)^n \leq 1$. But $\left(x^{1/n}\right)^n = x$. So $x \leq 1$.

Note: This result can also be derived by directly applying exercise 20 from Section 9.1.

b. *Proof 1*: Let p, q, r, and s be integers with q and s nonzero and $\frac{p}{q} > \frac{r}{s}$, and let x be any real number with $x > 1$. Since $\frac{p}{q} > \frac{r}{s}$, then $\frac{p}{q} - \frac{r}{s} > 0$, or, equivalently, $\frac{ps-qr}{qs} > 0$. Thus either both $ps - qr$ and qs are positive, or both are negative.

Case 1 (both $ps - qr$ and qs are positive): By part (a) since $x > 1$, $x^{1/qs} > 1$. Also since $ps - qr > 0$ and since $ps - qr$ is an integer, then $ps - qr \geq 1$. So by exercise 15 (with $n = ps - qr$ and $m = 1$), $\left(x^{1/qs}\right)^{pq-rs} \geq \left(x^{1/qs}\right)^1 = x^{1/qs} > 1$. But $\left(x^{1/qs}\right)^{pq-rs} = x^{\frac{ps-qr}{qs}} = x^{\frac{p}{q}-\frac{r}{s}} = \frac{x^{\frac{p}{q}}}{x^{\frac{r}{s}}}$.

So $\frac{x^{\frac{p}{q}}}{x^{\frac{r}{s}}} > 1$, or, equivalently, $x^{\frac{p}{q}} > x^{\frac{r}{s}}$.

Case 2 (both $ps - qr$ and qs are negative): Let $p' = -p$ and $q' = -q$. Then $\frac{p'}{q'} = \frac{-p}{-q} = \frac{p}{q}$, and so $\frac{p'}{q'} > \frac{r}{s}$. But $p's - q'r$ and $q's$ are both positive. So by case 1, $x^{\frac{p'}{q'}} > x^{\frac{r}{s}}$, and hence $x^{\frac{p}{q}} > x^{\frac{r}{s}}$.

Proof 2: Let p, q, r, and s be integers with q and s nonzero and $\frac{p}{q} > \frac{r}{s}$, and let x be any real number with $x > 1$. Since $\frac{p}{q} > \frac{r}{s}$, then $\frac{p}{q} - \frac{r}{s} > 0$, or, equivalently, $\frac{ps-qr}{qs} > 0$. Let f be the function defined by the formula $f(x) = x^{\frac{ps-qr}{qs}}$. By the result of exercise 20 in Section 9.1, f is increasing. Hence if $x > 1$, then $x^{\frac{ps-qr}{qs}} > 1^{\frac{ps-qr}{qs}} = 1$. But $x^{\frac{ps-qr}{qs}} = x^{\frac{p}{q}-\frac{r}{s}} = \frac{x^{\frac{p}{q}}}{x^{\frac{r}{s}}}$. So $\frac{x^{\frac{p}{q}}}{x^{\frac{r}{s}}} > 1$, or, equivalently, $x^{\frac{p}{q}} > x^{\frac{r}{s}}$.

17. c. Suppose $f: \mathbf{R} \to \mathbf{R}$ is a function, $f(x)$ is $O(g(x))$, and c is any real number. Since $f(x)$ is $O(g(x))$, there exist real numbers M and x_0 such that $|f(x)| \leq M|g(x)|$ for all $x > x_0$. In case $c = 0$, let $M' = M$. Then $cf(x) = 0$, and so $|cf(x)| = 0 \leq M'|g(x)|$. In case $c \neq 0$, let $M' = cM$. Then $M = M'/c$, and so

$$|f(x)| \leq \frac{M'}{c}|g(x)| \qquad \text{for all } x > x_0.$$

or, equivalently,

$$|cf(x)| \leq M'|g(x)| \qquad \text{for all } x > x_0.$$

Thus in either case $|cf(x)| \leq M'|g(x)|$ for all $x > x_0$, and so by definition of O-notation $cf(x)$ is $O(g(x))$.

d. By exercise 15, for all $x > 1$, $x^2 < x^5$ and $1 = x^0 < x^5$. Clearly also $x^5 \leq x^5$. So for all $x > 1$, $|x^5| \leq 1 \cdot |x^5|$, $|x^2| \leq 1 \cdot |x^5|$, and $|1| \leq 1 \cdot |x^5|$. Hence by definition of O-notation, x^5 is $O(x^5)$, x^2 is $O(x^5)$, and 1 is $O(x^5)$. By part (c), then, $12x^5$ is $O(x^5)$, $-34x^2$ is $O(x^5)$, and $7 = 7 \cdot 1$ is $O(x^5)$. So by part (a), $12x^5 + (-34)x^2 + 7 = 12x^5 - 34x^2 + 7$ is $O(x^5)$.

18. Let a nonnegative integer n be given, and suppose m is any integer with $m \geq n$. By exercise 15, for all integers k with $0 \leq k \leq n$ and for all real numbers $x > 1$, $1 = x^0 \leq x^m$, $x = x^1 \leq x^m$, $x^2 \leq x^m$, ..., and $x^n \leq x^m$. So since all expressions are positive, if $M = 1$ and $x_0 = 1$, then for all real numbers x with $x > x_0$, $|1| \leq M|x^m|$, $|x| \leq M|x^m|$, $|x^2| \leq M|x^m|$, ..., $|x^n| \leq M|x^m|$. Thus by definition of O-notation, 1 is $O(x^m)$, x is $O(x^m)$, x^2 is $O(x^m)$, ..., and x^n is $O(x^m)$. It follows by part (c) of exercise 17 that for any real numbers $a_0, a_1, a_2, \ldots, a_n$, $a_0 = a_0 \cdot 1$ is $O(x^m)$, $a_1 x$ is $O(x^m)$, $a_2 x^2$ is $O(x^m)$, ..., and $a_n x^n$ is $O(x^m)$. So by repeated application of part (a) of exercise 17, $a_n x^n + a_{n-1} x^{n-1} + \cdots + a_2 x^2 + a_1 x + a_0$ is $O(x^m)$.

20. $\frac{x}{3}(4x^2 - 1) = \frac{4}{3}x^3 - \frac{1}{3}x$ is $O(x^3)$ by the theorem on polynomial orders

22. $\left(\frac{n(n+1)}{2}\right)^2 = \frac{n^2(n^2 + 2n + 1)}{4} = \frac{1}{4}n^4 + \frac{1}{2}n^3 + \frac{1}{4}n^2$ is $O(n^4)$ by the theorem on polynomial orders

24. By exercise 10 of Section 4.2, $1^3 + 2^3 + 3^3 \ldots + n^3 = \left(\dfrac{n(n+1)}{2}\right)^2$, and by exercise 22 above this is $O(n^4)$. Hence $1^3 + 2^3 + 3^3 \ldots + n^3$ is $O(n^4)$.

26. Factoring out a 2 gives $2 + 4 + 6 + \ldots + 2n = 2(1 + 2 + 3 + \ldots + n) = 2\left(\dfrac{n(n+1)}{2}\right)$ [by Theorem 4.2.2] $= n^2 + n$, which is $O(n^2)$ by the theorem on polynomial orders.

28. Note that by the distributive law and the laws of exponents, $\sqrt{x}(38x^5 + 9) = 38x^{11/2} + 9^{1/2}$. By part (b) of exercise 16, for all $x > 1$, $x^{1/2} < x^{11/2}$, and certainly $x^{11/2} \leq x^{11/2}$. So for all $x > 1$, $|x^{1/2}| \leq 1 \cdot |x^{11/2}|$ and $|x^{11/2}| \leq 1 \cdot |x^{11/2}|$ (since when $x > 1$ all expressions are positive). Thus by definition of O-notation, $x^{11/2}$ is $O(x^{11/2})$ and $x^{1/2}$ is $O(x^{11/2})$. By part (c) of exercise 17, then, $38x^{11/2}$ is $O(x^{11/2})$ and $9x^{1/2}$ is $O(x^{11/2})$. Finally, by part (a) of exercise 17, $\sqrt{x}(38x^5 + 9) = 38x^{11/2} + 9^{1/2}$ is $O(x^{11/2})$.

30. *Proof*: Suppose f, g, and h are functions from **R** to **R** and $f(x)$ is $O(g(x))$ and $g(x)$ is $O(h(x))$. By definition of O-notation, there exist constants M_1, M_2, x_1, and x_2 such that

$$|f(x)| \leq M_1|g(x)| \quad \text{for all } x > x_1$$

and

$$|g(x)| \leq M_2|h(x)| \quad \text{for all } x > x_2.$$

Let $M = M_1 M_2$, and let x_0 be the greater of x_1 and x_2. Then if $x > x_0$,

$$|f(x)| \leq M_1|g(x)| \leq M_1(M_2|h(x)|) = M|h(x)|.$$

So by definition of O-notation, $f(x)$ is $O(h(x))$.

32. a. *Proof (by mathematical induction)*:

The formula holds for n = 1: For $n = 1$ the formula asserts that $1^{1/3} \leq 1^{4/3}$, which is certainly true.

If the formula holds for n = k then it holds for n = k+1: Let k be an integer with $k \geq 1$, and suppose

$$1^{1/3} + 2^{1/3} + \cdots + k^{1/3} \leq k^{4/3}.$$

[*This is the inductive hypothesis.*]

We must show that

$$1^{1/3} + 2^{1/3} + \cdots + (k+1)^{1/3} \leq (k+1)^{4/3}.$$

But by inductive hypothesis, $1^{1/3} + 2^{1/3} + \cdots + k^{1/3} \leq k^{4/3}$, and so

$$1^{1/3} + 2^{1/3} + \cdots + k^{1/3} + (k+1)^{1/3} \leq k^{4/3} + (k+1)^{1/3} \quad \text{by adding } (k+1)^{1/3} \text{ to both sides}$$

$$\Rightarrow 1^{1/3} + 2^{1/3} + \cdots + k^{1/3} + (k+1)^{1/3} \leq k^{1/3} \cdot k + (k+1)^{1/3} \quad \text{by factoring out } k \text{ from } k^{4/3}$$

$$\Rightarrow 1^{1/3} + 2^{1/3} + \cdots + k^{1/3} + (k+1)^{1/3} \leq (k+1)^{1/3} \cdot k + (k+1)^{1/3} \quad \text{because } k^{1/3} \leq (k+1)^{1/3}$$

$$\Rightarrow 1^{1/3} + 2^{1/3} + \cdots + k^{1/3} + (k+1)^{1/3} \leq (k+1)^{1/3}(k+1) \quad \text{by factoring out } (k+1)^{1/3}$$

$$\Rightarrow 1^{1/3} + 2^{1/3} + \cdots + k^{1/3} + (k+1)^{1/3} \leq (k+1)^{4/3} \quad \text{by the laws of exponents.}$$

b. Let $M = 1$ and $x_0 = 1$. By part (a) and because all quantities are positive,

$$\left|1^{1/3} + 2^{1/3} + \cdots + n^{1/3}\right| \leq M \cdot \left|n^{4/3}\right| \quad \text{for all } n > x_0.$$

Thus by definition of O-notation, $1^{1/3} + 2^{1/3} + \cdots + n^{1/3}$ is $O(n^{4/3})$.

33. *a.* Given any integer $n \geq 1$, for all integers k with $1 \leq k \leq \lfloor\sqrt{n}\rfloor$, $\frac{1}{k} \leq 1$. So since there are $\lfloor\sqrt{n}\rfloor$ terms in the sum $1 + \frac{1}{2} + \frac{1}{3} + \cdots + \frac{1}{\lfloor\sqrt{n}\rfloor}$, then

$$1 + \frac{1}{2} + \frac{1}{3} + \cdots + \frac{1}{\lfloor\sqrt{n}\rfloor} \leq \underbrace{1 + 1 + \cdots + 1}_{\lfloor\sqrt{n}\rfloor \text{ terms}} = \lfloor\sqrt{n}\rfloor \leq \sqrt{n}$$

by definition of floor.

b. Given any integer $n \geq 2$, for all integers k with $\lfloor\sqrt{n}\rfloor + 1 \leq k \leq n$, observe that

$$\frac{1}{k} \leq \frac{1}{\lfloor\sqrt{n}\rfloor + 1} \leq \frac{1}{\lfloor\sqrt{n}\rfloor},$$

where the second inequality holds because by definition of floor

$$\lfloor\sqrt{n}\rfloor \leq \sqrt{n} < \lfloor\sqrt{n}\rfloor + 1.$$

Thus

$$\frac{1}{\lfloor\sqrt{n}\rfloor + 1} + \frac{1}{\lfloor\sqrt{n}\rfloor + 2} + \cdots + \frac{1}{n} \leq \underbrace{\frac{1}{\lfloor\sqrt{n}\rfloor} + \frac{1}{\lfloor\sqrt{n}\rfloor} + \cdots + \frac{1}{\lfloor\sqrt{n}\rfloor}}_{n - (\lfloor\sqrt{n}\rfloor + 1) + 1 \,=\, n - \lfloor\sqrt{n}\rfloor \text{ terms}}.$$

Hence

$$\frac{1}{\lfloor\sqrt{n}\rfloor + 1} + \frac{1}{\lfloor\sqrt{n}\rfloor + 2} + \cdots + \frac{1}{n} \leq (n - \lfloor\sqrt{n}\rfloor)\frac{1}{\sqrt{n}} = \sqrt{n} - \frac{\lfloor\sqrt{n}\rfloor}{\sqrt{n}} \leq \sqrt{n}.$$

c. For all integers $n \geq 2$,

$$\begin{aligned}1 + \frac{1}{2} + \frac{1}{3} + \cdots + \frac{1}{n} &= \left(1 + \frac{1}{2} + \frac{1}{3} + \cdots + \frac{1}{\lfloor\sqrt{n}\rfloor}\right) + \left(\frac{1}{\lfloor\sqrt{n}\rfloor + 1} + \frac{1}{\lfloor\sqrt{n}\rfloor + 2} + \cdots + \frac{1}{n}\right) \\ &\leq \sqrt{n} + \sqrt{n} = 2\sqrt{n}\end{aligned}$$

by parts (a) and (b). Let $M = 2$ and $x_0 = 2$. Then because all terms are positive,

$$\left|1 + \frac{1}{2} + \frac{1}{3} + \cdots + \frac{1}{n}\right| \leq M |\sqrt{n}| \quad \text{for all integers } n > x_0.$$

So by definition of O-notation, $1 + \frac{1}{2} + \frac{1}{3} + \cdots + \frac{1}{n}$ is $O(\sqrt{n})$.

34. *Proof (by contradiction)*: Let r and s be rational numbers with $r > s$. Suppose that x^r is $O(x^s)$. We must show that this supposition leads to a contradiction. Then there exist real numbers M and x_0 so that $|x^r| \leq M|x^s|$ for all real numbers $x > x_0$. Let x be any real number greater than both x_0 and $M^{\frac{1}{r-s}}$. Then

$$x > M^{\frac{1}{r-s}}$$
$$\Rightarrow \quad x^{r-s} > \left(M^{\frac{1}{r-s}}\right)^{r-s} \quad \text{by exercise 20 of Section 9.1 and because } r - s > 0 \text{ since } r > s$$
$$\Rightarrow \quad \frac{x^r}{x^s} > M \quad \text{by the laws of exponents.}$$

Thus there is a real number x with $x > x_0$ such that

$$x^r > Mx^s,$$

which contradicts the fact that $|x^r| \leq M|x^s|$ for all real numbers $x > x_0$. Hence the supposition is false, and so x^r is not $O(x^s)$.

36. Note that $x(25x-4)(x^2+3) = (25x^2-4x)(x^2+3) = 25x^4 - 4x^3 + 75x^2 - 12x$. So by property (9.2.4), $x(25x-4)(x^2+3)$ is $O(x^r)$ for all rational numbers $r \geq 4$ and is not $O(x^s)$ for any rational number $s < 4$. Thus x^4 is the best big-oh approximation for $x(25x-4)(x^2+3)$ from among the set of all rational power functions.

38. Note that $\dfrac{(2x^{5/2}+1)(x-1)}{x^{1/2}} = \dfrac{2x^{7/2} - 2x^{5/2} + x - 1}{x^{1/2}} = 2x^3 - 2x^2 + x^{1/2} - \dfrac{1}{x^{1/2}}$. So by property (9.2.4), $\dfrac{(2x^{5/2}+1)(x-1)}{x^{1/2}}$ is $O(x^r)$ for all rational numbers $r \geq 3$ and is not $O(x^s)$ for any rational number $s < 3$. Thus x^3 is the best big-oh approximation for $\dfrac{(2x^{5/2}+1)(x-1)}{x^{1/2}}$ from among the set of all rational power functions.

39. *a. Proof*: Suppose that g is a best big-oh approximation for f in a set S of functions, and suppose $\lim\limits_{x \to \infty} \dfrac{h(x)}{k(x)} = c$, where c is a nonzero constant. By definition of O-notation and of limit, there exist constants M_1, x_1, and x_2 such that

$$|f(x)| \leq M_1 |g(x)| \quad \text{for all } x > x_1$$

and

$$\left| \dfrac{h(x)}{k(x)} - c \right| < 1 \quad \text{for all } x > x_2.$$

By the triangle inequality, then, if $x > x_2$,

$$\left| \dfrac{h(x)}{k(x)} \right| \leq \left| \left(\dfrac{h(x)}{k(x)} - c \right) + c \right| \leq \left| \dfrac{h(x)}{k(x)} - c \right| + |c| < 1 + c.$$

If, in addition, $x > x_1$, then

$$|f(x)| \cdot \left| \dfrac{h(x)}{k(x)} \right| \leq M_1(1+|c|)|g(x)|.$$

So let x_0 be the greater of x_1 and x_2, and let $M = M_1(1+|c|)$. Then

$$\left| \dfrac{f(x)h(x)}{k(x)} \right| \leq M|g(x)| \quad \text{for all } x > x_0.$$

Hence by definition of O-notation, $\dfrac{f(x)h(x)}{k(x)}$ is $O(g(x))$.

To show that $g(x)$ is the best big-oh approximation for $\dfrac{f(x)h(x)}{k(x)}$, suppose $\dfrac{f(x)h(x)}{k(x)}$ is $O(G(x))$ for another function G in S. *[We must show that $g(x)$ is $O(G(x))$.]* Then there exist real numbers M_3 and x_3 such that

$$\left| \dfrac{f(x)h(x)}{k(x)} \right| \leq M_3 |G(x)| \quad \text{for all } x > x_3.$$

Since $c \neq 0$,

$$\lim_{x \to \infty} \dfrac{k(x)}{h(x)} = \dfrac{1}{\lim\limits_{x \to \infty} \dfrac{h(x)}{k(x)}} = \dfrac{1}{c},$$

and so there exists a real number x_4 such that

$$\left| \dfrac{k(x)}{h(x)} - \dfrac{1}{c} \right| < 1 \quad \text{for all } x > x_4.$$

By the triangle inequality, then, if $x > x_4$,

$$\left|\frac{k(x)}{h(x)}\right| \leq \left|\left(\frac{k(x)}{h(x)} - \frac{1}{c}\right) + \frac{1}{c}\right| \leq \left|\frac{k(x)}{h(x)} - \frac{1}{c}\right| + \left|\frac{1}{c}\right| < 1 + \frac{1}{c}.$$

Thus if x is greater than both x_3 and x_4, then

$$|f(x)| = \left|\frac{k(x)}{h(x)}\right| \left|\frac{f(x)h(x)}{k(x)}\right| \leq M_3|G(x)|(1 + \left|\frac{1}{c}\right|).$$

So let x_5 be the greater of x_3 and x_4, and let $M_5 = M_3(1 + |\frac{1}{c}|)$. Then

$$|f(x)| \leq M_5|G(x)| \qquad \text{for all } x > x_5.$$

Hence by definition of O-notation, $f(x)$ is $O(G(x))$. But $g(x)$ is the best big-oh approximation for $f(x)$ in S. Thus $g(x)$ is also $O(G(x))$ [as was to be shown].

b. (ii) Note that

$$\lim_{x \to \infty} \frac{4x^3 + 5x + 2}{7x^3 - 3x^2 + 1} = \lim_{x \to \infty} \frac{4x^3 + 5x + 2}{7x^3 - 3x^2 + 1} \cdot \frac{\left(\frac{1}{x^3}\right)}{\left(\frac{1}{x^3}\right)} = \lim_{x \to \infty} \left(\frac{4x + \frac{5}{x^2} + \frac{2}{x^3}}{7 - \frac{3}{x} + \frac{1}{x^3}}\right) = \frac{4}{7},$$

and by property (9.2.4), $x^{3/2}$ is the best big-oh approximation for $25x^{3/2} + 4x^{1/2} + 7$ from among the set of all rational power functions. Hence by part (a), $x^{3/2}$ is a best big-oh approximation for $\dfrac{(25x^{3/2} + 4x^{1/2} + 7)(4x^3 + 5x + 2)}{7x^3 - 3x^2 + 1}$.

41. *Proof*: Suppose $f(x)$ and $g(x)$ are $o(h(x))$ and a and b are any real numbers. Then by properties of limits,

$$\lim_{x \to \infty} \frac{af(x) + bg(x)}{h(x)} = a \lim_{x \to \infty} \frac{f(x)}{h(x)} + b \lim_{x \to \infty} \frac{g(x)}{h(x)} = a \cdot 0 + b \cdot 0 = 0.$$

So $af(x) + bg(x)$ is $o(h(x))$.

42. *Proof*: Suppose a and b are any positive real numbers such that $a < b$. Then $b - a > 0$, and so since $\dfrac{1}{x^c} \to 0$ as $x \to \infty$ for any positive number c,

$$\lim_{x \to \infty} \frac{x^a}{x^b} = \lim_{x \to \infty} \frac{1}{x^{b-a}} = 0.$$

So x^a is $o(x^b)$.

Section 9.3

1. b. .0002 seconds   c. .00153 seconds   e. 8 seconds   f. 5.09×10^{46} years

3. a. When the input size is increased from m to $2m$, the number of operations increases from cm^3 to $c(2m)^3 = 8cm^3$.

   b. By part (a), the number of operations increases by a factor of $\dfrac{8cm^3}{cm^3} = 8$.

   c. When the input size is increased by a factor of 10 (from m to $10m$), the number of operations increases by a factor of $\dfrac{c(10m^3)}{cm^3} = \dfrac{1000cm^3}{cm^3} = 1000$.

5. *a.* The best big-oh approximation for the efficiency of algorithm A from among the set of power functions is n^2 and for algorithm B it is $n^{5/3}$. So algorithm A has order n^2 (or higher) and algorithm B has order $n^{5/3}$ (or higher).

b. Algorithm A is more efficient than algorithm B for values of n with $4n^2 < 200n^{5/3}$. This occurs exactly when $n^2 < 50n^{5/2} \Leftrightarrow \dfrac{n^2}{n^{5/3}} < 50 \Leftrightarrow \dfrac{n^{6/3}}{n^{5/3}} < 50 \Leftrightarrow n^{1/3} < 50 \Leftrightarrow n < 50^3 = 125{,}000$. So when $n < 125{,}000$, algorithm A is more efficient than algorithm B.

c. Algorithm B is 100 times more efficient than algorithm A for values of n with $100(200n^{5/3}) \leq 4n^2$. This occurs exactly when $20{,}000n^{5/3} \leq 4n^2 \Leftrightarrow 5000 \leq \dfrac{n^2}{n^{5/3}} \Leftrightarrow n^{1/3} \geq 5000 \Leftrightarrow n \geq 5000^3 = 1.25 \times 10^{11} = 125{,}000{,}000{,}000$.

7. *a.* For each iteration of the loop there is one comparison. The number of iterations of the loop is $n - 2 + 1 = n - 1$. Therefore, the total number of elementary operations that must be performed when the algorithm is executed is $n - 1$.

b. Since $n - 1$ is $O(n)$ [by the theorem on polynomial orders], the algorithm segment has order n.

9. *a.* For each iteration of the inner loop there are two multiplications and one addition. There are $2n$ iterations of the inner loop for each iteration of the outer loop, and there are n iterations of the outer loop. Therefore, the number of iterations of the inner loop is $2n \cdot n = 2n^2$. It follows that the total number of elementary operations that must be performed when the algorithm is executed is $3 \cdot 2n^2 = 6n^2$.

b. Since $6n^2$ is $O(n^2)$ [by the theorem on polynomial orders], the algorithm segment has order n^2.

11. *a.* For each iteration of the inner loop there is one comparison. For the ith iteration of the outer loop there are $n - (i+1) + 1 = n - i$ iterations of the inner loop, and there are $n - 1$ iterations of the outer loop. Therefore, the number of iterations of the inner loop is $(n-1)+(n-2)+\cdots+2+1 = n(n-1)/2$. It follows that the total number of elementary operations that must be performed when the algorithm is executed is $1 \cdot (n(n-1)/2) = \dfrac{1}{2}n^2 - \dfrac{1}{2}n$.

Since $\dfrac{1}{2}n^2 - \dfrac{1}{2}n$ is $O(n^2)$ [by the theorem on polynomial orders], the algorithm segment has order n^2.

13. *a.* For each iteration of the inner loop there are two multiplications. There are n iterations of the inner loop for each iteration of the middle loop; there are $2n$ iterations of the middle loop for each iteration of the outer loop; and there are n iterations of the outer loop. Therefore, there are $n \cdot 2n \cdot n = 2n^3$ iterations of the inner loop, and so the total number of elementary operations that must be performed when the algorithm is executed is $2 \cdot (2n^3) = 4n^3$.

b. Since $4n^4$ is $O(n^4)$ [by the theorem on polynomial orders], the algorithm segment has order n^4.

14. *a.* By the method of Example 6.5.4, the number of iterations of the inner loop is $\dfrac{n(n+1)(2n+1)}{6}$. Since there are two elementary operations (multiplications) for each iteration of the inner loop, the total number of elementary operations that must be performed when the algorithm is executed is $2\left(\dfrac{n(n+1)(2n+1)}{6}\right) = \dfrac{n(n+1)(2n+1)}{3} = \dfrac{1}{3}n^3 + n^2 + \dfrac{2}{3}n$.

b. Since $\frac{1}{3}n^3 + n^2 + \frac{2}{3}n$ is $O(n^3)$ [by the theorem on polynomial orders], the algorithm segment has order n^3.

16.

	a[1]	a[2]	a[3]	a[4]	a[5]
initial order	7	3	6	9	5
result of step 1	3	7	6	9	5
result of step 2	3	6	7	9	5
result of step 3	3	6	7	9	5
result of step 4	3	5	6	7	9

18. 1 *[from step 1]* + 2 *[from step 2]* + 3 *[from step 3]* + 2 *[from step 4]* = 8

19. One such array is $a[1] = 1,\ a[2] = 2,\ a[3] = 3,\ a[4] = 4,\ a[5] = 5$.

21. The result of each interchange is shown in a separate row.

a[1]	a[2]	a[3]	a[4]	a[5]
6	4	5	8	1
4	6	5	8	1
1	6	5	8	4
1	5	6	8	4
1	4	6	8	5
1	4	5	8	6
1	4	5	6	8

23. There are six interchanges, one less than the number of rows in the table constructed for exercise 21.

24. There are six comparisons; three to compare $a[1]$ to $a[2], a[3]$, and $a[4]$, two to compare $a[2]$ to $a[3]$ and $a[4]$, and one to compare $a[3]$ to $a[4]$.

25. a. $n - 1$

    c. $n - (i+1) + 1 = n - i$

    d. When $a[1]$ is compared to $a[2], a[3], \ldots, a[n]$, there are $n - 1$ comparisons. When $a[2]$ is compared to $a[3], a[4], \ldots, a[n]$, there are $n - 2$ comparisons. And so forth. In the second-to-last step, there are two comparisons: $a[n-2]$ is compared to $a[n-1]$ and $a[n]$. And in the final step, there is just one comparison: $a[n-1]$ is compared to $a[n]$. Therefore, the total number of comparisons is

    $$\begin{aligned}(n-1) + (n-2) + \cdots + 2 + 1 &= \frac{(n-1)[(n-1)+1]}{2} \quad \text{by Theorem 4.2.2} \\ &= \frac{n(n-1)}{2} \\ &= \frac{1}{2}n^2 + \frac{1}{2}n.\end{aligned}$$

    But $\frac{1}{2}n^2 + \frac{1}{2}n$ is $O(n^2)$) by the theorem on polynomial orders. So selection sort has order n^2.

26. One such array is $a[1] = 5,\ a[2] = 4,\ a[3] = 3,\ a[4] = 2,\ a[5] = 1$.

27. **Algorithm Insertion Sort**

    [The aim of this algorithm is to take an array of items $a[1], a[2], \ldots, a[n]$ (where $n \geq 1$) and to order it. The output array is also denoted $a[1], a[2], \ldots, a[n]$. It has the same values as the input array, but they are in ascending order. For each $k = 2, 3, \ldots, n$, the value of $a[k]$ is inserted into its correct position in the array $a[1], a[2], \ldots, a[k]$. This is achieved by successively comparing $a[k]$ with $a[i]$ for $i < k$ until a value of i is found for which $a[k] \leq a[i]$. At that point $a[i]$ is goven the vlue of $a[k]$ and each of the values of $a[i+1], a[i+2], \ldots a[k-1]$ is shifted up on position in the array.]

Input: n [a positive integer], $a[1], a[2], \ldots, a[n]$ [an array of data items that can be ordered]

Algorithm Body:
    for $k := 2$ to n
        $stillbigger := 1)$
        while$(i < k$ and $stillbigger = 1)$
            if $a[k] \leq a[i]$ then do
                $stillbigger := 0)$
                $temp := a[k], j := k - 1$
                while$(j \geq i)$
                      $a[j + 1] := a[j]$
                      $j := j - 1$
                end while
            end do
        end while
    next k

Output: $a[1], a[2], \ldots, a[n]$ [the array of data items in ascending order]
end Algorithm

28. **Algorithm   Selection Sort**

[The aim of this algorithm is to take an array of items $a[1], a[2], \ldots, a[n]$ (where $n \geq 1$) and to order it. The output array is also denoted $a[1], a[2], \ldots, a[n]$. It has the same values as the input array, but they are in ascending order. For each $i = 1, 2, \ldots, n - 1$, the value of $a[i]$ is compared to $a[k]$ for each $k = i+1, i+2, \ldots, n$. Each time the value of $a[i]$ is greater than than of $a[k]$, the two values are interchanged.]

Input: n [a positive integer], $a[1], a[2], \ldots, a[n]$ [an array of data items that can be ordered]

Algorithm Body:
    for $i := 1$ to $n - 1$
        for $k := i + 1$ to n
            if $a[k] < a[i]$ then do
                $temp := a[i]$
                $a[i] := a[k]$
                $a[k] := temp$
            end do
        next k
    next i

Output: $a[1], a[2], \ldots, a[n]$ [the array of data items in ascending order]
end Algorithm

30.

n	2				
$a[0]$	5				
$a[1]$	-1				
$a[2]$	2				
x	3				
$polyval$	5		2		20
i	1		2		
$term$	-1	-3	2	6	18
j	1		1	2	

32. By the result of exercise 31, $s_n = \frac{1}{2}n^2 + \frac{3}{2}n$, which is $O(n^2)$ by the theorem on polynomial orders.

34.

n	2		
$a[0]$	5		
$a[1]$	-1		
$a[2]$	2		
x	3		
$polyval$	2	5	20
i	1	2	

35. There are two operations (one addition and one multiplication) per iteration of the loop, and there are n iterations of the loop. Therefore, $t_n = 2n$.

36. By the result of exercise 35, $t_n = 2n$. This is $O(n)$ by the theorem on polynomial orders.

Section 9.4

2.

x	$g(x) = (1/3)^x$
0	1
1	1/3
2	1/9
-1	3
-2	9

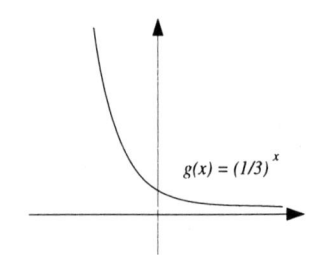

4.

x	$k(x) = \log_2 x$
1	0
2	1
4	2
8	3
1/2	-1
-1/4	-2

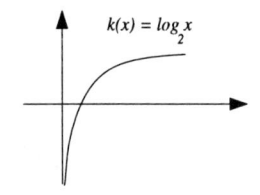

6.

x	$\lceil \log_2 x \rceil$
$1 < x \le 2$	1
$2 < x \le 4$	2
$4 < x \le 8$	3
$1/2 < x \le 0$	0
$1/4 < x \le 1/8$	-1
$1/8 < x \le 1/4$	-2

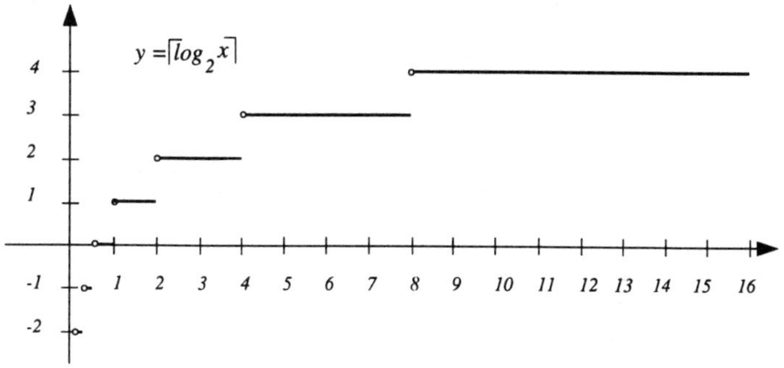

8.

x	$K(x) = x \log_{10} x$
1	0
10	10
20	~ 26.02
30	~ 44.31
1/10	-1/10
1/100	-1/50
1/1000	-3/1000

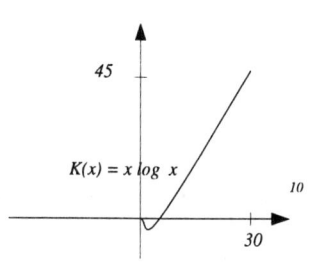

10. a. *Solution 1*: Let b be any positive real number. By definition of logarithm with base b, for any real number x, $\log_b(b^x)$ is the exponent to which b must be raised to obtain b^x. But this exponent is x. So $\log_b(b^x) = x$.

Solution 2: Let $\log_b(b^x) = y$. By definition of logarithm, $b^y = b^x$. It follows from property (7.3.4) that $y = x$. So $\log_b(b^x) = x$.

c. Let $f : \mathbf{R} \to \mathbf{R}^+$ be the exponential function with base b: $f(x) = \exp_b(x) = b^x$ for all real numbers x. Let $g : \mathbf{R}^+ \to \mathbf{R}$ be the logarithmic function with base b: $g(x) = \log_b(x)$ for all positive real numbers x. Then for all $x \in \mathbf{R}$, $(g \circ f)(x) = g(f(x)) = g(b^x) = \log_b(b^x) = x$ by part (a), and for all $x \in \mathbf{R}^+$, $(f \circ g)(x) = f(g(x)) = f(\log_b(x)) = b^{\log_b x} = x$ by part (b). So $f \circ g = i_{\mathbf{R}^+}$ and $g \circ f = i_{\mathbf{R}}$, and hence $g = \log_b$ and $f = \exp_b$ are inverse functions.

11. a. Suppose (u, v) lies on the graph of the logarithmic function with base b. Then by definition, $v = \log_b u$. But by definition of logarithm, this equation is equivalent to $b^v = u$. So (v, u) lies on the graph of the exponential function with base b.

c.

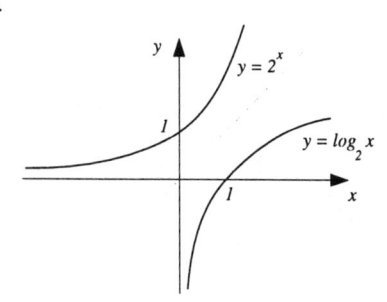

The graphs of $y = 2^x$ and $y = \log_2 x$ are symmetric about the line $y = x$. That is, if the two graphs are drawn on a piece of paper using the same scale on both axes and if the paper is folded along the line $y = x$, then the two graphs will coincide exactly.

12.

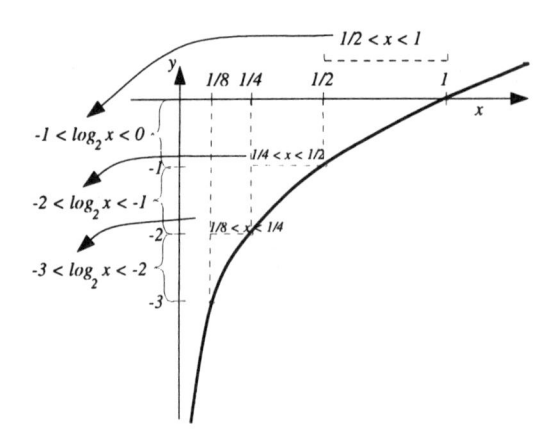

13. If $10^m \leq x < 10^{m+1}$, where m is an integer, then $m = \lfloor \log_{10} x \rfloor$.

    Proof: Suppose that m is an integer and x is a real number with $10^m \leq x < 10^{m+1}$. Because the logarithmic function with base 10 is increasing, this inequality implies that

    $$\log_{10}(10^m) \leq \log_{10} x < \log_{10}(10^{m+1}).$$

    But by definition of logarithm, $\log_{10}(10^m) = m$ and $\log_{10}(10^{m+1}) = m + 1$. Hence

    $$m \leq \log_{10} x < m + 1.$$

    It follows by definition of floor that $m = \lfloor \log_{10} x \rfloor$.

14. *Proof*: Suppose n is a positive integer, k is a nonnegative integer, and $2^k < n \leq 2^{k+1}$. *[We must show that $\lceil \log_2 n \rceil = k + 1$.]* Since the logarithm with base 2 is an increasing function, taking the logarithm with base 2 of all parts of this inequality preserves the directions of the inequality signs. Thus $\log_2(2^k) < \log_2(n) \leq \log_2(2^{k+1})$. But by definition of logarithm, $\log_2(2^k) = k$ and $\log_2(2^{k+1}) = k + 1$. Hence $k < \log_2(n) \leq k + 1$. Since k is an integer, by definition of ceiling, $\lceil \log_2 n \rceil = k + 1$ *[as was to be shown]*.

15. If x is a positive real number that lies strictly between two consecutive integer powers of two, the exponent of the higher power of two is the ceiling of the logarithm with base 2 of x.

17. If n is an odd integer and $n > 1$, then $\lceil \log_2 n \rceil = \lceil \log_2(n+1) \rceil$.

    Proof: Suppose n is an odd integer and $n > 1$. Since n is odd, n is not an integer power of 2 and so n lies strictly between two successive integer powers of 2. In other words, there is an integer k such that $2^k < n < 2^{k+1}$. Consequently, $2^k < n + 1 \leq 2^{k+1}$. By exercise 14, then, $\lceil \log_2 n \rceil = k + 1$ and $\lceil \log_2(n+1) \rceil = k + 1$ also. Hence $\lceil \log_2 n \rceil = \lceil \log_2(n+1) \rceil$.

18. No. *Counterexample*: Let $n = 3$. Then $\lfloor \log_2(n+1) \rfloor = \lfloor \log_2(4) \rfloor = 2$ whereas $\lfloor \log_2 n \rfloor = \lfloor \log_2 3 \rfloor = 1$.

19. *b.* $\lfloor \log_2(5,067,329) \rfloor + 1 = 23$

20. No. If $\log_2 n$ is not an integer, then the two formulas give identical answers. But if $\log_2 n$ is an integer, then $\lfloor \log_2 n \rfloor + 1 = \log_2 n + 1 > \log_2 n = \lceil \log_2 n \rceil$. Consider $n = 4$ for instance. Since $4_{10} = 100_2$, three binary digits are needed to represent n. This agrees with the answer obtained from the formula $\lfloor \log_2 n \rfloor + 1 = \lfloor \log_2 4 \rfloor + 1 = 2 + 1 = 3$. On the other hand, $\lceil \log_2 n \rceil = \lceil \log_2 4 \rceil = \lceil 2 \rceil = 2$, which is too small.

22. *a.*

$$\begin{aligned}
b_1 &= 1 \\
b_2 &= 1
\end{aligned}$$

$$\begin{aligned}
b_3 &= 3b_{\lfloor 3/3 \rfloor} = 3b_1 = 3 \cdot 1 = 3 \\
b_4 &= 3b_{\lfloor 4/3 \rfloor} = 3b_1 = 3 \cdot 1 = 3 \\
b_5 &= 3b_{\lfloor 5/3 \rfloor} = 3b_1 = 3 \cdot 1 = 3 \\
b_6 &= 3b_{\lfloor 6/3 \rfloor} = 3b_2 = 3 \cdot 1 = 3 \\
b_7 &= 3b_{\lfloor 7/3 \rfloor} = 3b_2 = 3 \cdot 1 = 3 \\
b_8 &= 3b_{\lfloor 8/3 \rfloor} = 3b_2 = 3 \cdot 1 = 3 \\
b_9 &= 3b_{\lfloor 9/3 \rfloor} = 3b_3 = 3 \cdot 3 = 3^2
\end{aligned}$$

$$\begin{aligned}
b_{17} &= 3b_{\lfloor 17/3 \rfloor} = 3b_5 = 3 \cdot 3 = 3^2 \\
b_{18} &= 3b_{\lfloor 18/3 \rfloor} = 3b_6 = 3 \cdot 3 = 3^2
\end{aligned}$$

$$\begin{aligned}
b_{26} &= 3b_{\lfloor 26/3 \rfloor} = 3b_8 = 3 \cdot 3 = 3^2 \\
b_{27} &= 3b_{\lfloor 27/3 \rfloor} = 3b_9 = 3 \cdot 3^2 = 3^3
\end{aligned}$$

Guess: $b_n = 3^{\lfloor \log_3 n \rfloor}$

b. Lemma: For any integer $m \geq 1$, $\lfloor \log_3(3m) \rfloor = \lfloor \log_3(3m+1) \rfloor = \lfloor \log_3(3m+2) \rfloor$.

Proof: First note that if x is any real number with $3^a \leq x < 3^{a+1}$, then $\lfloor \log_3(x) \rfloor = a$. The justification for this statement is identical to the proof of property (9.4.2) given in Example 9.4.1 except that the number 2 is everywhere replaced by the number 3.

Now suppose m is an integer with $m \geq 1$. Then for some integer $a \geq 1$, $3^a \leq 3m < 3^{a+1}$. Note that if $a \geq 1$, then $3^a + 2 < 3 \cdot 3^a = 3^{a+1}$. Thus

$$3^a \leq 3m < 3m+1 < 3m+2 < 3^{a+1}.$$

Applying the logarithmic function with base 3 to all parts of these inequalities and using the fact that the logarithmic function with base 3 is increasing gives

$$\log_3(3^a) \leq \log_3(3m) < \log_3(3m+1) < \log_3(3m+2) < \log_3(3^{a+1}).$$

Since $\log_3(3^a) = a$ and $\log_3(3^{a+1}) = a+1$,

$$a \leq \log_3(3m) < \log_3(3m+1) < \log_3(3m+2) < a+1.$$

So by definition of floor, $\lfloor \log_3(3m) \rfloor = \lfloor \log_3(3m+1) \rfloor = \lfloor \log_3(3m+2) \rfloor$.

Proof that the formula derived in part (a) is correct: Suppose the sequence $b_1, b_2, b_3, \ldots$ is defined recursively as follows: $b_1 = 1$, $b_2 = 1$, and $b_k = 3b_{\lfloor k/3 \rfloor}$ for all integers $k \geq 3$. We will show by strong mathematical induction that for all integers $n \geq 1$, $b_n = 3^{\lfloor \log_3 n \rfloor}$.

The formula holds for n = 1 and n = 2: When $n = 1$, $3^{\lfloor \log_3 n \rfloor} = 3^{\lfloor \log_3 1 \rfloor} = 3^0 = 1$, which is the value of b_1. And when $n = 2$, $3^{\lfloor \log_3 n \rfloor} = 3^{\lfloor \log_3 2 \rfloor} = 3^0 = 1$, which is the value of b_2.

If the formula holds for all integers i with $1 \leq i < k$ then it holds for n = k: Let $k > 2$ be an integer and suppose that $b_i = 3^{\lfloor \log_3 i \rfloor}$ for all i with $1 \leq i < k$. *[This is the inductive hypothesis.]* We must show that $b_k = 3^{\lfloor \log_3 k \rfloor}$. But

$$\begin{aligned}
b_k &= 3b_{\lfloor k/3 \rfloor} &&\text{by definition of } b_1, b_2, b_3, \ldots \\
&= 3 \cdot 3^{\lfloor \log_3 \lfloor k/3 \rfloor \rfloor} &&\text{by inductive hypothesis} \\
&= \begin{cases} 3 \cdot 3^{\lfloor \log_3(k/3) \rfloor} & \text{if } k \bmod 3 = 0 \\ 3 \cdot 3^{\lfloor \log_3((k-1)/3) \rfloor} & \text{if } k \bmod 3 = 1 \\ 3 \cdot 3^{\lfloor \log_3((k-2)/3) \rfloor} & \text{if } k \bmod 3 = 2 \end{cases} &&\text{by exercise 17 of Section 3.5} \\
&= \begin{cases} 3 \cdot 3^{\lfloor \log_3 k - \log_3 3 \rfloor} & \text{if } k \bmod 3 = 0 \\ 3 \cdot 3^{\lfloor \log_3(k-1) - \log_3 3 \rfloor} & \text{if } k \bmod 3 = 1 \\ 3 \cdot 3^{\lfloor \log_3(k-2) - \log_3 3 \rfloor} & \text{if } k \bmod 3 = 2 \end{cases} &&\text{by exercise 24 of Section 7.3.} \\
&= \begin{cases} 3 \cdot 3^{\lfloor \log_3 k - 1 \rfloor} & \text{if } k \bmod 3 = 0 \\ 3 \cdot 3^{\lfloor \log_3(k-1) - 1 \rfloor} & \text{if } k \bmod 3 = 1 \\ 3 \cdot 3^{\lfloor \log_3(k-2) - 1 \rfloor} & \text{if } k \bmod 3 = 2 \end{cases} &&\text{because } \log_3 3 = 1 \\
&= \begin{cases} 3 \cdot 3^{\lfloor \log_3 k \rfloor - 1} & \text{if } k \bmod 3 = 0 \\ 3 \cdot 3^{\lfloor \log_3(k-1) \rfloor - 1} & \text{if } k \bmod 3 = 1 \\ 3 \cdot 3^{\lfloor \log_3(k-2) \rfloor - 1} & \text{if } k \bmod 3 = 2 \end{cases} &&\text{by exercise 15 of Section 3.5} \\
&= \begin{cases} 3^{(\lfloor \log_3 k \rfloor - 1) + 1} & \text{if } k \bmod 3 = 0 \\ 3^{(\lfloor \log_3(k-1) \rfloor - 1) + 1} & \text{if } k \bmod 3 = 1 \\ 3^{(\lfloor \log_3(k-2) \rfloor - 1) + 1} & \text{if } k \bmod 3 = 2 \end{cases} &&\text{by the laws of exponents} \\
&= \begin{cases} 3^{\lfloor \log_3 k \rfloor} & \text{if } k \bmod 3 = 0 \\ 3^{\lfloor \log_3(k-1) \rfloor} & \text{if } k \bmod 3 = 1 \\ 3^{\lfloor \log_3(k-2) \rfloor} & \text{if } k \bmod 3 = 2 \end{cases}
\end{aligned}$$

Now if $k \bmod 3 = 1$ and $k \geq 3$, then $k = 3m + 1$ for some integer m, and so by the lemma above, $\lfloor \log_3 k \rfloor = \lfloor \log_3(3m+1) \rfloor = \lfloor \log_3(3m) \rfloor = \lfloor \log_3(k-1) \rfloor$. Similarly, if $k \bmod 3 = 2$ and $k \geq 3$, then by the lemma $\lfloor \log_3 k \rfloor = \lfloor \log_3(k-2) \rfloor$. Hence regardless of the value of $k \bmod 3$, $b_k = 3^{\lfloor \log_3 k \rfloor}$ *[as was to be shown]*.

[Since we have proved both the basis and the inductive steps, we conclude that the given formula holds for all integers $n \geq 1$.]

23. *Proof:* Suppose the sequence $c_1, c_2, c_3, \ldots$ is defined recursively as follows: $c_1 = 0$ and $c_k = 2 \cdot c_{\lfloor k/2 \rfloor} + k$ for all integers $k \geq 2$. Consider the inequality $c_n \leq n^2$. We will show by strong mathematical induction that this inequality is true for all integers $n \geq 1$.

The inequality is true for n = 1: When $n = 1$, $c_1 = 0$, and $0 \leq 1^2$.

If the inequality is true for all integers i with $1 \leq i < k$ then it is true for k: Let k be an integer with $k > 1$ and suppose $c_i \leq i^2$ for all integers i with $1 \leq i < k$. *[This is the inductive hypothesis.]* We must show that $c_k \leq k^2$.

Case 1 (k is even):

$$\begin{aligned}
c_k &= 2 \cdot c_{\lfloor k/2 \rfloor} + k &&\text{by the recursive definition of } c_1, c_2, c_3, \ldots \\
\Rightarrow c_k &= 2 \cdot c_{k/2} + k &&\text{because } k \text{ is even} \\
\Rightarrow c_k &\leq 2 \cdot (k/2)^2 + k &&\text{by inductive hypothesis} \\
\Rightarrow c_k &\leq k^2/2 + k^2/2 &&\text{because } 2 \cdot (k/2)^2 = k^2/2, \text{ and since} \\
& &&k \geq 2, \text{ then } k^2 \geq 2k \text{ and so } k^2/2 \geq k \\
\Rightarrow c_k &\leq k^2 &&\text{by algebra.}
\end{aligned}$$

Case 2 (k is odd):

$$\begin{aligned}
c_k &= 2 \cdot c_{\lfloor k/2 \rfloor} + k && \text{by the recursive definition of } c_1, c_2, c_3, \ldots \\
\Rightarrow c_k &= 2 \cdot c_{(k-1)/2} + k && \text{by Theorem 3.5.2 and because } k \text{ is odd} \\
\Rightarrow c_k &\leq 2 \cdot ((k-1)/2)^2 + k && \text{by inductive hypothesis} \\
\Rightarrow c_k &\leq (k-1)^2/2 + k^2/2 && \text{because } 2 \cdot ((k-1)/2)^2 = (k-1)^2/2, \text{ and since} \\
& && k \geq 2, \text{ then } k^2 \geq 2k \text{ and so } k^2/2 \geq k \\
\Rightarrow c_k &\leq k^2/2 + k^2/2 && \text{because } (k-1)^2/2 \leq k^2/2 \\
\Rightarrow c_k &\leq k^2 && \text{by algebra.}
\end{aligned}$$

Thus in either case, $c_k \leq k^2$ *[as was to be shown]*.

[Since we have proved both the basis and inductive steps, we conclude that the inequality holds for all integers $n \geq 1$.]

24. *Proof:* Suppose the sequence $c_1, c_2, c_3, \ldots$ is defined recursively as follows: $c_1 = 0$ and $c_k = 2 \cdot c_{\lfloor k/2 \rfloor} + k$ for all integers $k \geq 2$. Consider the inequality $c_n \leq n \cdot \log_2 n$. We will show by strong mathematical induction that this inequality is true for all integers $n \geq 1$.

The inequality is true for n = 1: When $n = 1$, $c_1 = 0$ and $1 \cdot \log_2 1 = 0$. So $c_1 \leq 1 \cdot \log_2 1$ and the inequality is true for $n = 1$.

If the inequality is true for all integers i with $1 \leq i < k$ then it is true for k: Let k be an integer with $k > 1$ and suppose $c_i \leq i \cdot \log_2 i$ for all integers i with $1 \leq i < k$. *[This is the inductive hypothesis.]* We must show that $c_k \leq k \cdot \log_2 k$.

Case 1 (k is even):

$$\begin{aligned}
c_k &= 2 \cdot c_{\lfloor k/2 \rfloor} + k && \text{by the recursive definition of } c_1, c_2, c_3, \ldots \\
\Rightarrow c_k &= 2 \cdot c_{k/2} + k && \text{by Theorem 3.5.3 and because } k \text{ is even} \\
\Rightarrow c_k &\leq 2 \cdot [(k/2) \cdot \log_2(k/2)] + k && \text{by inductive hypothesis} \\
\Rightarrow c_k &\leq k(\log_2 k - \log_2 2) + k && \text{by algebra and the identity } \log_b(x/y) = \log_b x - \log_b y \\
& && \text{(exercise 24 of Section 7.3)} \\
\Rightarrow c_k &\leq k(\log_2 k - 1) + k && \text{because } \log_2 2 = 1 \\
\Rightarrow c_k &\leq k \cdot \log_2 k && \text{by algebra.}
\end{aligned}$$

Case 2 (k is odd):

$$\begin{aligned}
c_k &= 2 \cdot c_{\lfloor k/2 \rfloor} + k & &\text{by the recursive definition} \\
& & & \text{of } c_1, c_2, c_3, \ldots \\
\Rightarrow c_k &= 2 \cdot c_{(k-1)/2} + k & &\text{because } k \text{ is odd} \\
\Rightarrow c_k &\leq 2 \cdot \left(\frac{k-1}{2} \cdot \log_2\left(\frac{k-1}{2}\right)\right) + k & &\text{by inductive hypothesis} \\
\Rightarrow c_k &\leq (k-1) \cdot (\log_2(k-1) - \log_2 2) + k & &\text{by algebra and the identity} \\
& & & \log_b(x/y) = \log_b x - \log_b y \\
& & & (\text{exercise 24 of Section 7.3}) \\
\Rightarrow c_k &\leq (k-1) \cdot (\log_2(k-1) - 1) + k & &\text{because } \log_2 2 = 1 \\
\Rightarrow c_k &\leq (k-1) \cdot \log_2(k-1) - (k-1) + k & &\text{by algebra} \\
\Rightarrow c_k &\leq (k-1) \cdot \log_2(k-1) + 1 & &\text{by algebra} \\
\Rightarrow c_k &\leq (k-1) \cdot \log_2(k-1) + \log_2(k-1) & &\text{because } k \geq 3 \text{ and so,} \\
& & & \text{since the logarithmic function with base 2} \\
& & & \text{is increasing, } \log_2(k-1) \geq \log_2 2 = 1 \\
\Rightarrow c_k &\leq ((k-1)+1) \cdot \log_2(k-1) & &\text{by factoring out } \log_2(k-1) \\
\Rightarrow c_k &\leq k \cdot \log_2(k-1) & &\text{by algebra} \\
\Rightarrow c_k &\leq k \cdot \log_2 k & &\text{because the logarithmic} \\
& & & \text{function with base 2} \\
& & & \text{is increasing and } k-1 < k.
\end{aligned}$$

Thus in either case, $c_k \leq k \cdot \log_2 k$ *[as was to be shown]*.

[Since we have proved both the basis and inductive steps, we conclude that the inequality holds for all integers $n \geq 1$.]

26. With some computer graphing programs (but not most graphing calculators) it is possible to find the point of intersection of $y = 2^x$ and $y = x^{50}$ by making the viewing window include very large values of y. However, because the values of 2^x and of x^{50} are so large in the region where the two are equal, probably the easiest way to solve this problem is to use logarithms. Note that $2^x > x^{50} \Leftrightarrow \log_2(2^x) > \log_2(x^{50}) \Leftrightarrow x > 50\log_2 x$. By computing values of x and of $50\log_2 x$ for various values of x or by using a graphing calculator or computer graphing program to graph $y = x$ and $y = 50\log_2 x$, one finds that the given inequality holds for values of x greater than approximately 438.884. So one answer would be $x = 440$.

    Another approach would involve numerical exploration using properties of exponents. For instance, if $x = 2^{10}$, then $2^x = 2^{(2^{10})} = 2^{1024}$ whereas $x^{50} = (2^{10})^{50} = 2^{500}$. So $x = 2^{10} = 1024$ would be another possible answer.

28. The values of x for which $x = 1.0001^x$ are approximately 1.0001 and 116677.5257. Furthermore, $x > 1.0001^x$ on the approximate interval $1.0001 < x < 1116677.5257$ and $x < 1.0001^x$ on the approximate intervals $1.0001 > x$ and $x > 1116677.5257$.

31. It is clear from the graphs of $y = \log_2 x$ and $y = x$ that $\log_2 x < x$ for all $x > 0$. So if $x > 0$, we may multiply both sides of $\log_2 x < x$ by $5x$ to obtain $5x \log_2 x < 5x \cdot x = 5x^2$. Adding x^2 to both sides gives $x^2 + 5x \log_2 x < x^2 + 5x^2 = 6x^2$. If $x > 1$, then all quantities are positive, and so $|x^2 + 5x \log_2 x| < 6|x^2|$. So let $M = 6$ and $x_0 = 1$. Then $|x^2 + 5x \log_2 x| < M|x^2|$ for all $x > x_0$. Thus by definition of O-notation, $x^2 + 5x \log_2 x$ is $O(x^2)$.

33. By factoring out an n and using the formula for the sum of a geometric sequence (Theorem 4.2.3),

$$n + \frac{n}{2} + \frac{n}{4} + \ldots + \frac{n}{2^n} = n(1 + \frac{1}{2} + \frac{1}{4} + \ldots + \frac{1}{2^n})$$

$$= n\left(\frac{(\frac{1}{2})^{n+1} - 1}{\frac{1}{2} - 1}\right)$$

$$= n\left(\frac{1 - 2^{n+1}}{-2^n}\right)$$

$$= n(2 - \frac{1}{2^n})$$

$$\leq 2n$$

for all integers $n \geq 1$. Let $M = 2$ and $x_0 = 1$. Then $\left|n + \frac{n}{2} + \frac{n}{4} + \ldots + \frac{n}{2^n}\right| \leq M \cdot |n|$ for all $n > x_0$. So by definition of O-notation, $n + \frac{n}{2} + \frac{n}{4} + \ldots + \frac{n}{2^n}$ is $O(n)$.

34. By factoring out $\frac{2n}{3}$ and using the formula for the sum of a geometric sequence (Theorem 4.2.3),

$$\frac{2n}{3} + \frac{2n}{3^2} + \frac{2n}{3^3} + \ldots + \frac{2n}{3^n} = \frac{2n}{3}(1 + \frac{1}{3} + \frac{1}{3^2} + \ldots + \frac{1}{3^{n-1}})$$

$$= \frac{2n}{3}\left(\frac{(\frac{1}{3})^n - 1}{\frac{1}{3} - 1}\right)$$

$$= \frac{2n}{3}\left(\frac{1 - 3^n}{(-2) \cdot 3^{n-1}}\right)$$

$$= 2n\left(\frac{1 - 3^n}{(-2) \cdot 3^n}\right)$$

$$= n\left(\frac{3^n - 1}{3^n}\right)$$

$$\leq n$$

for all integers $n \geq 1$. Let $M = 1$ and $x_0 = 1$. Then $\left|\frac{2n}{3} + \frac{2n}{3^2} + \frac{2n}{3^3} + \ldots + \frac{2n}{3^n}\right| \leq M \cdot |n|$ for all $n > 1$. So by definition of O-notation, $\frac{2n}{3} + \frac{2n}{3^2} + \frac{2n}{3^3} + \ldots + \frac{2n}{3^n}$ is $O(n)$.

35. Let k be any positive integer. If $n > 2$, then $1 < \log_2 n$ because the logarithmic function with base 2 is increasing and $\log_2 2 = 1$. Multiplying both sides of $1 < \log_2 n$ by kn (which is positive) gives $kn < kn \log_2 n$. Adding $kn \log_2 n$ to both sides gives $kn + kn \log_2 n < 2kn \log_2 n$. If $n > 2$, then all quantities are positive, and we have $|kn + kn \log_2 n| \leq 2k|n \log_2 n|$. Let $M = 2k$ and $x_0 = 2$. Then $|kn + kn \log_2 n| \leq M \cdot |n \log_2 n|$ for all $n > 2$. So by definition of O-notation, $kn + kn \log_2 n$ is $O(n \log_2 n)$.

37. $1 + \frac{1}{2} = \frac{3}{2}$, $1 + \frac{1}{2} + \frac{1}{3} = \frac{11}{6}$, $1 + \frac{1}{2} + \frac{1}{3} + \frac{1}{4} = \frac{50}{24} = \frac{25}{12}$, $1 + \frac{1}{2} + \frac{1}{3} + \frac{1}{4} + \frac{1}{5} = \frac{137}{60}$

39. By Example 9.4.7(c), $1 + \frac{1}{2} + \frac{1}{3} + \ldots + \frac{1}{n} \leq 2 \cdot \ln n$ for all integers $n \geq 3$. But by (7.3.6), $\log_2 n = \frac{\log_e n}{\log_e 2} = \frac{\ln n}{\ln 2}$, or, equivalently, $\ln n = (\ln 2)(\log_2 n)$. Substituting for $\ln n$ in the above inequality gives $1 + \frac{1}{2} + \frac{1}{3} + \ldots + \frac{1}{n} \leq 2 \cdot (\ln 2)(\log_2 n)$. Let $M = 2 \cdot \ln 2$ and $x_0 = 3$. Then

$$\left|1 + \frac{1}{2} + \frac{1}{3} + \ldots + \frac{1}{n}\right| \leq M \cdot |\log_2 n| \quad \text{for all integers } n > x_0.$$

So by definition of O-notation, $1 + \frac{1}{2} + \frac{1}{3} + \ldots + \frac{1}{n}$ is $O(\log_2 n)$.

40. a. *Proof:* If n is any positive integer, then $\log_2 n$ is defined and by definition of floor, $\lfloor \log_2 n \rfloor \leq \log_2 n$. If, in addition, n is greater than 1, then since the logarithmic function with base 2 is increasing $\log_2 n > 0$, and so, letting $M = 1$ and $x_0 = 1$, we have $|\lfloor \log_2 n \rfloor| \leq M \cdot |\log_2 n|$ for all integers $n > x_0$. Thus by definition of O-notation, $\lfloor \log_2 n \rfloor$ is $O(\log_2 n)$.

b. *Proof:* If n is any positive real number, then $\log_2 n$ is defined and by definition of floor, $\lfloor \log_2 n \rfloor \leq \log_2 n$. If, in addition, n is greater than 2, then since the logarithmic function with base 2 is increasing $\log_2 n > \log_2 2 = 1$. Adding the inequalities $\lfloor \log_2 n \rfloor \leq \log_2 n$ and $1 < \log_2 n$ gives $\lfloor \log_2 n \rfloor + 1 \leq 2\log_2 n$. Thus if we let $M = 2$ and $x_0 = 2$ and use the fact that for $n > 2 \log_2 n$ is positive, we have $|\lfloor \log_2 n \rfloor + 1| \leq M \cdot |\log_2 n|$ for all integers $n > x_0$. It follows by definition of O-notation that $\lfloor \log_2 n + 1 \rfloor$ is $O(\log_2 n)$.

41. a. *Proof:* Suppose n is a variable that takes positive integer values. Then

$$n! = \underbrace{n \cdot (n-1) \cdot (n-2) \cdot \ldots \cdot 2 \cdot 1}_{n \text{ factors}} \leq \underbrace{n \cdot n \cdot n \cdot n \cdot \ldots \cdot n}_{n \text{ factors}} = n^n.$$

because $(n-1) \leq n$, $(n-2) \leq n, \ldots$, and $1 \leq n$. Let $M = 1$ and $x_0 = 1$. It follows from the displayed inequality and the fact that $n!$ and n^n are positive that $|n!| \leq M \cdot |n^n|$ for all integers $n > x_0$. Hence by definition of O-notation $n!$ is $O(n^n)$.

b. By part (a), for all integers $n \geq 1$, $n! \leq n^n$. Since the logarithmic function with base 2 is increasing, we have $\log_2(n!) \leq \log_2(n^n) = n \log_2(n)$. Let $M = 1$ and $x_0 = 1$. Then $|\log_2(n!)| \leq M \cdot |n \log_2(n)|$ for all $n > x_0$, and so by definition of O-notation $\log_2(n!)$ is $O(n \log_2 n)$.

42. *Proof:* Suppose n is a variable that takes positive integer values. For all $n \geq 2$,

$$2^n = \underbrace{2 \cdot 2 \cdot 2 \cdot \ldots \cdot 2}_{n \text{ factors}} \leq \underbrace{2 \cdot 2 \cdot 3 \cdot 4 \cdot \ldots \cdot n}_{n \text{ factors}} = 2n!.$$

Let $M = 2$ and $x_0 = 2$. Since 2^n is positive for all n, $|2^n| \leq M \cdot |n!|$ for all integers $n > 2$. Hence by definition of O-notation 2^n is $O(n!)$.

44. *Proof (by mathematical induction):*

The inequality is true for n = 1: When $n = 1$ the inequality is $1 \leq 10^1$, which is true.

If the inequality is true for n = k then it is true for n = k+1: Let k be an integer with $k \geq 1$ and suppose that $k \leq 10^k$. *[This is the inductive hypothesis.]* We must show that $k+1 \leq 10^{k+1}$. By inductive hypothesis, $k \leq 10^k$. Adding 1 to both sides gives $k+1 \leq 10^k + 1$. But $10^k + 1 \leq 10^k + 9 \cdot 10^k = 10 \cdot 10^k = 10^{k+1}$. Hence $k+1 \leq 10^{k+1}$ *[as was to be shown]*.

45. a. *Lemma:* For all integers $n \geq 1$, $(1 + \frac{1}{n})^n < 3$.

Proof: Consider the function F defined by the rule $F(x) = (1 + \frac{1}{x})^x = e^{x \ln(1 + \frac{1}{x})}$ for all real numbers $x \geq 1$. The following argument shows that F is increasing for all $x \geq 1$. First note that

$$F'(x) = (1 + \frac{1}{x})^x (x \cdot \frac{1}{1 + \frac{1}{x}} \cdot (-\frac{1}{x^2}) + \ln(1 + \frac{1}{x})) = (1 + \frac{1}{x})^x (-\frac{1}{x+1} + \ln(1 + \frac{1}{x})).$$

Thus

$$F(x) \Leftrightarrow \ln(1 + \frac{1}{x}) > \frac{1}{x+1}$$

$$\Leftrightarrow \ln(x+1) - \ln(x) > \frac{1}{x+1}$$

$$\Leftrightarrow \int_x^{x+1} \frac{1}{t} dt > \frac{1}{x+1}.$$

But this last inequality is true because when $x \geq 1$ then $\dfrac{1}{t} \geq \dfrac{1}{x+1}$ for all t with $x \leq t \leq x+1$. Hence $F'(x) > 0$ for all x with $x \geq 1$, and so F is increasing for all $x \geq 1$.

Now from the study of calculus we know that $\lim_{n \to \infty} ((1 + \dfrac{1}{n})^n) = e < 3$. In addition, it follows from the argument in the paragraph above that numbers of the form $(1 + \dfrac{1}{n})^n$ form an increasing sequence for $n = 1, 2, 3, \ldots$. Hence $(1 + \dfrac{1}{n})^n < 3$ for all integers $n \geq 1$.

Theorem: For all integers $n \geq 2$, $n \log_2 n \leq 2 \log_2(n!)$.

Proof (by mathematical induction):

The inequality is true for n = 2: We must show that $2 \log_2 2 \leq 2 \log_2(2!)$. By properties of logarithms (see exercise 26 of Section 7.3) and since the logarithmic function with base 2 is increasing, this inequality is true if, and only if, $2^2 \leq (2!)^2$, or equivalently, $4 \leq 4$, which is a true statement. Hence the inequality holds for $n = 2$.

If the inequality is true for n = k then it is true for n = k + 1: Let k be an integer with $k \geq 2$ and suppose that $k \log_2 k \leq 2 \log_2(k!)$. *[This is the inductive hypothesis.]* We must show that $(k+1) \log_2(k+1) \leq 2 \log_2((k+1)!)$. But

$$\begin{aligned}
(k+1) \log_2(k+1) &= k \log_2(k+1) + \log_2(k+1) \\
&= k \log_2(\dfrac{k+1}{k} \cdot k) + \log_2(k+1) \\
&= k[\log_2(1 + \dfrac{1}{k}) + \log_2 k] + \log_2(k+1) \\
&= k \log_2(1 + \dfrac{1}{k}) + k \log_2 k + \log_2(k+1) \\
&= \log_2\left((1 + \dfrac{1}{k})^k\right) + k \log_2 k + \log_2(k+1) \\
&\leq \log_2 3 + k \log_2 k + \log_2(k+1)
\end{aligned}$$

[by the lemma above] $\leq \log_2 3 + 2 \log_2(k!) + \log_2(k+1)$ *[by inductive hypothesis]* $= \log_2 3 + 2(\log_2(k!) + \log_2(k+1)) - \log_2(k+1) = \log_2 3 + 2 \log_2((k+1)!) - \log_2(k+1) = 2 \log_2((k+1)!) - (\log_2(k+1) - \log_2 3) \leq 2 \log_2((k+1)!)$ *[because when $k \geq 2$, $k+1 \geq 3$ and so $\log_2(k+1) \geq \log_2 3$ since the logarithmic function with base 2 is increasing]*. *[This is what was to be shown.]*

[Since we have proved the basis and inductive steps, we conclude that the theorem is true.]

b. Let $M = 2$ and $x_0 = 2$. By part (a), $|n \log_2 n| \leq M \cdot |\log_2(n!)|$ for all integers $n \geq x_0$. Hence by definition of O-notation, $n \log_2 n$ is $O(\log_2(n!))$.

46. *Theorem*: For all integers $n \geq 4$, $\log_2(n!) \geq \dfrac{1}{4}(n \log_2 n)$.

    Proof 1: We first prove that $i(n - i + 1) \geq n$ for all integers n and i with $1 \leq i \leq n$. Suppose n is an integer with $n \geq 1$. Let x be a real variable and let $y = x(n - x + 1) - n = -x^2 + (n+1)x - n$. The graph of this equation is a parabola that opens downwards *[because the coefficient of x^2 is negative]*, and by inspection its zeros occur at $x = 1$ and $x = n$. Consequently, $x(n - x + 1) - n \geq 0$ for all x with $1 \leq x \leq n$. It follows that $i(n - i + 1) \geq n$ for all integers i with $1 \leq i \leq n$.

    We next show that for all integers $n \geq 1$, $n! \geq n^{n/2}$. Suppose n is an integer with $n \geq 1$. If n is even then

$$n! = [n \cdot 1] \cdot [(n-1) \cdot 2] \cdot [(n-2) \cdot 3] \cdot \ldots \cdot \left[\left(n - \left(\dfrac{n}{2} - 1\right)\right) \cdot \left(\dfrac{n}{2} - 1\right)\right] \geq \underbrace{n \cdot n \cdot n \cdot \ldots \cdot n}_{n/2 \text{ factors}} = n^{n/2}.$$

If n is odd, then

$$n! = [n \cdot 1] \cdot [(n-1) \cdot 2] \cdot [(n-2) \cdot 3] \cdot \ldots \cdot \left[\left(n - \left(\frac{n-1}{2} - 1\right)\right) \cdot \left(\frac{n-1}{2}\right)\right] \cdot \left(\frac{n+1}{2}\right)$$

$$\geq \underbrace{n \cdot n \cdot n \cdot \ldots \cdot n}_{(n-1)/2 \text{ factors}} \cdot \frac{n+1}{s^2}$$

$$\geq n^{(n-1)/2} \cdot \left(\frac{n+1}{2}\right) = n^{n/2} \cdot \left(\frac{n+1}{\sqrt{n} \cdot 2}\right)$$

$$\geq n^{n/2}$$

[because

$$\frac{n+1}{\sqrt{n} \cdot 2} \geq 1 \Leftrightarrow n+1 \geq 2\sqrt{n} \Leftrightarrow n^2 + 2n + 1 \geq 4n \Leftrightarrow n^2 - 2n + 1 \geq 0 \Leftrightarrow (n-1)^2 \geq 0$$

which is true]. Hence for all integers $n \geq 1$, $n! \geq n^{n/2}$.

Since the logarithmic function with base 2 is increasing for all positive real numbers, we may take the logarithm with base 2 of both sides of the above inequality to obtain

$$\log_2(n!) \geq \log_2(n^{n/2}) = \frac{n}{2} \log_2 n \geq \frac{1}{4}(n \log_2 n)$$

for all integers $n \geq 1$. So by definition of O-notation, $n \log_2 n$ is $O(\log(n!))$.

Proof 2: Let n be any integer with $n \geq 4$. Then

$$n! \geq \underbrace{n \cdot (n-1) \cdot (n-2) \cdot \ldots \cdot \lceil n/2 \rceil}_{\lfloor n/2 \rfloor + 1 \text{ factors}}.$$

Since each of these factors is at least $\lceil n/2 \rceil$, $n! \geq (\lceil n/2 \rceil)^{\lfloor n/2 \rfloor + 1} \geq (n/2)^{n/2}$ [because $\lceil n/2 \rceil \geq n/2$ and $\lfloor n/2 \rfloor + 1 \geq n/2$ for all integers $n \geq 1$]. Taking the logarithm with base 2 of both sides gives $\log_2(n!) \geq \log_2((n/2)^{n/2}) = \frac{n}{2}\log_2(n/2) \geq \frac{n}{2}(\log_2 n - 1) \geq \frac{n}{4}\log_2 n$ for all integers $n \geq 4$ [because if $n \geq 4$, then $\log_2 n \geq \log_2 4 = 2$, and so $\frac{n}{4}\log_2 n \geq \frac{n}{2}$, which implies that $\frac{n}{2}(\log_2 n - 1) \geq \frac{n}{4}\log_2 n$]. Let $M = 4$ and $x_0 = 4$. Then $|n \log_2 n| \leq M \cdot |\log(n!)|$ for all $n \geq x_0$, and so by definition of O-notation $n \log_2 n$ is $O(\log(n!))$.

47. By the binomial theorem with $a = 1$ and $b = r$,

$$(1+r)^n = 1 + \binom{n}{1}r + \binom{n}{2}r^2 + \cdots + \binom{n}{n-1}r^{n-1} + \binom{n}{n}r^n = 1 + nr + \text{other positive terms}.$$

Therefore, since $r \geq 1$, $(1+r)^n > 1 + n + \text{other positive terms} > n$.

49. a. Let n be a positive integer. Then for any real number x [because $u < 2^u$ for all real numbers u],

$$\frac{x}{n} < 2^{x/n} \Rightarrow x < n2^{x/n} \Rightarrow x^n < (n2^{x/n})^n = n^n \cdot 2^x.$$

So $x^n < n^n 2^x$.

51. Let n be a positive integer and x a real number with $x > (2n)^{2n}$. By exercise 50, $\log_2 x < x^{1/m}$ for all positive integers m. But if $n \geq 2$, then $x^{1/n} < x^{1/2}$ (by exercise 20 of Section 9.1). So, in particular,

$$\log_2 x < x^{1/2}.$$

But since $x > (2n)^{2n}$, then by properties of inequalities and exercise 20 of Section 9.1,

$$x > n^2 \Rightarrow \sqrt{x} > n \Rightarrow \frac{1}{n}\sqrt{x} > 1 \Rightarrow \frac{1}{n}\sqrt{x} \cdot \sqrt{x} > 1 \cdot \sqrt{x} \Rightarrow \frac{1}{n}x > x^{1/2}.$$

Putting the inequalities $\log_2 x < x^{1/2}$ and $x^{1/2} < \frac{1}{n}x$ together gives

$$\log_2 x < \frac{1}{n}x.$$

Applying the exponential function with base 2 to both sides results in

$$2^{\log_2 x} < 2^{\frac{1}{n}x} \Rightarrow x < (2^x)^{1/n} \Rightarrow x^n < 2^x.$$

53. a. Let b be a real number with $b > 1$, and let n be any integer with $n \geq 1$. By L'Hôpital's rule,

$$\lim_{x \to \infty} \frac{\log_b x}{x^{1/n}} = \lim_{x \to \infty} \frac{\frac{\ln x}{\ln b}}{x^{1/n}} = \lim_{x \to \infty} \frac{\frac{1}{x \ln b}}{\frac{1}{n} x^{1/n - 1}}$$

$$= \frac{n}{\ln b} \lim_{x \to \infty} \frac{\frac{1}{x}}{(x^{1/n})\frac{1}{x}} = \frac{n}{\ln b} \cdot \lim_{x \to \infty} \frac{1}{x^{1/n}} = \frac{n}{\ln b} \cdot 0 = 0.$$

b. By the result of part (a) and the definition of limit, given any real number ϵ, say $\epsilon = 1$, there exists a real number N so that

$$\left| \frac{\log_b x}{x^{1/n}} - 0 \right| < \epsilon = 1 \quad \text{for all } x > N.$$

It follows that

$$|\log_b x| < |x^{1/n}| \quad \text{for all } x > N.$$

Let $M = 1$ and $x_0 = N$. Then

$$|\log_b x| < M|x^{1/n}| \quad \text{for all } x > x_0,$$

and so by definition of O-notation, $\log_b x$ is $O(x^{1/n})$.

Section 9.5

2. b. If $m = 2^k$, where k is a positive integer, then the algorithm requires $c\lfloor \log_2(2^k) \rfloor = c\lfloor k \rfloor = ck$ operations. If the input size is increased to $m^{10} = (2^k)^{10} = 2^{10k}$, then the number of operations required is $c\lfloor \log_2(2^{10k}) \rfloor = c\lfloor 10k \rfloor = c \cdot 10k = 10 \cdot ck$. So in going from ck to $10 \cdot ck$, the number of operations increases by a factor of 10.

c. When the input size is increased from 2^7 to 2^{28}, the factor by which the number of operations increases is $\dfrac{c\lfloor \log_2(2^{28}) \rfloor}{c\lfloor \log_2(2^7) \rfloor} = \dfrac{28c}{7c} = 4.$

4. To answer this question, we need to find where the graph of $y = \lfloor n^2/10 \rfloor$ crosses above the graph of $y = \lfloor n \log_2 n \rfloor$. A little numerical exploration reveals that when $n = 20$, $\lfloor n^2/10 \rfloor = 40$ and $\lfloor n \log_2 n \rfloor = 86$, and thus for $n = 20$, $\lfloor n^2/10 \rfloor < \lfloor n \log_2 n \rfloor$. However, for $n = 100$, $\lfloor n^2/10 \rfloor = 1000$ and $\lfloor n \log_2 n \rfloor = 664$, and thus for $n = 100$, $\lfloor n^2/10 \rfloor > \lfloor n \log_2 n \rfloor$. So to find the crossing point of the two graphs, we can use an initial window going from $n = 20$ to $n = 100$ and from $y = 40$ to $y = 1000$. Zooming in shows that the crossing point occurs at approximately $n = 58$. Indeed, for $n = 58$, $\lfloor n^2/10 \rfloor = 336 < 339 = \lfloor n \log_2 n \rfloor$, and for $n = 59$, $\lfloor n^2/10 \rfloor = 348 > 347 = \lfloor n \log_2 n \rfloor$. So if $n \leq 58$, an algorithm that requires $\lfloor n^2/10 \rfloor$ operations is more efficient than an algorithm that requires $\lfloor n \log_2 n \rfloor$ operations.

6. *a.*

index	0			
bot	1	1	1	1
top	10	4	1	0
mid	5	2	1	

b.

index	0		8
bot	1	6	
top	10		
mid		5	8

7. *c.* Suppose there is an even number of elements in the array $a[bot], a[bot+1], \ldots, a[top]$. Then $top - bot + 1$ is an even number, and so $top - bot + 1 = 2k$ for some integer k. Solving for top gives $top = 2k + bot - 1$, and hence $bot + top = bot + (2k + bot - 1) = 2 \cdot bot + 2k - 2 + 1 = 2(bot + k - 1) + 1$, which is odd because $bot + k - 1$ is an integer.

12.

| n | 424 | 141 | 47 | 15 | 5 | 1 | 0 |

13. For each integer $k \geq 3$, $n \ div\ 3 = \lfloor k/3 \rfloor$. Thus when the algorithm segment is run for a particular k and the **while** loop has iterated one time, the input to the next iteration is $\lfloor k/3 \rfloor$. It follows that the number of iterations of the loop for k is one more than the number of iterations for $\lfloor k/3 \rfloor$. That is, $b_k = 1 + b_{\lfloor k/3 \rfloor}$ for all $k \geq 3$. Also $b_1 = 1$ and $b_2 = 1$ because $\lfloor 1/3 \rfloor = 0$ and $\lfloor 2/3 \rfloor = 0$, and so when k equals 1 or 2, the **while** loop iterates just one time.

14.
$$b_1 = 1$$
$$b_2 = 1$$
$$b_3 = 1 + b_{\lfloor 3/3 \rfloor} = 1 + b_1 = 1 + 1 = 2$$
$$b_4 = 1 + b_{\lfloor 4/3 \rfloor} = 1 + b_1 = 1 + 1 = 2$$
$$b_5 = 1 + b_{\lfloor 5/3 \rfloor} = 1 + b_1 = 1 + 1 = 2$$
$$b_6 = 1 + b_{\lfloor 6/3 \rfloor} = 1 + b_2 = 1 + 1 = 2$$
$$b_7 = 1 + b_{\lfloor 7/3 \rfloor} = 1 + b_2 = 1 + 1 = 2$$
$$b_8 = 1 + b_{\lfloor 8/3 \rfloor} = 1 + b_2 = 1 + 1 = 2$$
$$b_9 = 1 + b_{\lfloor 9/3 \rfloor} = 1 + b_3 = 1 + 2 = 3$$

.
.
.

$$b_{26} = 1 + b_{\lfloor 26/3 \rfloor} = 1 + b_8 = 1 + 2 = 3$$
$$b_{27} = 1 + b_{\lfloor 27/3 \rfloor} = 1 + b_9 = 1 + 3 = 4$$

.
.
.

Guess: If n satisfies the inequality $3^r \leq n < 3^{r+1}$ for some integer $r \geq 0$, then $b_n = 1 + r$. In this case $r \leq \log_3 n < r + 1$, and so $r = \lfloor \log_3 n \rfloor$ and the formula is

$$b_n = 1 + \lfloor \log_3 n \rfloor$$

for all integers $n \geq 1$.

To prove the correctness of this formula, we first establish the following lemma.

Lemma: For any integer $m \geq 1$, $\lfloor \log_3(3m) \rfloor = \lfloor \log_3(3m+1) \rfloor = \lfloor \log_3(3m+2) \rfloor$.

Proof: First note that if x is any real number with $3^a \leq x < 3^{a+1}$, then $\lfloor \log_3(x) \rfloor = a$. The justification for this statement is identical to the proof of property (9.4.2) given in Example 9.4.1 except that the number 2 is everywhere replaced by the number 3.

Now suppose m is an integer with $m \geq 1$. Then for some integer $a \geq 1$, $3^a \leq 3m < 3^{a+1}$. Note that if $a \geq 1$, then $3^a + 2 < 3 \cdot 3^a = 3^{a+1}$. Thus

$$3^a \leq 3m < 3m+1 < 3m+2 < 3^{a+1}.$$

Applying the logarithmic function with base 3 to all parts of these inequalities and using the fact that the logarithmic function with base 3 is increasing gives

$$\log_3(3^a) \leq \log_3(3m) < \log_3(3m+1) < log_3(3m+2) < log_3(3^{a+1}).$$

Since $\log_3(3^a) = a$ and $\log_3(3^{a+1}) = a+1$,

$$a \leq \log_3(3m) < \log_3(3m+1) < log_3(3m+2) < a+1.$$

So by definition of floor, $\lfloor \log_3(3m) \rfloor = \lfloor \log_3(3m+1) \rfloor = \lfloor \log_3(3m+2) \rfloor$.

Proof that the formula for b_n is correct:

The formula holds for n = 1 and n = 2: By part (a) we know that $b_1 = b_2 = 1$. The formula gives $1 + \lfloor \log_3 1 \rfloor = 1 + 0 = 1$ and $1 + \lfloor \log_3 2 \rfloor = 1 + 0 = 1$ respectively, which agree.

If the formula holds for all i with $1 \leq i < k$ then it holds for i = k: Let $k > 2$ be an integer and suppose that $b_i = 1 + \lfloor \log_3 i \rfloor$ for all i with $1 \leq i < k$. *[This is the inductive hypothesis.]* We must show that $b_k = 1 + \lfloor \log_3 k \rfloor$. But

$$
\begin{aligned}
b_k &= 1 + b_{\lfloor k/3 \rfloor} &&\text{by part (a)} \\
&= 1 + (1 + \lfloor \log_3 \lfloor k/3 \rfloor \rfloor) &&\text{by inductive hypothesis} \\
&= \begin{cases} 2 + \lfloor \log_3(k/3) \rfloor & \text{if } k \bmod 3 = 0 \\ 2 + \lfloor \log_3((k-1)/3) \rfloor & \text{if } k \bmod 3 = 1 \\ 2 + \lfloor \log_3((k-2)/3) \rfloor & \text{if } k \bmod 3 = 2 \end{cases} &&\text{by exercise 17 of Section 3.5} \\
&= \begin{cases} 2 + \lfloor \log_3 k - \log_3 3 \rfloor & \text{if } k \bmod 3 = 0 \\ 2 + \lfloor \log_3(k-1) - \log_3 3 \rfloor & \text{if } k \bmod 3 = 1 \\ 2 + \lfloor \log_3(k-2) - \log_3 3 \rfloor & \text{if } k \bmod 3 = 2 \end{cases} &&\text{by exercise 24 of Section 7.3.} \\
&= \begin{cases} 2 + \lfloor \log_3 k - 1 \rfloor & \text{if } k \bmod 3 = 0 \\ 2 + \lfloor \log_3(k-1) - 1 \rfloor & \text{if } k \bmod 3 = 1 \\ 2 + \lfloor \log_3(k-2) - 1 \rfloor & \text{if } k \bmod 3 = 2 \end{cases} &&\text{because } \log_3 3 = 1 \\
&= \begin{cases} 2 + \lfloor \log_3 k \rfloor - 1 & \text{if } k \bmod 3 = 0 \\ 2 + \lfloor \log_3(k-1) \rfloor - 1 & \text{if } k \bmod 3 = 1 \\ 2 + \lfloor \log_3(k-2) \rfloor - 1 & \text{if } k \bmod 3 = 2 \end{cases} &&\text{by exercise 15 of Section 3.5} \\
&= \begin{cases} 1 + \lfloor \log_3 k \rfloor & \text{if } k \bmod 3 = 0 \\ 1 + \lfloor \log_3(k-1) \rfloor & \text{if } k \bmod 3 = 1 \\ 1 + \lfloor \log_3(k-2) \rfloor & \text{if } k \bmod 3 = 2 \end{cases}
\end{aligned}
$$

Now if $k \bmod 3 = 1$ and $k \geq 3$, then $k = 3m+1$ for some integer m, and so by the lemma $\lfloor \log_3 k \rfloor = \lfloor \log_3(3m+1) \rfloor = \lfloor \log_3(3m) \rfloor = \lfloor \log_3(k-1) \rfloor$. Similarly, if $k \bmod 3 = 2$ and $k \geq 3$, then by the lemma $\lfloor \log_3 k \rfloor = \lfloor \log_3(k-2) \rfloor$. Hence regardless of the value of $k \bmod 3$, $b_k = 1 + \lfloor \log_3 k \rfloor$ *[as was to be shown]*.

[Since we have proved both the basis and the inductive steps, we conclude that the given formula holds for all integers $n \geq 1$.]

15. If $n \geq 3$, then $b_n = 1 + \lfloor \log_3 n \rfloor \leq 1 + \log_3 n$ [by definition of floor] $\leq \log_3 n + \log_3 n$ [because if $n \geq 3$ then $\log_3 n \geq 1$] $= 2 \cdot \log_3 n$. Let $M = 2$ and $x_0 = 3$. Then for all integers $n \geq x_0$, $\log_3 n \geq 0$, and so $|1 + \lfloor \log_3 n \rfloor| \leq M \cdot |\log_3 n|$. Hence by definition of O-notation, $b_n = 1 + \lfloor \log_3 n \rfloor$ is $O(\log_3 n)$. Thus the algorithm segment has order $\log_3 n$.

16. Let $w_1, w_2, w_3...$ be defined as follows: $w_1 = 1$ and $w_k = 1 + w_{\lfloor k/2 \rfloor}$ for all integers $k \geq 1$. Let k be an even integer, and suppose $w_i = \lfloor \log_2 i \rfloor + 1$ for all integers i with $1 \leq i < k$. [This is the inductive hypothesis.] [We will show that
$w_k = \lfloor \log_2 k \rfloor + 1$.] Since k is even, $\lfloor k/2 \rfloor = k/2$. Then $w_k = 1 + w_{\lfloor k/2 \rfloor} = 1 + w_{k/2} = 1 + (\lfloor \log_2(k/2) \rfloor + 1)$ [by inductive hypothesis] $= 1 + \lfloor \log_2 k - \log_2 2 \rfloor + 1)$ [by (9.2.11)] $= 1 + \lfloor \log_2 k - 1 \rfloor + 1)$ [because $\log_2 2 = 1$] $= 1 + (\lfloor \log_2 k \rfloor - 1 + 1)$ [by exercise 15 of Section 3.5] $= \lfloor \log_2 k \rfloor + 1$ [as was to be shown].

17. a.

index	0				7
bot	1	7			
top	10		8	7	
mid		6	9	8	7

b.

index	0			
bot	1	7	9	
top	10		8	
mid		6	9	8

18. Suppose an array of length k is input to the **while** loop and the loop is iterated one time. The elements of the array can be matched with the integers from 1 to k as shown below:

$$\underbrace{\begin{array}{ccccc} a[bot] & a[bot+1] & \ldots & a[mid-1] & a[mid] \\ \updownarrow & \updownarrow & & \updownarrow & \updownarrow \\ 1 & 2 & & m-1 & m \end{array}}_{\text{left subarray}} \quad \underbrace{\begin{array}{cccc} a[mid+1] & \ldots & a[top-1] & a[top] \\ \updownarrow & & \updownarrow & \updownarrow \\ m+1 & & k-1 & k \end{array}}_{\text{right subarray}}$$

where $m = \left\lceil \dfrac{k+1}{2} \right\rceil$.

Case 1 (k is even): In this case $m = \left\lceil \dfrac{k+1}{2} \right\rceil = \left\lceil \dfrac{k}{2} + \dfrac{1}{2} \right\rceil = \dfrac{k}{2} + 1$, and so the number of elements in the left subarray equals $m - 1 = (\dfrac{k}{2} + 1) - 1 = \dfrac{k}{2} = \left\lfloor \dfrac{k}{2} \right\rfloor$. The number of elements in the right subarray equals $k - (m+1) - 1 = k - m = k - (\dfrac{k}{2} + 1) = \dfrac{k}{2} - 1 < \left\lfloor \dfrac{k}{2} \right\rfloor$. Hence both subarrays (and thus the new input array) have length at most $\left\lfloor \dfrac{k}{2} \right\rfloor$.

Case 2 (k is odd): In this case $m = \left\lceil \dfrac{k+1}{2} \right\rceil = \dfrac{k+1}{2}$, and so the number of elements in the left subarray equals $m - 1 = \dfrac{k+1}{2} - 1 = \dfrac{k-1}{2} = \left\lfloor \dfrac{k}{2} \right\rfloor$. The number of elements in the right subarray equals $k - m = k - \dfrac{k+1}{2} = \dfrac{k-1}{2} = \left\lfloor \dfrac{k}{2} \right\rfloor$ also. Hence both subarrays (and thus the new input array) have length $\left\lfloor \dfrac{k}{2} \right\rfloor$.

The arguments in cases 1 and 2 show that the length of the new input array to the next iteration of the **while** loop has length at most $\lfloor k/2 \rfloor$.

19. By exercise 18, given an input array of length k to the **while** loop of the modified binary search algorithm, the worst that can happen is that the next iteration of the loop will have to search an array of length $\lfloor k/2 \rfloor$. Hence the maximum number of iterations of the loop is one more than the maximum number necessary to execute it for an input array of length $\lfloor k/2 \rfloor$. Thus $w_k = w_{\lfloor k/2 \rfloor} + 1$ for $k \geq 2$.

21.

23.

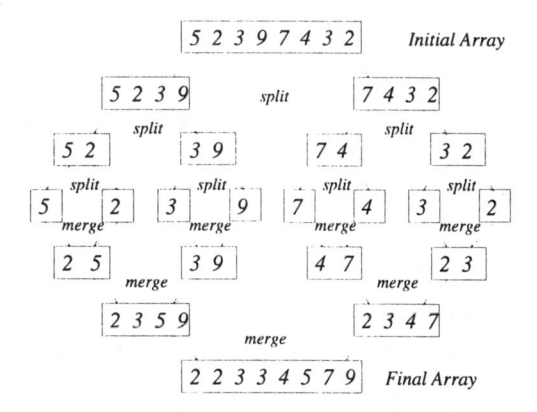

24. *a.* Refer to Figure 9.5.3. Observe that when k is odd, the subarray $a[mid+1], a[mid+2], \ldots a[top]$ has length $k - \left(\dfrac{k+1}{2} + 1\right) + 1 = \dfrac{k-1}{2} = \lfloor k/2 \rfloor$. And when k is even, the subarray $a[mid+1], a[mid+2], \ldots a[top]$ has length $k - \left(\dfrac{k}{2} + 1\right) + 1 = \dfrac{k}{2} = \lfloor k/2 \rfloor$. So in either case the subarray has length $\lfloor k/2 \rfloor$.

25. *Proof*:

    The inequality holds for n = 1 and n = 2: For $n = 1$ the inequality asserts that $m_1 \leq 2 \cdot 1 \log_2(1)$. But $m_1 = 0$ and $2 \cdot 1 \log_2(1) = 2 \cdot 0 = 0$ also. So the inequality holds for $n = 1$. For $n = 2$ the inequality asserts that $m_1 \leq 2 \cdot 2 \log_2(2)$. But $m_2 = m_{\lfloor 2/2 \rfloor} + m_{\lceil 2/2 \rceil} + 2 - 1 = m_1 + m_1 + 1 = 0 + 0 + 1 = 1$ and $2 \cdot 2 \log_2(2) = 2 \cdot 2 \cdot 1 = 4$, and $1 \leq 4$. So the inequality also holds for $n = 2$.

    If the inequality holds for all integers i with $1 \leq i < k$ then it holds for k: Let k be an integer with $k \geq 1$, and suppose that $m_i \leq 2i \cdot \log_2 i$ for all integers i with $1 \leq i < k$. [*This is the inductive hypothesis.*] We must show that $m_k \leq 2k \cdot \log_2 k$.

227

Case 1 (k is even): In this case,

$$\begin{aligned} m_k &= m_{\lfloor k/2 \rfloor} + m_{\lceil k/2 \rceil} + k - 1 \quad &&\text{by definition of } m_1, m_2, m_3, \ldots \\ &= m_{k/2} + m_{k/2} + k - 1 \quad &&\text{since } k \text{ is even} \\ &= 2m_{k/2} + k - 1. \end{aligned}$$

$$\begin{aligned} \text{Hence} \quad m_k &\leq 2 \cdot 2 \cdot \frac{k}{2} \log_2\left(\frac{k}{2}\right) + k - 1 \quad &&\text{by inductive hypothesis} \\ \Rightarrow \quad m_k &\leq 2k(\log_2 k - \log_2 2) + k - 1 \quad &&\text{by the identity } \log_b (x/y) = \log_b x - \log_b y \\ & &&\text{(exercise 24 of Section 7.3)} \\ \Rightarrow \quad m_k &\leq 2k(\log_2 k - 1) \quad &&\text{because } \log_2 2 = 1 \\ & +k - 1 \\ \Rightarrow \quad m_k &\leq 2k\log_2 k - 1 \\ \Rightarrow \quad m_k &\leq 2k\log_2 k. \end{aligned}$$

Case 2 (k is odd): In this case, first note that because the logarithmic function with base 2 is increasing,

(1) $\log_2((k-1)/2) = \log_2 k - 1$ *[because since $\frac{k-1}{2} < \frac{k}{2}$, then $\log_2\left(\frac{k-1}{2}\right) < \log_2\left(\frac{k}{2}\right) = \log_2 k - \log_2 2 = \log_2 k - 1$].*

(2) $\log_2((k+1)/2) \leq \log_2 k$ *[because when $1 \leq k$, then $k+1 \leq 2k$, and so $(k+1)/2 \leq k$.]*

$$\begin{aligned} m_k &= m_{\lfloor k/2 \rfloor} + m_{\lceil k/2 \rceil} + k - 1 \quad &&\text{by definition of } m_1, m_2, m_3, \ldots \\ &= m_{(k-1)/2} + m_{(k+1)/2} + k - 1 \quad &&\text{by Theorem 3.5.2 and exercise 21} \\ & &&\text{of Section 3.5 since } k \text{ is odd} \end{aligned}$$

$$\begin{aligned} \Rightarrow \quad m_k &\leq 2 \cdot \frac{k-1}{2} \log_2\left(\frac{k-1}{2}\right) + 2 \cdot \frac{k+1}{2} \log_2\left(\frac{k+1}{2}\right) \\ & +k - 1 \quad &&\text{by inductive hypothesis} \\ \Rightarrow \quad m_k &\leq (k-1)(\log_2 k - 1) + (k+1)\log_2 k + k - 1 \quad &&\text{by notes (1) and (2) above} \\ \Rightarrow \quad m_k &\leq k\log_2 k - k - \log_2 k + 1 + k\log_2 k + \log_2 k + k - 1 \quad &&\text{by multiplying out} \\ \Rightarrow \quad m_k &\leq k\log_2 k \quad &&\text{by cancelling common terms.} \end{aligned}$$

Thus in either case, $m_k \leq k\log_2 k$ *[as was to be shown]*.

26. **Algorithm   Fast Computation of Integral Powers**

    [Given a real number x and a positive integer n, this algorithm computes x^n. It first calls Algorithm 4.1.1 to find the binary representation of n : $(r[k]r[k-1]\ldots r[1]r[0])_2$. Then it computes x^n using the fact that $x^n = x^{r[k]2^k} \cdot x^{r[k-1]2^{k-1}} \cdot \ldots \cdot x^{r[1]2^1} \cdot x^{r[0]2^0}$. Initially, factor is set equal to x and answer is either set equal to x if $r[0] = 1$ (which means that $x^{r[0]2^0} = x$) or it is set equal to 1 if $r[0] = 0$ (which means that $x^{r[0]2^0} = 1$). For each $i = 1$ to k, the value of factor is squared to obtain the numbers x^{2^i}, and each such number is made a factor of the answer provided $r[i] = 1$ (which means that $x^{r[i]2^i} = x^{2^i}$.)]

    Input: x *[a real number]*, n *[a positive integer]*

    Algorithm Body:

    Call Algorithm 4.1.1 with input n to obtain the output $r[0], r[1], \ldots, r[k]$.

    factor := x

    if $r[0] = 1$ **then** *answer* := x **else** *answer* := 1

    for i := 1 **to** k

        factor := *factor*·*factor*

        if $r[i] = 1$ **then** *answer* := *answer*·*factor*

    next i

Output: *answer [a real number]*

end Algorithm

b. There are at most two multiplications per iteration of the **for-next** loop and there are k iterations of this loop. Hence the total number of multiplications is $2k$. Now $n = 2^k + $ *lower powers of* 2. Consequently, $2^k \leq n < 2^{k+1}$, and so $k = \lfloor \log_2 n \rfloor$ by property (9.4.2). Thus the number of multiplications is at most $2 \cdot \lfloor \log_2 n \rfloor$.

Chapter 10 Relations

The first section of this chapter is an introduction to concepts and notation. The emphasis is on understanding equivalent ways to specify and to represent relations, both finite and infinite. In Section 10.2 the reflexivity, symmetry, and transitivity properties of binary relations are introduced and explored, and in Section 10.3 equivalence relations are discussed. At this point in the course, students are generally able to handle this level of abstraction fairly well. Some still need to be reminded of the logic discussed in the first two chapters and the proof methods covered in Chapter 3. For instance, even at this late stage of the course, I find that a few students rephrase the definition of symmetric as "xRy and yRx". But even these students are responsive to correction and are generally able to succeed with some assistance. In fact, it is often only at this point in the course that one can count on virtually all students understanding that the same proof outline is used to prove universal conditional statements no matter what the mathematical context.

Class participation in the discussion of these sections is very helpful. For instance, each time you write a definition of a property on the board, you can ask what it means for something not to have that property. And instead of just working examples to prove or disprove the properties in various instances, you can pose the questions on the board and get the class to solve them for you, preferably by having several students each contribute a part of the solution. After all, once the definitions have been written down, it is just a question of thinking things through to apply them in any given instance, and students are supposed to have learned how to think things through or at least to be making significant progress in doing so. If time allows, it is desirable to have students present solutions to homework problems for the rest of the class to critique or to work on solving problems together in groups.

The equivalence and simplification of finite-state automata, discussed in Section 10.4, is not used later in the book, but it is a nice application of equivalence relations. Since equivalence of digital logic circuits is defined in Section 10.3, when covering Section 10.4 you can draw parallels between the simplification of digital logic circuits discussed in Section 1.4 and the simplification of finite-state automata developed in this section. Both kinds of simplification have obvious practical use.

Partial order relations are not discussed until the last section of this chapter so as not to confuse students by presenting the definitions of symmetry and antisymmetry side-by-side. Another reason for placing the discussion of partial order relations in Section 10.5 is that the flavor of partial order relations is rather different from that of equivalence relations.

Comments on Exercises:

Section 10.2: #6, #7, #8, #50, #51: It would seem as if finite sets would provide the simplest examples of relations, and by and large they do. But in these problems, a few properties are vacuously true (or "true by default"), which is a mode of reasoning that some students still find hard to grasp. Actually, the idea of vacuous truth has occurred frequently enough throughout the book that there are generally students who relish encountering it in new situations. And usually even the students who still find the idea mindboggling are amused by it.

Section 10.3: #27-31: These exercises are designed to give students practice doing the kind of reasoning that is used to prove the main theorem of the section. These are good exercises to go over with the class as a whole, having a different student supply each step of the solution. #38: This exercise is somewhat whimsical, and students often ask what is its point. It is intended to provide an occasion to discuss the fact that a given equivalence class may have many different names, at least as many as there are representatives in the class, and that one needs to distinguish between what such a mathematical object *is* and what it is *called*.

Section 10.1

2. a. No, $2 \not\geq 4$. Yes, $4 \geq 3$. Yes, $4 \geq 4$. No, $2 \notin D$.

   b. $S = \{(3,3), (4,3), (5,3), (4,4), (5,4)\}$

3. b. *Proof*: Let n be any even integer. Then $n - 0 = n$ is also even, and so $n \, E \, 0$ by definition of E.

4. The following is a rather complete proof. Shorter versions that have a correct flow and feel should certainly be acceptable.

   Proof: We first observe that for all integers m and n, if $m - n$ is even then both m and n are even or both m and n are odd. To do this, we prove the logically equivalent statement: for all integers m and n, if $m - n$ is even and at least one of m or n is odd, then both m and n are odd. So suppose m and n are any integers so that $m - n$ is even and at least one of m or n is odd. Since $m - n$ is even, $m - n = 2r$ for some integer r. In case m is odd, then $m = 2s + 1$ for some integer s, and so $n = m - (m - n) = (2s + 1) - 2r = 2(s - r) + 1$, which is odd *[because $s - r$ is an integer]*. In case n is odd, then $n = 2s + 1$ for some integer s, and so $m = (m - n) + n = 2r + (2s + 1) = 2(r + s) + 1$, which is odd *[because $r + s$ is an integer]*. Hence in either case, both m and n are odd *[as was to be shown]*. To finish the proof, observe that for all integers m and n, if both m and n are even or both m and n are odd, then $m - n$ is even. But this was proved in the solutions to exercises 27 and 30 of Section 3.1.

5. c. *One possible answer*: 4, 7, 10, -2, -5

   d. *One possible answer*: 5, 8, 11, -1, -4

   e. *Theorem*: 1. All integers of the form $3k$ are related by T to 0.
   2. All integers of the form $3k + 1$ are related by T to 1.
   3. All integers of the form $3k + 2$ are related by T to 2.

   Proof of (2): Let n be any integer of the form $n = 3k + 1$ for some integer k. By substitution, $n - 1 = (3k + 1) - 1 = 3k$, and so by definition of divisibility, $3 \mid (n - 1)$. Hence by definition of T, $n \, T \, 1$.

   The proofs of (1) and (3) are identical to the proof of (2) with 0 and 2 respectively substituted in place of 1.

6. a. Yes, $2 \geq 1$. Yes, $2 \geq 2$. No, $2 \not\geq 3$. Yes, $-1 \geq -2$.

7. b.

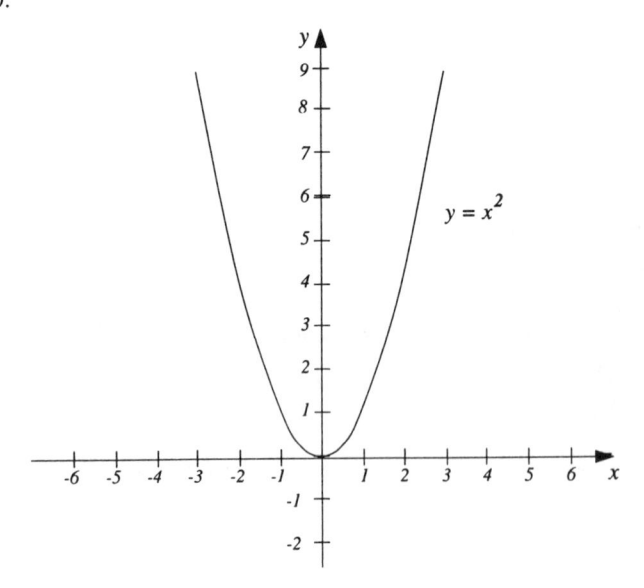

[Graph of $y = x^2$ parabola shown on coordinate axes with x from -6 to 6 and y from -2 to 9]

8. c. Yes, both have the prime factor 5. d. Yes, both have the prime factor 2.

9. b. No, $\{a\}$ has one element and $\{a,b\}$ has two. c. Yes, both have one element.

10. b. Yes, $\{a,b\} \cap \{b,c\} = \{b\} \neq \emptyset$. c. Yes, $\{a,b\} \cap \{a,b,c\} = \{a,b\} \neq \emptyset$.

11. b. No, $aa \neq bb$. c. Yes, $aa = aa$.

12. a.

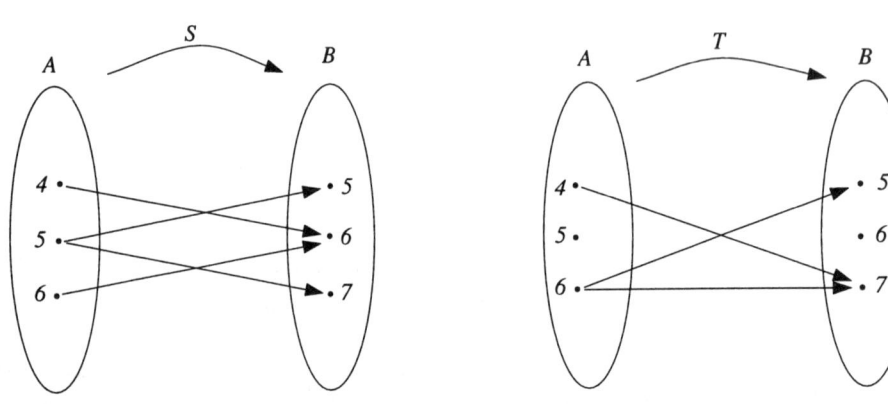

[Two diagrams: Left shows relation S from A={4,5,6} to B={5,6,7} with arrows; Right shows relation T from A={4,5,6} to B={5,6,7} with arrows]

b. S is not a function because $(5,5) \in S$ and $(5,7) \in S$ and $5 \neq 7$. So S does not satisfy property (2) of the definition of function. T is not a function both because $(5,x) \notin T$ for any x in B and because $(6,5) \in T$ and $(6,7) \in T$ and $5 \neq 7$. So T does not satisfy either property (1) or property (2) of the definition of function.

14. The following sets are all the relations from $\{a,b\}$ to $\{x,y\}$ that are not functions:

$\emptyset$, $\{(a,x)\}$, $\{(a,y)\}$, $\{(b,x)\}$, $\{(b,y)\}$, $\{(a,x),(a,y)\}$, $\{(b,x),(b,y)\}$, $\{(a,x),(a,y),(b,x)\}$, $\{(a,x),(a,y),(b,y)\}$, $\{(b,x),(b,y),(a,x)\}$, $\{(b,x),(b,y),(a,y)\}$, $\{(a,x),(a,y),(b,x),(b,y)\}$.

15. *a.* There are 2^{mn} binary relations from A to B because a binary relation from A to B is any subset of $A \times B$, $A \times B$ is a set with mn elements (since A has m elements and B has n elements), and the number of subsets of a set with mn elements is 2^{mn} (by Theorem 5.3.6).

b. In order to define a function from A to B we must specify exactly one image in B for each of the elements in A. But there are n choices of image for each of the m elements in A. So by the multiplication rule, the total number of functions is mn.

c. $mn/2^{mn}$

18. $S = \{(3,6),(4,4),(5,5)\}$ $\quad S^{-1} = \{(6,3),(4,4),(5,5)\}$

19. *c.* Yes, *aba* is the concatenation of *a* with *ba*.

d. No, $abb\ T^{-1}\ bba \Leftrightarrow bba\ T\ abb$, but $bba\ \not{T}\ abb$ because *abb* is not the concatenation of *a* with *bba*.

f. Yes, $abba\ T^{-1}\ bba \Leftrightarrow bba\ T\ abba$, and $bba\ T\ abba$ because *abba* is the concatenation of *a* with *bba*.

20.

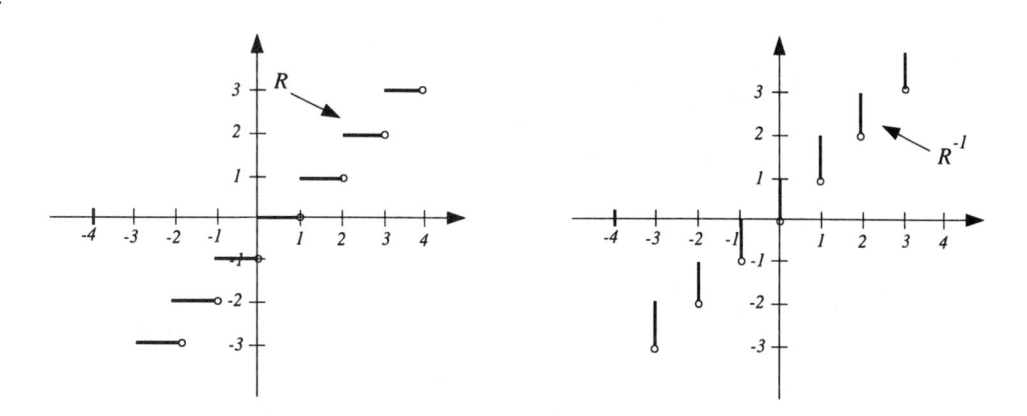

21. *b.* A function $F: X \to Y$ is onto if, and only if, for all $y \in Y$, $\exists x \in X$ such that $(x, y) \in F$.

22. *b.* No. If $F: X \to Y$ is not one-to-one, then there exist x_1 and x_2 in X and y in Y such that $(x_1, y) \in F$ and $(x_2, y) \in F$ and $x_1 \neq x_2$. But this implies that there exist x_1 and x_2 in X and y in Y such that $(y, x_1) \in F^{-1}$ and $(y, x_2) \in F^{-1}$ and $x_1 \neq x_2$. Consequently, F^{-1} does not satisfy property (2) of the definition of function.

24.

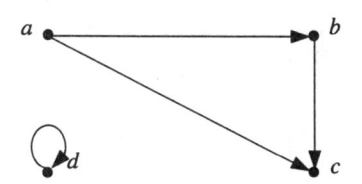

26.

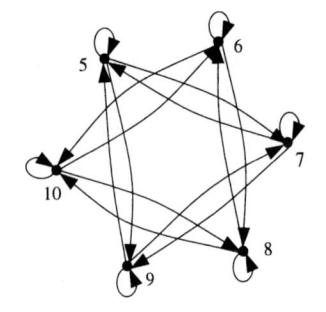

27.

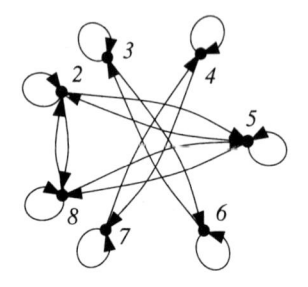

28. b.

  466581 Mary Lazars
  778400 Jamal Baskers

30. $A \times B = \{(-1,1),(1,1),(2,1),(4,1),(-1,2),(1,2),(2,2),(4,2)\}$

  $R = \{(-1,1),(1,1),(2,2)\}$

  $S = \{(-1,1),(1,1),(2,2),(4,2)\}$

  $R \cup S = S \quad R \cap S = R$

32.

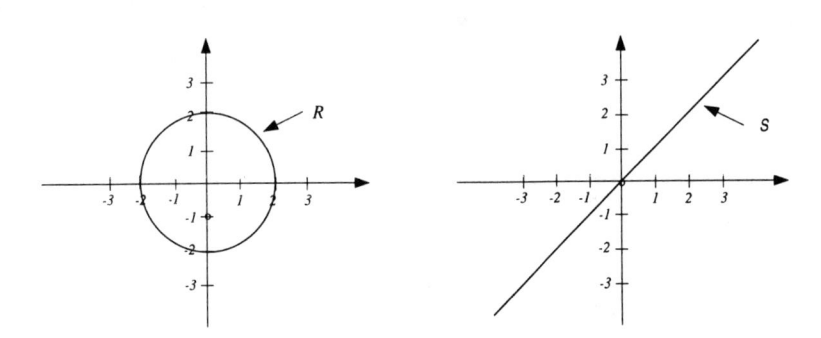

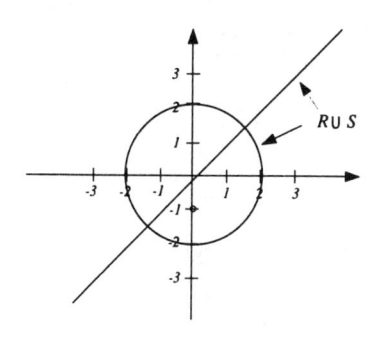

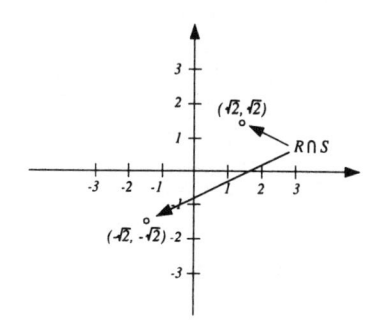

To obtain $R \cap S$, solve the system of equations:

$$x^2 + y^2 = 4$$
$$x = y$$

Substituting the second equation into the first gives $x^2 + y^2 = 4 \Rightarrow 2x^2 = 4 \Rightarrow x^2 = 2 \Rightarrow x = \pm\sqrt{2}$. Hence $x = y = \sqrt{2}$ or $x = y = -\sqrt{2}$.

33.

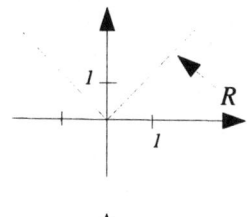

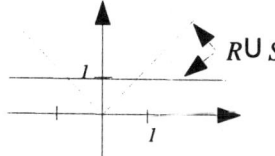

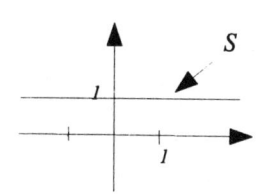

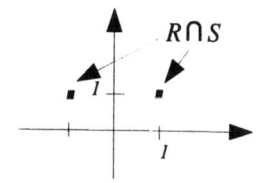

Section 10.2

2. a.

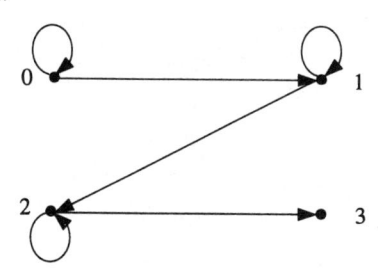

b. R_2 is not reflexive because $(3,3) \notin R_2$.
c. R_2 is not symmetric because, for example, $(0,1) \in R_2$ but $(1,0) \notin R_2$.
d. R_2 is not transitive because, for example, $(0,1) \in R_2$ and $(1,2) \in R_2$ but $(0,2) \notin R_2$.

235

4. *a.*

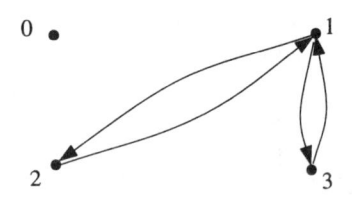

[Diagram with points 0, 1, 2, 3: arrows between 1 and 2 (both directions), and between 1 and 3 (both directions).]

b. R_4 is not reflexive because, for instance, $(0,0) \notin R_4$. (In fact, $(x,x) \notin R_4$ for any x in A.)

c. R_4 is symmetric.

d. R_4 is not transitive because, for example, $(3,1) \in R_4$ and $(1,2) \in R_4$ but $(3,2) \notin R_4$.

5. *a.*

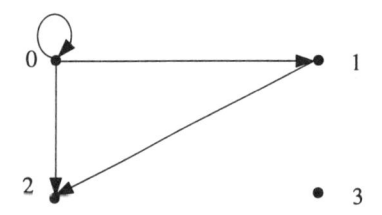

[Diagram with points 0, 1, 2, 3: self-loop at 0, arrow from 0 to 1, arrow from 0 to 2.]

b. R_5 is not reflexive because, for example, $(1,1) \notin R_2$.

c. R_5 is not symmetric because, for example, $(0,1) \in R_5$ but $(1,0) \notin R_5$.

d. R_5 is transitive.

7. *a.*

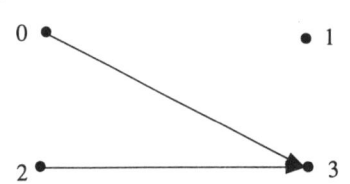

[Diagram with points 0, 1, 2, 3: arrow from 0 to 3, arrow from 2 to 3.]

b. R_7 is not reflexive because, for example, $(0,0) \notin R_7$.

c. R_7 is not symmetric because, for example, $(0,3) \in R_7$ but $(3,0) \notin R_7$.

d. R_7 is transitive.

8. *a.*

[Diagram with points 0, 1, 2, 3: self-loop at 0, self-loop at 1.]

b. R_8 is not reflexive because, for example, $(2,2) \notin R_8$.

c. R_8 is symmetric.

d. R_8 is transitive.

10. $S^t = \{(0,0),(0,3),(1,0),(1,2),(2,0),(3,2),(0,2),(1,3),(2,2),(2,3),(3,3),(3,0)\}$

11. $T^t = \{(0,2),(1,0),(2,3),(3,1),(0,3),(1,2),(2,1),(3,2),(3,0),(0,0),(0,1),(1,1),(1,3),(3,3),$
    $(2,2),(2,0)\}$

13. *C is not reflexive*: C is reflexive $\Leftrightarrow$ for all real numbers x, $x\,C\,x$. By definition of C this means that for all real numbers x, $x^2 + x^2 = 1$. But this is false. As a counterexample, take $x = 0$. Then $x^2 + x^2 = 0^2 + 0^2 = 0 \neq 1$.

    C is symmetric: C is symmetric $\Leftrightarrow$ for all real numbers x and y, if $x\,C\,y$ then $y\,C\,x$. By definition of C this means that for all real numbers x and y, if $x^2 + y^2 = 1$ then $y^2 + x^2 = 1$. But this is true because by commutativity of addition, $x^2 + y^2 = y^2 + x^2$ for all real numbers x and y.

    C is not transitive: C is transitive $\Leftrightarrow$ for all real numbers x, y, and z, if $x\,C\,y$ and $y\,C\,z$ then $x\,C\,z$. By definition of C this means that for all real numbers x, y and z, if $x^2 + y^2 = 1$ and $y^2 + z^2 = 1$ then $x^2 + z^2 = 1$. But this is false. As a counterexample, take $x = 0$, $y = 1$, and $z = 0$. Then $x^2 + y^2 = 1$ because $0^2 + 1^2 = 1$ and $y^2 + z^2 = 1$ because $1^2 + 0^2 = 1$, but $x^2 + z^2 \neq 1$ because $0^2 + 0^2 = 0 \neq 1$.

16. *F is reflexive*: Suppose m is any integer. Since $m - m = 0$ and $5 \mid 0$, we have that $5 \mid (m - m)$. Consequently, $m\,F\,m$ by definition of F.

    F is symmetric: Suppose m and n are any integers such that $m\,F\,n$. By definition of F this means that $5 \mid (m - n)$, and so by definition of divisibility $m - n = 5k$ for some integer k. Now $n - m = -(m - n)$. Hence by substitution, $n - m = -(5k) = 5 \cdot (-k)$. It follows that $5 \mid n - m$ by definition of divisibility (since $-k$ is an integer), and thus $n\,F\,m$ by definition of F.

    F is transitive: Suppose m, n and p are any integers such that $m\,F\,n$ and $n\,F\,p$. By definition of F this means that $5 \mid (m - n)$ and $5 \mid (n - p)$, and so by definition of divisibility $m - n = 5k$ for some integer k and $n - p = 5l$ for some integer l. Now $m - p = (m - n) + (n - p)$. Hence by substitution, $m - p = 5k + 5l = 5(k + l)$. It follows that $5 \mid m - p$ by definition of divisibility (since $k + l$ is an integer), and thus $m\,F\,p$ by definition of F.

17. *O is not reflexive*: O is reflexive $\Leftrightarrow$ for all integers m, $m\,O\,m$. By definition of O this means that for all integers m, $m - m$ is odd. But this is false. As a counterexample, take any integer m. Then $m - m = 0$, which is even, not odd.

    O is symmetric: Suppose m and n are any integers such that $m\,O\,n$. By definition of O this means that $m - n$ is odd, and so by definition of odd $m - n = 2k + 1$ for some integer k. Now $n - m = -(m - n)$. Hence by substitution, $n - m = -(2k + 1) = 2(k - 1) + 1$. It follows that $n - m$ is odd by definition of odd (since $k - 1$ is an integer), and thus $n\,O\,m$ by definition of O.

    O is not transitive: O is transitive $\Leftrightarrow$ for all integers m, n, and p, if $m\,O\,n$ and $n\,O\,p$ then $m\,O\,p$. By definition of O this means that for all integers m, n, and p, if $m - n$ is odd and $n - p$ is odd then $m - p$ is odd. But this is false. As a counterexample, take $m = 1$, $n = 0$, and $p = 1$. Then $m - n = 1 - 0 = 1$ is odd and $n - p = 0 - 1 = -1$ is also odd, but $m - p = 1 - 1 = 0$ is not odd. Hence $m\,O\,n$ and $n\,O\,p$ but $m\,\cancel{O}\,p$.

19. *A is reflexive*: A is reflexive $\Leftrightarrow$ for all real numbers x, $\mid x \mid = \mid x \mid$. But this is true by the reflexive property of equality.

    A is symmetric: [We must show that for all real numbers x and y, if $\mid x \mid = \mid y \mid$ then $\mid y \mid = \mid x \mid$.] But this is true by the symmetric property of equality.

    A is transitive: A is transitive $\Leftrightarrow$ for all real numbers x, y, and z, if $\mid x \mid = \mid y \mid$ and $\mid y \mid = \mid z \mid$, then $\mid x \mid = \mid z \mid$. But this is true by the transitive property of equality.

20. *P is not reflexive*: P is reflexive $\Leftrightarrow$ for all integers n, $n\,P\,n$. By definition of P this means that for all integers n, $\exists$ a prime number p such that $p \mid n$ and $p \mid n$. But this is false. As a counterexample, take $n = 1$. There is no prime number that divides 1.

P is symmetric: *[We must show that for all integers m and n, if m P n then n P m.]* Suppose m and n are integers such that m P n. By definition of P this means that there exists a prime number p such that $p \mid m$ and $p \mid n$. But to say that "$p \mid m$ and $p \mid n$" is logically equivalent to saying that "$p \mid n$ and $p \mid m$." Hence there exists a prime number p such that $p \mid n$ and $p \mid m$, and so by definition of P, n P m.

P is not transitive: P is transitive $\Leftrightarrow$ for all integers m, n, and p, if m P n and n P p then m P p. But this is false. As a counterexample, take $m = 2$, $n = 6$, and $p = 9$. Then m P n because the prime number 2 divides both 2 and 6 and n P p because the prime number 3 divides both 6 and 9, but $m \not{P} n$ because the numbers 2 and 9 have no common prime factor.

22. *G is not reflexive:* G is reflexive $\Leftrightarrow$ for all strings $s \in \Sigma^*$, s G s. By definition of G this means that for all strings s in Σ^*, the number of 0's in s is greater than the number of 0's in s. But this is false for every string in Σ^*. For instance, let $s = 00$. Then the number of 0's in s is 2 which is not greater than 2.

    G is not symmetric: For G to be symmetric would mean that for all strings s and t in Σ^*, if s G t then t R s. By definition of G, this would mean that for all strings s and t in Σ^*, if the number of 0's in s is greater than the number of 0's in t, then the number of 0's in t is greater than the number of 0's in s. But this is false for all strings s and t in Σ^*. For instance, take $s = 100$ and $t = 10$. Then the number of 0's in s is 2 and the number of 0's in t is 1. It follows that s G t (since $2 > 1$), but $t \not{G} s$ (since $1 \not> 2$).

    G is transitive: To prove transitivity of G, we must show that for all strings s, t, and u in Σ^*, if s G t and t G u then s G u. By definition of G this means that for all strings s, t, and u in Σ^*, if the number of 0's in s is greater than the number of 0's in t and the number of 0's in t is greater than the number of 0's in u, then the number of 0's in s is greater than the number of 0's in u. But this is true by the transitivity property of order (Appendix A, T17).

24. *R is not reflexive:* $\mathcal{R}$ is reflexive $\Leftrightarrow$ for all sets $A \in \mathcal{P}(X)$, A $\mathcal{R}$ A. By definition of $\mathcal{R}$ this means that for all sets A in $\mathcal{P}(X)$, $n(A) < n(A)$. But this is false for every set in $\mathcal{P}(X)$. For instance, let $A = \emptyset$. Then $n(A) = 0$, and 0 is not less than 0.

    R is not symmetric: For $\mathcal{R}$ to be symmetric would mean that for all sets A and B in $\mathcal{P}(X)$, if A $\mathcal{R}$ B then B $\mathcal{R}$ A. By definition of $\mathcal{R}$, this would mean that for all sets A and B in $\mathcal{P}(X)$, if $n(A) < n(B)$, then $n(B) < n(A)$. But this is false for all sets A and B in $\mathcal{P}(X)$. For instance, take $A = \emptyset$ and $B = \{a\}$. Then $n(A) = 0$ and $n(B) = 1$. It follows that A is related to B by $\mathcal{R}$ (since $0 < 1$), but B is not related to A by $\mathcal{R}$ (since $1 \not< 0$).

    R is transitive: To prove transitivity of $\mathcal{R}$, we must show that for all sets A, B, and C in $\mathcal{P}(X)$, if A $\mathcal{R}$ B and B $\mathcal{R}$ C then A $\mathcal{R}$ C. By definition of $\mathcal{R}$ this means that for all sets A, B, and C in $\mathcal{P}(X)$, if $n(A) < n(B)$ and $n(B) < n(C)$, then $n(A) < n(C)$. But this is true by the transitivity property of order (Appendix A, T17).

25. *N is not reflexive:* $\mathcal{N}$ is reflexive $\Leftrightarrow$ for all sets $A \in \mathcal{P}(X)$, A $\mathcal{N}$ A. By definition of $\mathcal{N}$ this means that for all sets A in $\mathcal{P}(X)$, $n(A) \neq n(A)$. But this is false for every set in $\mathcal{P}(X)$. For instance, let $A = \emptyset$. Then $n(A) = 0$. And it is not true that $0 \neq 0$.

    N is symmetric: $\mathcal{N}$ is symmetric $\Leftrightarrow$ for all sets A and B in $\mathcal{P}(X)$, if A $\mathcal{N}$ B then B $\mathcal{N}$ A. By definition of $\mathcal{N}$, this means that for all sets A and B in $\mathcal{P}(X)$, if $n(A) \neq n(B)$, then $n(B) \neq n(A)$. But this is true.

    N is not transitive: $\mathcal{N}$ is transitive $\Leftrightarrow$ for all sets A, B, and C in $\mathcal{P}(X)$, if A $\mathcal{N}$ B and B $\mathcal{N}$ C then A $\mathcal{N}$ C. By definition of $\mathcal{N}$ this means that for all sets A, B, and C in $\mathcal{P}(X)$, if $n(A) \neq n(B)$ and $n(B) \neq n(C)$, then $n(A) \neq n(C)$. But this is false. As a counterexample, let $A = \{a\}$, $B = \{a, b\}$, and $C = \{b\}$. Then $n(A) = 1$, $n(B) = 2$, and $n(C) = 1$. So $n(A) \neq n(B)$ and $n(B) \neq n(C)$, but $n(A) = n(C)$.

27. **$\mathcal{R}$ is not reflexive:** $\mathcal{R}$ is reflexive $\Leftrightarrow$ for all sets $X \in \mathcal{P}(A)$, $X \mathcal{R} X$. By definition of $\mathcal{R}$ this means that for all sets X in $\mathcal{P}(A)$, $X \neq X$. But this is false for every set in $\mathcal{P}(A)$. For instance, let $X = \emptyset$. It is not true that $\emptyset \neq \emptyset$.

$\mathcal{R}$ is symmetric: $\mathcal{R}$ is symmetric $\Leftrightarrow$ for all sets X and Y in $\mathcal{P}(A)$, if $X \mathcal{R} Y$ then $Y \mathcal{R} X$. By definition of $\mathcal{R}$, this means that for all sets X and Y in $\mathcal{P}(A)$, if $X \neq Y$, then $Y \neq X$. But this is true.

$\mathcal{R}$ is not transitive: $\mathcal{R}$ is transitive $\Leftrightarrow$ for all sets X, Y, and Z in $\mathcal{P}(A)$, if $X \mathcal{R} Y$ and $Y \mathcal{R} Z$ then $X \mathcal{R} Z$. By definition of $\mathcal{R}$ this means that for all sets X, Y, and Z in $\mathcal{P}(A)$, if $X \neq Y$ and $Y \neq Z$, then $X \neq Z$. But this is false as the following counterexample shows. Since $A \neq \emptyset$, there exists an element x in A. Let $X = \{x\}$, $Y = \emptyset$, and $Z = \{x\}$. Then $X \neq Y$ and $Y \neq Z$, but $X = Z$.

28. **$\mathcal{C}$ is not reflexive:** $\mathcal{C}$ is reflexive $\Leftrightarrow$ for all sets $X \in \mathcal{P}(A)$, $X \mathcal{C} X$. By definition of $\mathcal{C}$ this means that for all sets X in $\mathcal{P}(A)$, $X = A - X$. But this is false for every set in $\mathcal{P}(A)$ because $A \neq \emptyset$. For instance, let $X = \emptyset$. It is not true that $\emptyset = A - \emptyset$ because $A - \emptyset = A$ and $A \neq \emptyset$.

$\mathcal{C}$ is symmetric: *[We must show that for all sets X and Y in $\mathcal{P}(A)$, if $X \mathcal{C} Y$ then $Y \mathcal{C} X$.]* Suppose X and Y are sets in $\mathcal{P}(A)$ and $X \mathcal{C} Y$. By definition of $\mathcal{C}$, this means that $Y = A - X$. By the properties of sets given in Sections 5.2 and 5.3, $A - Y = A - (A - X) = A \cap (A \cap X^c)^c = A \cap (A^c \cup X) = (A \cap A^c) \cup (A \cap X) = \emptyset \cup (A \cap X) = (A \cap X) \cup \emptyset = A \cap X = X$ (because $X \subseteq A$). Hence $Y \mathcal{C} X$.

$\mathcal{C}$ is not transitive: $\mathcal{C}$ is transitive $\Leftrightarrow$ for all sets X, Y, and Z in $\mathcal{P}(A)$, if $X \mathcal{R} Y$ and $Y \mathcal{C} Z$ then $X \mathcal{C} Z$. By definition of $\mathcal{C}$ this means that for all sets X, Y, and Z in $\mathcal{P}(A)$, if $Y = A - X$ and $Z = A - Y$, then $Z = A - X$. However this is false, as the following counterexample shows. Since $A \neq \emptyset$, there exists an element x in A. Let $X = \{x\}$, $Y = A - X$, and $Z = A - Y$. Then by substitution, $Z = A - (A - X) = X$ *[by the same argument as in the proof of symmetry above]*. Suppose $Z = A - X$. Then we would have $A - X = X$. But this is impossible because $x \in \{x\} = X$ and $x \notin A - \{x\} = A - X$. Therefore it cannot be the case that $Z = A - X$. Consequently $X \mathcal{C} Y$ and $Y \mathcal{C} Z$ but X is not related to Z by $\mathcal{C}$.

29. **$\mathcal{R}$ is reflexive:** *[We must show that for all sets X in $\mathcal{P}(A)$, $X \mathcal{R} X$.]* Suppose X is a set in $\mathcal{P}(A)$. By definition of subset, we know that $X \subseteq X$. By definition of $\mathcal{R}$, then, $X \mathcal{R} X$.

$\mathcal{R}$ is symmetric: *[We must show that for all sets X and Y in $\mathcal{P}(A)$, if $X \mathcal{R} Y$ then $Y \mathcal{R} X$.]* Suppose X and Y are sets in $\mathcal{P}(A)$ and $X \mathcal{R} Y$. By definition of $\mathcal{R}$, this means that $X \subseteq Y$ or $Y \subseteq X$. In case $X \subseteq Y$, then the statement "$Y \subseteq X$ or $X \subseteq Y$" is true, and so by definition of $\mathcal{R}$, $Y \mathcal{R} X$. In case $Y \subseteq X$ then the statement "$Y \subseteq X$ or $X \subseteq Y$" is also true, and so by definition of $\mathcal{R}$, $Y \mathcal{R} X$. Therefore in either case $Y \mathcal{R} X$.

If A has at least two elements, then $\mathcal{R}$ is not transitive: $\mathcal{R}$ is transitive $\Leftrightarrow$ for all sets X, Y, and Z in $\mathcal{P}(A)$, if $X \mathcal{R} Y$ and $Y \mathcal{R} Z$ then $X \mathcal{R} Z$. By definition of $\mathcal{R}$ this means that for all sets X, Y, and Z in $\mathcal{P}(A)$, if either $X \subseteq Y$ or $Y \subseteq X$ and either $Y \subseteq Z$ or $Z \subseteq Y$, then either $X \subseteq Z$ or $Z \subseteq X$. However this is false, as the following counterexample shows. Since A has at least two elements, there exist elements x and y in A with $x \neq y$. Let $X = \{x\}$, $Y = A$, and $Z = \{y\}$. Then $X \subseteq Y$ and so $X \mathcal{R} Y$ and $Z \subseteq Y$ and so $Y \mathcal{R} Z$. But $X \not\subseteq Z$ and $Z \not\subseteq X$ because $x \neq y$. Hence X is not related to Z by $\mathcal{R}$.

If A has a single element, then $\mathcal{R}$ is transitive: In this case, given any two subsets of A, either one is a subset of the other or the other is a subset of the one. Hence regardless of the choice of X, Y, and Z, it must be the case that $X \subseteq Z$ or $Z \subseteq X$ and so $X \mathcal{R} Z$.

32. **$\mathcal{R}$ is reflexive:** $\mathcal{R}$ is reflexive $\Leftrightarrow$ for all elements (x, y) in $\mathbf{R} \times \mathbf{R}$, $(x, y) \mathcal{R}(x, y)$. By definition of $\mathcal{R}$ this means that for all elements (x, y) in $\mathbf{R} \times \mathbf{R}$, $y = y$. But this is true.

$\mathcal{R}$ is symmetric: *[We must show that for all elements (x_1, y_1) and (x_2, y_2) in $\mathbf{R} \times \mathbf{R}$, if $(x_1, y_1) \mathcal{R}(x_2, y_2)$ then $(x_2, y_2) \mathcal{R}(x_1, y_1)$.]* Suppose (x_1, y_1) and (x_2, y_2) are elements of $\mathbf{R} \times \mathbf{R}$.

such that $(x_1, y_1)\mathcal{R}(x_2, y_2)$. By definition of $\mathcal{R}$ this means that $y_1 = y_2$. By symmetry of equality, $y_2 = y_1$. So by definition of $\mathcal{R}$, $(x_2, y_2)\mathcal{R}(x_1, y_1)$.

$\mathcal{R}$ is transitive: [We must show that for all elements (x_1, y_1), (x_2, y_2) and (x_3, y_3) in $\mathbf{R} \times \mathbf{R}$, if $(x_1, y_1)\mathcal{R}(x_2, y_2)$ and $(x_2, y_2)\mathcal{R}(x_3, y_3)$ then $(x_1, y_1)\mathcal{R}(x_3, y_3)$.] Suppose (x_1, y_1), (x_2, y_2), and (x_3, y_3) are elements of $\mathbf{R} \times \mathbf{R}$ such that $(x_1, y_1)\mathcal{R}(x_2, y_2)$ and $(x_2, y_2)\mathcal{R}(x_3, y_3)$. By definition of $\mathcal{R}$ this means that $y_1 = y_2$ and $y_2 = y_3$. By transitivity of equality, $y_1 = y_3$. Hence by definition of $\mathcal{R}$, $(x_1, y_1)\mathcal{R}(x_3, y_3)$.

33. R is reflexive: R is reflexive $\Leftrightarrow$ for all points p in A, $p\,R\,p$. By definition of R this means that for all elements p in A, p and p both lie on the same half line emanating from the origin. But this is true.

    R is symmetric: [We must show that for all points p_1 and p_2 in A, if $p_1 R\, p_2$ then $p_2 R\, p_1$.] Suppose p_1 and p_2 are points in A such that $p_1 R\, p_2$. By definition of R this means that p_1 and p_2 lie on the same half line emanating from the origin. But this implies that p_2 and p_1 lie on the same half line emanating from the origin. So by definition of R, $p_2 R\, p_1$.

    R is transitive: [We must show that for all points p_1, p_2 and p_3 in A, if $p_1 R\, p_2$ and $p_2 R\, p_3$ then $p_1 R\, p_3$.] Suppose p_1, p_2, and p_3 are points in A such that $p_1 R\, p_2$ and $p_2 R\, p_3$. By definition of R, this means that p_1 and p_2 lie on the same half line emanating from the origin and p_2 and p_3 lie on the same half line emanating from the origin. Since two points determine a line, it follows that both p_1 and p_3 lie on the same half line determined by the origin and p_2. Thus p_1 and p_3 lie on the same half line emanating from the origin, and so by definition of R, $p_1 R\, p_3$.

35. R is reflexive: R is reflexive $\Leftrightarrow$ for all lines l in A, $l\, R\, l$. By definition of R this means that for all lines l in the plane, l is parallel to itself. But this is true.

    R is symmetric: [We must show that for all lines l_1 and l_2 in A, if $l_1 R\, l_2$ then $l_2 R\, l_1$.] Suppose l_1 and l_2 are lines in A such that $l_1 R\, l_2$. By definition of R this means that l_1 is parallel to l_2. But this implies that l_2 is parallel to l_1. So by definition of R, $l_2 R\, l_1$.

    R is transitive: [We must show that for all lines l_1, l_2 and l_3 in A, if $l_1 R\, l_2$ and $l_2 R\, l_3$ then $l_1 R\, l_3$.] Suppose l_1, l_2, and l_3 are lines of A such that $l_1 R\, l_2$ and $l_2 R\, l_3$. By definition of R this means that l_1 is parallel to l_2 and l_2 is parallel to l_3. Since two lines each parallel to a third line are parallel to each other, it follows that l_1 is parallel to l_3 because both are parallel to l_2. Hence by definition of R, $l_1 R\, l_3$.

36. R is not reflexive: R is reflexive $\Leftrightarrow$ for all lines l in A, $l\, R\, l$. By definition of R this means that for all lines l in the plane, l is perpendicular to itself. But this is false for every line in the plane.

    R is symmetric: [We must show that for all lines l_1 and l_2 in A, if $l_1 R\, l_2$ then $l_2 R\, l_1$.] Suppose l_1 and l_2 are lines in A such that $l_1 R\, l_2$. By definition of R this means that l_1 is perpendicular to l_2. But this implies that l_2 is perpendicular to l_1. So by definition of R, $l_2 R\, l_1$.

    R is not transitive: R is transitive $\Leftrightarrow$ for all lines l_1, l_2, and l_3 in A, if $l_1 R\, l_2$ and $l_2 R\, l_3$ then $l_1 R\, l_3$. But this is false. As a counterexample, take l_1 and l_3 to be the horizontal axis and l_2 to be the vertical axis. Then $l_1 R\, l_2$ and $l_2 R\, l_3$ because the horizontal axis is perpendicular to the vertical axis and the vertical axis is perpendicular to the horizontal axis. But $l_1 \not{R}\, l_3$ because the horizontal axis is not perpendicular to itself.

37. b. A reflexive relation must contain (a,a) for all eight elements a in A. Any subset of the remaining 56 elements of $A \times A$ (which has a total of 64 elements) can be combined with these eight to produce a reflexive relation. Therefore, there are as many reflexive binary relations as there are subsets of a set of 56 elements, namely 2^{56}.

    d. Form a relation that is both reflexive and symmetric by a two-step process: (1) pick all eight elements of the form (x, x) where $x \in A$, (2) pick a set of pairs of elements of the form (a, b) and (b, a). There is just one way to perform step 1, and, as explained in the answer to part (c), there are 2^{28} ways to perform step 2. Therefore, there are 2^{28} binary relations on A that are reflexive and symmetric.

39. **Algorithm   Test for Symmetry**

    [*The input for this algorithm consists of a binary relation R defined on a set A which is represented as the one-dimensional array $a[1], a[2], \ldots, a[n]$. To test whether R is symmetric, the variable answer is initially set equal to "yes" and then each pair of elements $a[i]$ and $a[j]$ of A is examined in turn to see whether the condition "if $(a[i], a[j]) \in R$ then $(a[j], a[i]) \in R$" is satisfied. If not, then answer is set equal to "no", the while loop is not repeated, and processing terminates.*]

    Input: n [*a positive integer*], $a[1], a[2], \ldots, a[n]$ [*a one-dimensional array representing a set A*], R [*a subset of $A \times A$*]

    Algorithm Body:

    $i := 1$, answer := "yes"

    while (answer = "yes" and $i \leq n$)

    $\quad j := 1$

    $\quad$**while** (answer = "yes" and $j \leq n$)

    $\qquad$**if** $(a[i], a[j]) \in R$ and $(a[j], a[i]) \notin R$ **then** answer := "no"

    $\qquad j := j + 1$

    $\quad$**end while**

    $\quad i := i + 1$

    end while

    Output: answer [*a string*]

    end Algorithm

40. **Algorithm   Test for Transitivity**

    [*The input for this algorithm consists of a binary relation R defined on a set A which is represented as the one-dimensional array $a[1], a[2], \ldots, a[n]$. To test whether R is transitive, the variable answer is initially set equal to "yes" and then each triple of elements $a[i]$, $a[j]$, and $a[k]$ of A is examined in turn to see whether the condition "if $(a[i], a[j]) \in R$ and $(a[j], a[k]) \in R$ then $(a[i], a[k]) \in R$" is satisfied. If not, then answer is set equal to "no", the while loop is not repeated, and processing terminates.*]

    Input: n [*a positive integer*], $a[1], a[2], \ldots, a[n]$ [*a one-dimensional array representing a set A*], R [*a subset of $A \times A$*]

    Algorithm Body:

    $i := 1$, answer := "yes"

    while (answer = "yes" and $i \leq n$)

    $\quad j := 1$

    $\quad$**while** (answer = "yes" and $j \leq n$)

    $\qquad k := 1$

    $\qquad$**while** (answer = "yes" and $k \leq n$)

    $\qquad\quad$**if** $(a[i], a[j]) \in R$ and $(a[j], a[k]) \in R$ and $(a[i], a[k]) \notin R$ **then** answer := "no"

    $\qquad\quad k := k + 1$

    $\qquad$**end while**

    $\qquad j := j + 1$

        end while

        $i := i + 1$

    end while

    Output: *answer [a string]*

end Algorithm

41. *Proof:* Suppose xRy or there is a sequence of elements of A, $x_1, x_2, \ldots, x_n$ such that $x = x_1$, $x_1 R x_2$, $x_2 R x_3, \ldots, x_{n-1} R x_n$ and $x_n = y$. *[We will show that $xR^t y$.]* In case xRy then $xR^t y$ because $R \subseteq R^t$. In case there is a sequence of elements of A, $x_1, x_2, \ldots, x_n$ such that $x = x_1$, $x_1 R x_2$, $x_2 R x_3, \ldots, x_{n-1} R x_n$ and $x_n = y$, then $x_1 R^t x_2$, $x_2 R^t x_3, \ldots, x_{n-1} R^t x_n$ because $R \subseteq R^t$, and since R^t is transitive, repeated application of the transitivity property gives that $x_1 R^t x_n$, or equivalently $xR^t y$ *[as was to be shown]*. On the other hand, suppose $xR^t y$ and suppose that $x\cancel{R}y$ and there does not exist a sequence of elements of A, $x_1, x_2, \ldots, x_n$ such that $x = x_1$, $x_1 R x_2$, $x_2 R x_3, \ldots, x_{n-1} R x_n$ and $x_n = y$. *[We will show that this supposition leads to a contradiction.]* Let $S = R^t - \{(x, y)\}$. Then S is a transitive relation that contains R and is a proper subset of R^t. This contradicts the fact that R^t is the "smallest" transitive relation containing R. Hence the supposition is false, and so if $xR^t y$, then xRy or there is a sequence of elements of A, $x_1, x_2, \ldots, x_n$ such that xRx_1, $x_1 R x_2$, $x_2 R x_3, \ldots, x_{n-1} R x_n$ and $x_n = y$.

    Note: A complete justification of the statements in the above proof requires the use of mathematical induction. The induction is on the length of the sequence joining x and y.

42. The following is an easily understood algorithm to construct the transitive closure of a relation. In fact, by keeping the variable names in the statements in the inner loop the same and interchanging the variable names governing the two outer loops, the algorithm can be made significantly more efficient. The resulting algorithm, known as Warshall's algorithm, is discussed in most books on the design and analysis of algorithms. See, for example, *Computer Algorithms* (Second Edition) by Sara Baase, Reading, Massachusetts: Addison-Wesley Publishing Company, 1988, pp.287-90.

    Algorithm   Constructing a Transitive Closure

    [The input for this algorithm consists of a binary relation R defined on a set A which is represented as the one-dimensional array $a[1], a[2], \ldots, a[n]$. The transitive closure is constructed by modifying the procedure used to test for transitivity described in the answer to exercise 39. Initially, the transitive closure, R^t, is set equal to R. Then a check is made through all triples of elements of A. In all cases for which $(a[i], a[j]) \in R^t$, $(a[j], a[k]) \in R^t$, and $(a[i], a[k]) \notin R^t$, the pair $(a[i], a[k])$ is added to R^t. After R^t has been enlarged in this way, it still may not equal the actual transitive closure of R because some of the added pairs may combine with pairs already present to necessitate the presence of additional pairs. So if any pairs have been added, an additional pass through all triples of elements of A may be necessary. Therefore, a variable flag is used to signal the addition of a pair. Initially, flag is set equal to 1 and if any pair is added, flag is set equal to 0. After all triples of elements of A have been examined, if the value of flag is 0, another pass is made through all triples of elements of A. If after all triples of elements of A have been examined the value of flag is 1, the enlarged relation R^t is transitive and thus equals the actual transitive closure of R.]

    Input: *n [a positive integer], $a[1], a[2], \ldots, a[n]$ [a one-dimensional array representing a set A], R [a subset of $A \times A$]*

    Algorithm Body:

        $R^t := R$

        $repeat := \text{``yes''}$, $flag := 1$

        $i := 1$

```
        while (repeat = "yes" and i ≤ n)
            j := 1
            while (j ≤ n)
                k := 1
                while (k ≤ n)
                    if (a[i], a[j]) ∈ R and (a[j], a[k]) ∈ R and (a[i], a[k]) ∉ R
                        then do flag := 0, R^t := R^t ∪ {(a[i], a[k])} end do
                    k := k + 1
                end while
                j := j + 1
            end while
            i := i + 1
            if flag = 1 then repeat := "no"
        end while
```
Output: R^t [a subset of $A \times A$]

end Algorithm

43. b. $R \cap S$ is symmetric: Suppose R and S are symmetric. [To show that $R \cap S$ is symmetric, we must show that $\forall x, y \in A$, if $(x, y) \in R \cap S$ then $(y, x) \in R \cap S$.] So suppose x and y are elements of A such that $(x, y) \in R \cap S$. By definition of intersection, $(x, y) \in R$ and $(x, y) \in S$. It follows that $(y, x) \in R$ because R is symmetric and $(x, y) \in R$, and also $(y, x) \in S$ because S is symmetric and $(x, y) \in S$. Thus by definition of intersection $(y, x) \in R \cap S$.

c. $R \cap S$ is transitive: Suppose R and S are transitive. [To show that $R \cap S$ is transitive, we must show that $\forall x, y, z \in A$, if $(x, y) \in R \cap S$ and $(y, z) \in R \cap S$ then $(x, z) \in R \cap S$.] So suppose x, y, and z are elements of A such that $(x, y) \in R \cap S$ and $(y, z) \in R \cap S$. By definition of intersection, $(x, y) \in R$, $(x, y) \in S$, $(y, z) \in R$, and $(y, z) \in S$. It follows that $(x, z) \in R$ because R is transitive and $(x, y) \in R$ and $(y, z) \in R$. Also $(x, z) \in S$ because S is transitive and $(x, y) \in S$ and $(y, z) \in S$. Thus by definition of intersection $(x, z) \in R \cap S$.

44. a. $R \cup S$ is reflexive: Suppose R and S are reflexive. [To show that $R \cup S$ is reflexive, we must show that $\forall x \in A$, $(x, x) \in R \cup S$.] So suppose $x \in A$. Since R is reflexive, $(x, x) \in R$, and since S is reflexive, $(x, x) \in S$. Thus by definition of union $(x, x) \in R \cup S$ [as was to be shown].

c. $R \cup S$ is not necessarily transitive: As a counterexample, let $R = \{(0, 1)\}$ and $S = \{(1, 2)\}$. Then both R and S are transitive (by default), but $R \cup S = \{(0, 1), (1, 2)\}$ is not transitive because $(0, 1) \in R \cup S$ and $(1, 2) \in R \cup S$ but $(0, 2) \notin R \cup S$. As another counterexample, let $R = \{(x, y) \in \mathbf{R} \times \mathbf{R} \mid x < y\}$ and let $S = \{(x, y) \in \mathbf{R} \times \mathbf{R} \mid x > y\}$. Then both R and S are transitive because of the transitivity of order for the real numbers. But $R \cup S = \{(x, y) \in \mathbf{R} \times \mathbf{R} \mid x \neq y\}$ is not transitive because, for instance, $(1, 2) \in R \cup S$ and $(2, 1) \in R \cup S$ but $(1, 1) \notin R \cup S$.

45. R_1 is not irreflexive because $(0, 0) \in R_1$. R_1 is not asymmetric because $(0, 1) \in R_1$ and $(1, 0) \in R_1$. R_1 is not intransitive because $(0, 1) \in R_1$ and $(1, 0) \in R_1$ and $(0, 0) \in R_1$.

46. R_2 is not irreflexive because $(0, 0) \in R_2$. R_2 is not asymmetric because $(0, 0) \in R_2$ and $(0, 0) \in R_2$. R_2 is not intransitive because $(0, 0) \in R_2$ and $(0, 1) \in R_2$ and $(0, 1) \in R_2$.

47. R_3 is irreflexive. R_3 is not asymmetric because $(2, 3) \in R_3$ and $(3, 2) \in R_3$. R_3 is intransitive.

48. R_4 is irreflexive. R_4 is not asymmetric because $(1, 2) \in R_4$ and $(2, 1) \in R_4$. R_4 is intransitive.

49. R_5 is not irreflexive because $(0,0) \in R_5$. R_5 is not asymmetric because $(0,0) \in R_5$ and $(0,0) \in R_5$. R_5 is not intransitive because $(0,1) \in R_5$ and $(1,2) \in R_5$ and $(0,2) \in R_5$.

50. R_6 is irreflexive. R_6 is asymmetric. R_6 is intransitive (by default).

51. R_7 is irreflexive. R_7 is asymmetric. R_7 is intransitive (by default).

52. R_8 is not irreflexive because $(0,0) \in R_8$. R_8 is not asymmetric because $(0,0) \in R_8$ and $(0,0) \in R_8$. R_8 is not intransitive because $(0,0) \in R_8$ (so a counterexample is $x = y = z = 0$).

Section 10.3

1. b. $\{(0,0), (1,1), (1,3), (1,4), (3,1), (3,3), (3,4), (4,1)(4,3), (4,4), (2,2)\}$

   c.
   $\{(0,0),(1,1),(1,2),(1,3),(1,4), (2,1),(2,2),(2,3),(2,4), (3,1),(3,2),(3,3),(3,4), (4,1),(4,2),(4,3),(4,4)\}$

3. distinct equivalence classes: $\{a\}, \{b,d\}, \{c\}$

5. distinct equivalence classes: $\{0,3,-3\}, \{1,4,-2\}, \{2,5,-1,-4\}$

7. distinct equivalence classes: $\{\emptyset\}, \{\{a\},\{b\},\{c\}\}, \{\{a,b\},\{a,c\},\{b,c\}\}, \{\{a,b,c\}\}$

9. distinct equivalence classes: $\{0,3,-3\}, \{1,4,-2,-5,5,-1,-4\}$

10. b. false   c. true   d. true

11. b. $[35] = [-7] = [0], [3] = [17], [12] = [-2]$

12. b. Proof: Suppose that m and n are integers such that $m \equiv n \pmod{d}$. [We must show that $m \bmod d = n \bmod d$.] By definition of congruence, $d \mid (m-n)$, and so by definition of divisibility $m - n = dk$ for some integer k. Let $m \bmod d = r$. Then $m = dl + r$ for some integer l. Since $m - n = dk$, then by substitution, $(dl + r) - n = dk$, or, equivalently, $n = d(l-k) + r$. Since $l - k$ is an integer, it follows by definition of mod, that $n \bmod d = r$ also, and so $m \bmod d = n \bmod d$ [as was to be shown].

    Suppose that m and n are integers such that $m \bmod d = n \bmod d$. [We must show that $m \equiv n \pmod{d}$.] Let $r = m \bmod d = n \bmod d$. Then by definition of mod, $m = dp + r$ and $n = dq + r$ for some integers p and q. By substitution, $m - n = (dp + r) - (dq + r) = d(p - q)$. Since $p - q$ is an integer, it follows that $d \mid (m - n)$, and so by definition of congruence, $m \equiv n \pmod{d}$.

13. b. Let $A_1 = \{1,2\}, A_2 = \{2,3\}, x = 1, y = 2$, and $z = 3$. Then both x and y are in A_1 and both y and z are in A_2, but x and z are not both in either A_1 or A_2.

14. b. There is one equivalence class for each age that is represented in the student body of the college. Each equivalence class consists of all the students of a given age.

16. Four distinct classes: $\{x \in \mathbf{Z} \mid x = 4k \text{ for some integer } k\}, \{x \in \mathbf{Z} \mid x = 4k + 1 \text{ for some integer } k\}, \{x \in \mathbf{Z} \mid x = 4k + 2 \text{ for some integer } k\}, \{x \in \mathbf{Z} \mid x = 4k + 3 \text{ for some integer } k\}$

17. There are $2^8 = 256$ distinct equivalence classes, one for each distinct truth table for a statement form in three variables. Each equivalence class consists of all statement forms with a given truth table.

18. There are as many distinct equivalence classes as there are distinct ordered pairs of the form (n, s) where n is a part number and s is a supplier name and s supplies a part with the number n. Each equivalence class consists of all parts that have the same part number and are shipped from the same supplier.

21. Since $m^2 - n^2 = (m-n)(m+n)$ for all integers m and n, $mDn \Leftrightarrow 3 \mid (m-n)$ or $3 \mid (m+n)$. Then by examining cases, one sees that $mDn \Leftrightarrow$ for some integers k and l either $(m = 3k$ and $n = 3l)$ or $(m = 3k+1$ and $n = 3l+1)$ or $(m = 3k+1$ and $n = 3l+2)$ or $(m = 3k+2$ and $n = 3l+1)$ or $(m = 3k+2$ and $n = 3l+2)$. Therefore, there are two distinct equivalence classes $\{m \in \mathbf{Z} \mid m = 3k$ for some integer $k\}$ and $\{m \in \mathbf{Z} \mid m = 3k+1$ or $m = 3k+2$ for some integer $k\}$.

23. There are as many distinct equivalence classes as there are distinct memory locations that are used to store variables during execution of the program. Each equivalence class consists of all variables that are stored in the same location.

25. There are as many distinct equivalence classes as there are points on a circle centered at the origin. Each equivalence class consists of all points that lie on the same half-line emanating from the origin.

26. The distinct equivalence classes can be identified with the points on a geometric figure, called a *torus*, that has the shape of the surface of a doughnut. Each point in the interior of the rectangle $\{(x,y) \mid 0 < x < 1$ and $0 < y < 1\}$ is only equivalent to itself. Each point on the top edge of the rectangle is in the same equivalence class as the point vertically below it on the bottom edge of the rectangle (so we can imagine identifying these points by gluing them together — this gives us a cylinder), and each point on the left edge of the rectangle is in the same equivalence class as the point horizontally across from it on the right edge of the rectangle (so we can also imagine identifying these points by gluing them together — this brings the two ends of the cylinder together to produce a torus).

27. *Proof:* Suppose R is an equivalence relation on a set A and $a \in A$. Because R is an equivalence relation, R is reflexive, and because R is reflexive each element of A is related to itself by R. In particular, $a\,R\,a$. Hence by definition of equivalence class, $a \in cl(a)$.

28. *Proof:* Suppose R is an equivalence relation on a set A, a and b are in A, and $b \in [a]$. By definition of equivalence class, $b\,R\,a$. But since R is an equivalence relation, R is symmetric; hence $a\,R\,b$.

30. *Proof:* Suppose R is an equivalence relation on a set A, a and b are in A, and $[a] = [b]$. Since R is reflexive, $a\,R\,a$, and so by definition of class, $a \in [a]$. *[Alternately, one could reference exercise 27 here.]* Since $[a] = [b]$, by definition of set equality, $a \in [b]$. But then by definition of equivalence class, $a\,R\,b$.

31. *Proof:* Suppose R is an equivalence relation on a set A, a and b are in A, and $a \in [b]$. By definition of class, $a\,R\,b$. We must show that $[a] = [b]$. To show that $[a] \subseteq [b]$, suppose $x \in [a]$. *[We must show that $x \in [b]$.]* By definition of class, $x\,R\,a$. By transitivity of R, since $x\,R\,a$ and $a\,R\,b$ then $x\,R\,b$. Thus by definition of class, $x \in [b]$ *[as was to be shown]*. To show that $[b] \subseteq [a]$, suppose $x \in [b]$. *[We must show that $x \in [a]$.]* By definition of class, $x\,R\,b$. But also $a\,R\,b$, and so by symmetry, $b\,R\,a$. Thus since R is transitive and since $x\,R\,b$ and $b\,R\,a$, then $x\,R\,a$. Therefore, by definition of class, $x \in [a]$ *[as was to be shown]*. Since we have proved both subset relations $[a] \subseteq [b]$ and $[b] \subseteq [a]$, we conclude that $[a] = [b]$.

32. There are, of course, an infinite number of answers to this exercise. One simple approach is to use one of the circuits given in the exercise, adding two NOT-gates that cancel each other out, or adding an AND-gate and a circuit for, say, $Q \vee \sim Q$, or an OR-gate and a circuit for, say, $Q \wedge \sim Q$. Another approach is to play with the input-output table directly, looking at it as, for instance, the table for $\sim((P \wedge Q) \vee (P \wedge \sim Q) \vee (\sim P \wedge Q))$. Some possible answers are shown below.

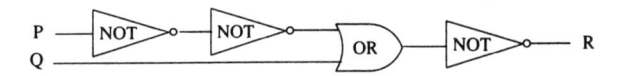

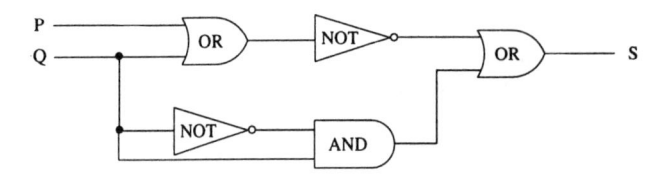

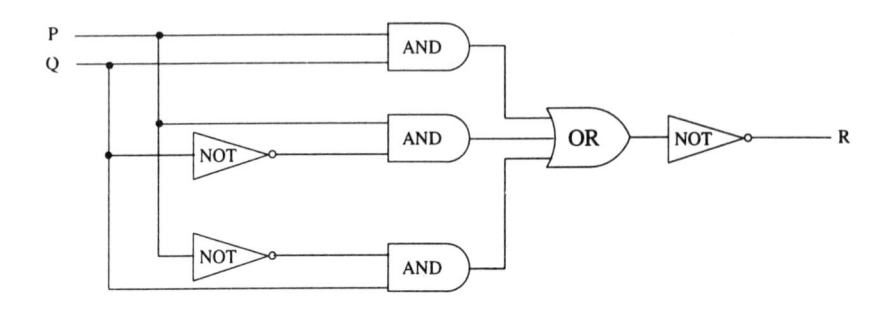

33. *a.* Suppose $(a,b) \in A$. By commutativity of multiplication for the real numbers, $ab = ba$. But then by definition of R, $(a,b)R(a,b)$.

b. Suppose $(a,b), (c,d) \in A$ and $(a,b)R(c,d)$. By definition of R, $ad = bc$, and so by commutativity of multiplication for the real numbers and symmetry of equality, $cb = da$. But then by definition of R, $(c,d)R(a,b)$.

d. For example, (2,5), (4,10), (-2,-5), and (6,15) are all in $[(2,5)]$.

34. *b. Proof:* Suppose (a,b), (a',b'), (c,d), and (c',d') are any elements of A such that $[(a,b)] = [(a',b')]$ and $[(c,d)] = [(c',d')]$. By definition of the relation, a, a', c, and c' are integers and b, b', d, and d' are nonzero integers, and $ab' = a'b$ (*) and $cd' = c'd$ (**). We must show that $[(a,b)] \cdot [(c,d)] = [(a',b')] \cdot [(c',d')]$. By definition of the multiplication, this equation holds if, and only if, $[(ac,bd)] = [(a'c',b'd')]$. And by definition of the relation, this equation holds if, and only if, $ac \cdot b'd' = bd \cdot a'c'$. (***) But multiplying equations (*) and (**) gives $ab' \cdot cd' = a'b \cdot c'd$. And by the associative and commutative laws for real numbers, this equation is equivalent to (***). Hence $[(a,b)] \cdot [(c,d)] = [(a',b')] \cdot [(c',d')]$. *[as was to be shown].*

d. The identity element for multiplication is $[(1,1)]$. To prove this, suppose (a,b) is any element of A. We must show that $[(a,b)] \cdot [(1,1)] = [(a,b)]$. But by definition of the multiplication this equation holds if, and only if, $[(a \cdot 1, b \cdot 1)] = [(a,b)]$. By definition of the relation, this equation holds if, and only if, $a \cdot 1 \cdot b = b \cdot 1 \cdot a$, which is true for all integers a and b. Thus $[(a,b)] \cdot [(1,1)] = [(a,b)]$ *[as was to be shown]*.

e. Given any $(a,b) \in A$ with $a \neq 0$, $[(b,a)]$ is an inverse for multiplication for $[(a,b)]$. To prove this, we must show that $[(a,b)] \cdot [(b,a)] = [(1,1)]$, which (by part (d)) is the identity element for multiplication. But by definition of the multiplication, $[(a,b)] \cdot [(b,a)] = [(ab,ba)]$. And by definition of the relation, $[(ab,ba)] = [(1,1)]$ if, and only if, $ab \cdot 1 = ba \cdot 1$, which is true for all integers a and b. Thus $[(a,b)] \cdot [(b,a)] = [(1,1)]$ *[as was to be shown]*.

35. *b.* Let (a,b) and (c,d) be any elements of $A = \mathbf{Z}^+ \times \mathbf{Z}^+$, and suppose $(a,b) \ R \ (c,d)$. *[We must show that $(c,d) \ R \ (a,b)$].* By definition of R, $a + d = c + b$, and so by the symmetry property of equality, $c + b = a + d$. But then by definition of R, $(c,d) \ R \ (a,b)$ *[as was to be shown]*.

c. Proof: Let $(a,b), (c,d),$ and (e,f) be any elements of $A = \mathbf{Z}^+ \times \mathbf{Z}^+$, and suppose $(a,b) \ R \ (c,d)$ and $(c,d) \ R \ (e,f)$. *[We must show that $(a,b) \ R \ (e,f)$].* By definition of R, $a + d = c + b$ (*) and $c + f = e + d$ (**). Adding (*) and (**) together gives $a + d + c + f = c + b + e + d$, and subtracting $c + d$ from both sides gives $a + f = b + e$. Then by definition of R, $(a,b) \ R \ (e,f)$ *[as was to be shown]*.

e. One possible answer: $(4,2), (5,3), (6,4), (7,5), (8,6)$

f. One possible answer: $(2,3), (3,4), (4,5), (5,6), (6,7)$

36. The given argument assumes that from the fact that the statement "$\forall x$ in A, if $x \ R \ y$ then $y \ R \ x$" is true, it follows that given any element x in R, there must exist an element y in R such that $x \ R \ y$ and $y \ R \ x$. This is false. For instance, consider the following binary relation R defined on $A = \{1, 2\} : R = \{(1, 1)\}$. This relation is symmetric and transitive, but it is not reflexive. Given $2 \in A$, there is no element y in A such that $(2, y) \in R$. Thus we cannot go on to use symmetry to say that $(y, 2) \in R$ and transitivity to conclude that $(2, 2) \in R$.

37. *Proof*: Suppose R is a binary relation on a set A, R is symmetric and transitive, and for every x in A there is a y in A such that $x \ R \ y$. Suppose x is any particular but arbitrarily chosen element of A. By hypothesis, there is a y in A such that $x \ R \ y$. By symmetry, $y \ R \ x$, and so by transitivity $x \ R \ x$. Therefore, R is reflexive. Since we already know that R is symmetric and transitive, we conclude that R is an equivalence relation.

38. *a.* Haddock's Eyes   *b.* The Aged Aged Man   *c.* Ways and Means   *d.* A-sittin on a Gate

Section 10.4

2. *a.* 0-equivalence classes: $\{s_0, s_1, s_3, s_4\}, \{s_2, s_5, s_6\}$

   1-equivalence classes: $\{s_0, s_1\}, \{s_3\}, \{s_4\}, \{s_2, s_5\}, \{s_6\}$

   2-equivalence classes: $\{s_0, s_1\}, \{s_3\}, \{s_4\}, \{s_2, s_5\}, \{s_6\}$

   b. transition diagram for $\bar{A}$:

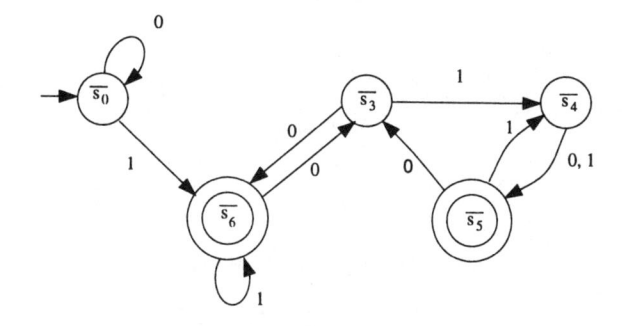

3. *a.* 0-equivalence classes: $\{s_1, s_3\}, \{s_0, s_2\}$

   1-equivalence classes: $\{s_1, s_3\}, \{s_0, s_2\}$

b. transition diagram for $\bar{A}$:

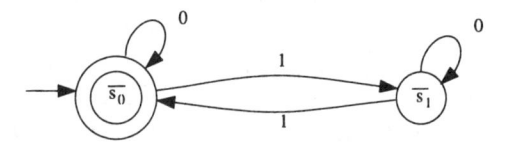

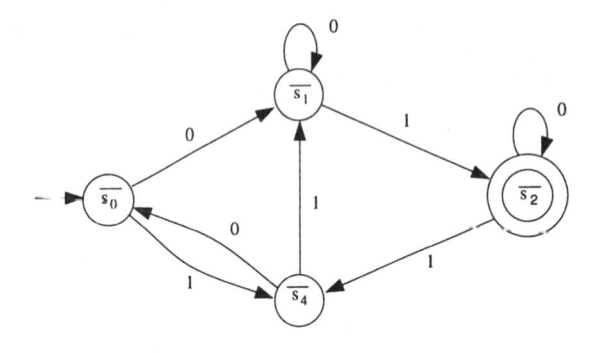

5. a. 0-equivalence classes: $\{s_0, s_1, s_3, s_4\}, \{s_2, s_5\}$
   1-equivalence classes: $\{s_0, s_3, s_4\}, \{s_1\}, \{s_2, s_5\}$
   2-equivalence classes: $\{s_0, s_3\}, \{s_4\}, \{s_1\}, \{s_2, s_5\}$
   3-equivalence classes: $\{s_0, s_3\}, \{s_4\}, \{s_1\}, \{s_2, s_5\}$
   b. transition diagram for $\bar{A}$:

6. a. 0-equivalence classes: $\{s_0, s_1, s_3, s_4, s_5\}, \{s_2, s_6\}$
   1-equivalence classes: $\{s_0, s_4, s_5\}, \{s_1, s_3\}, \{s_2\}, \{s_6\}$
   2-equivalence classes: $\{s_0, s_4\}, \{s_5\}, \{s_1\}, \{s_3\}, \{s_2\}, \{s_6\}$
   3-equivalence classes: $\{s_0\}, \{s_4\}, \{s_5\}, \{s_1\}, \{s_3\}, \{s_2\}, \{s_6\}$
   b. The transition diagram for $\overline{A}$ is the same as the one given for A except that the states are denoted $\overline{s_0}, \overline{s_1}, \overline{s_2}, \overline{s_3}, \overline{s_4}, \overline{s_5}, \overline{s_6}$.

8. For A:
   0-equivalence classes: $\{s_2, s_4\}, \{s_0, s_1, s_3\}$
   1-equivalence classes: $\{s_2, s_4\}, \{s_0, s_1\}, \{s_3\}$
   2-equivalence classes: $\{s_2, s_4\}, \{s_0, s_1\}, \{s_3\}$
   Therefore, the states of $\overline{A}$ are the 2-equivalence classes of A.
   For A':
   0-equivalence classes: $\{s'_4\}, \{s'_0, s'_1, s'_2, s'_3\}$
   1-equivalence classes: $\{s'_4\}, \{s'_1, s'_3\}, \{s'_0, s'_2\}$
   2-equivalence classes: $\{s'_4\}, \{s'_1, s'_3\}, \{s'_0, s'_2\}$
   Therefore, the states of $\overline{A'}$ are the 2-equivalence classes of A'.

   According to the text, two automata are equivalent if, and only if, their quotient automata are isomorphic, provided inaccessible states have first been removed. Now A and A' have no inaccessible states, and $\overline{A}$ has one accepting state and two nonaccepting states as does $\overline{A'}$. But the labels on the arrows connecting the states are different. For instance, in both quotient automata, there is one nonaccepting state which has an arrow going out from it to the accepting state and an arrow going back from the accepting state to it. But for $\overline{A}$, the label on the arrow going to the accepting state is labeled 0 whereas for $\overline{A'}$ it is labeled 1.

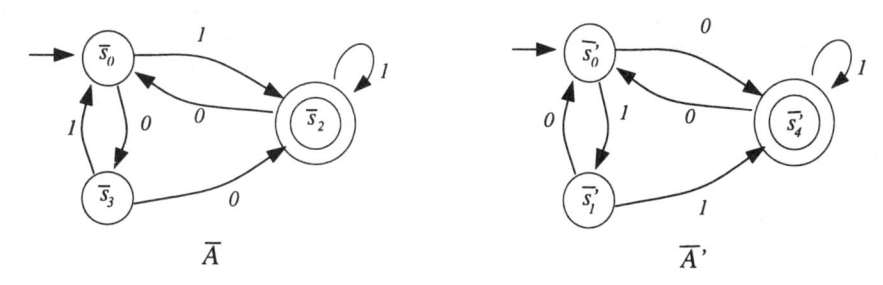

The nonequivalence of A and A' can also be seen by noting, for example, that the string 00 is accepted by A but not by A'.

10. For A:

0-equivalence classes: $\{s_0, s_1, s_2, s_3\}, \{s_4\}$

1-equivalence classes: $\{s_0, s_1, s_2\}, \{s_3\}, \{s_4\}$

2-equivalence classes: $\{s_0\}, \{s_1, s_2\}, \{s_3\}, \{s_4\}$

3-equivalence classes: $\{s_0\}, \{s_1, s_2\}, \{s_3\}, \{s_4\}$

Therefore, the states of $\overline{A}$ are the 3-equivalence classes of A.

For A':

0-equivalence classes: $\{s'_0, s'_1, s'_3\}, \{s'_2, s'_4\}$

1-equivalence classes: $\{s'_0\}, \{s'_1\}, \{s'_3\}, \{s'_2, s'_4\}$

2-equivalence classes: $\{s'_0\}, \{s'_1\}, \{s'_3\}, \{s'_2\}, \{s'_4\}$

Therefore, the states of $\overline{A'}$ are the 2-equivalence classes of A'.

According to the text, two automata are equivalent if, and only if, their quotient automata are isomorphic, provided inaccessible states have first been removed. Now A and A' have no inaccessible states, and $\overline{A}$ has four states whereas $\overline{A'}$ has five states. Therefore A and A' are not equivalent. This result can also be obtained by noting, for example, that the string 10 is accepted by A' but not by A.

11. *Proof:* Suppose A is a finite-state automaton with set of states S and relation R_* of $*$-equivalence of states. We will show that R_* is reflexive, symmetric, and transitive.

R_* *is reflexive*: Suppose that s is a state of A. It is certainly true that for all input strings w, $N^*(s, w)$ is an accepting state $\Leftrightarrow N^*(s, w)$ is an accepting state. So by definition of R_*, $s\ R_* s$.

R_* *is symmetric*: This is proved in Appendix B of the text.

R_* *is transitive*: Suppose that s, t, and u are states of A such that $s\ R_* t$ and $t\ R_* u$. By definition of R_*, for all input strings w, $N^*(s, w)$ is an accepting state $\Leftrightarrow N^*(t, w)$ is an accepting state and $N^*(t, w)$ is an accepting state $\Leftrightarrow N^*(u, w)$ is an accepting state. It follows by transitivity of the $\Leftrightarrow$ relation that $N^*(s, w)$ is an accepting state $\Leftrightarrow N^*(u, w)$ is an accepting state. Hence by definition of R_*, $s\ R_* u$.

Since R_* is reflexive, symmetric, and transitive, R_* is an equivalence relation.

14. *Proof:* Suppose k is an integer such that $k \geq 1$ and states s and t are k-equivalent. Then for all input strings w of length less than or equal to k, $N^*(s, w)$ is an accepting state $\Leftrightarrow N^*(t, w)$ is an accepting state. Since $k - 1 < k$, it follows that for all input strings w of length less than

or equal to $k-1$, $N^*(s,w)$ is an accepting state $\Leftrightarrow N^*(t,w)$ is an accepting state. Hence s and t are $(k-1)$-equivalent.

15. *Proof:* Suppose k is an integer such that $k \geq 1$ and C_k is a k-equivalence class. We must show that there is a $k-1$ equivalence class, C_{k-1}, such that $C_k \subseteq C_{k-1}$. By property (10.4.3), the $(k-1)$-equivalence classes partition the set of all states of A in to a union of mutually disjoint subsets. Let s be any state in C_k. Then s is in *some* $(k-1)$-equivalence class; call it C_{k-1}. Let t be any other state in C_k. *[We will show that $t \in C_{k-1}$ also.]* Then $t\ R_k\ s$, and so for all input strings of length k, $N^*(t,w)$ is an accepting state $\Leftrightarrow N^*(s,w)$ is an accepting state. Since $k-1 < k$, it follows that for all input strings of length $k-1$, $N^*(t,w)$ is an accepting state $\Leftrightarrow N^*(s,w)$ is an accepting state. Consequently, $t\ R_{k-1}\ s$, and so t and s are in the same $(k-1)$-equivalence class. But $s \in C_{k-1}$. Hence $t \in C_{k-1}$ also. We, therefore, conclude that $C_k \subseteq C_{k-1}$.

16. *Proof:* Suppose s and t are states that are k-equivalent for all integers $k \geq 0$. Let w be any *[particular but arbitrarily chosen]* input string and let the length of w be l. Then $l \geq 0$ and so by hypothesis, s and t are R_l-equivalent. By definition of R_l, $N^*(s,w)$ is an accepting state $\Leftrightarrow N^*(t,w)$ is an accepting state. Since the choice of w was arbitrary, we conclude that for all input strings w, $N^*(s,w)$ is an accepting state $\Leftrightarrow N^*(t,w)$ is an accepting state. Thus by definition of $*$-equivalence, s and t are $*$-equivalent.

17. *Proof:* Suppose k is an integer such that states s and t are k-equivalent and suppose that m is a nonnegative integer less than k. Then for all input strings w of length less than or equal to k, $N^*(s,w)$ is an accepting state $\Leftrightarrow N^*(t,w)$ is an accepting state. Now since m is a nonnegative integer and $m < k$, then any string of length less than or equal to m has length less than or equal to k. Consequently, for all input strings w of length less than or equal to m, $N^*(s,w)$ is an accepting state $\Leftrightarrow N^*(t,w)$ is an accepting state. Hence s and t are m-equivalent.

18. *Proof:* Suppose A is an automaton and C is a $*$-equivalence class of states of A. By Theorem 10.4.2, for some integer $K \geq 0$, C is a K-equivalence class of A. Suppose C contains both an accepting state s and a nonaccepting state t of A. Since both s and t are in the same K-equivalence class, s is K-equivalent to t (by exercise 30 of Section 10.3), and so by exercise 17, s is 0-equivalent to t. But this is impossible because there are only two 0-equivalence classes, the set of all accepting states and the set of all nonaccepting states, and these two sets are disjoint. Hence the supposition that C contains both an accepting and a nonaccepting state is false: C consists entirely of accepting states or entirely of nonaccepting states.

19. *Proof:* Suppose A is an automaton and states s and t of A are $*$-equivalent. Let m be any input symbol and let w be any input string. By definition of the next-state and eventual-state functions, $N^*(N(s,m),w) = N^*(s,wm)$ and $N^*(N(t,m),w) = N^*(t,wm)$, where wm is the concatenation of w and m. But since s and t are $*$-equivalent, $N^*(s,wm)$ is an accepting state $\Leftrightarrow N^*(t,wm)$ is an accepting state. Hence $N^*(N(s,m),w)$ is an accepting state $\Leftrightarrow N^*(N(t,m),w)$ is an accepting state. So by definition of $*$-equivalence, $N(s,m)$ is $*$-equivalent to $N(t,m)$.

Section 10.5

1. c.

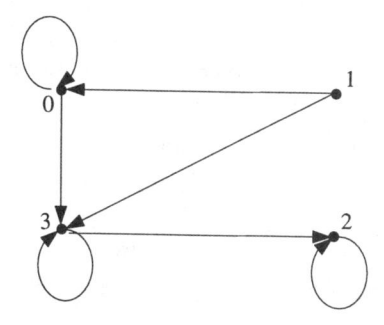

   R_3 is antisymmetric: there are no cases where $a\ R\ b$ and $b\ R\ a$ and $a \neq b$.

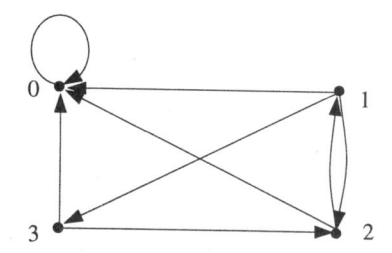

   d.

   R_4 is not antisymmetric: $1\ R_4\ 2$ and $2\ R_4\ 1$ and $1 \neq 2$.

3. R is not antisymmetric. *Counterexample*: Let $s = 0$ and $t = 1$. Then $\ell(s) \leq \ell(t)$ and $\ell(t) \leq \ell(s)$, because both $\ell(s)$ and $\ell(t)$ equal 1, but $s \neq t$.

4. R is antisymmetric.

   Proof 1: The statement "For all real numbers x and y, if $x < y$ and $y < x$, then $x = y$" is vacuously true. By the trichotomy law (Appendix A, T16), there are no real numbers x and y such that $x < y$ and $y < x$.

   Proof 2 (by contradiction): Suppose R is not antisymmetric. Then there exist real numbers x and y such that $x < y$ and $y < x$ (and $x \neq y$). But this contradicts the trichotomy law (Appendix A, T16) which says that both $x < y$ and $y < x$ are not simultaneously true. *[Hence the supposition is false and so R is antisymmetric.]*

6. R is a partial order relation.

   Proof:

   R is reflexive: Suppose $r \in P$. Then $r = r$, and so by definition of R, $r\ R\ r$.

   R is antisymmetric: Suppose $r, s \in P$ and $r\ R\ s$ and $s\ R\ r$. *[We must show that $r = s$.]* By definition of R, either r is an ancestor of s or $r = s$ and either s is an ancestor of r or $s = r$. Now it is impossible for both r to be an ancestor of s and s to be an ancestor of r. Hence one of these conditions must be false, and so $r = s$ *[as was to be shown]*.

   R is transitive: Suppose $r, s, t \in P$ and $r\ R\ s$ and $s\ R\ t$. *[We must show that $r\ R\ t$.]* By definition of R, either r is an ancestor of s or $r = s$ and either s is an ancestor of t or $s = t$. In case r is an ancestor of s and s is an ancestor of t, then r is an ancestor of t, and so $r\ R\ t$. In case r is an ancestor of s and $s = t$, then r is an ancestor of t, and so $r\ R\ t$. In case $r = s$ and s

is an ancestor of t, then r is an ancestor of t, and so $r\ R\ t$. In case $r = s$ and $s = t$, then $r = t$, and so $r\ R\ t$. Thus in all four possible cases, $r\ R\ t$ [as was to be shown].

Since R is reflexive, antisymmetric, and transitive, R is a partial order relation.

7. R is not a partial order relation because R is not antisymmetric. *Counterexample*: Let $m = 2$ and $n = 4$. Then $m\ R\ n$ because every prime factor of 2 is a prime factor of 4, and $n\ R\ m$ because every prime factor of 4 is a prime factor of 2. But $m \neq n$ because $2 \neq 4$.

9. R is not a partial order relation because R is not antisymmetric. *Counterexample*: Let $x = 2$ and $y = -2$. Then $x\ R\ y$ because $(-2)^2 \leq 2^2$, and $y\ R\ x$ because $2^2 \leq (-2)^2$. But $x \neq y$ because $2 \neq -2$.

10. No. *Counterexample*: Let $A = \{1, 2\}$, $R = \{(1, 2)\}$, and $S = \{(2, 1)\}$. Then R and S are antisymmetric (by default), but $R \cup S$ is not antisymmetric because $(1, 2) \in R \cup S$ and $(2, 1) \in R \cup S$ and $1 \neq 2$.

11. *c.* True, by (3).   *d.* True, by (2).   *e.* False. By (2), $bbaa \preceq bbab$.
    f. True, by (1).   *g.* True, by (2).

12. *Proof:*

    $\preceq$ *is reflexive*: Suppose s is in Σ^*. If $s = \epsilon$, then $s \preceq s$ by (3). If $s \neq \epsilon$, then $s \preceq s$ by (1). Hence in either case, $s \preceq s$.

    $\preceq$ *is symmetric*: Suppose s and t are in Σ^* and $s \preceq t$ and $t \preceq s$. [We must show that $s = t$.] By definition of Σ^*, either $s = \epsilon$ or $s = a_1 a_2 \ldots a_m$ and either $t = \epsilon$ or $t = b_1 b_2 \ldots b_n$ for some positive integers m and n and elements $a_1, a_2, \ldots, a_m$ and $b_1, b_2, \ldots, b_n$ in Σ. It is impossible to have $s \preceq t$ by virtue of condition (2) because in that case there is no condition that would give $t \preceq s$. [For suppose $s \preceq t$ by virtue of condition (2). Then for some integer k with $k \leq m$, $k \leq n$, and $k \geq 1$, $a_i = b_i$ for all $i = 1, 2, \ldots, k-1$, and $a_k R\ b_k$ and $a_k \neq b_k$. In this situation, it is clearly impossible for $t \preceq s$ by virtue either of condition (1) or (3), and so if $t \preceq s$, then it must be by virtue of condition (2). But in that case, since $a_k \neq b_k$, it must follow that $b_k R\ a_k$, and so since R is a partial order relation, $a_k = b_k$. However, this contradicts the fact that $a_k \neq b_k$. Hence it cannot be the case that $s \preceq t$ by virtue of condition (2).] Similarly, it is impossible for $t \preceq s$ by virtue of condition (2). Hence $s \preceq t$ and $t \preceq s$ by virtue either of condition (1) or of condition (3). In case $s \preceq t$ by virtue of condition (1), then neither s nor t is the null string and so $t \preceq s$ by virtue of condition (1) also. Then by (1) $m \leq n$ and $a_i = b_i$ for all $i = 1, 2, \ldots, m$ and $n \leq m$ and $b_i = a_i$ for all $i = 1, 2, \ldots, m$, and so in this case $s = t$. In case $s \preceq t$ by virtue of condition (3), then $s = \epsilon$, and so since $t \preceq s$, $t \preceq \epsilon$. But the only condition that can give this result is (3) with $t = \epsilon$. Hence in this case, $s = t = \epsilon$. Thus in all possible cases, if $s \preceq t$ and $t \preceq s$, then $s = t$ [as was to be shown].

    $\preceq$ *is transitive*: Suppose s and t are in Σ^* and $s \preceq t$ and $t \preceq u$. [We must show that $s = u$.] By definition of Σ^*, either $s = \epsilon$ or $s = a_1 a_2 \ldots a_m$, either $t = \epsilon$ or $t = b_1 b_2 \ldots b_n$, and either $u = \epsilon$ or $u = c_1 c_2 \ldots c_o$ for some positive integers m, n, and o and elements $a_1, a_2, \ldots, a_m$, $b_1, b_2, \ldots, b_n$, and $c_1, c_2, \ldots, c_o$ in Σ.

    Case 1 ($s = \epsilon$): In this case, $s\ R\ u$ by (3).

    Case 2 ($s \neq \epsilon$): In this case, since $s\ R\ t$, $t \neq \epsilon$ either, and since $t\ R\ u$, $u \neq \epsilon$ either.

    Subcase a ($s\ R\ t$ and $t\ R\ u$ by condition (1)): Then $m \leq n$ and $n \leq o$ and $a_i = b_i$ for all $i = i, 2, \ldots, m$ and $b_j = c_j$ for all $j = 1, 2, \ldots, n$. It follows that $a_i = c_i$ for all $i = i, 2, \ldots, m$, and so by (1), $s\ R\ u$.

    Subcase b ($s\ R\ t$ by condition (1) and $t\ R\ u$ by condition (2)): Then $m \leq n$ and $a_i = b_i$ for all $i = i, 2, \ldots, m$, and for some integer k with $k \leq n$, $k \leq o$, and $k \geq 1$, $b_j = c_j$ for all $j = 1, 2, \ldots, k-1$, $b_k R\ c_k$, and $b_k \neq c_k$. If $k \leq m$, then s and u satisfy condition (2) [because

Ship returns to:

THOMSON LEARNING
DISTRIBUTION CENTER
7625 EMPIRE DRIVE
FLORENCE, KY 41042

The enclosed materials are sent to you for your review by
MICHAEL LEE 800 8762350X7007

SALES SUPPORT

SHIP TO: THOMAS PETERS
UNIV OF CONNECTICUT
COMPUTER SCIENCE DEPT.
191 AUDITORIUM RD UNIT 3155
STORRS CT 06269

WAREHOUSE INSTRUCTIONS

SLA: 7   BOX: Staple

LOCATION	QTY	ISBN	AUTHOR/TITLE
K-35H-003-02	1	0-534-94449-3	EPP INST S.M. DISC MATH 2E

INV# 47066805SM
PO# 33917323
DATE: 09/05/02
CARTON: 1 of 1
ID# 1323670

PRIME-INDCT-UPSTRS
-SLSB

VIA: U2

PAGE 1 OF 1

BATCH: 1441132

001/016

$a_i = b_i$ for all $i = 1, 2, \ldots, m$ and so $k \le m$, $k \le o$, $k \ge 1$, $a_i = b_i$ for all $k = 1, 2, \ldots, k-1$, $a_k R\, b_k$, and $a_k \ne b_k$]. If $k > m$, then s and u satisfy condition (1) [because $a_i = b_i = c_i$ for all $i = 1, 2, \ldots, m$]. Thus in either case $s\ R\ u$.

Subcase c ($s\ R\ t$ *by condition (2) and* $t\ R\ u$ *by condition (1)*): Then for some integer k with $k \le m$, $k \le n$, $k \ge 1$, $a_i = b_i$ for all $i = 1, 2, \ldots, k-1$, $a_k R\, b_k$, and $a_k \ne b_k$, and $n \le o$ and $b_j = c_j$ for all $j = i, 2, \ldots, n$. Then s and u satisfy condition (2) [because $k \le n$, $k \le o$ (since $k \le n$ and $n \le o$), $k \ge 1$, $a_i = b_i = c_i$ for all $i = 1, 2, \ldots, k-1$ (since $k-1 < n$), $a_k R c_k$ (since $b_k = c_k$ because $k \le n$), and $a_k \ne c_k$ (since $b_k = c_k$ and $a_k \ne b_k$)]. Thus $s\ R\ u$.

Subcase d ($s\ R\ t$ *by condition (2) and* $t\ R\ u$ *by condition (2)*): Then for some integer k with $k \le m$, $k \le n$, $k \ge 1$, $a_i = b_i$ for all $i = 1, 2, \ldots, k-1$, $a_k R\, b_k$, and $a_k \ne b_k$, and for some integer l with $l \le n$, $l \le o$, and $l \ge 1$, $b_j = c_j$ for all $j = 1, 2, \ldots, l-1$, $b_l R\, c_l$, and $b_l \ne c_l$. If $k < l$, then $a_i = b_i = c_i$ for all $i = 1, 2, \ldots, k-1$, $a_k R\, b_k$, $b_k = c_k$ (in which case $a_k R\, c_k$), and $a_k \ne c_k$ (since $a_k \ne b_k$). Thus if $k < l$, then $s \preceq u$ by condition (2). If $k = l$, then $b_k R\, c_k$ (in which case $a_k R\, c_k$ by transitivity of R) and $b_k \ne c_k$. It follows that $a_k \ne c_k$ [for if $a_k = c_k$, then $a_k R\, b_k$ and $b_k R\, a_k$, which implies that $a_k = b_k$ (since R is a partial order) and contradicts the fact that $a_k \ne b_k$]. Thus if $k = l$, then $s \preceq u$ by condition (2). If $k > l$, then $a_i = b_i = c_i$ for all $i = 1, 2, \ldots, l-1$, $a_l R\, c_l$ (because $b_l R\, c_l$ and $a_l = b_l$), $a_l \ne c_l$ (because $b_l \ne c_l$ and $a_l = b_l$). Thus if $k > l$, then $s \preceq u$ by condition (2). Hence in all cases $s \preceq u$.

The above arguments show that in all possible cases, $s \preceq u$ [as was to be shown]. Hence $\preceq$ is transitive.

Since $\preceq$ *is* reflexive, antisymmetric, and transitive, $\preceq$ *is* a partial order relation.

14. b. $\{(a,a),(b,b),(c,c)\}$, $\{(a,a),(b,b),(c,c),(a,b)\}$,
$\{(a,a),(b,b),(c,c),(a,c)\}$, $\{(a,a),(b,b),(c,c),(a,b),(a,c)\}$,
$\{(a,a),(b,b),(c,c),(a,b),(b,c)\}$, $\{(a,a),(b,b),(c,c),(a,c),(c,b)\}$,
$\{(a,a),(b,b),(c,c),(b,c)\}$, $\{(a,a),(b,b),(c,c),(c,b)\}$,
$\{(a,a),(b,b),(c,c),(a,b),(c,b)\}$, $\{(a,a),(b,b),(c,c),(a,c),(b,c)\}$

16. b.

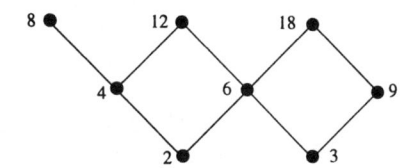

17. *b*.

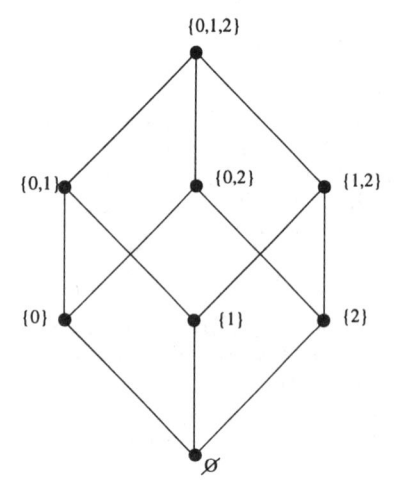

```
                    {0,1,2}
                   /   |   \
                  /    |    \
            {0,1}   {0,2}   {1,2}
             / \   / \ / \   / \
            /   \ /   X   \ /   \
           /    / \  / \  / \    \
          {0}    {1}    {2}
            \     |     /
             \    |    /
                  ∅
```

19.

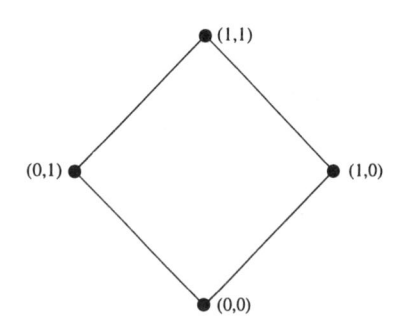

```
           (1,1)
          /     \
      (0,1)    (1,0)
          \     /
           (0,0)
```

20.

```
                 (1,1,1)
                /   |   \
          (0,1,1) (1,0,1) (1,1,0)
            / \   / \ / \   / \
          (0,0,1) (0,1,0) (1,0,0)
             \     |     /
                (0,0,0)
```

254

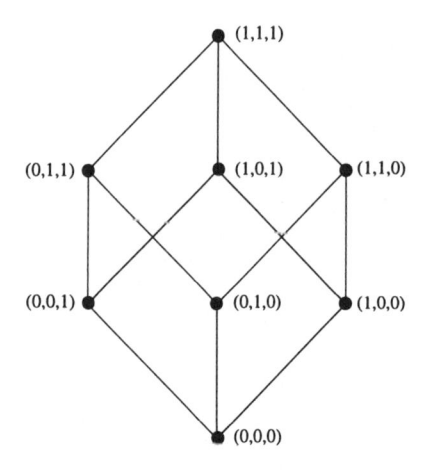

21. b.

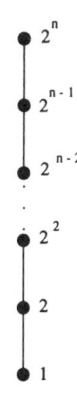

```
• 2^n
|
• 2^{n-1}
|
• 2^{n-2}
.
.
• 2^2
|
• 2
|
• 1
```

23. greatest element: none   least element: none

    maximal elements: 8, 12, 18   minimal elements: 2, 3

25. greatest element: $\{0, 1, 2\}$ least element: $\emptyset$

    maximal elements: $\{0, 1, 2\}$ minimal elements: $\emptyset$

27. greatest element: (1,1)   least element: (0,0)

    maximal elements: (1,1)   minimal elements: (0,0)

28. greatest element: (1,1,1)   least element: (0,0,0)

    maximal elements: (1,1,1)   minimal elements: (0,0,0)

29. greatest element: 2^n least element: 1

    maximal elements: 2^n minimal elements: 1

30. c. no greatest element and no least element

    d. greatest element: 9   least element: 1

31. There are $n!$ total orderings on a set with n elements because $n!$ is the number of ways to write n elements in a row.

32. *Proof:* Let R' be the restriction of R to B. Then R' is reflexive because given any x in B, $(x, x) \in R$ since R is reflexive and $B \subseteq A$, and so $(x, x) \in R'$ by definition of R'. Furthermore, R' is antisymmetric because given any x and y in B such that $(x, y) \in R'$ and $(y, x) \in R'$, then $(x, y) \in R$ and $(y, x) \in R$ by definition of R' and since R is antisymmetric, $x = y$. Finally, R' is transitive because given any x, y, and z in B such that $(x, y) \in R'$ and $(y, z) \in R'$, then by definition of R', $(x, y) \in R$ and $(y, z) \in R$. Since R is transitive, $(x, z) \in R$, and so by definition of R', $(x, z) \in R'$. Since R' is reflexive, antisymmetric, and transitive, R' is a partial order relation on B.

34. 2, 4, 12, 24 or 3, 6, 12, 24

35. (0,0), (0,1), (1,1) or (0,0), (1,0), (1,1)

37. b. This proof is identical to that given in part (a) provided the following changes are made: (1) Change "minimal" to "maximal" throughout the entire proof; (2) Change "$\preceq$" to "$\succeq$" and "$\succeq$" to "$\preceq$" throughout step 2 of the proof.

38. *a. Proof:* Suppose A is any finite partially ordered set, ordered with respect to a relation $\preceq$, and a and b are greatest elements of A. By definition of greatest element, $x \preceq a$ for all x in A; in particular, $b \preceq a$. Similarly, $x \preceq b$ for all x in A, and so $a \preceq b$. Since $\preceq$ is a partial order, it is antisymmetric, and thus $a = b$. Hence A has at most one greatest element.

b. Proof: Suppose A is any finite partially ordered set, ordered with respect to a relation $\preceq$, and a and b are least elements of A. By definition of least element, $a \preceq x$ for all x in A; in particular, $a \preceq b$. Similarly, $b \preceq x$ for all x in A, and so $b \preceq a$. Since $\preceq$ is a partial order, it is antisymmetric, and thus $a = b$. Hence A has at most one least element.

40.

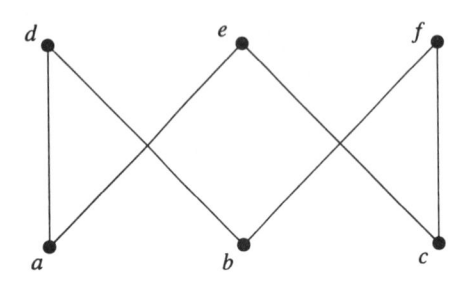

42. One such total order is 3,9,2,6,18,4,12,8.

44. One such total order is (0,0,0),(0,0,1),(0,1,0),(0,1,1),(1,0,0),(1,0,1),(1,1,0),(1,1,1).

45. One such total order is $\emptyset, \{a\}, \{b\}, \{c\}, \{d\}, \{a,b\}, \{a,c\}, \{a,d\}, \{b,c\}, \{b,d\}, \{c,d\},$
$\{a,b,c\}, \{a,b,d\}, \{a,c,d\}, \{b,c,d\}, \{a,b,c,d\}.$

46. *a.* One such order is 1,6,9,10,5,7,2,8,4,3.

    b. (i) Annotate the given Hasse diagram by indicating in boxes the least number of days needed to accomplish each job, taking into account the time needed to perform prerequisite jobs.

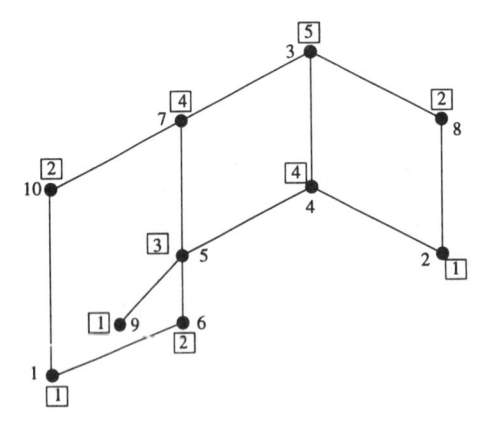

    Therefore, at least five days are needed to perform all ten jobs.

    (ii) At most four jobs can be performed at the same time. For instance, 10, 9, 6 and 2 could be performed simultaneously.

47. *a.* 33 hours

Chapter 11 Graphs and Trees

The first section of this chapter introduces the terminology of graph theory, illustrating it in a variety of different instances. Several exercises are designed to clarify the distinction between a graph and a drawing of a graph. You might point out to students the advantage of the formal definition over the informal drawing for computer representation of graphs. Other exercises explore the use of graphs to solve problems of various sorts. In some cases, students may be able to solve the given problems, such as the wolf, the goat, the cabbage and the ferryman, more easily without using graphs than using them. The point to make is that such problems *can* be solved using graphs and that for more complex problems involving, say, hundreds of possible states, a graphical representation coupled with a computer path-finding algorithm makes it possible find a solution that could not be discovered by trial-and-error alone. The rest of the exercises in this section give students practice in applying the theorem that relates the total degree of a graph to the number of its edges, especially for exploring properties of simple graphs, complete graphs, and bipartite graphs.

In Section 11.2 the general topic of paths and circuits is discussed, including the notion of connectedness and Euler and Hamiltonian circuits. As throughout the chapter, an attempt is made to balance the presentation of theory and application so that you can create whatever mix seems most appropriate for your students. Thus while many exercises are designed to develop facility with terminology and the use of theorems, quite a few others provide opportunities for students to engage in the kind of reasoning that lies behind the theorems.

Section 11.3 introduces the concept of the adjacency matrix of a graph. The main theorem of the section states that the ijth entry of the kth power of the adjacency matrix equals the number of walks of length k from the ith to the jth vertices in the graph. Matrix multiplication is defined and explored in this section in a way that is intended to be adequate for students who have never seen the definition before but also provide some challenge to students who were exposed to the topic in high school.

The concept of graph isomorphism is discussed in Section 11.4. In this section the main theorem gives a list of isomorphic invariants that can be used to determine the non-isomorphism of two graphs. The theoretical exercises at the end of this section give students an opportunity to fill in the parts of the proof of this theorem that are not included in the text.

The last two sections of the chapter deal with the subject of trees. Section 11.5 is rather long. In addition to definitions, examples, and theorems giving necessary and sufficient conditions for graphs to be trees, the section also contains the definition of rooted tree, binary tree, and the theorems that relate the number of internal to the number of terminal vertices of a full binary tree and the maximum height of a binary tree to the number of its terminal vertices. Section 11.6 on spanning trees contains Kruskal's and Prim's algorithms and proofs of their correctness, as well as applications of minimal spanning trees.

Section 11.1

2. $V(G) = \{v_1, v_2, v_3, v_4\}$, $E(G) = \{e_1, e_2, e_3, e_4, e_5\}$

   edge-endpoint function:

edge	endpoints
e_1	$\{v_1, v_2\}$
e_2	$\{v_2, v_3\}$
e_3	$\{v_2, v_3\}$
e_4	$\{v_2, v_4\}$
e_5	$\{v_4\}$

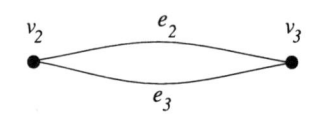

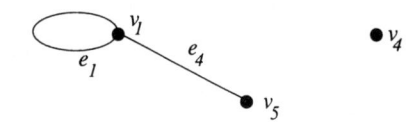

4.

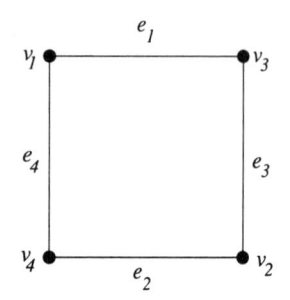

6.

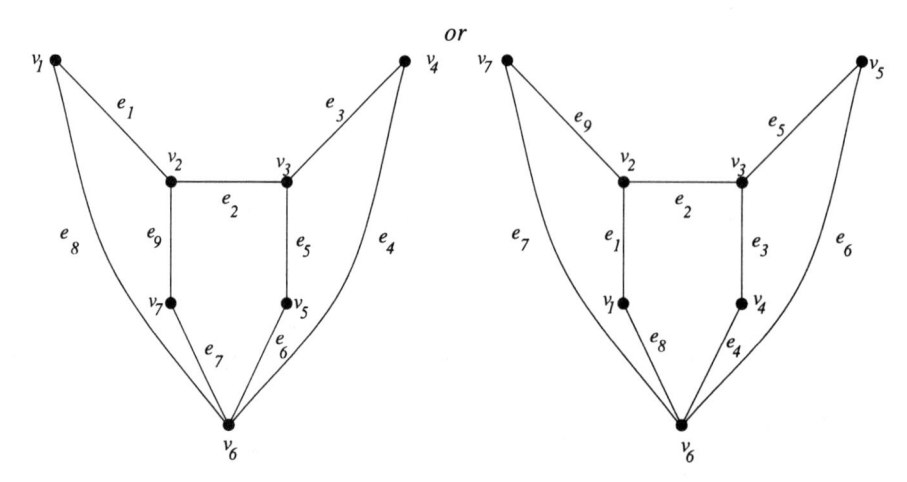

7.   or

9. (i) e_1, e_2, e_7 are incident on e_1.
   (ii) v_1 and v_2 are adjacent to v_3.
   (iii) e_2 and e_7 are adjacent to e_1.
   (iv) e_1 and e_3 are loops.
   (v) e_4 and e_5 are parallel.
   (vi) v_4 is an isolated vertex.
   (vii) degree of $v_3 = 2$
   (viii) total degree of the graph $= 14$

10. *b.* Yes. According to the graph, *Poetry Magazine* is an instance of a literary journal which is a scholarly journal and, therefore, contains long words.

11.

$$(vvccB/) \to (vc/Bvc) \to (vvcB/c) \to (c/Bvvc) \to (vcB/vc) \to (/Bvvcc)$$
$$(vvccB/) \to (vv/Bcc) \to (vvcB/c) \to (c/Bvvc) \to (vcB/vc) \to (/Bvvcc)$$
$$(vvccB/) \to (vv/Bcc) \to (vvcB/c) \to (c/Bvvc) \to (ccB/vv) \to (/Bvvcc)$$

13.

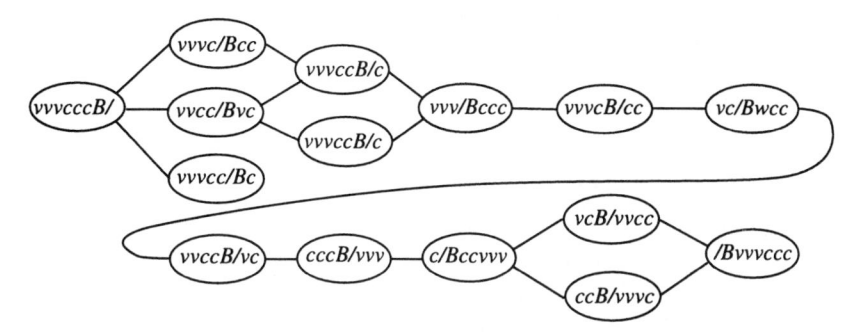

The diagram shows several solutions. Among them is $(vvvccB/) \to (vvcc/Bvc) \to (vvvccB/c)$ $\to (vvv/Bccc) \to (vvvcB/cc) \to (vc/Bvvcc) \to (vvccB/vc) \to (cc/Bvvvc) \to (cccB/vvv)$ $\to (c/Bvvvcc) \to (vcB/vvcc) \to (/Bvvvccc)$, or one can end with $(c/Bvvvcc) \to (ccB/vvvc)$ $\to (/Bvvvccc)$, or one can start with $(vvvcccB/) \to (vvvc/Bcc) \to (vvvccB/c)$ or with $(vvvcccB/) \to (vvcc/Bvc) \to (vvvccB/c) \to (vvv/Bccc)$.

14. Represent possible amounts of water in jugs A and B by ordered pairs with, say, the ordered pair (1,3) indicating that there is one quart of water in jug A and three quarts in jug B. Starting with (0,0), draw an edge from one ordered pair to another if it is possible to go from the situation represented by the one pair to that represented by the other and back by either filling a jug from the tap, emptying a jug into the drain, or transferring water from one jug to another. The resulting graph is shown below.

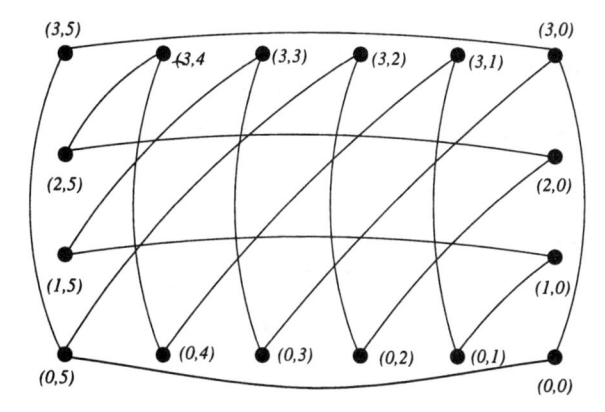

It is clear from the graph that one solution is $(0,0) \to (3,0) \to (0,3) \to (3,3) \to (1,5) \to (1,0)$ and another solution is $(0,0) \to (0,5) \to (3,2) \to (0,2) \to (2,0) \to (2,5) \to (3,4) \to (0,4) \to (3,1) \to (0,1)$.

Note that it would be possible to add arrows to the above graph from each reachable state to each other state that could be obtained from it either by filling one of the jugs to the top or by emptying the entire contents of one of the jugs. For instance, one could draw an arrow from $(0,3)$ to $(0,5)$ or from $(0,3)$ to $(0,0)$. Inspection shows that all such arrows would point to states already reachable by other means, so that it is not necessary to add such additional arrows to find solutions to the problem (and it makes the diagram look more complicated). However, if the problem were to find all possible solutions, the arrows would have to be added.

17. If there were a graph with four vertices of degrees 1, 1, 1 and 4, then its total degree would be 7, which is odd. Yet by Corollary 11.1.2, its total degree must be even. *[This is a contradiction.]* Thus there is no such graph.

18.

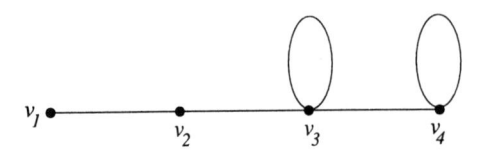

20. Since a simple graph has no loops or parallel edges, the maximum number of edges incident on a vertex equals the number of other vertices in the graph (because the vertex can only be connected to these and only once each). In a simple graph with five vertices, therefore, the maximum degree of any vertex is four, and so there can be no vertex of degree 5. Thus there is no simple graph with five vertices of degrees 2, 3, 3, 3, and 5..

21.

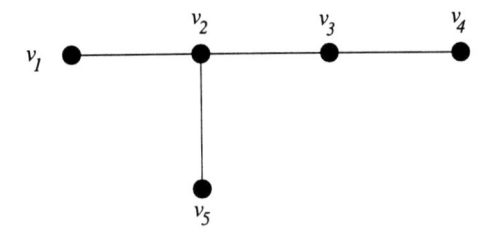

23. Let us first deduce what we can about such a graph. Its total degree would be two times the number of edges, or 18, and since each vertex would have degree 3, the number of vertices would be 18/3, or 6. A graph that satisfies the given properties is shown below.

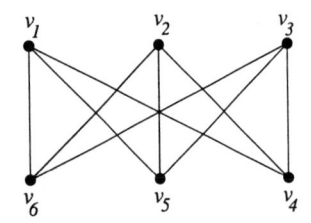

24. *b.* There are 17 nonempty subgraphs.

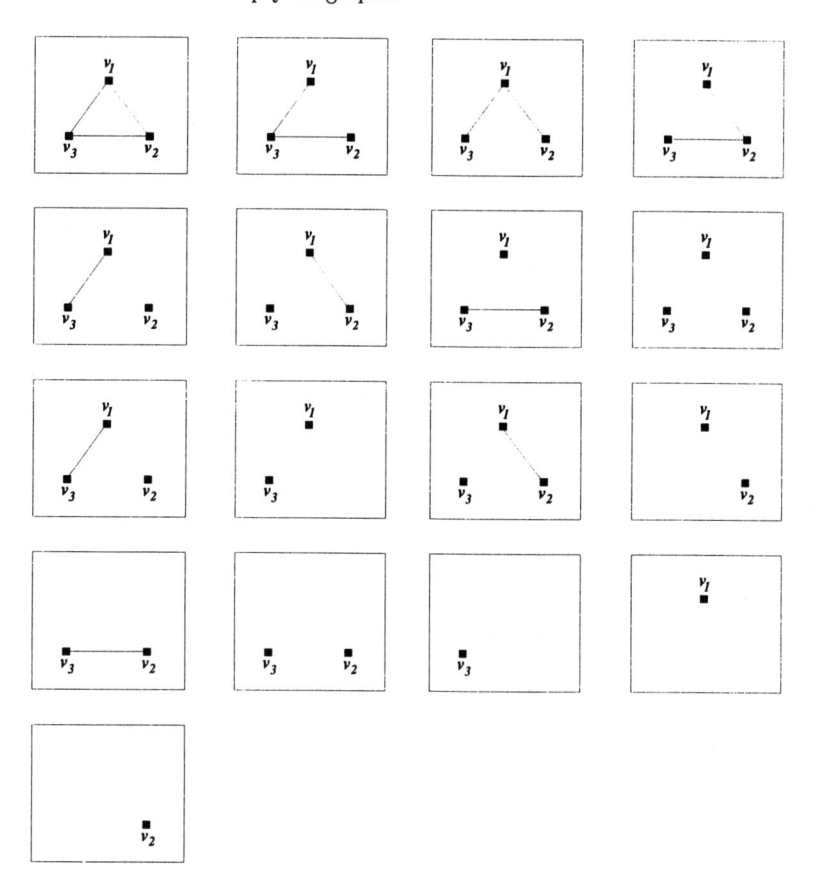

c. There are 7 nonempty subgraphs.

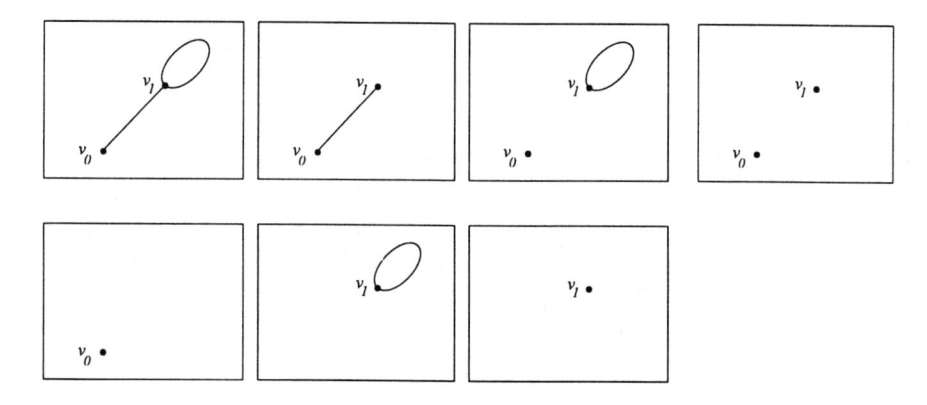

25. *b.* Yes. Each could be friends with all three others.

26. No. If the people were represented by vertices of a graph and each handshake were represented by an edge joining two vertices, the result would be a graph with a total degree of 25, which is odd. But this is impossible because the total degree of a graph must be even.

27. Yes. For example, the following graph satisfies this condition.

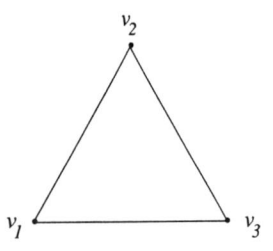

29. The total degree of the graph is $1+1+4+4+6 = 16$. So by Theorem 11.1.1, the number of edges is $16/2$, or 8.

30. Let t be the total degree of the graph. Since the degree of each vertex is at least d_{min} and at most d_{max}, $d_{min} \cdot v \leq t \leq d_{max} \cdot t$. But by Theorem 11.1.1, t equals twice the number of edges. So by substitution, $d_{min} \cdot v \leq 2e \leq d_{max} \cdot v$.

32. By the parity theorem, any integer that is not even is odd. So the given statement is the contrapositive of the statement proved in exercise 31. Thus it is true by the equivalence between a conditional statement and its contrapositive.

33. b. *Proof 1*: Suppose n is an integer with $n \geq 1$ and K_n is a complete graph on n vertices. If $n = 1$, then K_n has one vertex and 0 edges and $\frac{n(n-1)}{2} = \frac{1(1-1)}{2} = 0$, and so K_n has $\frac{n(n-1)}{2}$ edges. If $n \geq 2$, then since each pair of distinct vertices of K_n is connected by exactly one edge, there are as many edges in K_n as there are subsets of size two of the set of n vertices. By Theorem 6.4.1, there are $\binom{n}{2}$ such sets. But $\binom{n}{2} = \frac{n!}{2! \cdot (n-2)!} = \frac{n(n-1)}{2}$. Hence there are $\frac{n(n-1)}{2}$ edges in K_n.

Proof 2 (by mathematical induction):

The formula is true for n = 1: K_1 has one vertex and 0 edges. And if $n = 1$, then $\frac{n(n-1)}{2} = \frac{1(1-1)}{2} = 0$ also. Hence the formula holds for $n = 1$.

If the formula is true for n = k then it is true for n = k + 1: Let k be an integer with $k \geq 1$, and suppose that a complete graph on k vertices has $\frac{k(k-1)}{2}$ edges. *[This is the inductive hypothesis.]* Let K_{k+1} be a complete graph on $k + 1$ vertices. *[We must show that K_{k+1} has $\frac{(k+1)((k+1)-1)}{2}$ vertices.]* Note that K_{k+1} can be obtained from a complete graph on k vertices, K_k, by adding one vertex, say v, and connecting v to each of the k other vertices. But by inductive hypothesis, K_k has $\frac{k(k-1)}{2}$ edges. Connecting v to each of the k other vertices adds another k edges. Hence the total number of edges of K_{k+1} is $\frac{k(k-1)}{2} + k = \frac{k(k-1)}{2} + \frac{2k}{2} = \frac{k^2-k+2k}{2} = \frac{k(k+1)}{2} = \frac{(k+1)((k+1)-1)}{2}$ *[as was to be shown]*.

Proof 3: Suppose n is an integer with $n \geq 1$ and K_n is a complete graph on n vertices. Since each vertex of K_n is connected by an edge to each of the other $n-1$ vertices of K_n by exactly one edge, the degree of each vertex of K_n is $n-1$. Thus the total degree of K_n equals the number of vertices times the degree of each vertex, or $n(n-1)$. But by Theorem 11.1.1, the total degree of K_n equals twice the number e of edges of K_n, and so $n(n-1) = 2 \cdot e$. Equivalently, $e = n(n-1)/2$, *[as was to be shown]*.

34. *Proof:* Let n be a positive integer and let G be any simple graph with n vertices. Add edges to G to connect any pairs of vertices not already connected by an edge of G. The result is a complete graph on n vertices which has $n(n-1)/2$ edges by exercise 33. Hence the number of edges of G is at most $n(n-1)/2$.

36. *b.*

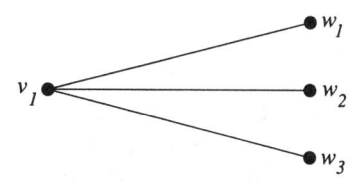

c.

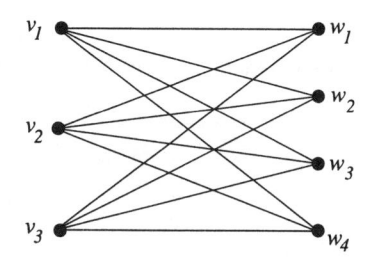

d, e, and *f.* The number of edges of $K_{m,n} = m \cdot n$. The reason is that $K_{m,n}$ has m vertices of degree n and n vertices of degree m, making its total degree $2mn$. By Theorem 11.1.1, therefore, it has $2mn/2 = mn$ edges. Alternatively, $K_{m,n}$ has n edges coming out of each of m vertices (leading to the n vertices) for a total of mn edges, or, equivalently, $K_{m,n}$ has m edges coming out of each of n vertices (leading to the m vertices) for a total of mn edges.

37. *c.*

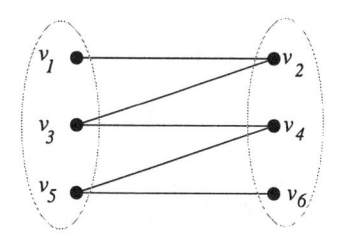

d. Suppose the graph were bipartite with disjoint vertex sets V_1 and V_2, where no vertices within either V_1 or V_2 are connected by edges. Then v_1 would be in one of the sets, say V_1, and so v_2 and v_6 would be in V_2 (because each is connected by an edge to v_1). Furthermore, v_3, v_4, and v_5 would be in V_1 (because all are connected by edges to v_2). But v_4 is connected by an edge to v_5, and so both cannot be in V_1. This contradiction shows that the supposition is false, and so the graph is not bipartite.

e.

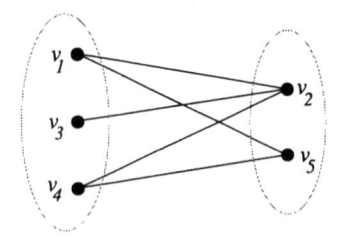

[Bipartite graph with $V_1 = \{v_1, v_3, v_4\}$ and $V_2 = \{v_2, v_5\}$ and edges connecting them.]

f. Suppose the graph were bipartite with disjoint vertex sets V_1 and V_2, where no vertices within either V_1 or V_2 are connected by edges. Then v_1 would be in one of the sets, say V_1, and so v_2 and v_5 would be in V_2 (because each is connected by an edge to v_1). Furthermore, v_3 and v_4 would be in V_1 (because v_3 is connected by an edge to v_2 and v_4 is connected by an edge to v_5). But v_3 is connected by an edge to v_4, and so both cannot be in V_1. This contradiction shows that the supposition is false, and so the graph is not bipartite.

38. Yes. Given positive integers r and s, let G be the bipartite graph $K_{r,s}$. Then G has r vertices of degree s and s vertices of degree r.

39. b.

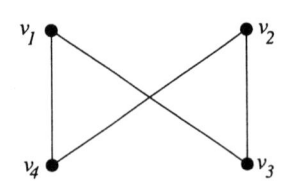

[Graph with vertices v_1, v_2, v_3, v_4 forming a complete bipartite $K_{2,2}$ drawn as an X between $\{v_1, v_4\}$ and $\{v_2, v_3\}$.]

40. a.

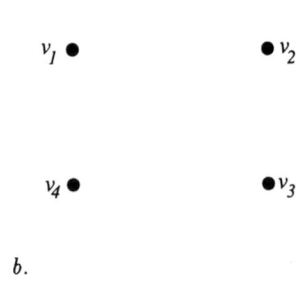

[Four isolated vertices v_1, v_2, v_3, v_4.]

b.

[Graph with vertices v_1, v_2, v_3 on the left (with edges) and w_1, w_2 on the right connected by an edge.]

264

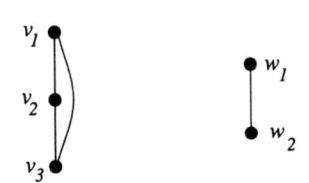

41. *a.*

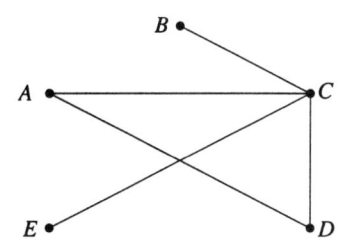

```
        B •
A •            • C
        
E •            • D
```

42. The graph obtained by taking all the vertices and edges of G together with all the edges of G' is K_n. Therefore, by exercise 33b, the number of edges of G plus the number of edges of G' equals $n(n-1)/2$.

43. Represent each person at the party by a vertex of an acquaintance graph and draw an edge connecting each pair of acquaintances. By assumption, the graph has at least two vertices. If the acquaintance graph has at least one edge, then the people represented by the endpoints of that edge are acquaintances. If the acquaintance graph has no edges, then its complement has at least one edge (because the graph has at least two vertices), and we may choose such an edge. Then the people represented by the endpoints of that edge are mutual strangers.

44. *a.* Yes. Let G be a simple graph with n vertices and let v be a vertex of G. Since G has no parallel edges, v can be joined by at most a single edge to each of the $n-1$ other vertices of G, and since G has no loops, v cannot be joined to itself. Therefore, the maximum degree of v is $n-1$.

b. No. Suppose there is a simple graph with four vertices each of which has a different degree. By part (a), no vertex can have degree greater than three, and, of course, no vertex can have degree less than 0. Therefore, the only possible degrees of the vertices are 0, 1, 2, and 3. Since all four vertices have different degrees, there is one vertex with each degree. But then the vertex of degree 3 is connected to all the other vertices, which contradicts the fact that one of the vertices has degree 0. Hence the supposition is false, and there is no simple graph with four vertices each of which has a different degree.

c. No. Suppose there is a simple graph with n vertices (where $n \geq 2$) each of which has a different degree. By part (a), no vertex can have degree greater than $n-1$, and, of course, no vertex can have degree less than 0. Therefore, the only possible degrees of the vertices are 0, 1, 2,..., $n-1$. Since the vertices all have different degrees, there are n vertices, and there are n integers from 0 to $n-1$ inclusive, there is one vertex with each degree. But then the vertex of degree $n-1$ is connected to all the other vertices, which contradicts the fact that one of the vertices has degree 0. Hence the supposition is false, and there is no simple graph with n vertices each of which has a different degree.

45. Yes. Suppose that in a group of two or more people, each person is acquainted with a different number of people. Then the acquaintance graph representing the situation is a simple graph in which all the vertices have different degrees. But by exercise 44(c) such a graph does not exist. Hence the supposition is false, and so in a group of two or more people there must be at least two people who are acquainted with the same number of people within the group.

46. In the graph below each committee name is represented as a vertex and labeled with the first letter of the name of the committee. Vertices are joined if, and only if, the corresponding committees have a member in common.

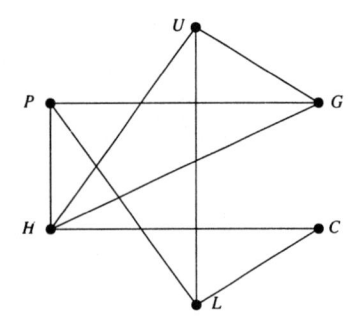

To the first time slot, assign a committee whose vertex has maximal degree. There is only one choice, the hiring committee. Since the library committee has no members in common with the hiring committee, assign it to meet in the first time slot also. Every other committee shares a member with the hiring committee and so cannot meet during the first time slot. To the second time slot, assign a committee that has not already been scheduled and whose vertex has next highest degree. This will be either the personnel, undergraduate education, or graduate education committee. Say the personnel committee is selected. The committees that have not already been scheduled and that do not share a member with the personnel committee are the undergraduate education and the colloquium committees. So assign these to the second time slot also. To the third time slot, assign a committee that has not already been scheduled and whose vertex has next highest degree. Only the graduate education committee satisfies this condition. The time slots of all the committee meetings are as follows.

Time 1: hiring, library

Time 2: personnel, undergraduate education, colloquium

Time 3: graduate education

Note that if the graduate education committee is chose in step 2, the result is as follows.

Time 1: hiring, library

Time 2: graduate education, colloquium

Time 3: personnel, undergraduate education

Section 11.2

2. *a.* just a walk, not a path or a circuit

   b. simple circuit

   c. just a walk, not a path or a circuit (has a repeated edge)

   d. circuit, not a simple circuit

   e. path, not a simple path, not a circuit

   f. simple path

3. *b.* No, because $e_1 e_2$ could refer either to $v_1 e_1 v_2 e_2 v_1$ or to $v_2 e_1 v_1 e_2 v_2$.

5. *a.* The number of simple paths from a to c is 4 *[the number of ways to choose an edge from a to b]*.

   b. The number of paths from a to c is $4 + 4 \cdot 3 \cdot 2 = 28$ *[the number of ways to choose an edge from a to b plus the number of ways to choose three distinct edges from a to b]*.

   c. There are infinitely many walks from a to c because it is possible to travel back and forth from a to b or from b to c an arbitrarily large number of times before ending up at c.

6. b. $\{v_8, v_7\}$, $\{v_1, v_2\}$, $\{v_3, v_4\}$

   c. $\{v_2, v_3\}$, $\{v_6, v_7\}$, $\{v_7, v_8\}$, $\{v_9, v_{10}\}$

7. Yes. For instance, for any positive integer n, consider the graph with distinct vertices v_0, v_1, $v_2, \ldots, v_n$ and edges $\{v_0, v_1\}$, $\{v_1, v_2\}, \ldots, \{v_{i-1}, v_i\}, \ldots, \{v_{n-1}, v_n\}$. Removal of any of these edges disconnects the graph.

8. b. Two connected components:

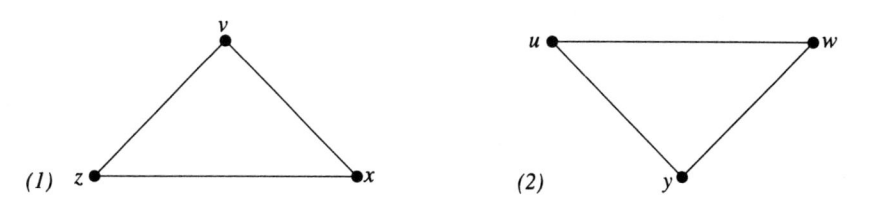

   c. Three connected components:

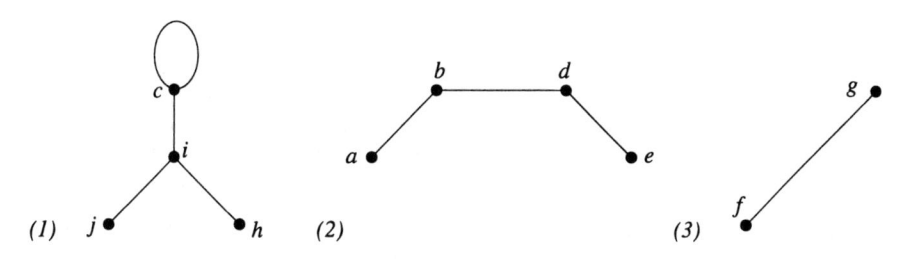

   d. Two connected components:

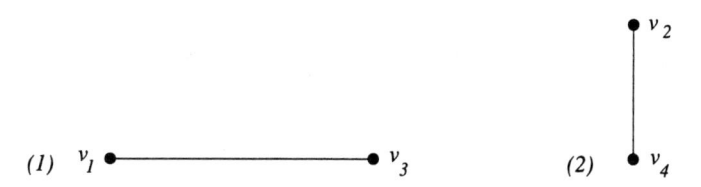

9. b. Yes, by Theorem 11.2.3 since G is connected and every vertex has even degree.

   c. Not necessarily. It is not specified that G is connected. For instance, the following graph satisfies the given conditions but does not have an Euler circuit:

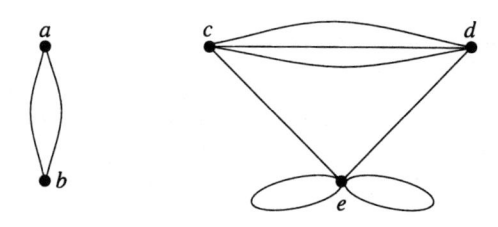

10. One such example is given in the answer to exercise 9c above. A simpler example is the graph shown below. Its vertices have degrees 2, 2, 2, and 0 which are all even numbers, but the graph does not have an Euler circuit.

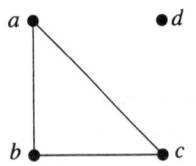

11. Yes. The graph that models the situation in which each bridge is crossed twice is the following.

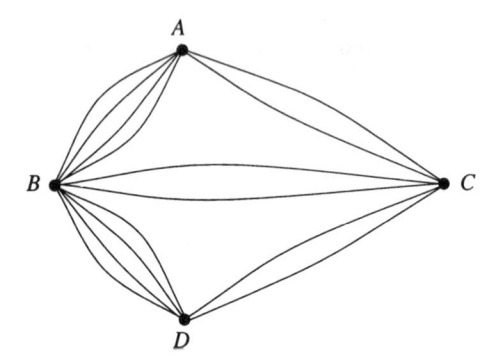

This graph is connected and its vertices have degrees 6, 6, 6, and 10, all of which are even numbers. Therefore, by Theorem 11.2.3, the graph has an Euler circuit, and so it is possible for a citizen of Königsberg to make a tour of the city and cross each bridge exactly twice.

13. This graph does not have an Euler circuit because vertices v_1, v_8, v_9, and v_7 have odd degree.

15. One Euler circuit is the following: *stuvwxyzrsuwyuzs*.

16. This graph does not have an Euler circuit because it is not connected.

17. This graph does not have an Euler circuit because vertices C and D have odd degree.

18. Yes. One Euler circuit is $ABDEACDA$.

20. There is not an Euler path from u to w because w has even degree and e, f, and h have odd degree.

21. One Euler path from u to w is $uv_1v_2v_3uv_0v_7v_6v_3v_4v_6wv_5v_4w$.

22. Yes. One such path is $AHGBCDGFE$.

24. One Hamiltonian circuit is *balkjedcfihgb*.

25. Call the given graph G and suppose G has a Hamiltonian circuit. Then G has a subgraph H that satisfies conditions (1) – (4) of Proposition 11.2.6. Since the degree of c in G is five and every vertex in H has degree two, three edges incident on c must be removed from G to create H. Edge $\{c, d\}$ cannot be removed because doing so would result in vertex d having degree less than two in H. Similar reasoning shows that edges $\{c, f\}$, $\{c, b\}$, and $\{c, g\}$ cannot be removed either. It follows that the degree of c in H must be at least four, which contradicts the condition that every vertex in H has degree two in H. Hence no such subgraph H can exist, and so G does not have a Hamiltonian circuit.

27. Call the given graph G and suppose G has a Hamiltonian circuit. Then G has a subgraph H that satisfies conditions (1) – (4) of Proposition 11.2.6. Since the degree of B in G is five and every vertex in H has degree two, three edges incident on B must be removed from G to create

H. Edge $\{B,C\}$ cannot be removed because doing so would result in vertex C having degree less than two in H. Similar reasoning shows that edges $\{B,E\}$, $\{B,F\}$, and $\{B,A\}$ cannot be removed either. It follows that the degree of B in H must be at least four, which contradicts the condition that every vertex in H has degree two in H. Hence no such subgraph H can exist, and so G does not have a Hamiltonian circuit.

28. Call the given graph G and suppose G has a Hamiltonian circuit. Then G has a subgraph H that satisfies conditions (1) – (4) of Proposition 11.2.6. Since the degree of b in G is three and every vertex in H has degree two, one edge incident on b must be removed from G to create H. Edge $\{b,a\}$ cannot be removed because doing so would result in vertex a having degree less than two in H. Similar reasoning shows that edges $\{b,d\}$ and $\{b,c\}$ cannot be removed either. It follows that the degree of b in H must be at least three, which contradicts the condition that every vertex in H has degree two in H. Hence no such subgraph H can exist, and so G does not have a Hamiltonian circuit.

29. One Hamiltonian circuit is $abcdefga$.

30. One Hamiltonian circuit is $v_0v_1v_5v_4v_7v_6v_2v_3v_0$.

31. Call the given graph G and suppose G has a Hamiltonian circuit. Then G has a subgraph H that satisfies conditions (1) – (4) of Proposition 11.2.6. Edges $\{a,b\}$ and $\{a,d\}$ must be part of H because otherwise vertices b and d would have degree less than two in H. Similarly, edges $\{c,b\}$ and $\{c,d\}$ must be in H. Therefore, edges $\{a,e\}$ and $\{f,c\}$ are not in H because otherwise vertices a and c would have degrees greater than two in H. But removal of $\{a,e\}$ and $\{f,c\}$ disconnects G, which implies that H is not connected. This contradicts the condition that H is connected. Hence no such subgraph H can exist, and so G does not have a Hamiltonian circuit.

32. Other such graphs are those shown in exercise 12 and Example 11.2.6.

33. Other such graphs are those shown in exercises 17, 21, 23, 24, 29 and 30.

34. In the graph below, $abcda$ is both an Euler and a Hamiltonian circuit.

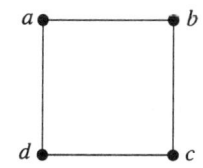

35. In the graph below, ve_1we_2v is a Hamiltonian circuit that is not an Euler circuit, and $ve_1we_2ve_3we_4v$ is an Euler circuit that is not a Hamiltonian circuit.

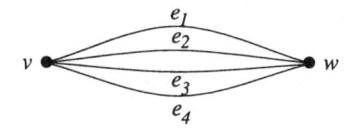

36. It is clear from the map that only a few routes have a chance of minimizing the distance. These are shown below.

Route	Total Distance (in km.)
Bru-Lux-Düss-Ber-Mun-Par-Bru	$219 + 224 + 564 + 585 + 832 + 308 = 2732$
Bru-Düss-Ber-Mun-Par-Lux-Bru	$223 + 564 + 585 + 832 + 375 + 219 = 2798$
Bru-Düss-Lux-Ber-Mun-Par-Bru	$223 + 224 + 764 + 585 + 832 + 308 = 2936$
Bru-Düss-Ber-Mun-Lux-Par-Bru	$223 + 564 + 585 - 517 + 375 + 308 = 2572$

The route that minimizes distance, therefore, is the bottom one shown in the table.

37. *b.* This statement is the contrapositive of the statement proved in part (a). So since a statement is logically equivalent to its contrapositive, this statement is true.

39. *Proof:* Suppose vertices v and w are part of a circuit in a graph G and one edge e is removed from the circuit. Without loss of generality, we may assume the v occurs before the w in the circuit, and we may denote the circuit by $v_0 e_1 v_1 e_2 \ldots e_{n-1} v_{n-1} e_n v_0$ with $v_i = v$, $v_j = w$, $i < j$, and $e_k = e$. If either $k \leq i$ or $k > j$, then $v = v_i e_{i+1} v_{i+1} \ldots v_{j-1} e_j v_j = w$ is a path in G from v to w that does not include e. If $i < k \leq j$, then $v = v_i e_i v_{i-1} e_{i-1} \ldots e_{j+1} v_j = w$ is a path in G from v to w that does not include e. These possibilities are illustrated by examples (1) and (2) in the diagram below. In either case there is a path in G from v to w that does not include e.

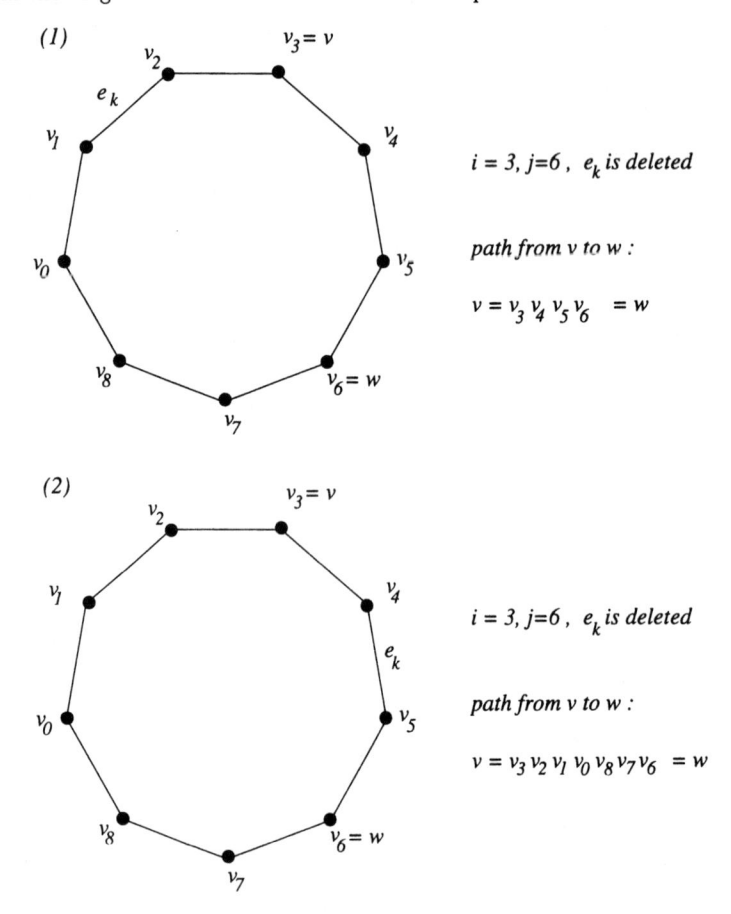

(1)

$i = 3, j=6$, e_k is deleted

path from v to w:

$v = v_3\, v_4\, v_5\, v_6 = w$

(2)

$i = 3, j=6$, e_k is deleted

path from v to w:

$v = v_3\, v_2\, v_1\, v_0\, v_8\, v_7\, v_6 = w$

41. *Proof:* Suppose there is a path P in a graph G from a vertex v to a vertex w. By definition of path from v to w, P has the form $v = v_0 e_1 v_1 e_2 v_2 \ldots e_{n-1} v_{n-1} e_n v_n = w$ for some vertices $v_0, v_1, \ldots, v_n$ and distinct edges $e_1, e_2, \ldots, e_n$. Then $w = v_n e_n v_{n-1} e_{n-1} \ldots v_2 e_2 v_1 e_1 v_0 = v$ is a path from w to v.

43. *Proof:* Suppose C is a circuit in a graph G that starts and ends at a vertex v, and suppose w is another vertex in the circuit. By definition the circuit has the form $v e_1 v_1 e_2 v_2 \ldots e_{n-1} v_{n-1} e_n v$ where $v_1, v_2, \ldots, v_{n-1}$ are vertices of G, $e_1, e_2, \ldots, e_n$ are distinct edges of G, and $v_i = w$ for some i with $1 \leq i \leq n-1$. Then $w = v_i e_{i+1} v_{i+1} \ldots e_n v e_1 v_1 e_2 v_2 \ldots v_i = w$ is a circuit that starts and ends at w.

46. *Proof:* Let G be a graph and let v and w be two distinct vertices of G.

($\Rightarrow$) Suppose there is an Euler path in G from v to w. Form a new graph G' from G by adding an edge e from v to w. To the end of the Euler path in G from v to w, add edge e and vertex v. The result is an Euler circuit in G' from v to v. It follows from Theorem 11.1.1 that every vertex in G' has even degree. Now the degrees of all vertices in G' except v and w are the same as their degrees in G. So all such vertices have even degree in G. Also the degrees of v and w in G are one less than their degrees in G'; so since one less than an even number is odd, both v and w have odd degree in G. Furthermore, G' is connected because it has an Euler circuit, and since removing an edge from a circuit does not disconnect a graph (Lemma 11.2.1c) and since G is obtained from G' by removing edge e (which is an edge of a circuit), G is also connected.

($\Leftarrow$) Suppose G is connected, v and w have odd degree, and all other vertices of G have even degree. Form a new graph G' from G by adding an edge e from v to w. This increases the degrees of v and w by one each, so that every vertex of G' has even degree. Therefore, by Theorem 11.2.3, G' has an Euler circuit. Construct a (possibly different) Euler circuit for G' by starting at w, following e to v, and continuing from v using the method outlined in the proof of Theorem 11.2.3, eventually to return to w having traversed every edge of G'. Removing the initial vertex w and edge e from this circuit gives an Euler path in G from v to w.

47. *a.* Let n be a positive integer and let K_n be a complete graph on n vertices. Since K_n is connected, by Theorem 11.2.4 it has an Euler circuit if, and only if, every vertex has even degree. But the degree of each vertex of K_n is $n-1$, and $n-1$ is even exactly when n is odd. So K_n has an Euler circuit if, and only if, n is odd.

b. Let n be a positive integer and let K_n be a complete graph on n vertices. Arrange the vertices of K_n in any order, and construct a Hamiltonian circuit by starting at any vertex, visiting each other vertex in the order listed, and returning to the starting vertex. This is possible because each pair of vertices is connected by an edge. Hence K_n has a Hamiltonian circuit.

48. *a.* Let m and n be positive integers and let $K_{m,n}$ be a complete bipartite graph on (m,n) vertices. Since $K_{m,n}$ is connected, by Theorem 11.2.4 it has an Euler circuit if, and only if, every vertex has even degree. But $K_{m,n}$ has m vertices of degree n and n vertices of degree m. So $K_{m,n}$ has an Euler circuit if, and only if, both m and n are even.

b. Let m and n be positive integers, let $K_{m,n}$ be a complete bipartite graph on (m,n) vertices, and suppose $V_1 = \{v_1, v_2, \ldots, v_m\}$ and $V_2 = \{w_1, w_2, \ldots, w_n\}$ are the disjoint sets of vertices such that each vertex in V_1 is joined by an edge to each vertex in V_2 and no vertex within V_1 or V_2 is joined by an edge to any other vertex within the same set. If $m = n \geq 2$, then $K_{m,n}$ has the following Hamiltonian circuit: $v_1 w_1 v_2 w_2 \ldots v_m w_m v_1$. If $K_{m,n}$ has a Hamiltonian circuit, then $m = n$ because the vertices in any Hamiltonian circuit must alternate between V_1 and V_2 (since no edges connect vertices within either set) and because no vertex, except the first and last, appears twice in a Hamiltonian circuit. If $m = n = 1$, then $K_{m,n}$ does not have a Hamiltonian circuit because $K_{1,1}$ contains just one edge joining two vertices. Therefore, $K_{m,n}$ has a Hamiltonian circuit if, and only if, $m = n \geq 2$.

49. *Proposition:* If n is an integer with $n \geq 2$, then a simple disconnected graph with n vertices has a maximum of $(n-1)(n-2)/2$ edges.

Proof: Let n be an integer with $n \geq 2$, and let G be a simple disconnected graph with n vertices and a maximum number of edges. Then G consists of just two connected components because if G had more than two components, an edge could be added between two vertices in two separate components, giving a graph that would still be disconnected but would have more edges than G. Suppose one connected component contains k vertices ($1 \leq k \leq n-1$). Then the other connected component contains $n - k$ vertices. By exercise 34 of Section 11.1, the maximum number of edges in the two components are $k(k-1)/2$ and $(n-k)(n-k-1)/2$ respectively.

Therefore, the total maximum number of edges is $k(k-1)/2 + (n-k)(n-k-1)/2$. Now

$$\frac{k(k-1)}{2} + \frac{(n-k)(n-k-1)}{2} = \frac{k^2 - k + n^2 - nk - n - nk + k^2 + k}{2} = k^2 - nk + \frac{n^2 - n}{2}.$$

We wish to find an integer k with $1 \leq k \leq n-1$ that maximizes $k^2 - nk + \frac{n^2-n}{2}$. Let f be the function defined by the rule $f(x) = x^2 - nx + \frac{n^2-n}{2}$ on the interval $1 \leq x \leq n-1$. Since $f'(x) = 2x - n$, (1) $f'(x) > 0 \Leftrightarrow 2x - n > 0 \Leftrightarrow x > n/2$, (2) $f'(x) = 0 \Leftrightarrow 2x - n = 0 \Leftrightarrow x = n/2$, and (3) $f'(x) < 0 \Leftrightarrow 2x - n < 0 \Leftrightarrow x < n/2$. Therefore, f is decreasing on $x \leq n/2$, attains a minimum at $x = n/2$, and is increasing for $x \geq n/2$. It follows that f achieves its maximum values at the endpoints of the interval: $x = 1$ and $x = n-1$. These both correspond to the situation in which one component of G has one vertex and no edges and the other component is a complete graph on $n-1$ vertices. Consequently, the total number of edges for the graph is the same as the total number of edges of a complete graph on $n-1$ vertices, namely $(n-1)(n-2)/2$ (by exercise 33b of Section 11.1).

Note that it is possible to solve this problem without using calculus by completing the square to show that $f(x) = (x - \frac{n}{2})^2 + (\frac{n^2 - 2n}{4})$ for all real x. It follows that the graph of f is a parabola with minimum value at $x = n/2$.

50. *Proof:*

($\Rightarrow$) Suppose a graph G is bipartite. Let V_1 and V_2 be disjoint sets of vertices such that vertices in V_1 are joined by edges to vertices in V_2 but no edges join vertices within either V_1 or V_2, and suppose that $v \in V_1$. Let C be any circuit in G. Since no edges join vertices within either V_1 or V_2, the only way that C can start from a vertex and return to the same vertex is for it to go back and forth from V_1 to V_2. Thus adjacent edges of C can be divided into pairs, one leading from V_1 to V_2 and the other leading back. It follows that the total number of edges in C is even.

($\Leftarrow$) Suppose that no circuit in a graph G has an odd number of edges. Build disjoint sets of vertices V_1 and V_2 of G by the following repetitive procedure. Start by letting V_1 and V_2 be empty. Pick any vertex u to place in V_1 and place in V_2 all vertices connected to u by an edge. If any vertices remain, pick one and add it to V_1, adding to V_2 all vertices connected to the newly added vertex by an edge. Repeat this process until all vertices have been assigned either to V_1 or V_2. Here is a more formal description of the procedure.

Let $V_1 = V_2 = \emptyset$.

while (there exists a vertex u such that $u \notin V_1$ and $u \notin V_2$)

Find u such that $u \notin V_1$ and $u \notin V_2$.

Let $V_1 := V_1 \cup \{u\}$ and $V_2 := V_2 \cup \{v \in V(G) \mid \text{there is an edge joining } v \text{ and } u\}$

end while

Note that as each vertex is added to V_1, all vertices connected to it by an edge are added to V_2. Thus after execution of the steps inside the **while** loop, if u is any vertex not in V_2, then u is not connected by an edge to any vertex in V_1. Consequently, after execution of the **while** loop, no two vertices in V_1 are connected by an edge. Furthermore, no two vertices in V_2 are connected by an edge because given any vertex v in V_2, by construction of V_2 there is an edge joining u to a vertex v in V_1, and so if there were an edge joining v to another vertex w in V_2, then $vuwv$ would be a circuit of length three in G, which is impossible by hypothesis. Hence it is possible to produce vertex sets V_1 and V_2 with the properties needed to show that G is bipartite.

Section 11.3

1. b. By equating corresponding entries, we see that $2a = 4$, $b+c = 3$, $c-a = 1$, and $2b-a = -2$. Now $2a = 4 \Rightarrow a = 2$, $c - a = c - 2 = 1 \Rightarrow c = 3$, and $b + c = b + 3 = 3 \Rightarrow b = 0$. Substituting these solutions into the last equation to check for consistency gives $2b - a = 2 \cdot 0 - 2 = -2$, which agrees. Therefore, $a = 2$, $b = 0$, and $c = 3$.

2. b.

$$\begin{array}{c c} & \begin{array}{cccc} v_1 & v_2 & v_3 & v_4 \end{array} \\ \begin{array}{c} v_1 \\ v_2 \\ v_3 \\ v_4 \end{array} & \left[\begin{array}{cccc} 1 & 0 & 1 & 0 \\ 0 & 0 & 1 & 0 \\ 1 & 0 & 0 & 1 \\ 0 & 0 & 1 & 0 \end{array} \right] \end{array}$$

3. b.

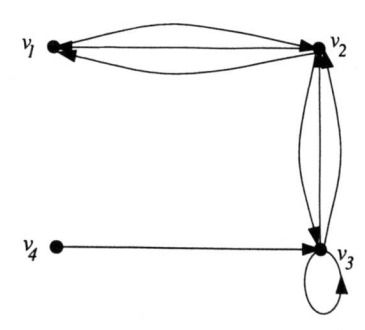

Any labels may be applied to the edges because the adjacency matrix does not determine edge labels.

4. b.

$$\begin{array}{c c} & \begin{array}{cccc} v_1 & v_2 & v_3 & v_4 \end{array} \\ \begin{array}{c} v_1 \\ v_2 \\ v_3 \\ v_4 \end{array} & \left[\begin{array}{cccc} 0 & 0 & 0 & 0 \\ 0 & 1 & 1 & 2 \\ 0 & 1 & 1 & 0 \\ 0 & 2 & 0 & 0 \end{array} \right] \end{array}$$

5. b.

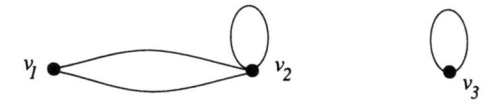

Any labels may be applied to the edges because the adjacency matrix does not determine edge labels.

6. b. The graph is not connected; the matrix shows that there are no edges joining the vertices from the set $\{v_1, v_2\}$ to those in the set $\{v_3, v_4\}$.

7. If, for all integers $i \geq 1$, all entries in the ith row and ith column of the adjacency matrix of a graph are zero, then the graph has no loops.

8. b. 2

9. b.
$$\begin{bmatrix} 0 & 8 \\ -5 & 4 \end{bmatrix}$$

c.
$$\begin{bmatrix} -2 & 3 \\ 4 & 6 \end{bmatrix}$$

10. c. no product (**A** does not have the same number of columns as rows)

    d. no product (**B** has two columns and **C** has three rows)

    e.
    $$\begin{bmatrix} -2 & -6 \\ -5 & 3 \\ -2 & 0 \end{bmatrix}$$

    g.
    $$\begin{bmatrix} -8 & 0 \\ 7 & 27 \end{bmatrix}$$

    h. no product (**C** does not have the same number of columns as rows)

    i.
    $$\begin{bmatrix} 0 & 4 & -2 \\ 3 & 1 & -2 \\ 1 & 1 & -1 \end{bmatrix}$$

11. Let $\mathbf{A} = \begin{bmatrix} 1 & 2 \\ 0 & 0 \end{bmatrix}$ and $\mathbf{B} = \begin{bmatrix} 1 & 0 \\ 2 & 0 \end{bmatrix}$. Then $\mathbf{AB} = \begin{bmatrix} 5 & 0 \\ 0 & 0 \end{bmatrix}$ whereas $\mathbf{BA} = \begin{bmatrix} 1 & 2 \\ 2 & 4 \end{bmatrix}$.

13. Let $\mathbf{A} = \begin{bmatrix} 1 & 1 \\ 1 & 1 \end{bmatrix}$ and $\mathbf{B} = \begin{bmatrix} 0 & 0 \\ 1 & -1 \end{bmatrix}$. Then $\mathbf{AB} = \begin{bmatrix} 1 & -1 \\ 1 & -1 \end{bmatrix} \ne 0$ whereas $\mathbf{BA} = \begin{bmatrix} 0 & 0 \\ 0 & 0 \end{bmatrix} = 0$.

14. *Proof:* Let **I** be the $m \times m$ identity matrix and let $\mathbf{A} = (a_{ij})$ be any $m \times n$ matrix. Then for all $i, j = 1, 2, \ldots, m$, the ijth entry of **IA** is $\Sigma_{k=1}^{m} \delta_{ik} a_{kj} = \delta_{ii} a_{ij} = a_{ij}$ because by definition of **I**, $\delta_{ik} = 0$ for all k with $i \ne k$ and $\delta_{ii} = 1$. But a_{ij} is the ijth entry of **A**. So $\mathbf{IA} = \mathbf{A}$.

16. *Proof:* Let $\mathbf{A} = (a_{ij})$, $\mathbf{B} = (b_{ij})$, and $\mathbf{C} = (c_{ij})$ be any $m \times k$, $k \times r$, and $r \times n$ matrices, respectively, and suppose that the entries of **A**, **B** and **C** are real numbers. The numbers of rows and columns agree so that **AB**, **BC**, (**AB**)**C**, and **A**(**BC**) are all defined. Let $\mathbf{AB} = (d_{ij})$ and $\mathbf{BC} = (e_{ij})$. Then for all integers i and j with $1 \le i \le m$ and $1 \le j \le n$,

$$\begin{aligned}
\text{the } ij\text{th entry of } (\mathbf{AB})\mathbf{C} &= \Sigma_{p=1}^{r} d_{ip} c_{pj} \\
&= \Sigma_{p=1}^{r} \left(\Sigma_{q=1}^{k} a_{iq} b_{qp} \right) c_{pj} \\
&= \Sigma_{p=1}^{r} \Sigma_{q=1}^{k} a_{iq} b_{qp} c_{pj} \\
&= \Sigma_{q=1}^{k} \Sigma_{p=1}^{r} a_{iq} b_{qp} c_{pj} \\
&= \Sigma_{q=1}^{k} a_{iq} \left(\Sigma_{p=1}^{r} b_{qp} c_{pj} \right) \\
&= \Sigma_{q=1}^{k} a_{iq} e_{qj} \\
&= \text{the } ij \text{ th entry of } \mathbf{A}(\mathbf{BC}).
\end{aligned}$$

Since all corresponding entries are equal, $(\mathbf{AB})\mathbf{C} = \mathbf{A}(\mathbf{BC})$.

18. *Proof (by mathematical induction)*:

The property holds for n = 1: For $n = 1$, $\mathbf{A}^n = \mathbf{A}^1 = \mathbf{A}$, which is symmetric by hypothesis.

If the property holds for n = k then it holds for n = k + 1: Let k be an integer with $k \geq 1$, and suppose that $\mathbf{A}^k$ is symmetric. *[This is the inductive hypothesis.]* We must show that $\mathbf{A}^{k+1}$ is symmetric. Let $\mathbf{A}^k = (b_{ij})$. Then for all $i, j = 1, 2, \ldots, m$, the ijth entry of $\mathbf{A}^{k+1}$ = the ijth entry of $\mathbf{A}\mathbf{A}^k$ *[by definition of matrix power]* $= \sum_{r=1}^{m} a_{ir} b_{rj}$ *[by definition of matrix multiplication]* $= \sum_{r=1}^{m} a_{ri} b_{jr}$ *[because $\mathbf{A}$ is symmetric by hypothesis and $\mathbf{A}^k$ is symmetric by inductive hypothesis]* $= \sum_{r=1}^{m} b_{jr} a_{ri}$ *[because multiplication of real numbers is commutative]* = the jith entry of $\mathbf{A}^k \mathbf{A}$ *[by definition of matrix multiplication]* = the jith entry of $\mathbf{A}\mathbf{A}^k$ *[by part (a)]* = the jith entry of $\mathbf{A}^{k+1}$ *[by definition of matrix power]*. Therefore, $\mathbf{A}^{k+1}$ is symmetric *[as was to be shown]*.

19. *b*. There are three walks of length two from v_1 to v_3 because the entry in row 1 column 3 (and row 3 column 1) of $\mathbf{A}^2$ is 3. There are 15 walks of length three from v_1 to v_3 because the entry in row 1 column 3 (and row 3 column 1) of $\mathbf{A}^3$ is 15.

c. The calculations are $\begin{bmatrix} 2 & 1 & 0 \end{bmatrix} \begin{bmatrix} 2 \\ 1 \\ 0 \end{bmatrix} = 2 \cdot 2 + 1 \cdot 1 + 0 \cdot 0 = 5$. In this sum

$2 \cdot 2 = \begin{bmatrix} \text{number of edges} \\ \text{from } v_3 \text{ to } v_1 \end{bmatrix} \cdot \begin{bmatrix} \text{number of edges} \\ \text{from } v_1 \text{ to } v_3 \end{bmatrix} = \begin{bmatrix} \text{number of walks of length 2} \\ \text{from } v_3 \text{ to } v_3 \text{ that go via } v_1 \end{bmatrix}$

$1 \cdot 1 = \begin{bmatrix} \text{number of edges} \\ \text{from } v_3 \text{ to } v_2 \end{bmatrix} \cdot \begin{bmatrix} \text{number of edges} \\ \text{from } v_2 \text{ to } v_3 \end{bmatrix} = \begin{bmatrix} \text{number of walks of length 2} \\ \text{from } v_3 \text{ to } v_3 \text{ that go via } v_2 \end{bmatrix}$

$0 \cdot 0 = \begin{bmatrix} \text{number of edges} \\ \text{from } v_3 \text{ to } v_3 \end{bmatrix} \cdot \begin{bmatrix} \text{number of edges} \\ \text{from } v_3 \text{ to } v_3 \end{bmatrix} = \begin{bmatrix} \text{number of walks of length 2} \\ \text{from } v_3 \text{ to } v_3 \text{ that go via } v_3 \end{bmatrix}$

Since any walk of length two from v_3 to v_3 must go via either v_1, v_2, or v_3, $2 \cdot 2 + 1 \cdot 1 + 0 \cdot 0 = 5$ is the total number of walks of length two from v_3 to v_3.

In the diagram below, the five walks of length two from v_3 to v_3 can be seen to be $v_3 e_1 v_1 e_1 v_3$, $v_3 e_1 v_1 e_2 v_3$, $v_3 e_2 v_1 e_1 v_3$, $v_3 e_2 v_1 e_2 v_3$, and $v_3 e_5 v_2 e_5 v_3$.

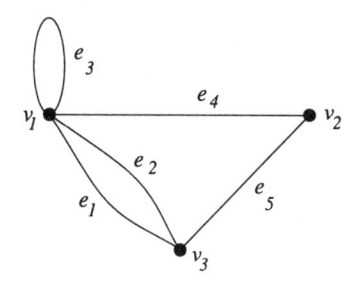

20.

21. *Proof (by mathematical induction)*:

The property holds for n = 1: Since $\mathbf{A}$ is the adjacency matrix for K_3, all the entries along the main diagonal of $\mathbf{A}$ are 0 (because K_3 has no loops) and all the entries off the main diagonal are 1 (because each pair of vertices is connected by exactly one edge).

If the property holds for n = k then it holds for n = k + 1 : Let k be an integer with $k \geq 1$, and suppose all the entries along the main diagonal of $\mathbf{A}$ are equal to each other and all the entries off the main diagonal are also equal to each other. Then

$$\mathbf{A}^k = \begin{bmatrix} b & c & c \\ c & b & c \\ c & c & b \end{bmatrix} \text{ for some integers } b \text{ and } c.$$

It follows that

$$\mathbf{A}^{k+1} = \mathbf{A}\mathbf{A}^k = \begin{bmatrix} 0 & 1 & 1 \\ 1 & 0 & 1 \\ 1 & 1 & 0 \end{bmatrix} \begin{bmatrix} b & c & c \\ c & b & c \\ c & c & b \end{bmatrix} = \begin{bmatrix} 2c & b+c & b+c \\ b+c & 2c & b+c \\ b+c & b+c & 2c \end{bmatrix}$$

As can be seen, all the entries of $\mathbf{A}^{k+1}$ along the main diagonal are equal to each other and all the entries off the main diagonal are equal to each other *[as was to be shown]*.

22. *a.*

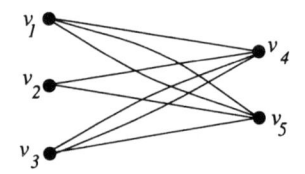

[Graph with vertices v_1, v_2, v_3 on the left and v_4, v_5 on the right, showing a complete bipartite graph $K_{3,2}$]

Any labels may be applied to the edges because the adjacency matrix does not determine edge labels. Regardless of edge labels, this graph is bipartite.

b. Proof:

($\Rightarrow$) Suppose that a graph G with n vertices is bipartite. Then its vertices can be partitioned into two disjoint sets V_1 and V_2 so that no two vertices within V_1 are connected to each other by an edge and no two vertices within V_2 are connected to each other by an edge. Label the vertices in V_1 as $v_1, v_2, \ldots, v_k$ and label the vertices in V_2 as $v_{k+1}, v_{k+2}, \ldots, v_n$. Then the adjacency matrix of G relative to this vertex labeling is

$$\begin{array}{c} \\ v_1 \\ v_2 \\ \vdots \\ v_k \\ v_{k+1} \\ \vdots \\ v_n \end{array} \begin{array}{c} \begin{array}{cccccc} v_1 & v_2 & \cdots & v_k & v_{k+1} & \cdots & v_n \end{array} \\ \left[\begin{array}{ccccccc} 0 & 0 & \cdots & 0 & a_{1,k+1} & \cdots & a_{1,n} \\ 0 & 0 & \cdots & 0 & a_{2,k+1} & \cdots & a_{2,n} \\ \vdots & \vdots & & \vdots & \vdots & & \vdots \\ 0 & 0 & \cdots & 0 & a_{k,k+1} & \cdots & a_{k,n} \\ a_{k+1,1} & a_{k+1,2} & \cdots & a_{k+1,k} & 0 & \cdots & 0 \\ \vdots & \vdots & & \vdots & \vdots & & \vdots \\ a_{n,1} & a_{n,2} & \cdots & a_{n,k} & 0 & \cdots & 0 \end{array} \right] \end{array}$$

Let $\mathbf{A}$ be the $k \times (n-k)$ matrix whose ijth entry is $a_{i,k+j}$ for all $i = 1, 2, \ldots, k$ and $j = 1, 2, \ldots, n-k$, and let $\mathbf{B}$ be the $(n-k) \times k$ matrix whose jith entry is $a_{k+j,i}$ for all $i = 1, 2, \ldots, k$ and $j = 1, 2, \ldots, n-k$. For all i and j, the ijth entry of $\mathbf{A} = a_{i,k+j}$ = the number of edges from v_i to v_{k+j} = the number of edges from v_{k+j} to v_i = $a_{k+j,i}$ = the jith entry of $\mathbf{B}$, and so $\mathbf{B} = \mathbf{A}^t$ and the adjacency matrix of G has the required form.

($\Leftarrow$) Suppose that for some labeling of the vertices of a graph G, its adjacency matrix has the given form. Denote the labeling of the vertices of G that gives rise to this adjacency matrix by $v_1, v_2, \ldots, v_n$. Let $V_1 = \{v_1, v_2, \ldots, v_k\}$ and $V_2 = \{v_{k+1}, v_{k+2}, \ldots, v_n\}$. For all i and j with

$1 \leq i, j \leq k$, the ijth entry of the adjacency matrix is zero. This implies that there is no edge that connects two vertices in V_1. Similarly, for all i and j with $k+1 \leq i, j \leq n$, the ijth entry in the adjacency matrix is zero, and so there is no edge that connects two vertices in V_2. Therefore, G is bipartite.

23. *a. Proof:* Suppose G is a graph with n vertices, v and w are distinct vertices of G, and there is a walk in G from v to w. If this walk has length greater than or equal to n, then it contains a repeated vertex, say u, because there are only n vertices in G and there are $n-1$ edges in a walk with n vertices. Replace the section of the walk from u to u by the vertex u alone; the result is still a walk from v to w but it has shorter length than the given walk *[because a walk consists of an alternating sequence of vertices and edges so that the section of a walk between two vertices contains at least one edge]*. If this walk has length greater than or equal to n, then repeat the replacement process described above. Continue repeating the replacement process until a walk from v to w with no more than $n-1$ edges is found. This must happen eventually because the total number of edges and vertices in the initial walk is finite and each repetition of the process results in the removal of at least one edge.

    b. Proof: Let G be a graph with n vertices (where $n > 1$) and let $\mathbf{A}$ be the adjacency matrix of G relative to the vertex labeling $v_1, v_2, \ldots, v_n$.

    ($\Rightarrow$) Suppose G is connected. Let integers i and j with $1 \leq i < j \leq n$ be given. We will show that the ijth entry in the matrix sum $\mathbf{A} + \mathbf{A}^2 + \ldots + \mathbf{A}^{n-1}$ is positive. Since G is connected, there is a walk from v_i to v_j, and so by part (a), there is a walk of length at most $n-1$ from v_i to v_j. Let the length of such a walk be k. Then the ijth entry of $\mathbf{A}^k$, which equals the number of walks of length k from v_i to v_j (by Theorem 11.3.2), is at least one. Now all entries in all powers of $\mathbf{A}$ are nonnegative *[because each equals the number of walks from one vertex to another]*, and if one term of a sum of nonnegative numbers is positive then the entire sum is positive. Hence the ijth entry in $\mathbf{A} + \mathbf{A}^2 + \ldots + \mathbf{A}^{n-1}$ *[which is the sum of the ijth entries in all powers of $\mathbf{A}$ from 1 to $n-1$]* is positive.

    ($\Leftarrow$) Suppose every entry in $\mathbf{A} + \mathbf{A}^2 + \ldots + \mathbf{A}^{n-1}$ is positive. Let v_i and v_j be any two vertices of G. We must show that there is a walk from v_i to v_j. For each $k = 1, 2, \ldots, n-1$, the ijth entry of $\mathbf{A}^k$ equals the number of walks of length k from v_i to v_j (by Theorem 11.3.2) and is therefore nonnegative. By supposition, when these nonnegative numbers are added together, the sum is positive. Now the only way that a sum of nonnegative real numbers can be positive is for at least one of the numbers to be positive. Hence for some k, the ijth entry of $\mathbf{A}^k$ is positive. It follows that the number of walks of length k from v_i to v_j is positive, and so there is at least one walk joining v_i to v_j.

Section 11.4

3. The graphs are isomorphic. One way to define to isomorphism is as follows.

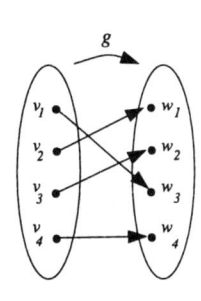

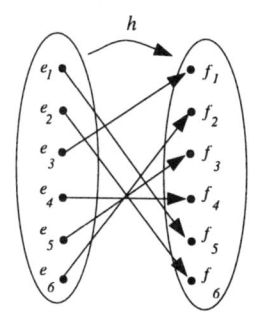

4. The graphs are not isomorphic: G has eight edges whereas G' has seven.

5. The graphs are not isomorphic: G has a vertex of degree five whereas G' does not.

7. The graphs are isomorphic. One way to define to isomorphism is as follows.

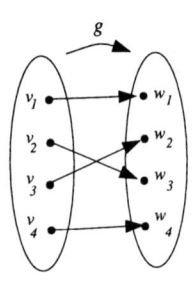

9. The graphs are isomorphic. One way to define to isomorphism is as follows.

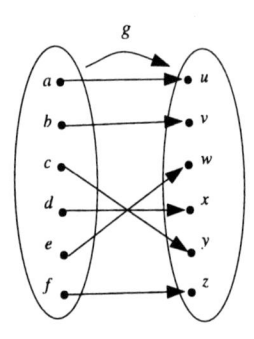

11. The graphs are not isomorphic: G' has a circuit of length six whereas G does not. Also G' is connected whereas G is not.

13. The graphs are not isomorphic: G has a simple circuit of length five whereas G' does not.

15. There is one such graph with 0 edges, one with 1 edge, and there are two with 2 edges, three

with 3 edges, two with 4 edges, one with 5 edges, and one with 6 edges. These are shown below.

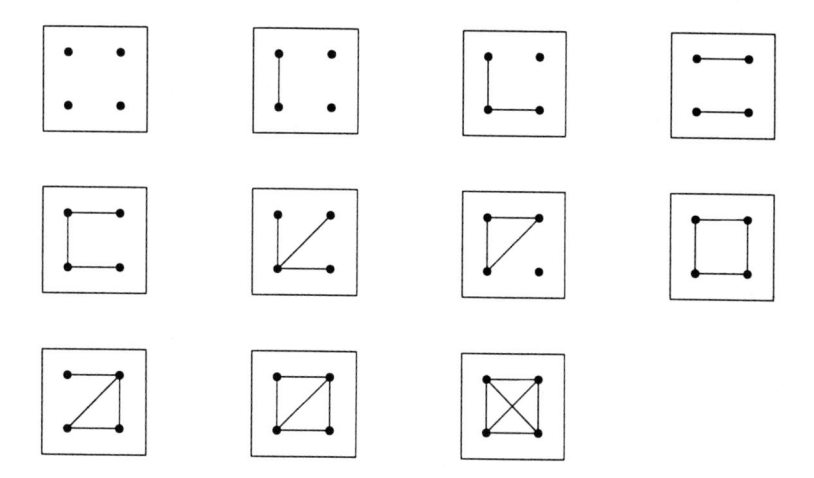

17. There is one such graph with 0 edges, and there are two with 1 edge (one in which the edge is a loop and one in which it is not), and six with 2 edges (two simple graphs, one with two parallel edges, two in which one of the edges is a loop and the other is not a loop, and two in which both edges are loops). These are shown below.

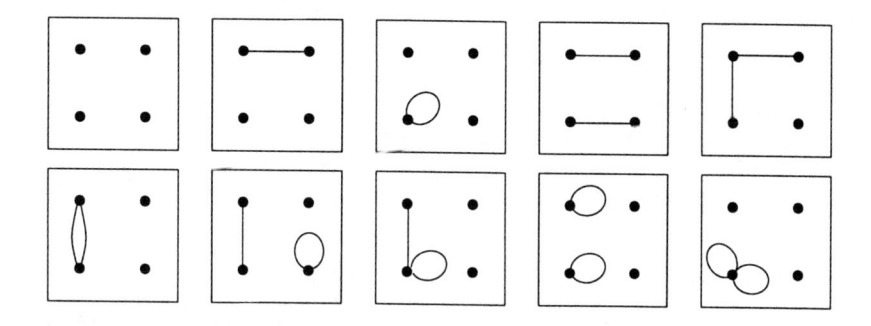

18. There are three such graphs in which all 3 edges are loops, five in which 2 edges are loops and 1 is not a loop, six in which 1 edge is a loop and 2 edges are not loops, and five in which none of the 3 edges is a loop. These are shown below.

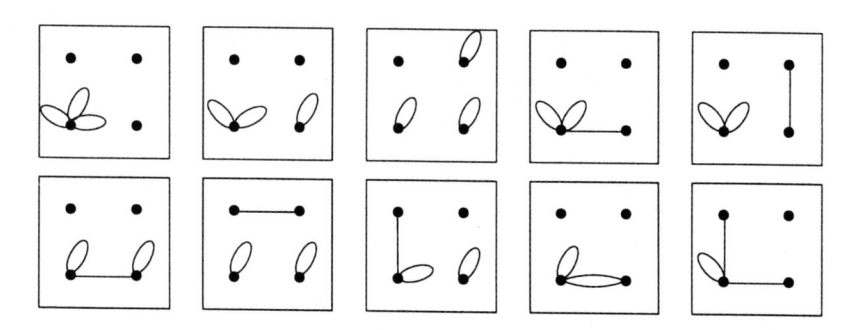

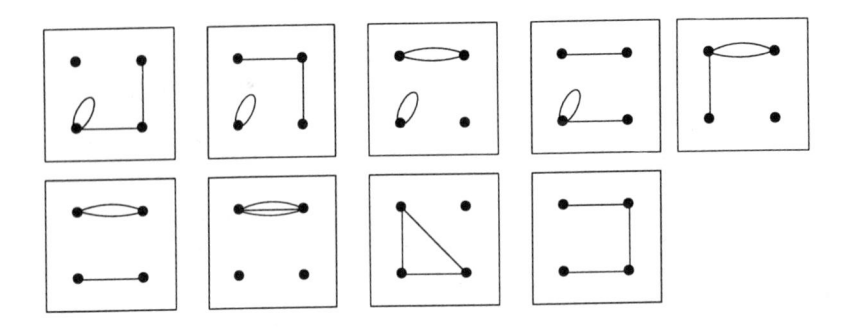

20. Four (of many) such graphs are shown below.

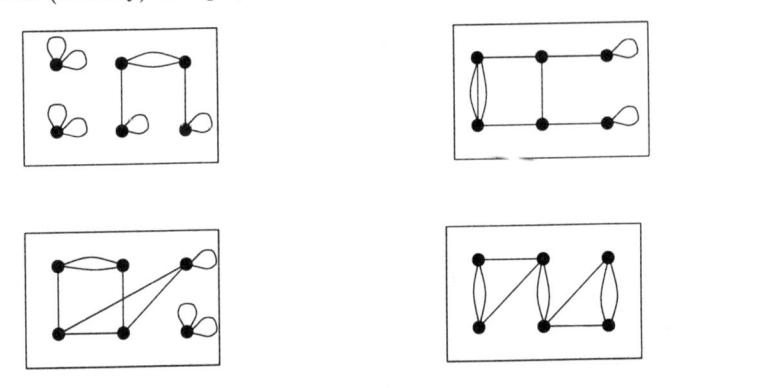

22. *Proof:* Suppose G and G' are isomorphic graphs and G has m edges where m is a nonnegative integer. By definition of graph isomorphism, there is a one-to-one correspondence $h: E(G) \to E(G')$. But $E(G)$ is a finite set and two finite sets in one-to-one correspondence have the same number of elements. Therefore, there are as many edges in $E(G')$ as there are in $E(G)$, and so G and G' have the same number of edges.

24. *Proof:* Suppose G and G' are isomorphic graphs and suppose G has a simple circuit C of length k, where k is a nonnegative integer. Let C be $v_0 e_1 v_1 e_2 \ldots e_k v_k (= v_0)$, and let C' be $g(v_0) h(e_1) g(v_1) h(e_2) \ldots h(e_k) g(v_k) (= g(v_0))$. By the same reasoning as in the solution to exercise 23, C' is a circuit of length k in G'. Suppose C' is not a simple circuit. Then C' has a repeated vertex, say $g(v_i) = g(v_j)$ for some $i, j = 0, 1, 2, \ldots, k-1$ with $i \neq j$. But since g is a one-to-one correspondence this implies that $v_i = v_j$, which is impossible because C is a simple circuit. Hence the supposition is false, C' is a simple circuit, and therefore G' has a simple circuit of length k.

25. *Proof:* Suppose G and G' are isomorphic graphs and suppose G has m vertices of degree k, $v_1, v_2, \ldots, v_m$, where m and k are nonnegative integers. By definition of graph isomorphism, there are one-to-one correspondences $g: V(G) \to V(G')$ and $h: E(G) \to E(G')$ that preserve the edge-endpoint functions in the sense that for all v in $V(G)$ and e in $E(G)$, v is an endpoint of $e \Leftrightarrow g(v)$ is an endpoint of $h(e)$. Consider the vertices $g(v_1), g(v_2), \ldots, g(v_m)$ in G'. Because g is a one-to-one correspondence, these vertices are all distinct. And applying the same argument as that used in Example 11.4.4 to each vertex $g(v_i)$ enables us to conclude that each has degree k. Hence G' has m vertices of degree k.

26. *Proof:* Suppose G and G' are isomorphic graphs and suppose G has m distinct simple circuits of length k, where k is a nonnegative integer. The same argument used in the answer to exercise

24 can be used to show that each such circuit gives rise to a corresponding simple circuit of length k in G'. For each pair of distinct circuits in G, the corresponding circuits in G' are also distinct. The reason is that if C_1 and C_2 are distinct circuits of length k in G,, then there is an edge e in C_1 that is not in C_2. Let C'_1 and C'_2 be the corresponding circuits in G'. It follows from the fact that h is one-to-one that $h(e)$ is in C'_1 but not in C'_2. Hence C'_1 and C'_2 are distinct. Therefore, G' has m distinct simple circuits of length k.

27. *Proof:* Suppose G and G' are isomorphic graphs and suppose G is connected. By definition of graph isomorphism, there are one-to-one correspondences $g: V(G) \to V(G')$ and $h: E(G) \to E(G')$ that preserve the edge-endpoint functions in the sense that for all v in $V(G)$ and e in $E(G)$, v is an endpoint of $e \Leftrightarrow g(v)$ is an endpoint of $h(e)$. Suppose w and x are any two vertices of G'. Then $u = g^{-1}(w)$ and $v = g^{-1}(x)$ are distinct vertices in G (because g is a one-to-one correspondence). Since G is connected, there is a walk in G connecting u and v. Say this walk is $ue_1v_1e_2v_2\ldots e_nv$. Because g and h preserve the edge-endpoint functions, $w = g(u)h(e_1)g(v_1)h(e_2)g(v_2)\ldots h(e_n)g(v) = x$ is a walk in G' connecting w and x.

28. *Proof:* Suppose G and G' are isomorphic graphs and suppose G has an Euler circuit C. Let m be the number of edges in G. Then C has length m because it includes every edge of G. By the same argument as in the answer to exercise 23, G' has a corresponding circuit C' of length m, and by exercise 22, G' also has m edges. Since all the edges of a circuit are distinct, C' includes all of the m edges of G'. Hence C' is an Euler circuit for G'.

29. *Proof:* Suppose G and G' are isomorphic graphs and suppose G has an Hamiltonian circuit C. Let the number of vertices of G be n. Since C is a Hamiltonian circuit, it is a simple circuit that has length n (because it includes every vertex of G exactly once, except for the first and last which are repeated). By the same argument as in the answer to exercise 23, G' has a corresponding circuit C' of length n, and by exercise 21, G' also has n vertices. Now all the edges and vertices of a simple circuit are distinct except for the repetition of the first and last vertex, the vertices of C' are by construction the images under g of the vertices of C, and since g is a one-to-one correspondence, g sends the n distinct vertices of C to n distinct vertices in G'. Hence C' is a simple circuit of length n in G', and so C' includes all of the n vertices of G'. Therefore, C' is a Hamiltonian circuit for G'.

30. Suppose that G and G' are isomorphic via one-to-one correspondences $g: V(G) \to V(G')$ and $h: E(G) \to E(G')$, where g and h preserve the edge-endpoint functions. Now w_6 has degree one in G', and so by the argument given in Example 11.4.4, w_6 must correspond to one of the vertices of degree one in G : either $g(v_1) = w_6$ or $g(v_6) = w_6$. Similarly, since w_5 has degree three in G', w_5 must correspond to one of the vertices of degree three in G : either $g(v_3) = w_5$ or $g(v_4) = w_5$. Because g and h preserve the edge-endpoint functions, edge f_6 with endpoints w_5 and w_6 must correspond to an edge in G with endpoints v_1 and v_3, or v_1 and v_4, or v_6 and v_3, or v_6 and v_4. But this contradicts the fact that none of these pairs of vertices are connected by edges in G. Hence the supposition is false, and G and G' are not isomorphic.

Section 11.5

1. *b.* Math 110

2. b.

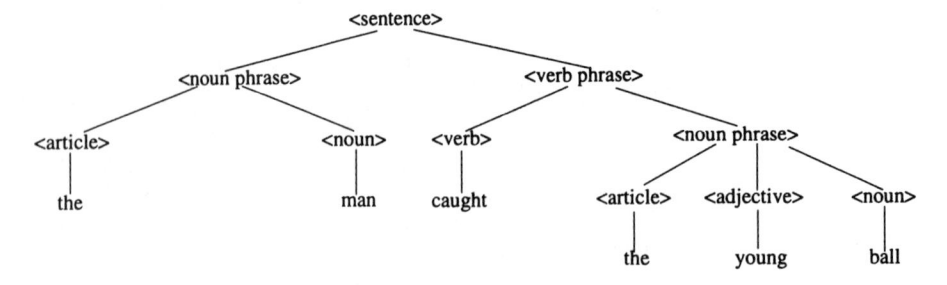

```
                              <sentence>
                     /                        \
              <noun phrase>                <verb phrase>
              /        \                  /            \
         <article>   <noun>          <verb>         <noun phrase>
             |         |               |           /     |       \
            the       man           caught    <article> <adjective> <noun>
                                                  |         |        |
                                                 the      young     ball
```

3. By Theorem 11.5.2, a tree with n vertices (where $n \geq 1$) has $n-1$ edges, and so by Theorem 11.1.1, its total degree is twice the number of edges, or $2 \cdot (n-1) = 2n-2$.

4. b.

```
        H   H   H   H   H
        |   |   |   |   |
   H—C—C—C—C—C—H
        |   |   |   |   |
        H   H   H   H   H


        H   H   H   H
        |   |   |   |
   H—C—C—C—C—H
        |   |   |   |
        H   |   H   H
        H—C—H
            |
            H


        H       H       H
        |       |       |
   H—C———C———C—H
        |       |       |
   H—C—H   H   H—C—H
        |       |       |
        H       H       H
```

c. *Proof:* Let G be the graph of a hydrocarbon molecule with the maximum number of atoms for the number of its carbon atoms, and suppose G has k carbon atoms and m hydrogen atoms. By Example 11.5.4, G is a tree with $k+m$ vertices. By exercise 3, the total degree of this tree is $2(k+m) - 2 = 2k + 2m - 2 = 2(k+m-1)$.

d. *Proof:* Let G be the graph of a hydrocarbon molecule with the maximum number of atoms for the number of its carbon atoms, and suppose G has k carbon atoms and m hydrogen atoms. Each carbon atom is bonded to four other atoms because otherwise an additional hydrogen atom could be bonded to it, which would contradict the assumption that the number of hydrogen atoms is maximal for the given number of carbon atoms. Hence each of the k carbon atom vertices has degree four in the graph. Also each hydrogen atom is bonded to exactly one carbon atom because otherwise the molecule would not be connected. Hence each of the m hydrogen atom vertices has degree one in the graph. It follows that the total degree of the graph is $4 \cdot k + 1 \cdot m = 4k + m$.

e. Equating the results of parts (c) and (d) above gives $2k + 2m - 2 = 4k + m$. Solving for m gives $m = 2k + 2$. In other words, a hydrocarbon molecule with k carbon atoms and a maximal number of hydrogen atoms has $2k + 2$ hydrogen atoms.

5. *Proof:* Let T be a particular but arbitrarily chosen tree that has more than one vertex, and consider the following algorithm. For justification of the various steps, see the proof of Lemma 11.5.1.

   Step 1: Pick a vertex v_0 of T and let e_0 be an edge incident on v_0.

   [The starting vertex and edge are given the names v_0 and e_0 that will not be changed during execution of the algorithm.]

   Step 2: if $\deg(v_0) > 1$

       then choose e_1 to be an edge incident on v_0 such that $e_1 \neq e_0$

       else let $v_1 := v_0$ and let $e_1 := e_0$

   [The name v_1 is given to the first vertex of degree 1 that is found. The second vertex of degree 1 will be named v_2. The name e_1 is given to the edge adjacent to the starting vertex along which the search for a vertex of degree 1 begins.]

   Step 3: Let v' be the vertex at the other end of e_1 from v_0, and let $e := e_1$ and $v := v'$.

   [The values of v and e may change many times during the search outward from v_0 toward a vertex of degree 1.]

   Step 4: while $(\deg(v) > 1)$

       Choose e' to be an edge incident on v such that $e' \neq e$.

       Let v' be the vertex at the other end of e' from v.

       Let $e := e'$ and $v := v'$.

   end while

   Step 5: if v_1 does not have a value

       then let $v_1 := v$ and $e_1 := e_0$, and go to step 3

   [If $\deg(v_0) \neq 1$, a vertex v_1 of degree 1 was first sought by moving away from v_0 along an edge other than e_0. Now a return is made to v_0, and the search for a second vertex of degree 1 is made by moving away from v_0 starting along e_0.]

       else let $v_2 := v$

   [If $\deg(v_0) = 1$, Step 5 is executed just once; otherwise it is executed twice, once to give v_1 a value and again to give v_2 a value.]

   After execution of this algorithm, v_1 and v_2 are distinct vertices of degree 1.

6. Define an infinite graph G as follows: $V(G) = \{v_i \mid i \in \mathbf{Z}\} = \{\ldots, v_{-2}, v_{-1}, v_0, v_1, v_2, \ldots\}$, $E(G) = \{e_i \mid i \in \mathbf{Z}\} = \{\ldots, e_{-2}, e_{-1}, e_0, e_1, e_2, \ldots\}$, and the edge-endpoint function is defined by the rule $f(e_i) = \{v_{i-1}, v_i\}$ for all $i \in \mathbf{Z}$. Then G is circuit-free, but each vertex has degree two. G is illustrated below.

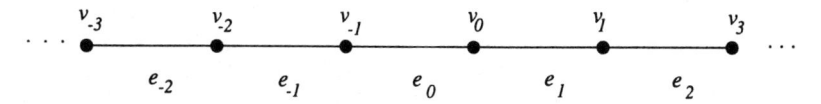

7. b. terminal vertices: v_1, v_2, v_5, v_6, v_8 internal vertices: v_3, v_4, v_7

15. One such graph is shown below.

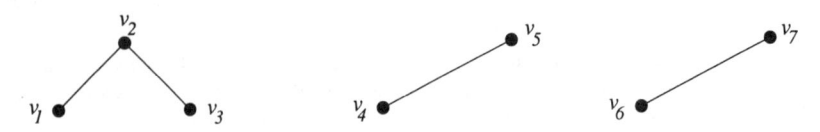

16. Any tree with twelve vertices has eleven edges, not fifteen. Thus there is no such graph.

17. One such graph is shown below.

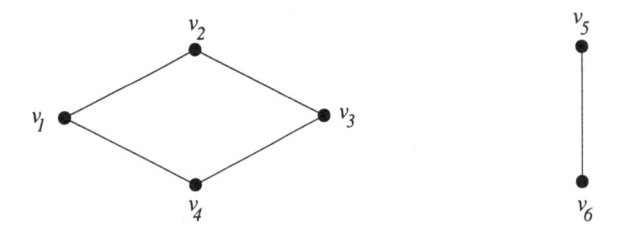

18. Any tree with five vertices has four edges. By Theorem 11.1.1, the total degree of such a graph is eight, not ten. Hence there is no such graph.

19. By Theorem 11.5.4, a connected graph with ten vertices and nine edges is a tree. By definition of tree, a tree cannot have a nontrivial circuit. Hence there is no such graph.

20. One such graph is shown below.

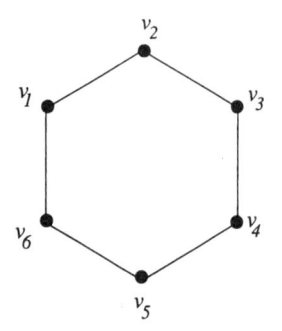

21. Any tree with ten vertices has nine edges. By Theorem 11.1.1, the total degree of such a tree is 18, not 24. Hence there is no such graph.

23. Yes, because a connected graph with no nontrivial circuits is a tree, and a tree with nine vertices has eight edges, not twelve.

24. Yes. Given any two vertices u and w of G', then u and w are vertices of G neither equal to v. Since G is connected, there is a walk in G from u to w, and so by Lemma 11.2.1, there is a path in G from u to w. This path does not include edge e or vertex v because a path does not have a repeated edge, and e is the unique edge incident on v. [If a path from u to w leads into v, then it must do so via e. But then it cannot emerge from v to continue on to w because no edge other than e is incident on v.] Thus this path is a path in G'. It follows that any two vertices of G' are connected by a walk in G', and so G' is connected.

26. No. Suppose there were a connected graph with n vertices and $n-2$ or fewer edges. Either the graph itself would be a tree or edges could be eliminated from its circuits to obtain a tree. In either case, there would be a tree with n vertices and $n-2$ or fewer edges. This would contradict the result of Theorem 11.5.2, which says that a tree with n vertices has $n-1$ edges. So there is no graph with n vertices and $n-2$ or fewer edges.

28. Yes. Suppose G is a circuit-free graph with n vertices and $n-1$ edges. Let $G_1, G_2, \ldots, G_k$ be the connected components of G. Each G_i is a tree because each is connected and circuit-free. For each $i = 1, 2, \ldots, k$, let n_i be the number of vertices in G_i. Since G has n vertices in all, $n_1 + n_2 + \ldots + n_k = n$. By Theorem 11.5.2, G_i has $n_i - 1$ edges for all $i = 1, 2, \ldots, k$. So the number of edges in G is $(n_1 - 1) + (n_2 - 1) + \ldots + (n_k - 1) = (n_1 + n_2 + \ldots + n_k) - k = n - k$. But by hypothesis, G has $n-1$ edges. So $n - k = n - 1$. It follows that $k = 1$, and so G has just one connected component. Therefore, G is connected.

29. *Proof:* Let T be a nonempty, nontrivial tree and let S be the set of all paths from one vertex to another of T. Among all the paths in S, choose a path P with the most edges. *[This is possible because since the number of vertices and edges of a graph is finite, there are only finitely many paths in T.]* The initial and final vertices of P both have degree one. For suppose that one of these vertices, say v, does not have degree one. Let e be the edge of the path that leads into v. Since $\deg(v) > 1$, there is an edge e' of T with $e \neq e'$. Add e' and the vertex at the other end of e' from v to the path. The result is a path that is longer than P (the path of maximal length), which is a contradiction. Hence the supposition is false, and so both the initial and final vertices of P have degree one.

30. Such a tree must have 4 edges and, therefore, a total degree of 8. Since at least two vertices have degree 1 and no vertex has degree greater than 4, the possible degrees of the five vertices are as follows: 1,1,1,1,4; 1,1,1,2,3; and 1,1,2,2,2. The corresponding trees are shown below.

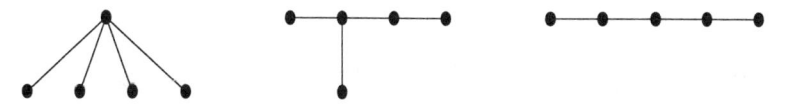

31. *a. Proof:* Suppose G and G' are isomorphic graphs and G has a vertex v of degree i that is adjacent to a vertex w of degree j (where i and j are positive integers). Since G and G' are isomorphic, there are one-to-one correspondences $g: V(G) \to V(G')$ and $h: E(G) \to E(G')$ that preserve the edge-endpoint functions in the sense that for all vertices v and edges e in G, e is an endpoint of $v \Leftrightarrow h(e)$ is an endpoint of $g(v)$. It follows that since v and w are adjacent vertices of G, $g(v)$ and $g(w)$ are adjacent vertices of G'. Let $e_1, e_2, \ldots, e_i$ be the i edges incident on v and let $f_1, f_2, \ldots, f_j$ be the j edges incident on w. Then since g and h preserve the edge-endpoint functions and h is one-to-one, $h(e_1), h(e_2), \ldots, h(e_i)$ are i distinct edges incident on $g(v)$, and $h(f_1), h(f_2), \ldots, h(f_j)$ are j distinct edges incident on $g(w)$. There are no more than i edges incident on $g(w)$ because any such edge would have to be the image under h of an edge incident on v *[because h is onto]*. Similarly, there are no more than j edges incident on $g(w)$. Hence $g(v)$ has degree i and $g(w)$ has degree j, and so G' has a vertex of degree i that is adjacent to a vertex of degree j.

b. The six nonisomorphic trees with six vertices are shown below. Note that a tree with six vertices has five edges and hence a total degree of ten. Also any such tree has at least two vertices of degree one, and it has no vertex of degree greater than five. Thus the possible degrees of the vertices of such a tree are the following: 2, 2, 2, 2, 1, 1; 3, 2, 2, 1, 1, 1; 3, 3, 1, 1, 1, 1; 4, 2, 1, 1, 1, 1; or 5, 1, 1, 1, 1, 1. Furthermore, by part (a) there are two nonisomorphic trees whose vertices have degrees 1, 1, 1, 2, 2, and 3. In one, the vertex of degree 3 is adjacent to vertices of degrees 1, 1, and 2, whereas in the other, the vertex of degree 3 is adjacent to vertices of degrees 2, 2, and 1.

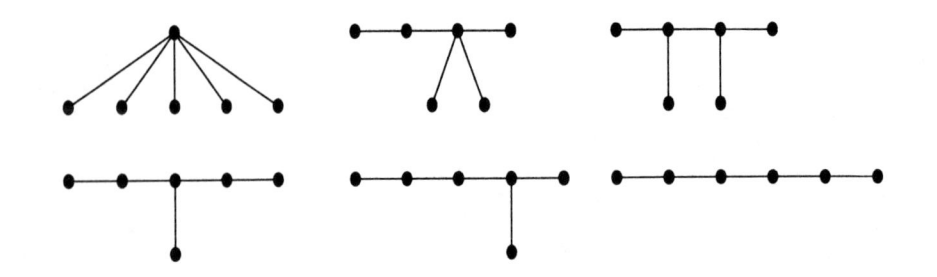

33. a. 3   b. 0   c. 5   d. v_{14}, v_{15}, v_{16} e. v_1 f. v_2 g. v_{17}, v_{18}, v_{19}

34. b.

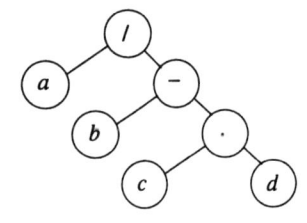

43. There is no tree with the given properties because any full binary tree with eight internal vertices has nine terminal vertices, not seven.

44. One such tree is the following.

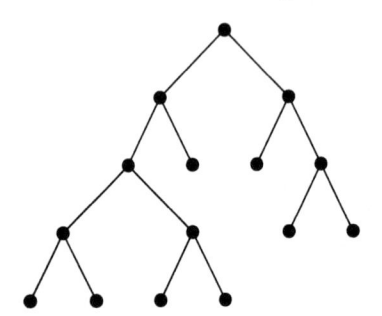

45. A full binary tree with k internal vertices has $2k + 1$ vertices in all. If $2k + 1 = 7$, then $k = 3$. Thus such a tree would have three internal and four terminal vertices. Two such trees are shown below.

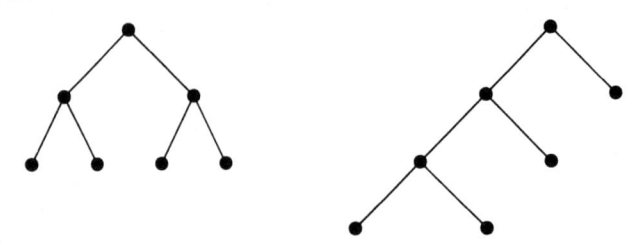

46. There is no tree with the given properties because a full binary tree with five internal vertices has $2 \cdot 5 + 1$ or eleven vertices in all, not nine.

47. A full binary tree with four internal vertices has five terminal vertices. One such tree is shown below.

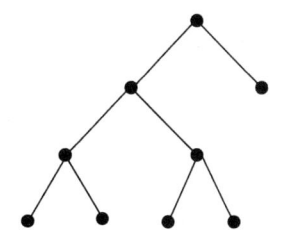

48. There is no tree with the given properties because a binary tree of height four has at most $2^4 = 16$ terminal vertices.

49. There is no tree with the given properties because a full binary tree has $2k + 1$ vertices, where k is the number of internal vertices, and $16 \neq 2k + 1$ for any integer k.

50. A full binary tree with seven terminal vertices has six internal vertices. One such tree of height three is shown below.

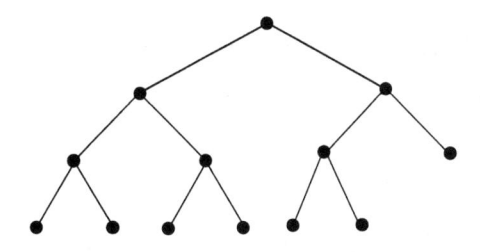

51. *b.* height $\geq \log_2 40 \cong 5.322$. Since the height of a tree is an integer, the height must be at least 6.

   c. height $\geq \log_2 60 \cong 5.907$. Since the height of a tree is an integer, the height must be at least 6.

Section 11.6

2.

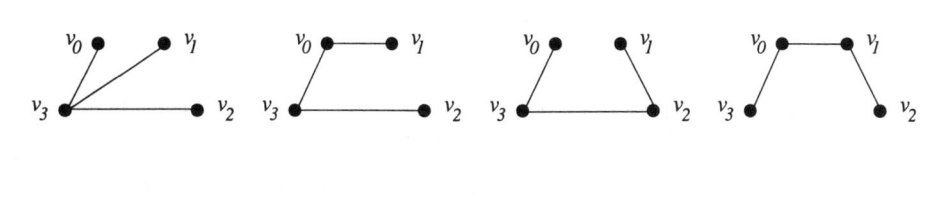

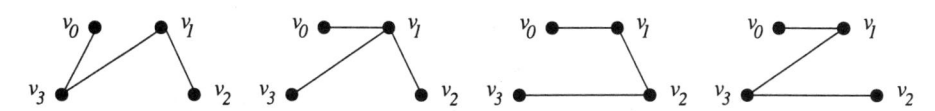

287

4. One of many spanning trees is the following.

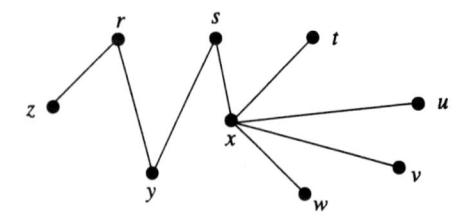

6. Minimal spanning tree:

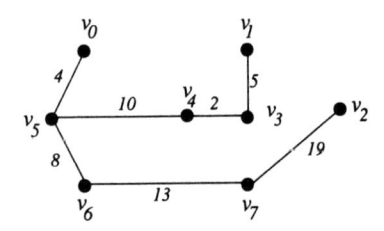

   Order of adding the edges: $\{v_4, v_3\}$, $\{v_0, v_5\}$, $\{v_3, v_1\}$, $\{v_5, v_6\}$, $\{v_5, v_4\}$, $\{v_6, v_7\}$, $\{v_7, v_2\}$

8. Minimal spanning tree: same as in exercise 6.

   Order of adding the edges: $\{v_0, v_5\}$, $\{v_5, v_6\}$, $\{v_5, v_4\}$, $\{v_4, v_3\}$, $\{v_3, v_1\}$, $\{v_6, v_7\}$, $\{v_7, v_2\}$

10. There are four minimal spanning trees:

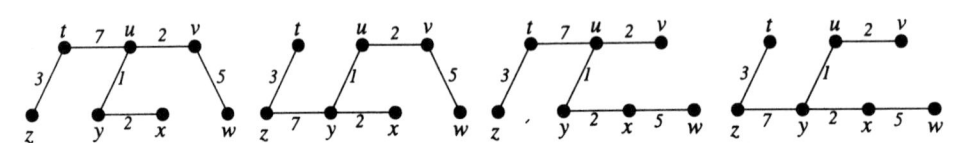

    When Kruskal's algorithm is used, edges are added in any of the orders obtained by following one of the eight paths from left to right across the diagram below.

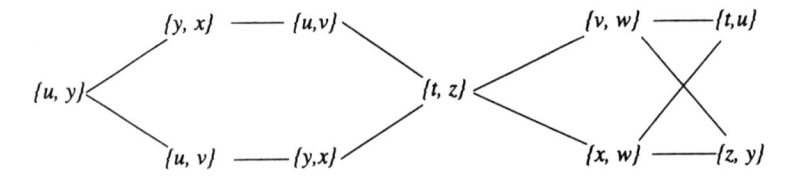

    When Prim's algorithm is used, edges are added in any of the orders obtained by following one of the eight paths from left to right across the diagram below.

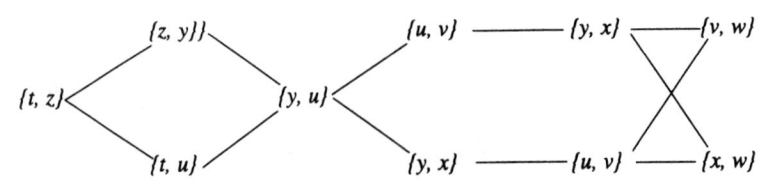

11.

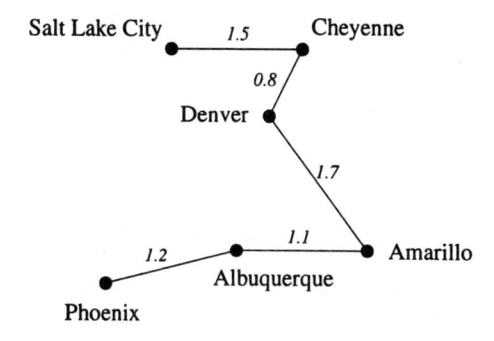

Salt Lake City —1.5— Cheyenne
          0.8
       Denver
           1.7
   1.2   1.1   Amarillo
      Albuquerque
Phoenix

12. *Proof:* Suppose T_1 and T_2 are spanning trees for a graph G with n vertices. By definition of spanning tree, both T_1 and T_2 contain all n vertices of G, and so by Theorem 11.5.2, both T_1 and T_2 have $n-1$ edges.

13. *a.* If there were two distinct paths from one vertex of a tree to another, they (or pieces of them) could be patched together to obtain a nontrivial circuit. But a tree cannot have a nontrivial circuit.

14. *Proof:* Suppose that G is a graph with n vertices and with spanning tree T, and suppose e is an edge of G that is not in T.

    (1) Let H be the graph obtained by adding e to T. Then H has n vertices and n edges and is connected, and so H contains a nontrivial circuit *[because if H were both connected and circuit-free, then H would be a tree and would therefore have $n-1$ edges which it does not]*.

    (2) Suppose H contains two distinct nontrivial circuits C_1 and C_2. Then one of the circuits, say C_1, would contain an edge, say e_1, not contained in C_2. Remove e_1 from H to obtain a graph H'. By Lemma 11.5.3, H' is connected. Since e_1 is not in C_2, and e_1 was the only edge removed from H to obtain H', C_2 is a circuit in H'. Remove an edge from C_2 to obtain a graph H'. Again, by Lemma 11.5.3, H' is connected. Then H' is a connected graph with $n-2$ edges that contains all n vertices of G. By exercise 26 of Section 11.5, this is impossible. Hence the supposition is false, and so the graph obtained by adding e to T contains at most one circuit.

    By (1) and (2) above, T contains one and only one nontrivial circuit.

16. *b. Counterexample:* Let G be the following simple graph.

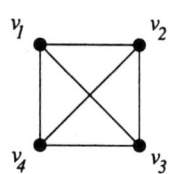

Then G has the spanning trees shown below.

These trees have no edge in common.

17. *Proof:*

    ($\Rightarrow$) Let G be a graph and e an edge that is contained in every spanning tree for G. Suppose that removal of e does not disconnect G. Let G' be the graph obtained by removing e from G. Then G' is connected, and so it has a spanning tree T' (by Proposition 11.6.1). But T' contains every vertex of G (because no vertices were removed from G to create G'), and every edge in T' is also an edge in G (by construction). Hence T' is a spanning tree for G that does not contain e. This contradicts the fact that e is contained in every spanning tree for G. Hence the supposition is false, and so removal of e disconnects G.

    ($\Leftarrow$) Let G be a graph and e an edge of G such that removal of e disconnects G. Suppose there is a spanning tree T of G that does not contain e. Then T is a connected subgraph of G that does not contain e. Hence removal of e plus (possibly) other edges from G does not disconnect G, which implies that removal of e alone from G does not disconnect G, a contradiction. Hence the supposition is false, and so e is contained in every spanning tree T of G.

18. *Proof:* Since T_2 is obtained from T_1 by removing e' and adding e, $w(T_2) = w(T_1) - w(e') + w(e)$. Now according to the proof of Theorem 11.6.3, $w(e') \geq w(e)$. Hence $w(e') - w(e) \geq 0$, and so $w(T_2) = w(T_1) - (w(e') - w(e)) \leq w(T_1)$.

20. *Proof:* Let G be a connected, weighted graph, and let e be an edge of G (not a loop) that has smaller weight than any other edge of G. Suppose there is a minimal spanning tree T that does not contain e. Since T is a spanning tree, T contains the endpoints v and w of e. By exercise 13, there is a unique path in T from v to w. Let e' be any edge along this path. By exercise 19, $w(e') \leq w(e)$. This contradicts the fact that $w(e)$ has smaller weight than any other edge of G. Hence the supposition is false: every minimal spanning tree contains e.

22. *Proof:* Suppose not. Suppose G is a connected, weighted graph, e is an edge of G that (1) has larger weight than any other edge of G and (2) is in a circuit of G, and suppose that there is a minimal spanning tree T for G that contains e. Construct another tree T' as follows: Let v and w be the endpoints of e. Because e is part of a circuit in G, there is a path in G that joins v and w. Also there is an edge e' of this path such that e' is not an edge of T (for otherwise T would contain a circuit). Form T' by removing e from T and adding e'. Then T' contains all n vertices of G, has $n-1$ edges, and is connected *[because T is connected and contains all the vertices of G, and so some edges of T must be incident on the endpoints of e']*. By Theorem 11.6.4, therefore, T' is a tree. But T' also contains all the vertices of G *[because T' is formed from T by adding and deleting only edges]*, and so T' is a spanning tree for G. Now $w(T') = w(T) - w(e) + w(e') = w(T) - (w(e) - w(e')) < w(T)$ because $w(e) > w(e')$. Thus T' is a spanning tree of smaller weight than a minimal spanning tree for G, which is a contradiction. Hence the supposition is false, and the given statement is true.

24. The output will be a minimal spanning tree for the connected component of the graph that contains the starting vertex input to Prim's algorithm.

25. *Proof:* Suppose that G is a connected, weighted graph with n vertices and that T is the output graph produced when G is input to Algorithm 11.6.3. Clearly T is a subgraph of G and T is connected because no edge is removed from T as T is being constructed if its removal would disconnect T. Also T is circuit-free because if T had a circuit then the circuit would contain edges $e_1, e_2, \ldots, e_k$ of maximal weight. At some point during execution of the algorithm, each of these edges would be examined (since all edges are examined eventually). Let e_i be the first such edge to be examined. When examined, e_i must be removed because deletion of an edge from a circuit does not disconnect a graph and at the time e_i is examined no other edge of the circuit would have been removed. But this contradicts the supposition that e_i was one of the edges in the output graph T. Thus T is circuit-free. Furthermore, T contains every vertex of G since only edges, not vertices, are removed from G in the construction of T. Hence T is a spanning tree for G.

Next we show that T has minimal weight. Let T_1 be any minimal spanning tree for G. If $T = T_1$, we are done. If $T \neq T_1$, then there is an edge e of T that is not in T_1. Now adding e to T_1 produces a graph with a unique circuit C (by exercise 14). Let e' be an edge of C that is not in T, and let T_2 be the graph obtained from T_1 by removing e' and adding e. Note that T_2 has $n-1$ edges and n vertices and that T_2 is connected. Consequently, T_2 is a spanning tree for G. Now $w(e') \leq w(e)$ because at the stage in Algorithm 11.6.3 when e' was deleted from T, e was also available to be deleted from T [since it was in T, and at that stage its deletion would not have disconnected T because e' was also in T and so were all the other edges in C which stayed in T throughout execution of the algorithm], and e would have been deleted from T if its weight had been greater than that of e'. Therefore, $w(T_2) = w(T_1) - w(e') + w(e) = w(T_1) - (w(e') - w(e)) \leq w(T_1)$. Since T_1 is a minimal spanning tree and T_2 is a spanning tree with weight less than or equal to T_1, T_2 is also a minimal spanning tree for G. In addition, T_2 has one more edge in common with T than does T_1. *[The remainder of the proof is identical to the last few lines of the proofs of Theorems 11.6.2 and 11.6.3.]* If T now equals T_2, then we are done. If not, then, as above, we can find another minimal spanning tree T_3 having one more edge in common with T than T_2. Continuing in this way produces a sequence of minimal spanning trees $T_1, T_2, T_3, \ldots$ each of which has one more edge in common with T than the preceding tree. Since T has only a finite number of edges, this sequence if finite, and so for some k, T_k is identical to T. This shows that T is itself a minimal spanning tree.